Biology, Ecology and Management of Crown-of-Thorns Starfish

Special Issue Editors

Morgan Pratchett
Sven Uthicke

MDPI • Basel • Beijing • Wuhan • Barcelona • Belgrade

MDPI

Special Issue Editors

Morgan Pratchett
James Cook University
Australia

Sven Uthicke
Australian Institute of Marine Science
Australia

Editorial Office
MDPI AG
St. Alban-Anlage 66
Basel, Switzerland

This edition is a reprint of the Special Issue published online in the open access journal *Diversity* (ISSN 1424-2818) from 2016–2017 (available at: http://www.mdpi.com/journal/diversity/special_issues/crown_thorns_starfish).

For citation purposes, cite each article independently as indicated on the article page online and as indicated below:

Author 1; Author 2. Article title. *Journal Name* **Year**, *Article number*, page range.

First Edition 2017

ISBN 978-3-03842-602-8 (Pbk)
ISBN 978-3-03842-603-5 (PDF)

Table of Contents

About the Special Issue Editors

Morgan Pratchett is chief investigator and research leader in the ARC Centre of Excellence for Coral Reef Studies, based at James Cook University. He also studied marine biology at James Cook University, but had several overseas postdoctoral appointments (University of Perpignan, Nova Southeastern University Oceanographic Centre, and the University of Oxford) before taking up his current position. Morgan is a marine field ecologist and has been studying crown-of-thorns starfish for over 20 years. For his PhD, he documented the population dynamics of crown-of-thorns starfish during the outbreak on the Great Barrier Reef in the 1990s. Morgan has now published more than 40 papers on crown-of-thorns starfish, but also studies the demography of reef-building corals, ecological reliance of reef fishes on coral habitats, and the effects of climate change on reef organisms and systems.

Sven Uthicke is a Principal Research Scientist at the Australian Institute of Marine Science. He was born in Germany and studied Marine Biology at the University of Hamburg. His current research activities include molecular and ecological research investigate cumulative impacts on the Great Barrier Reef. This work combines field-based ecological techniques with state-of-the-art molecular and other laboratory based methods. Sven has worked on echinoderm population dynamics for over 20 years in collaboration with national and international researchers. While focusing on coral reef environments, Sven has worked in several marine ecosystems stretching from Antarctica to the Deep Sea. He currently investigates causes for crown-of-thorns starfish outbreaks. In this research he is specifically interested in the development of new genetic tools (e.g., 'eDNA') to elucidate life-history parameters of CoTS larvae and early outbreak detection. He has strong expertise in experimental ecology, molecular ecology and population genetics.

Preface to "Biology, Ecology and Management of Crown-of-Thorns Starfish"

In 1986, Peter Moran wrote that, "the crown-of-thorns starfish has become one of the most well-known animals in coral reef ecosystems (Moran 1986, p. 379)". Crown-of-thorns starfish (CoTS) have attracted considerable public and political interest due to their propensity for population outbreaks, and the threat that outbreaks pose to coral reef ecosystems. However, Moran was referring, at least in part, to the plethora of scientific studies and research papers that describe and detail various aspects of the biology, ecology, and behaviour of CoTS, as well as the ecosystem impacts of CoTS outbreaks. Since Moran (1986), there has also been extensive research on crown-of-thorns starfish (CoTS) and many more scientific articles have been published.

The volume of scientific studies and research articles focussed on CoTS is a double-edged sword. On one hand, it seems incomprehensible that there would still be significant or major gaps in our knowledge of their biology and ecology, or that any persistent knowledge gaps will ever be effectively addressed. Accordingly, potential donors and funding agencies often question the merit of further investment in CoTS research. Indeed, I have been asked, both by funding committees and established researchers who have previously dabbled in CoTS research, "what can you possibly hope to achieve with thousands of dollars, given the millions of dollars already expended on CoTS research?" To redress this issue, we must periodically take stock of what we do know as well as what we do not. We also need to be increasingly pragmatic in conducting research that will affect meaningful changes in management of CoTS outbreaks and reef ecosystems.

The large body of literature pertaining to CoTS (and their impacts on reef ecosystems) also makes it very difficult for new and contemporary researchers to effectively capture and distil the cumulative knowledge acquired during multiple successive outbreak episodes, especially given that the research itself is punctuated by rapid and prolonged scarcity of CoTS following most major outbreaks. Many of the pioneers of CoTS research, including Robert Endean, Peter Moran, Peter Glynn, John Lucas, and Charles Birkeland, wrote extensively and provided phenomenal insights into the fundamental biology of CoTS. This early research provides a strong foundation for more informed and targeted research. The dual purpose of this edited monograph (and the associated Special Issue) is therefore, to synthesise prior research on topics of specific relevance, as well as highlight recent advances in the Biology, Ecology and Management of Crown-of-Thorns Starfish.

Morgan Pratchett and Sven Uthicke
Special Issue Editors

diversity

MDPI

Review

Thirty Years of Research on Crown-of-Thorns Starfish (1986–2016): Scientific Advances and Emerging Opportunities

Morgan S. Pratchett [1,*] ID, Ciemon F. Caballes [1] ID, Jennifer C. Wilmes [1], Samuel Matthews [1], Camille Mellin [2,3], Hugh P. A. Sweatman [2], Lauren E. Nadler [1,4], Jon Brodie [1], Cassandra A. Thompson [1], Jessica Hoey [5], Arthur R. Bos [6], Maria Byrne [7], Vanessa Messmer [1], Sofia A. V. Fortunato [1], Carla C. M. Chen [2] ID, Alexander C. E. Buck [1] ID, Russell C. Babcock [8] ID and Sven Uthicke [2] ID

[1] ARC Centre of Excellence for Coral Reef Studies, James Cook University, Townsville, QLD 4811, Australia; ciemon.caballes@my.jcu.edu.au (C.F.C.); wilmes.jennifer@gmail.com (J.C.W); sammatthews990@gmail.com (S.M.); lnadler@ucsd.edu (L.E.N.); jon.brodie@jcu.edu.au (J.B.); cassandra.thompson@my.jcu.edu.au (C.A.T.); vanessa.messmer@gmail.com (V.M.); sofia.valerofortunato@jcu.edu.au (S.A.V.F.); alexander.buck@my.jcu.edu.au (A.C.E.B.)
[2] Australian Institute of Marine Science, PMB No. 3, Townsville, QLD 4810, Australia; camille.mellin@adelaide.edu.au (C.M.); h.sweatman@aims.gov.au (H.P.A.S.); c.ewels@aims.gov.au (C.C.M.C.); s.uthicke@aims.gov.au (S.U.)
[3] The Environment Institute and School of Biological Sciences, University of Adelaide, Adelaide, SA 5005, Australia
[4] Scripps Institution of Oceanography, UC San Diego, La Jolla, CA 92037, USA
[5] Great Barrier Reef Marine Park Authority, Townsville, QLD 4810, Australia; jessica.hoey@gbrmpa.gov.au
[6] American University in Cairo, AUC Avenue, New Cairo 11835, Egypt; arbos@aucegypt.edu
[7] School of Medical Science and School of Life Science, University of Sydney, Sydney, NSW 2006, Australia; mbyrne@anatomy.usyd.edu.au
[8] CSIRO Oceans and Atmosphere, GPO Box 2583, Brisbane, QLD 4001, Australia; Russ.Babcock@csiro.au
* Correspondence: morgan.pratchett@jcu.edu.au; Tel.: +61-747-81-5747

Received: 22 August 2017; Accepted: 14 September 2017; Published: 21 September 2017

Abstract: Research on the coral-eating crown-of-thorns starfish (CoTS) has waxed and waned over the last few decades, mostly in response to population outbreaks at specific locations. This review considers advances in our understanding of the biology and ecology of CoTS based on the resurgence of research interest, which culminated in this current special issue on the *Biology, Ecology and Management of Crown-of-Thorns Starfish*. More specifically, this review considers progress in addressing 41 specific research questions posed in a seminal review by P. Moran 30 years ago, as well as exploring new directions for CoTS research. Despite the plethora of research on CoTS (>1200 research articles), there are persistent knowledge gaps that constrain effective management of outbreaks. Although directly addressing some of these questions will be extremely difficult, there have been considerable advances in understanding the biology of CoTS, if not the proximate and ultimate cause(s) of outbreaks. Moving forward, researchers need to embrace new technologies and opportunities to advance our understanding of CoTS biology and behavior, focusing on key questions that will improve effectiveness of management in reducing the frequency and likelihood of outbreaks, if not preventing them altogether.

Keywords: *Acanthaster*; coral reefs; disturbance; management; population outbreaks; research priorities

1. Background

Crown-of-thorns starfish (CoTS; *Acanthaster* spp., excluding *A. brevispinus*) are renowned for their ability to devastate coral reef ecosystems [1]. This is primarily because local densities of CoTS can increase from normally very low densities (<1 starfish per hectare) to extremely high densities (>1000 starfish per hectare) during periodic population outbreaks (e.g., [2]). Moreover, CoTS are one of the largest and most efficient predators on scleractinian corals [3]. Whereas most other individual coral-feeding organisms (e.g., *Chaetodon* butterflyfishes, and *Drupella* snails) cause only localized injuries or tissue-loss [4,5], adult CoTS can kill entire corals, including relatively large colonies. High densities of CoTS will, therefore, cause rapid and extensive short- to long-term coral depletion. In French Polynesia, for example, high densities of CoTS caused systematic coral loss around the entire circumference of the island of Moorea, killing > 96% of coral between 2005 and 2010 [6]. More broadly, outbreaks of *Acanthaster* spp. are a major contributor to sustained declines in coral cover and degradation of coral reefs at many locations throughout the Indo-West Pacific [7–9].

While there has been considerable research, and a large number of scientific articles (>940) focused on *Acanthaster* spp., extending back to the 1960s [10,11], research interest and funding has waxed and waned through this period (Figure 1). In Australia, a disproportionate number of research papers on CoTS have followed the initiation of each new wave of outbreaks (in 1962, 1979, 1993 and 2009), with apparent declines in publications as outbreaks subside. The main exception to this pattern was in 1979 to 1992, where the number of papers published on CoTS was lower than expected even after the initiation of the outbreak in 1979, whereas publication output was highest in 1992, immediately prior to the start of the third documented wave of outbreaks. The high number of publications in 1992 (46 publications) was partly due to two separate special issues on *Acanthaster* spp. [12,13], as well as an explicit recognition of the need to study CoTS during non-outbreak periods [14,15]. Scientific, management and political interest in CoTS did decline towards the end of the third documented wave of outbreaks in the early 2000s (Figure 1), with increasing concern about climate change and coral bleaching deflecting some attention from CoTS outbreaks. There has however, been a sustained increase in the number of scientific studies and publications on CoTS from 2010–2017 (Figure 1).

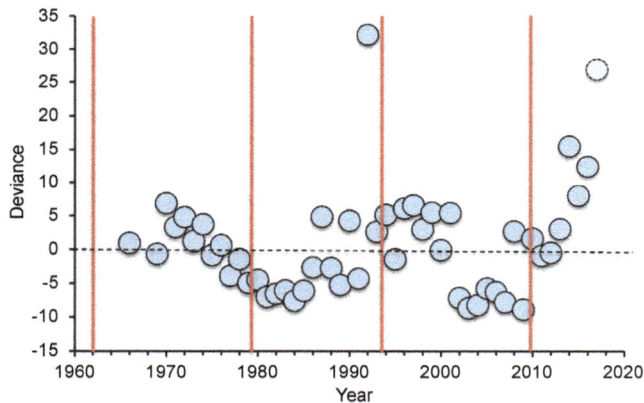

Figure 1. Interannual variation in the number of crown-of-thorns starfish (CoTS) publications relative to the start of successive outbreaks on Australia's Great Barrier Reef (as indicated by vertical red lines). The annual number of publications was determined based on a Web of Science search (topic = "*Acanthaster*" or "crown-of-thorns" and "Australia"), which was then detrended (showing the deviance from a linear regression between annual number of publications and year from 1965 to 2016) to account for increases in the number of publications through time. NB. Number of papers for 2017 is projected, based on number of papers published to end of April, 2017.

There is greatly renewed interest in *Acanthaster* spp. for two reasons. Firstly, renewed outbreaks of CoTS are occurring on the Great Barrier Reef (GBR), Australia, combined with unprecedented outbreaks at many locations throughout the Indo-Pacific [1]. Secondly, outbreaks of CoTS remain a major contributor to sustained decline in coral cover at many reefs throughout the Indo-Pacific (e.g., [16,17]) and scientists and managers alike recognize the critical need to halt and reverse this decline. Preventing and/or containing CoTS outbreaks is generally considered to be one of the most feasible management actions to reduce rates of coral mortality (e.g., [7]), thereby improving the capacity of reef systems to cope with threats due to climate change and other anthropogenic disturbances [18]. Despite persistent knowledge gaps regarding the ultimate cause(s) of outbreaks, and the considerable time and cost required for effective management of established outbreaks, outbreaks of CoTS are one of the principal causes of major coral loss (along with severe tropical storms and mass coral bleaching) that are amenable to direct and immediate intervention [7]. De'ath et al. [7] suggested that preventing outbreaks of *Acanthaster* spp. on the GBR could in itself reverse sustained declines in coral cover. The same is probably true for other reef regions, where outbreaks of CoTS are among the major causes of acute coral loss [9,17,19,20].

The purpose of this review is not to provide a comprehensive overview of CoTS biology or research, which are already available elsewhere [1,21,22], but to consider major advances (and apparent failings) in CoTS research over the past few decades. Our assessment of research progress is based on a critical judgment of the extent to which research (including original research presented within this special issue) has addressed the knowledge gaps highlighted by a seminal review in 1986 [21]. Ongoing research interest in *Acanthaster* spp. is largely driven by persistent controversy around the cause(s) of outbreaks and corresponding management actions [1]. This special issue (*Biology, Ecology and Management of Crown-of-Thorns Starfish*) reflects the latest resurgence in scientific interest and research on *Acanthaster* spp., unequivocally focused on better understanding the initiation and spread of outbreaks, as well as refining the capability and capacity for effective management. The papers presented in this special issue address: (i) environmental drivers of fertilization and early development [23,24]; (ii) larval nutrition, larval development and implications for the 'nutrient enrichment hypothesis' [25–29]; (iii) predation and implications for the 'predator removal hypothesis' [30–32]; (iv) factors influencing settlement [27,30]; (v) dispersal and genetics [33]; (vi) longevity, growth, and size-and-age relationship [34,35]; (vii) movement [36]; and (viii) control and management [16,37].

Aside from research to address current issues related to the effective management of CoTS outbreaks, there have also been some fundamentally new directions for CoTS research in the last decade, enabled by advances in research methods and analytical capabilities. Therefore, this review also considers some of the foremost new directions for CoTS research, related to (i) systematics and biogeography, (ii) genetic and genomic sampling, (iii) ecological modelling, and (iv) projected effects of environmental change. Notably, genetic sequencing of CoTS populations from throughout their geographic range (Red Sea to the eastern Pacific) has resolved that there are several distinct species [38,39]; *Acanthaster planci*, which is restricted to the northern Indian Ocean, is readily distinguishable from the Pacific species (*A.* cf. *solaris*) based on color, as well as a general lack of spines along the aboral distal portion of the arms (Figure 2). Overall, there are at least four distinct and geographically separated species [38,39], though most of the research and knowledge of CoTS biology and ecology comes from research in the western Pacific on *A.* cf. *solaris*. However, species-specific differences in behavior and biology may account for geographic variation in the occurrence of outbreaks, and their impacts on reef ecosystems [1]. Thus, there is a definite need for comparative studies across multiple species of *Acanthaster*, testing for differences in key demographic rates (e.g., growth and reproductive output) as well as feeding rates and dietary preferences.

Figure 2. Distinct species of crown-of-thorns starfish: (**a**) *Acanthaster planci* from northern Indian Ocean (Photo taken by M. Pratchett in Maldives), (**b**) *Acanthaster* cf. *solaris* from the Pacific (Photo taken by C. Caballes in Guam).

2. Advances in CoTS Research

It has been 30 years since the publication of a comprehensive review titled, "The *Acanthaster* Phenomenon" by Moran [21]. Moran [21] reported that crown-of-thorns starfish are probably the single most studied species (or species complex) on coral reefs, and yet there were many knowledge gaps that directly hampered effective management of population outbreaks. Accordingly, Moran [21] presented a list of 41 questions considered fundamental to understanding the causes and consequences of outbreaks. These questions were intended to guide research through the subsequent years and decades to improve both understanding and management of outbreaks of *Acanthaster* spp. They were presented within three broad categories (Larvae and Juveniles—21 questions, Adults—11 questions, and Effects on Communities and Processes—3 questions) together with a further six overarching questions. To assess the advances (or lack thereof) in our understanding of the biology of CoTS over the last three decades, as well as causes and consequences of outbreaks, we review progress against each of these 41 research questions, specifically highlighting the contributions of research articles presented in this issue. Where relevant, questions have been grouped together under a single section heading to minimize repetition.

2.1. Questions 1 and 3 (Larvae and Juveniles)—Are high nutrient conditions needed for the enhanced survival of larvae in the field? Can larvae develop and settle under 'non-bloom' nutrient conditions in the field? If so, can high densities of larvae be sustained under these conditions?

While receiving considerable attention, these questions are LARGELY UNRESOLVED.

One of the foremost hypotheses proposed to account for outbreaks of CoTS, the *larval starvation hypothesis*, is predicated on a link between rates of effective larval development and availability of suitable prey (mostly unicellular phytoplankton; e.g., [40–43]). Early studies by Lucas [40] suggested that rates of development and survivorship increased with increasing algal concentrations. Moreover, Okaji [41] demonstrated that CoTS larvae grew faster and had a higher survivorship with increasing chlorophyll-*a* (chl-*a*) levels (>2 μg chl-*a* L^{-1}), suggesting that enrichment of nearshore waters and phytoplankton assemblages would lead to increased densities of *A.* cf. *solaris* larvae [44,45]. Conversely, field-based experiments conducted by Olson [46], using an apparatus designed to rear larvae in situ, suggested that larvae grew well under low chl-*a* conditions. However, these results were potentially confounded by contamination and retention of phytoplankton within the experimental apparatus, resulting in higher than expected chl-*a* concentrations [41]. Nevertheless, in situ studies of larval development and survival, with simultaneous sampling of environmental parameters (e.g., cell-counts, chl-*a*, organic carbon content), are still needed.

Recent experimental studies have shown that elevated chlorophyll concentrations may not be necessary for CoTS larvae to complete development (reviewed in [47]). Laboratory experiments by Wolfe et al. [43,47] demonstrated high larval survival and settlement success across a broad range of nutrient levels, and most importantly, below the lower threshold levels previously suggested by Fabricius et al. [45]. The robust nature of larvae, even in oligotrophic conditions, may be driven by increased investment in provisioning of eggs by well-fed adults [25,48] and the morphological plasticity of larvae to respond to changes in the availability of exogenous prey [25,49]. Furthermore, there appears to be an upper threshold for optimal larval survival and development, whereby very high levels of nutrients have been shown to be deleterious to larval development for CoTS, and larval growth and development are maximized at close to normal background concentrations of chl-*a* [27,43]. Larval survival and size at set intervals (4, 7, and 10 days after fertilization), as well as settlement rate and size of newly settled juveniles after 18 days were significantly lower for treatments with the highest phytoplankton concentration (100,000 cells per mL; 10 μg chl-*a* per L) compared to intermediate food levels (10,000 cells per mL; 1 μg chl-*a* per L) [43]. Mortality rates for larvae under high algal concentration (100,000 cells per mL; 10 μg chl-*a* per L) were significantly higher from very early in larval development, compared to low (1000 cells per mL; 0.1 μg chl-*a* per L) and intermediate (10,000 cells per mL; 1 μg chl-*a* per L) food levels [27]. Taken together, these results suggest that chl-*a* levels between 0.4 and 1.0 μg per L may be the optimal range for larval survival and development.

The use of *Proteomonas sulcata* in both Wolfe et al. [43] and Pratchett et al. [27] was criticized by Brodie et al. [29] because of the very low chl-*a* concentration of this phytoplankton species. In addition, the high cell numbers (~100,000 cells per mL) are unlikely to occur in the field; even under flood conditions, algal cell numbers rarely exceed 1000 cells per ml [50]. Similarly, nutrient enrichment of natural seawater used in larval feeding experiments presented in Fabricius et al. [45] had maximum cell numbers of 4400 cells per ml (equivalent to ~5.2 μg chl-*a* per L in their study). As these numbers illustrate, it is unfortunate that the "nutrient hypothesis" (or "terrestrial runoff hypothesis") is currently expressed in terms of chl-*a* concentration. Although chl-*a* concentration is easy to measure, chl-*a* content varies widely among species and is not necessarily reflective of the nutritional value of the algae. Further studies need to focus on energy content (or at least organic carbon content) of food organisms (see [26]). Debate continues as to whether terrestrial runoff promotes primary outbreaks on the GBR [44,45] and/or plays an important role in fuelling subsequent secondary outbreaks [1,29].

One of the foremost limitations in understanding purported links between nutrient enrichment, increases in phytoplankton abundance or changes in phytoplankton composition, and the increased survivorship and settlement of *Acanthaster* spp. is the lack of systematic monitoring of relevant variables. On the GBR, for example, we lack the necessary information to explicitly compare nutrient availability and phytoplankton assemblages among reefs considered important for initiating primary outbreaks, versus those that support secondary outbreaks or are generally unaffected by CoTS outbreaks. It is also now understood that the algorithms used in the GBR to estimate chl-*a* from satellite remote sensing, in particular the MODIS satellite with the Aqua sensor, can give inaccurate and biased results; and therefore, usage should be avoided for absolute measures of chl-*a* [51]. In situ monitoring of nutrients, phytoplankton assemblages, and CoTS larvae, to relate size, abundance, and condition of larvae to local biological and environmental parameters (e.g., [49,52]) in the lead up to the next outbreak of *A.* cf. *solaris* on the GBR, would clearly establish whether CoTS larvae can develop and settle even in the absence of nutrient plumes and phytoplankton blooms. This would go a long way to resolving the extent to which sustained but gradual improvements in catchment management and water quality are an effective mechanism for limiting future outbreaks. In addition, new genetic methods (eDNA, next generation sequencing) to identify larvae [53,54] and potentially characterize phytoplankton abundance and community structure in the same water parcel provide opportunities to investigate the connection between nutrient inputs and changes in the quantity and quality of food for the larvae.

2.2. Question 2 (Larvae and Juveniles)—Do these types of conditions (i.e., high nutrient conditions and associated phytoplankton blooms) occur frequently in the field? If so, do they coincide with observed spawning periods and how long do they occur?

These questions are Largely Unresolved.

Spatial and temporal coincidence between high nutrient conditions and enhanced survivorship of CoTS larvae in the field would lend significant support to the *larval starvation hypothesis* [45,55]. However, establishing these links is complicated by aforementioned controversies about levels of food availability necessary to ensure development and survival of CoTS larvae (Section 2.1), as well as the inability to directly assess larval densities and survivorship in the wild. Fabricius et al. [45] argued that interannual variation in the cumulative annual discharge from the Burdekin river (one of five major rivers that discharge into the GBR) corresponds with the initiation of outbreaks of *A. cf. solaris* after allowing for inevitable lags in the timing of floods (and corresponding increases in larval densities) versus first reports of elevated densities of adult starfish in 1962, 1979, 1993 and 2009. However, the lag between major flood events and corresponding outbreaks of *A. cf. solaris* ranges from 2–5 years, possibly due to limitations in detecting the specific onset of outbreaks [1]. Moreover, major flood events may or may not initiate outbreaks, depending on the availability of coral food resources [25,48], the timing of floods relative to specific spawning periods and developmental rates for *A.* cf. *solaris* [45] and whether larval food supply is, in fact, a major limiting factor.

While there is a definite need for further research to resolve interannual variability in the specific pattern and occurrence of spawning by *A. solaris* on the GBR relative to particular flood events [52], the general onset of reproduction and larval development (December–March) broadly coincides with periods of heavy rainfall and increased likelihood of flooding [29]. Moreover, flood plumes and phytoplankton blooms occur frequently, almost annually within certain areas of the GBR (almost annually) and can persist for weeks to months [29]. These floodwaters provide increases in nutrient concentrations, especially from within heavily modified catchments with intensive agriculture [45,50]. However, it is the confluence of nutrient enrichment from flood plumes and relatively clean offshore waters that enable the proliferation of phytoplankton, potentially explaining why outbreaks of *A.* cf. *solaris* predominantly occur on mid-shelf reefs. If, however, phytoplankton blooms sufficient to sustain elevated densities of CoTS larvae occur almost annually, this cannot explain why initial outbreaks (primary outbreaks) occur relatively infrequently and at specific locations in the northern GBR [1]. Rather, persistent nutrient enrichment may be important in sustaining the proliferation and spread of outbreaks (secondary outbreaks) once they have become established at discrete reef locations [29]. If food is limiting, however, we might expect successive waves of outbreaks to peter out, especially in areas of the central GBR, which have lowest nutrient inputs (but see [56]).

Wooldridge and Brodie [57] explicitly acknowledge that high nutrient conditions and associated phytoplankton blooms occur quite commonly during summer in the northern GBR, but do not always initiate outbreaks of *A.* cf. *solaris*. At Green Island, for example chl-*a* concentrations exceeded 0.80 μg chl-*a* per L for prolonged periods at least six times between 1969 and 1998, though outbreaks developed only twice, in 1979 and 1993. Elevated nutrients may therefore, be a necessary precursor for outbreaks to become established, but there are other conditions that must also be met. Modelling studies by Wooldridge and Brodie [57] suggested that it is interannual variation in levels of larval retention (see [58]) that may explain when, and perhaps where, primary outbreaks become established. Importantly, hydrodynamic conditions would have promoted high levels of self-recruitment, rather than dispersal of larvae among widely separated reefs, in years prior to outbreaks arising in 1979 and 1993 [57]. However, it is yet to be tested whether primary outbreaks are directly attributable to iterative increases in population size, due to high levels of self-recruitment [59]. As stated previously, increased monitoring of relevant metrics for food availability is needed alongside fine-scale spatial and temporal sampling to document the initiation and spread of CoTS outbreaks.

2.3. Question 4 (Larvae and Juveniles)—How important is diet in influencing the survival of larvae? Is survival more dependent on the diversity rather than density of food species? What other factors influence the survival of larvae?

The first two components are MOSTLY RESOLVED, but larval survivorship is yet to be studied in field settings (Section 2.12).

The abundance of specific phytoplankton is much more important than overall abundance of phytoplankton (and corresponding chl-*a* concentrations) for promoting rapid development and survival of CoTS larvae. Larvae of *A.* cf. *solaris* feed predominantly on mid-sized phytoplankton (e.g., dinoflagellates and pennate diatoms >5 μm) [40,60–62], whereas tropical coral reef waters are generally dominated by picoplankton (e.g., pelagic cyanobacteria—*Synechococcus* and *Prochlorococcus* <2 μm), which typically make up >50% of total primary production in oligotrophic waters (e.g., [63–65]). As such, total chl-*a* concentrations may grossly overestimate food availability for CoTS larvae. Elevated nutrients due to terrestrial runoff and upwelling do promote increased dominance of phytoplankton species with larger cell sizes [65–67], potentially benefitting CoTS larvae, but it is nonetheless necessary to sample phytoplankton assemblages directly (rather than relying on chl-*a* concentrations) to assess the conditions that promote larval survival and the onset of outbreaks.

Very few studies have specifically considered the feeding selectivity of CoTS larvae, other than showing that there is strong size selectivity. Okaji et al. [68] demonstrated that clearance rates of larval feeding were significantly lower for smaller phytoplankton (1–2 μm) compared to phytoplankton with larger cell sizes (4–5 μm), even when smaller phytoplankton were overwhelmingly dominant. Also, heterotrophic bacteria appear to have a negligible role in larval nutrition [69]; though CoTS larvae are able to assimilate and utilize dissolved organic matter (DOM) in the water column [60,70], as well as coral-derived organic matter, such as mucus and associated microorganisms [28]. Mellin et al. [26] tested for selective feeding by CoTS larvae among phytoplankton species of similar size (>5 μm). CoTS larvae consistently preferred algal species with the highest energetic content (*Chaetoceros*, *Dunaliella*) over microalgae with lower energy content (*Pavlova lutheri*, *Phaeodactylum tricornutum*), which would presumably lead to higher growth rates and elevated survival of CoTS larvae, though this was not tested. Although algal species used in that study have been commonly used in aquaculture and are within the size-range preferred by CoTS, these species are not naturally present at high concentrations on the GBR (except for *Chaetoceros* spp.). It is important therefore, to establish feeding selectivity for algal species that are particularly dominant during flood events (e.g., *Skeletonema*; [50]) and DOM, as well as testing for variation in larval growth and survivorship with changes in the availability of specific prey types.

The extent to which development and survival rates of CoTS larvae are constrained by exogenous food availability is equivocal, and somewhat dependent on several other factors (e.g., maternal provisioning of larvae, and vulnerability to predation). Caballes et al. [48] demonstrated that differences in the nutritional condition of female *A.* cf. *solaris*, based on contrasting diets (e.g., *Acropora* versus *Porites*), have a major bearing on the growth and performance of their progeny. Well-fed females provision their offspring with increased levels of endogenous energetic reserves, which not only allows larvae to withstand prolonged periods of starvation, but also enables them to grow larger and feed more efficiently [25,49]. Larval survivorship in the wild will also be limited by competition, predation and environmental constraints on development (see Sections 2.13 and 2.14). If higher food concentrations also benefit predators of CoTS, this has the potential to suppress larval survival and outbreaks [71]. There are no empirical data on rates of predation for CoTS larvae in the wild, and such information will be very difficult to obtain, though CoTS larvae are certainly vulnerable to predation [72].

2.4. Question 5 (Larvae and Juveniles)—Do certain physical conditions occur in the field that cause the increased survival of larvae? Do these conditions act in conjunction with any other factors?

These questions are LARGELY RESOLVED, though there is scope for more work on synergistic effects of climate-induced changes in physical conditions.

The specific physical conditions that promote larval development and survival for *Acanthaster* spp. has received a great deal of attention (e.g., [40,73]) and there has been a recent resurgence in research on the environmental tolerances of CoTS larvae aimed at establishing the vulnerability of *Acanthaster* spp. to ocean warming and acidification [24,74–76]. Temperature is widely regarded as the foremost abiotic factor influencing development rates and survivorship of planktonic larvae [77], and *Acanthaster* spp. are sensitive to extreme temperatures at all stages of their life-cycle [24,61]. Fitness of CoTS larvae is generally highest at 26–30 °C [24,76]. There is increasing evidence that CoTS larvae are adversely affected when exposed to temperatures $\geq$ 30 °C [75], suggesting that ocean warming suppresses population outbreaks at low latitudes. Projected effects of ocean warming are further compounded by constraints on fertilization and larval development due to ocean acidification [24,74,75], though it is possible that *Acanthaster* spp. could acclimate or adapt to changing environmental conditions [78].

Aside from temperature, salinity is considered to have an important influence on reproductive success for *Acanthaster* spp. [24,79]. While echinoderms are generally very sensitive to changes in salinity, Lucas [79] showed that larval survival was 3-fold higher at 30‰ salinity compared to ambient conditions. If so, temporary declines in salinity could further enhance the beneficial effects of nutrient inputs, during flooding [44,55]. However, Caballes et al. [24] showed that rates of fertilization, gastrulation and cleavage were generally high between 30‰ and 34‰ and declined significantly at salinities <30‰.

2.5. Question 6 and 7 (Larvae and Juveniles)—How long do [CoTS] larvae spend in the plankton before settling? What is the maximum period of time they can spend in this phase and yet still be able to settle? How far can [CoTS] larvae be dispersed in the field?

These questions are LARGELY RESOLVED.

The time that larvae spend in the plankton, or planktonic larval duration (PLD), is constrained by the minimum pre-competency period, which is the necessary time for larvae to complete development before being capable of settling, and the maximum competency period, which is maximum time that larvae can spend in the plankton and still be capable of effective settlement [27]. The minimum time taken for CoTS to develop into late-stage brachiolaria larvae, which are assumed to be competent to settle, is just 9 days [80], though actual settlement has never been documented <14 days post-fertilization [46]. At the other end of the spectrum, Pratchett et al. [27] recorded settlement among larvae of *A*. cf. *solaris* up to 43 days post-fertilization. However, settlement rates peaked at 22 days for optimal (intermediate) food levels and declined through time due to limited survivorship >30 days post-fertilization and reduced settlement competency of surviving larvae [27].

While CoTS larvae may settle in as little as 9–14 days after fertilization, their maximum competency period and capacity for long-distance dispersal is largely unknown. The maximum recorded longevity for CoTS larvae is 50 days [27], and could be even longer under conditions of limited food availability and if deprived of opportunities to settle. However, CoTS larvae are planktotrophic and must maintain certain levels of energetic reserves to complete metamorphosis and settlement [27]. Moreover, rates of larval mortality in the wild are likely to be even higher than have been documented in experimental studies (e.g., due to predation), such that few larvae are likely to persist beyond 30 days, let alone 50 days. Assuming average daily rates of natural mortality for echinoderm larvae (~0.16 per day; [81]), CoTS larvae will have a survival probability of approximately 0.82% after 30 days, and 0.03% after 50 days. For the most part, CoTS larvae are expected to be dispersed only 10 s–100 s km between reefs [82], if not entrained within the confines of their natal reef [83,84]. Genetic sampling of CoTS populations demonstrated that there is effective connectivity (reflective of ecological significant levels of larval dispersal) between reefs separated by <1000 km [33,85,86]. However, there tends to be very strong genetic differentiation of CoTS populations among geographic provinces [87,88], not to mention distinct species in different ocean basins [39], suggesting that there is extremely limited connectivity, and therefore, negligible larval dispersal, at distances of >1000 km.

2.6. Question 8 and 14 (Larvae and Juveniles)—What factors are important in causing dispersal [of CoTS larvae]? Do larvae tend to settle on those reefs from which they were propagated or do they generally recruit to reefs other than the parent reef?

The first component is WELL RESOLVED, but there is limited empirical information regarding rates of dispersal versus retention.

CoTS larvae are, for the most part, passively dispersed by ocean currents [89]. Therefore, the primary factors that influence dispersal are (i) how long larvae can persist while still retaining the capacity to settle (see Section 2.5) and (ii) oceanographic conditions, specifically, the strength and direction of water movement, during spawning and larval development [85]. On the GBR, the extent to which larvae are retained and settle on their natal reefs (self-recruitment), rather than being dispersed, has been modelled for at least one small sub-region and may vary among years with changes in ocean current velocities driven by El Niño Southern Oscillation (ENSO) [57]. However, it is not clear whether such effects exist in other regions at other phases of the ENSO cycle, particularly those that are putative seed areas for outbreaks. Weak or variable along-shore currents, which occur during neutral phases of the Southern Oscillation Index (SOI) may promote strong larval retention or very limited dispersal, which is fundamental to the progressive accumulation of CoTS within a given location and is likely to give rise to primary outbreaks [57,59]. In contrast, strong directional (southerly) along-shore currents associated with strong El Niño or La Niña conditions will increase the likelihood of inter-reef dispersal, which could lead to proliferation of outbreaks once they become established [57]. These ideas are however, based on temporal autocorrelation in the initiation of outbreaks, rather than explicit empirical data on rates of self-recruitment versus larval dispersal.

High rates of self-recruitment by *Acanthaster* spp. may be reinforced by conspecific chemo attraction of settling larvae towards feeding aggregations of adult CoTS [2,30]. In static choice chamber experiments, Cowan et al. [30] showed that CoTS larvae were significantly attracted to adult conspecifics, which could lead to elevated rates of settlement on reefs already infested with high densities of CoTS. While settling in the presence of adult conspecifics may seem maladaptive due to ultimate competition for coral prey, this strategy may limit predation by sessile invertebrates [61,90] leading to overall increases in larval survivorship. However, the spread of population outbreaks, determined based on field surveys [56,91,92] and modelling [56,58,82–84,93,94] shows that at least some larvae must be dispersed and settle on non-natal reefs, regardless of the presence of adult CoTS.

Improvements in hydrodynamic models, combined with advances in computational power and new methods for analyzing patterns of particle dispersal, are providing increasingly resolved and tractable models to inform patterns of initiation and spread for CoTS outbreaks [58,82,93,95,96]. However, these models are potentially very sensitive to the precise timing of spawning and the relevant speed and direction of currents, and predictions arising from these models need explicit testing based on extensive spatial and temporal sampling to resolve the occurrence and timing of outbreaks. Alternatively, genetic approaches may be used to explicitly resolve actual connections among discrete populations to validate dispersal patterns. However, using genetics to track the spread of outbreaks has proved difficult for CoTS due to the low levels of genetic differentiation apparent when using existing markers [33,97,98]. On the GBR, for example, genetic sampling during outbreaks has failed to resolve any structure [33,97], indicating rapid expansion in population size from multiple, undifferentiated latent populations. Similarly, studies elsewhere in the Pacific have identified largely homogeneous populations within specific reef systems [86,99,100], though CoTS generally exhibit substantial regional, archipelagic genetic structuring [88], reflective of limited large-scale dispersal. Greater resolution in genetic structure among outbreak populations, enabling greater insight into the source and spread of outbreaks, may be possible using more extensive and comprehensive sampling (e.g., single-nucleotide polymorphisms or SNPs), but these are yet to be tested for any *Acanthaster* sp.

2.7. Question 9 (Larvae And Juveniles)—Is there a positive correlation between larval density, recruitment density, and adult density?

Given previous limitations in sampling larvae and recruits, this question has not been addressed and is UNRESOLVED.

High densities of CoTS larvae and high recruitment will intuitively lead to increased densities of adult CoTS, and are a fundamental precursor to manifest rapid population outbreaks (mostly, secondary outbreaks). However, the more important question is whether it is the local densities of larvae (larval supply) or effective rates of recruitment that generally limit adult densities? There are also important, and as yet unresolved issues, about how far and how fast CoTS can move within and among reef habitats after they have settled (Section 2.18), which will determine relevant scales of recruitment limitation. Most hypotheses that seek to explain the initiation of CoTS outbreaks assume that larval supply is generally limiting, such that outbreaks arise due to increased reproductive success and/or larval survivorship [1]. However, the inability to quantify larval supply, settlement, and recruitment has so far prevented explicit testing of such assumptions. New methods aimed at measuring these processes are being developed and tested [34,53] and will not only provide new opportunities to test questions pertaining to recruitment limitation, but may also provide an early warning system for detecting new and renewed outbreaks.

The relationship between adult densities of *Acanthaster* spp. and their corresponding reproductive output versus local densities of larvae or juveniles (stock-recruitment relationships) is important for understanding the role of adult biology and behavior in initiating outbreaks, as well as informing the effectiveness of population regulation based on culling of adult starfish. Given the high fecundity of individual CoTS [101], larval production is likely to asymptote at relatively low adult densities [102,103] such that larval production may be largely insensitive to changes in adult abundance above a certain threshold. If so, this would mean that local densities would need to be reduced to very low levels before adult culling would have any meaningful impact on reproductive output and the progression of outbreaks. Moreover, the distribution and proximity of spawning starfish may be more important than adult densities in determining reproductive success (though it would be expected that there must be some relationship between these factors). Babcock and Mundy [104] showed that *A.* cf. *solaris* achieve remarkably high rates of fertilization even when spawning male and female starfish were separated by >30 m. However, fertilization success is fundamentally dependent on spawning synchrony, which appears to be triggered via intrinsic cues (pheromones) and will therefore, be most effective when starfish are aggregated [52]. Still, it is unclear whether *Acanthaster* spp. actively aggregate to spawn, and what environmental cues induce spawning.

Recent modelling of fertilization success in CoTS suggests that both density and aggregation are important to reproductive success at low densities with a threshold density for enhanced reproductive success of 3 starfish per hectare when individuals are moderately aggregated [105]. Reproductive success increased linearly above this density. At the highest levels of density and aggregation fertilization success for individuals did not increase due the increases in unsuccessful polyspermic fertilizations although population level zygote production did continue to increase [105]. The reduced fertilization success at high levels of aggregation may in part explain the relative lack of aggregation by CoTS at the time of spawning [106].

2.8. Question 10 (Larvae and Juveniles)—Where do larvae occur in the water column? Does their position vary throughout their planktonic period? What factors are responsible for determining their position?

These questions are LARGELY UNRESOLVED and important for understanding the environmental and nutritional conditions to which larvae are exposed during their development.

In laboratory cultures, hatched gastrulae of *A.* cf. *solaris* swim upward and remain close to the surface (negative geotaxis) throughout much of the formative period of their development [61]. Larvae then become negatively buoyant at the late brachiolaria stage and actively orientate towards the bottom in search of potential settlement substrates [41]. It is assumed therefore, that CoTS larvae are likely to be largely concentrated in surface waters in the wild. Accordingly, CoTS DNA have been detected in near-surface plankton tows along 320 km of coastline in the northern GBR [53]. Similarly, plankton tows at 7 m below the surface along reefs in southern Japan have also yielded high concentrations of advanced-stage brachiolaria larvae [107]. CoTS larvae are also capable of swimming, albeit at limited speeds (~0.4 mm per second), via ciliary movement that causes the body to rotate on its long axis [61]. The horizontal transport of CoTS larvae has typically been considered a passive process that is mainly mediated by currents [58,87,89]. However, current velocity and direction can vary with depth, so the vertical position of larvae and larval behavior in response to thermoclines, haloclines, or pycnoclines will have important implications for distance and direction of larval transport (e.g., [108–111]). Explicit field sampling is still required to ascertain the specific position of CoTS larvae at different stages throughout their development and under varying conditions; but this may now be possible with the development of new genetic quantification methods [54].

2.9. Question 11 and 13 (Larvae and Juveniles)—Where do larvae settle in the field? Is it in shallow or deep water on reefs? Are there particular areas on reefs which are more suitable for settlement than others? Do they settle in high densities?

There have been significant insights on patterns and rates of settlement by specific, localized studies (e.g., [112]), but these questions are MOSTLY UNRESOLVED.

One of the foremost controversies surrounding settlement patterns for *Acanthaster* spp. is the extent to which larvae settle in deep-water (>20 m) versus shallow reef environments. Although earlier reports suggested that settlement occurred in shallow reef environments [2,113] and high densities of newly settled and juvenile *Acanthaster* spp. have never been recorded in deep water (>20 m depth), the deep-water recruitment hypothesis [114] has gained a lot of attention. CoTS larvae were suggested to settle mainly in deep water, at the base of reef slopes [114] because (i) highest rates of settlement and metamorphosis occurred on coral rubble encrusted with the sciaphilic crustose coralline algae (CCA), *Lithothamnium pseudosorum*, which was found predominantly in deep water habitats (but are also common in caves, crevices and overhangs in shallow water; S. Uthicke, personal observation); (ii) late-stage brachiolaria larvae are negatively buoyant and are expected to be concentrated along reef margins, such that larvae will ultimately fall out in deep water [83]; (iii) few newly settled (0+ year class) starfish have historically been found in shallow reef environments; and (iv) on some reefs (e.g., Davies Reef in the central GBR), high densities of adult CoTS were initially detected moving up from deep water [114]. To test whether *Acanthaster* spp. preferentially settle in shallow (<5 m depth) or deep-water habitats (>20 m depth), standardized settlement collectors (e.g., [115]) should be deployed across a range of depths, and this research is currently underway. It is possible however, that CoTS larvae settle across a broad range of depths, but settle preferentially in areas that provide best access to food and shelter. Alternatively, they may settle indiscriminately among habitats, but have vastly different rates of post-settlement survival depending on local abundance of prey, shelter, and/or predators (Section 2.11).

Field-based studies on patterns of CoTS settlement are largely focused on the detection of newly settled CoTS, rather than explicitly measuring settlement rates in different habitats (but see [115]). Even so, understanding the habitat preferences of newly settled CoTS has been greatly constrained by the small size and cryptic nature of newly settled individuals [112,114]. Until recently, few newly settled (0+ year class) *A.* cf. *solaris* had ever been detected on the GBR [11,114,116]. At Suva Reef in Fiji, however, high densities of very small (10–32 mm in diameter) *A.* cf. *solaris* were detected in July 1984 [112]. These individuals were presumed to have settled *en masse* in January 1984, but were not detected until they were ≥6 months old. Even so, the newly settled (0+ year class) starfish were mainly found on the encrusting coralline algae, *Porolithon onkodes*, on the underside of rubble and coral blocks [112], suggesting that they had settled in the area and habitat in which they were found. Significant densities of newly settled *A.* cf. *solaris* were found immediately behind the exposed reef crest, in very shallow habitat dominated by rubble and intact skeletons of robust corals dislodged during cyclones and tsunamis [112]. Newly settled CoTS were also sighted off the reef slope on the windward side of Suva Reef (6–8 m depth), indicating that their distribution extended subtidally [112]. At Iriomote Island, Japan, Habe et al. [117] detected highest densities (0.82 individuals per m^2) of newly settled CoTS on reef slopes (6.0–9.3 m depth), and lowest densities (0.06. individuals per m^2) on the reef flat (0.8–1.2 m).

During recent opportunistic sampling on the GBR, Wilmes et al. [34] collected 3532 juvenile *A.* cf. *solaris* ranging in size from 3 to 64 mm in diameter. Sampling was conducted across 64 reefs throughout the course of 2015, with searching concentrated on visible patches of CCA on dead corals or coral rubble. Newly settled (0+ year class) starfish were collected from a range of depths (up to 15 m depth) and habitats. However, collections were mainly intended to inform growth models (see Section 2.16) and so limited information was collected regarding the specific habitat conditions and exact densities. Despite limited success in the past [116,118], current work by Wilmes et al. [34] demonstrates that 0+ year juveniles can be effectively sampled in the field, and much more sampling is required to establish variation in rates of settlement across different reef areas, depths and habitats. There are also methods available for measuring settlement rates of *Acanthaster* spp. using settlement collectors constructed from high surface area plastic biospheres [115], which will be important to differentiate where larvae settle versus where they survive and are actually recorded several months after settlement (Sections 2.9–2.11). Previous constraints on the use of settlement collectors were the time and effort required to manual sort and visually detect newly settled CoTS, though modification of genetic sampling protocols used for larval detection [53,54] may overcome such constraints.

2.10. Question 12 (Larvae and Juveniles)—Do larvae tend to settle on a particular type of surface? What factors are important in determining the type of surface chosen by larvae for settlement?

These questions are LARGELY RESOLVED.

Settlement preferences of *Acanthaster* spp. are dictated by both physical and biological habitat structure. Larvae settle preferentially in habitats with fine-scale topographic complexity, so that the larvae are completely hidden within the carbonate matrix, or among coral rubble, prior to metamorphosis [80]. Ormond and Campbell [119] demonstrated that skeletons of dead *Acropora hyacinthus* were among the most preferred settlement substrates, probably owing to the fine-scale complexity provided by individual calices and branchlets. Conversely, CoTS will rarely settle on glass or ceramic tiles [119–121]. However, CCA is a strong settlement inducer, and biological stimuli may override physical microhabitat preferences [120]. Observations of newly settled *A.* cf. *solaris* in the field [34,112,122], have revealed a strong and consistent association with CCA, which is expected given that newly settled starfish feed almost exclusively on coralline algae [61,112,123].

Settlement experiments conducted under laboratory conditions [27,42,43,61,121,124] reaffirm that CCA is important for inducing CoTS settlement. When examined microscopically, the surface of CCA is roughly textured—this fine scale topographic complexity can provide a tactile stimulus for larvae to settle [80]. Conversely, Johnson et al. [121] argued that a tactile stimulus was unlikely given that settlement was high on live CCA as opposed to physically similar, but boiled, bleached, or autoclaved CCA fragments. They also observed high rates of settlement among larvae that were physically separated from CCA using a mesh, suggesting that settlement may be chemically mediated [121]. However, bioassays with common marine invertebrate settlement inducers, γ-amino butyric acid (GABA) and potassium chloride (KCl) at different concentrations, did not induce settlement and metamorphosis in CoTS larvae [121]. Antibiotic treatment of highly inductive shards of CCA significantly reduced settlement to low levels, indicating that settlement may be mediated by chemical cues produced by epiphytic bacteria [121]. Settlement was inhibited in the absence of bacteria and larvae always settled on sections of CCA thalli that had high densities of bacteria, but not where epiphytic bacteria were sparse [121,125]. However, surface bacteria were not inductive when isolated from soluble algal compounds, suggesting that bacteria require the algal substrate to produce inductive compounds or that compounds from both the bacteria and CCA are required to induce settlement [125]. It appears that both tactile and chemical stimuli may play a role in determining settlement preferences, though further field sampling is required to establish the extent to which these preferences determine settlement patterns in the wild (e.g., [119]).

2.11. Question 15 (Larvae and Juveniles)—Do juveniles tend to be in shallow or deep water on reefs? Does this location vary depending on whether or not the reef has recently suffered an outbreak of adults?

Given limited effective sampling of recruits, these questions are LARGELY UNRESOLVED.

The distribution of juvenile CoTS will be largely dictated by patterns of larval settlement (see Section 2.9), though these patterns may be greatly altered and obscured by differential rates of post-settlement mortality and movement [126]. Mortality rates for newly settled (post-metamorphosed) juvenile CoTS are unknown, but are expected to be very high due to the combination of predation, disease, and food limitation ([126]; Section 2.12). In terms of moderating the distribution and abundance of juvenile CoTS, the key question is whether rates of mortality vary spatially (e.g., among habitat types or with depth). Keesing and Halford [126] suggested that known predators on very small CoTS occur in particularly high abundance among dead coral rubble, though it is also possible that high complexity of these habitats moderates actual predation rates. Conversely, predators associated with coral-rich habitats may represent an even greater threat to survival of juvenile CoTS [30]. There will also be an inherent tradeoff between the risk of predation and the necessary settlement cues and food resources that will determine the extent to which juvenile CoTS are associated with different habitat types [30]. Another key factor that will potentially influence the survival and therefore relative abundance of juvenile CoTS in different habitats is the availability of suitable prey, and corresponding effects on the size and growth of individuals (Section 2.15).

The locomotor capacity of *A.* cf. *solaris* is size-dependent [127], such that very small CoTS move very slowly [61] and are unlikely to venture far from where they settle [126]. With sustained directional movement, which is unlikely given their generally cryptic behavior and limited energetic reserves, newly settled CoTS could travel only 5 m per week. As juvenile CoTS transition from feeding on CCA to coral, it is to be expected that their distribution and habitat-associations will also change. Notably, coral feeding juveniles are predominantly found in areas with moderate to high cover of scleractinian corals [112,116], whereas newly settled individuals targeting CCA will tend to be more abundant in habitats with low coral cover. In Fiji, for example, high densities of newly settled *A.* cf. *solaris* were first detected immediately behind the exposed reef crest in habitat with very limited coral cover [112]. The following year, the same cohort of starfish was concentrated along the reef crest, feeding on abundant coral within this zone [112,128]. There has not however, been equivalent sampling in other areas to establish the generalities of these ontogenetic shifts in habitat use. It is also very likely that the distribution of juvenile CoTS will be affected by depletion of coral prey by high densities of adults. Moran [21] suggested that juvenile starfish predominate in shallow-water habitats on reefs subject to outbreaks because large adults generally avoid habitats subject to high levels of wave action and surge. During the initiation of outbreaks, however, CoTS larvae may settle at the base of reef slopes and then gradually move to shallow reef environments as they grow and mature ([21,121]; see Section 2.9).

2.12. Question 16 (Larvae and Juveniles)—What are the mortality rates of larvae and juveniles in the field?

Field-based rates of survivorship/mortality are LARGELY UNRESOLVED, and represent a major limitation in understanding the population dynamics of *Acanthaster* spp.

While there is some preliminary information regarding mortality rates of CoTS larvae and juveniles from laboratory and experimental studies, estimating natural mortality rates for CoTS throughout the formative stages of their life history (Figure 3) is extremely challenging. In culture, concentrations of CoTS larvae steadily decline with time [25,27,43,48], indicative of significant rates of intrinsic larval mortality [81], which depend on food availability (Section 2.1) and food quality (Section 2.3). Mortality rates recorded for larvae in laboratory cultures are substantial [43]. However, there is increasing evidence that both larvae and juveniles are vulnerable to predation (Section 2.13), which will further increase rates of mortality. Pratchett et al. [27] showed that larval survivorship and settlement rates were highest for larvae maintained at intermediate food levels, while higher and lower levels of food availability resulted in higher mortality rates and delayed development. The recent feeding history and nutritional condition of maternal gamete sources of larvae may also interact with larval diet to affect larval survival, growth, and development [25].

The factors affecting survival during the settlement and metamorphosis are still poorly understood. Rearing of larvae in laboratory conditions and settlement assays show high mortality rates during settlement and metamorphosis [27,43,61,126]. Yamaguchi [61] suggested that ~50% of larvae are consumed by epibenthic fauna during settlement (see also [30]). However, settlement assays on CCA that were carefully cleaned of epifauna still resulted in mortality rates as high as 84% during settlement [126], which suggests that the physiological condition of competent larvae may influence intrinsic mortality rates. More recent laboratory assays have shown that larval nutritional history influences mortality rates during settlement [27,43]. The availability of suitable microhabitats for settlement (Section 2.10) may also influence mortality rates [126]; for example, mortality rates increased abruptly in larvae that have not settled after 30 days [27]. Minor changes in mortality rates before and during settlement can potentially have a significant impact on the eventual adult population size, particularly over the course of several generations, especially given that a single female starfish can produce over 100 million eggs in a single year [101]. Further investigations of factors influencing mortality rates during these critical phases, under natural conditions, are warranted.

Mortality rates of post-settlement juveniles are likely to be influenced by predation, food availability, and disease [129]. Keesing and Halford [130] recorded significant daily mortality rates (~6.5%) for small juveniles (~1-mm diameter) in the field. Mortality rates appear to decline with size, whereby mortality rates for ~3-mm and ~5-mm juveniles were 1.24% and 0.45% per day, respectively [130]. Visual predators (e.g., reef fishes) have also been implicated as a source of mortality due to the cryptic and nocturnal behavior of juvenile CoTS. However, in a field experiment where laboratory-reared juvenile starfish were placed in an area with suspected fish predators present, Sweatman [131] found that losses attributable to predation were low (0.13% per day)—much lower than the mortality rate due to predation (1.5% per day) predicted to have an impact on population regulation [132]. These field studies highlight the importance of epibenthic predators (e.g., [30]) in regulating population sizes during the post-settlement stage [133]. The influence of food availability may be more pronounced once CoTS shift their diet from CCA to coral, since herbivorous juveniles are unlikely to be food limited in the field (Section 2.14). Food availability and the conditions of juvenile CoTS may also influence susceptibility to predation [126] and disease [112,128]. Using hypothetical rates of post-settlement mortality, Keesing and Halford [126] argued that small changes in post-settlement mortality can have a disproportionate effect on the population size of adult starfish.

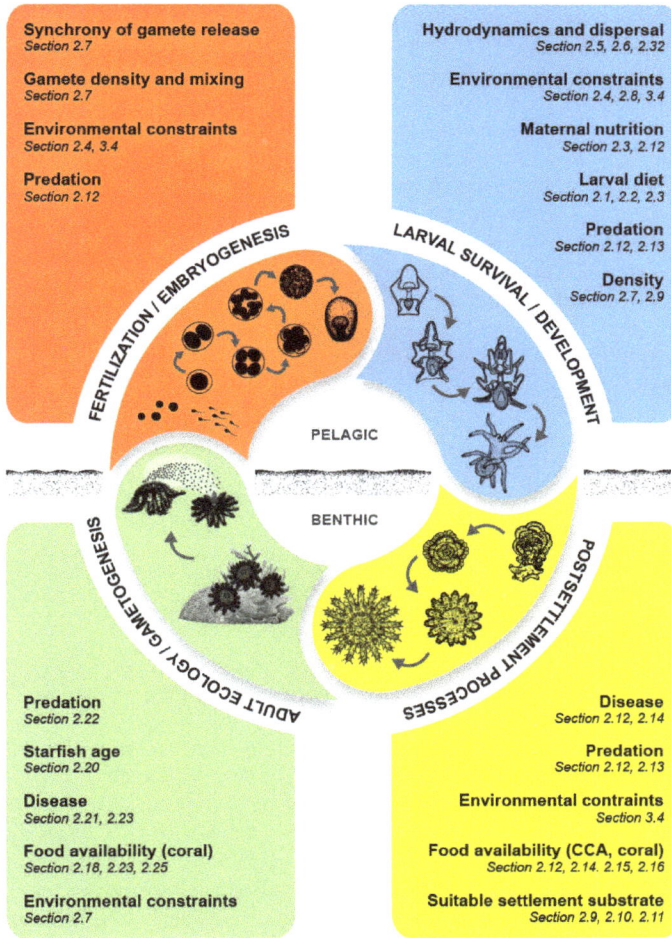

Figure 3. Generalized life cycle of *Acanthaster* spp. indicating potential sources of mortality and major bottlenecks in population replenishment, which are likely to be fundamental in understanding the occurrence of population outbreak. Late-stage larvae digitized from illustrations by M. Yamaguchi [61] and D. Engelhardt [134].

2.13. Question 17 (Larvae and Juveniles)—Is predation important in determining the density of larvae and juveniles? What are the main predators of each stage?

These questions are LARGELY UNRESOLVED.

Early field observations [11] and laboratory experiments [135] suggested that CoTS larvae are unpalatable to planktivorous fishes, such that predation was considered to exert limited influence on larval mortality. CoTS larvae contain steroidal saponins, which may have specific anti-predatory functions, as demonstrated by experimental assays showing planktivorous fishes discriminating against saponin-impregnated food pellets [136]. However, more recent experiments have found that planktivorous fishes readily consumed CoTS larvae [72]. Predation on larvae by scleractinian corals [2], predaceous zooplankton [137], and fishes [11] may therefore be an important determinant of larval survivorship. There have not, however, been any studies aimed at estimating predation or mortality rates of CoTS larvae in the field, and the relative contributions of different groups of predators is unknown. Cowan et al. [72] showed that planktivorous damselfish may be capable of consuming sufficient numbers of CoTS larvae (up to 158 larvae per hour) to effectively suppress larval settlement, especially when starfish are in low abundance. This level of predation may be sufficient to prevent the onset of outbreaks. However, reef-based predators are likely to be overwhelmed by extremely high densities of larvae [72], accounting for the propagation of outbreaks once they become established.

Unlike larvae, newly settled CoTS have long been considered to be extremely vulnerable to predation [21,138]. Indeed, the highly cryptic and generally nocturnal habits of newly settled *Acanthaster* spp. are considered to be adaptations to moderate natural predation rates. Keesing and Halford [126] demonstrated that epibenthic predators were the major source of mortality for captive reared *A.* cf. *solaris* that were deployed to field environments within boxes filled with freshly collected rubble. Predation rates were estimated to be 5.05% per day for 1-month old *A.* cf. *solaris*, which declined to 0.85% per day for 4-month old starfish [130]. These results suggest that epibenthic predators may be a major factor in regulating local densities of *Acanthaster* spp. However, natural predation rates may be moderated by selective settlement within microhabitats with relatively few benthic predators [30]. Cowan et al. [30] demonstrated that competent larvae of *A.* cf. *solaris* were able to detect some predators in the substrate and preferentially settled in microhabitats without predators, where possible. Variation in the abundance of benthic predators may therefore, influence settlement patterns of *A.* cf. *solaris*, even if they do not cause significant predation mortality in newly settled CoTS [30].

2.14. Question 18 (Larvae and Juveniles)—Apart from predation what other factors are important in causing the mortality of juveniles (e.g., disease, lack of nutrients)?

This question is LARGELY UNRESOLVED.

Food limitation and constraints on the physiological condition of juvenile CoTS will have consequences for survivorship, though the ultimate factors responsible for mortality will be predation and/or disease. Disease was shown to contribute to mass-mortality of juvenile CoTS in Fiji, which was attributed to an undescribed sporozoan pathogen (intracellular parasite), which ultimately resulted in the extirpation of the entire cohort [128]. The general susceptibility of juvenile CoTS to disease is largely unknown, but likely depends on the conspecific densities and individual condition, which in turn may be influenced by availability of specific prey. While certain coralline algae may promote higher growth or survivorship (Section 2.15), crustose coralline algae are considered to be ubiquitous within potential settlement habitats [139–141]. Therefore, it is expected to be coral prey that ultimately constrains the growth and survivorship of juvenile CoTS. In the absence of suitable coral prey, juvenile CoTS may continue to feed on a CCA for >2 years [123,126], though timely transition to coral prey leads to marked increases in growth. In the wild, it is expected that CoTS that have limited access to coral prey will experience high rates of mortality [126], though this is yet to be explicitly tested (Section 2.12). If newly settled CoTS can withstand temporary or localized depletion of coral prey by continuing to feed on coralline algae, these latent populations may proliferate following the recovery of coral populations and assemblages, potentially accounting for the sudden onset of population outbreaks.

2.15. Question 19 (Larvae and Juveniles)—What type of food do juveniles eat in the field? do they show any feeding preferences?

Feeding preferences of juvenile CoTS are POORLY RESOLVED, but the question also needs to be REPHRASED to focus on the fitness consequences associated with differential access to preferred versus non-preferred prey (see Section 2.16).

Newly-settled *Acanthaster* spp. have been reported to feed on a wide variety (at least 12 different species) of coralline algae [112,117,122], including *Lithothamnium pseudosorum* and *Porolithon onkodes* [125,142], though the fitness consequences of settling and feeding on differential species of coralline algae have not been considered. Johnson et al. [121] suggested that the most preferred coralline algae is *L. pseudosorum* but this was based on settlement (rather than feeding) preferences. Higher growth and survival rates as a result of feeding on the preferred species of coralline algae may be fundamental in understanding settlement preferences, as well as accounting for spatial and temporal variation in population dynamics.

Coral-feeding juvenile *Acanthaster* spp. do exhibit pronounced feeding preferences as shown by a recent laboratory study [143]. In this study, juvenile *A.* cf. *solaris* that were given the choice between eight species of coral (*Acropora formosa*, *A. millepora*, *A. tenuis*, *Pavona cactus*, *Echinopora lamellosa*, *Pocillopora damicornis*, *Stylophora pistillata* and *Porites lutea*) and preferred *Acropora tenuis* while avoiding *S. pistillata*, *E. lamellosa* and *P. lutea* [143]. While these laboratory studies show that juvenile CoTS do have distinct feeding preferences, field surveys of feeding preferences of juveniles (based on feeding scars on corals upon collection of juveniles; *sensu* [144]), are needed to evaluate the ecological impact of feeding by juvenile CoTS in the field, where prey choices are likely to be influenced by a wide variety of different factors [21]. For example, juvenile CoTS (1+ year old; mean size ~35 mm) sampled from Green Island and Fitzroy Island (Great Barrier Reef) were consistently observed feeding on bushy scleractinian corals such as *Acropora echinata* and *Stylophora mordax* [11], which may reflect preference for complex habitats to evade predators rather than inherent feeding preferences.

2.16. Question 20 (Larvae and Juveniles)—Wow fast do juveniles grow in the field? Is it similar to that recorded in the laboratory? How important is diet in determining the growth rate of juveniles?

General patterns of juvenile growth are WELL RESOLVED, but it is unknown whether different species of coralline algae significantly influence growth of algal-feeding juveniles.

Growth rates of *Acanthaster* spp. vary with ontogeny, but can also vary in response to environmental conditions (e.g., temperature), food availability and conspecific densities [123]. On the GBR, growth rates of newly-settled (0+ year) *A.* cf. *solaris* increase exponentially from 0.03–0.04 mm per day at 1-month to 0.11–0.22 mm per day at 12-months [34], which is comparable to laboratory-based growth estimates for *A.* cf. *solaris* [123]. Growth rates further accelerate after 12-months or as soon as individuals switch from feeding on CCA to scleractinian corals and peak at 20–30 mm per month when *A.* cf. *solaris* are ~100–200 mm diameter for both wild and captive individuals [61,123].

Broad changes in diet (from coralline algae to coral) and overall prey availability have a major impact on growth rates for juvenile CoTS [61,123]. Most notably, growth is relatively slow during the algal-feeding phase, but accelerates after switching to coral prey [41,61,112,117,123,145]. Accordingly, Lucas [123] showed that *A.* cf. *solaris* constrained to feeding on CCA for 2 years were up to 20 times smaller than counterparts from the same cohort provided with access to coral prey. Given that growth and survival of newly settled juveniles may represent a critical bottleneck to recruitment, more work is required to understand the relative importance of different types of coralline algae. After switching to coral prey, growth rates certainly vary according to availability of different coral prey; laboratory-reared juveniles maintained on an exclusive diet of *Acropora formosa* grew at 12.0 mm per month compared 0.1 mm per month for those maintained on a diet of *Porites lichen* [126]. Difference in growth rates may be due to variable nutritional content of specific coral diets. Laboratory-reared juveniles maintained under *Acropora* or *Pocillopora* feeding treatments grew at the same rate despite consumption of *Acropora* being twice as much compared to *Pocillopora* [145].

2.17. Question 21 (Larvae and Juveniles)—How far do juveniles move in the field?

Locomotor capacities of newly settled CoTS are WELL RESOLVED, but there has been limited consideration of rates and patterns of movement by larger juveniles (up to 10–15 cm total diameter).

Research on the movement of newly settled *Acanthaster* spp. is limited to short-term experimental studies during which individuals were deprived of access to prey and placed on petri dishes or bare sand. Two-week old juveniles (<2 mm total diameter) placed in a clean dish without food moved at 1 cm per minute [61], while larger juveniles (19–70 mm total diameter) moved over bare sand at rates of 2.34 to 6.67 cm per minute [11]. However, CoTS generally settle on or near their preferred prey (CCA) and spend most of their time feeding and hiding, rather than moving [61,126]. Keesing and

Halford [130] deployed pervious boxes to reef habitats to measure predation rates on captive reared *A.* cf. *solaris*. Through the course of these experiments and explicit escape controls, Keesing and Halford [130] demonstrated that 1-month old starfish move very little (<26 cm) on time frames of 1–2 weeks, though CoTS certainly become more mobile as they grow. Larger juveniles (up to 10–15 cm total diameter), which have increased capacity for movement, but are still very vulnerable to predation, may be expected to frequently move between feeding sites and predator refuges, and may also move over larger distances in search of more optimal habitats. If there are significant ontogenetic shifts in habitat use, it is likely that these occur once starfish attain sufficient size to maximize food intake and energetic reserves, but have not yet invested in reproduction.

2.18. Question 1 (Adults)—Are adults capable of moving between reefs?

Adult CoTS are certainly capable of moving between reefs under certain conditions, but the question needs to be REPHRASED to explicitly consider the maximum distances over which adult CoTS can and do move between reefs.

Rates of movement for CoTS have been extensively studied, mostly to inform the extent of their impacts on coral assemblages and reef ecosystems [127]. Over small distances (meters), CoTS are capable of moving at 33 to 51 cm per minute [11,127,146], with maximum rates of movement (which probably reflect escape responses) moderated by the size of the starfish and the complexity of the substrate [127,147]. If maximum rates of movement recorded on sand [127] can be sustained, large *A.* cf. *solaris* could travel up to 520 m per day. Even so, it would take weeks to months for *A.* cf. *solaris* to move several kilometers between reefs, and longer-term and larger-scale movement will be greatly constrained by habitat heterogeneity, resource acquisition, and diurnally restricted periods of activity. Ultimately, the likelihood of adult CoTS moving between reefs will depend on the distance separating adjacent reefs as well as the nature of the intervening habitat.

When tracked for periods of days to weeks (by relocating uniquely tagged individuals at regular intervals), CoTS move <35 m per day and mostly move only after they have depleted coral prey within the immediate area [2]. Adult CoTS also tend to avoid crossing open expanses of sand (e.g., [148]). Sigl and Laforsch [36] demonstrated that well-fed *A.* cf. *solaris* remain within shelter, whereas starved individuals more readily leave shelter and travel over sand, presumably in search of food. Suzuki et al. [149] reported large numbers of adult CoTS moving across shallow sand flats in Ishigaki Island, southern Japan, which were in very poor condition and ultimately became stranded at low tide. For the most part, adult CoTS in good condition and with reasonable access to coral prey will have limited impetus to move, whereas individuals that are starving are probably unlikely to succeed in traversing large distances between reefs, despite demonstrated capacity for detection of reef structures and selective migration toward coral-rich areas via "vision" or chemoreception [150–153]. It is very likely that CoTS can and do move between close positioned reefs, especially where there is contiguous reef habitat connecting reefs, but the limited temporal and spatial scales of previous movement studies (as well as the predominant focus on movement within coral habitats) do not really inform the capacity of CoTS to travel large distances between reefs. Acoustic tagging may provide new opportunities to assess the scale and occurrence of inter-reef movement and migrations by adult CoTS, assuming that small acoustic tags can be permanently affixed to the body of these starfish.

2.19. Question 2 and 3 (Adults)—How rapidly do [adult CoTS] grow in the field? Is their rate of growth similar to that recorded in the laboratory? Can the age of a starfish be determined from its size?

These questions are LARGELY RESOLVED, though there are some persistent controversies surrounding the ability to reliably age adult CoTS using biological proxies.

Growth rates of *Acanthaster* spp. have long been considered to be extremely plastic [123,154]. In the extreme, Lucas [123] reported a 20-fold difference in the size of *A.* cf. *solaris* at 2 years of age, depending on whether they did or did not transition from feeding on coralline algae to coral prey. Substantial differences in the size of CoTS within any given population [59] may, therefore, reflect individual differences in growth, rather than differences in age, whereby the latter would reflect the multiple cohorts within the same population. That said, direct comparisons of size-at-age data from previous studies, including laboratory-based measurements of Lucas [123], have revealed remarkable consistency in age-specific growth rates [1]. Growth rates of *A*. cf. *solaris* are maximized (100–150 mm per year) among 1+ and 2+ annual age classes, and thereafter, follow a von Bertalanffy growth function. There are however, apparent differences in the extent to which CoTS exhibit finite versus indeterminate growth [154], as well as variation in asymptotic size [35]. These differences are largely manifest when comparing between outbreaking and non-outbreak populations [155], whereby growth is increasingly constrained (and potentially finite) during severe population outbreaks due to strong intraspecific competition for food and scarcity of prey resources [156]. MacNeil et al. [35] compared the size-structure of *A*. cf. *solaris* populations across 17 reefs on the GBR and showed that the asymptotic size varied among reefs (from <300 mm to >400 mm total diameter, with smaller asymptotic sizes recorded on reefs with higher CoTS densities.

The sigmoidal growth exhibited by *A*. cf. *solaris*, combined with variation in their asymptotic size, may obscure the general relationship between size and age, especially for larger and older individuals. For this reason, the capacity to distinguish individual cohorts based on population size-structure and retrospectively establish interannual variation in rates of settlement based on size has been contentious (but see [157]). Therefore, various size-independent proxies of age have been explored for *Acanthaster* spp. including spine length and pigment bands on spines [158]. Care is needed to consistently sample the longest spines from the upper portion of non-adjacent arms, specifically avoiding regenerating spines and arms [159]. Stump [160] used mark-recapture to confirm that spine-banding couplets are deposited annually for *A*. cf. *solaris* at Davies Reef in the central GBR. While absolute age-estimates based on spine banding still need to be validated, this method enables coarse estimates and comparisons of age-structure (and therefore growth) among discrete populations [35]. Validation of age estimates is critically dependent upon determining the specific timing of initial band formation [158], which may vary spatially and especially among distinct species (Section 3.1).

2.20. Question 4 and 8 (Adults)—How long do adults survive in the field? Do adult starfish enter a senile phase in the field where their growth declines greatly and they become infertile?

Maximum longevity is still unclear, but predominant patterns of growth and longevity are WELL RESOLVED.

Demographics of CoTS populations are strongly dependent on food availability (e.g., [123,154]), and may also vary with environmental conditions (especially temperature). In captivity, Lucas [123] demonstrated that *A*. cf. *solaris* grew to >300 mm total diameter within 3 years, but then largely stopped growing and reproducing, and mostly died within 4–5 years. The limited size of these starfish and the early onset of apparent senescence were suggested to be experimental artefacts, and at least partly attributable to food limitation. In the wild, *Acanthaster* spp. can grow to >750 mm total diameter and live >8 years [157,161]. On the GBR, large individuals of *A*. cf. *solaris* (>600 mm total diameter) have been recorded before the onset of active outbreaks, potentially representing individuals that have lived throughout an entire outbreak cycle [157]. If so, these individuals may be >14 years old, though it is also possible that these starfish simply recruited during non-outbreak periods and grew quickly or survived longer due to abundant coral prey and limited food competition. For outbreak populations, the maximum size of *A*. cf. *solaris* is generally <400 mm (e.g., [35]), which probably reflects constraints on growth and longevity due to local depletion of prey resources and density-dependent mortality [156,162]. Even so, there is no evidence of size- or age-specific onset of senility among wild populations [101,129]. Babcock et al. [101] demonstrated that there was an exponential increase in egg and sperm production with increasing size, and no apparent reduction in reproductive investment for individuals up to 500 mm total diameter.

2.21. Question 5 and 7 (Adults)—What are the rates of mortality for adults in the field? Are there any other factors which are important in causing the mortality of adult starfish (e.g., disease)?

These questions are LARGELY UNRESOLVED though there is considerable evidence that CoTS are highly susceptible to disease.

Aside from documented declines in the local abundance of *Acanthaster* spp. at the end of outbreaks [59], there is very limited information on rates (or causes) of natural mortality for CoTS. Moran [21] proposed a comprehensive field study to explicitly quantify mortality and longevity for *Acanthaster* spp., but such studies are still hampered by the limited capacity to tag and/or identify individual starfish over necessary periods of several years, especially during major outbreaks. If or when conducted, such studies should attempt to discern density-dependent effects on population dynamics and individual demography, or at the very least, test for differences in key demographic rates between outbreak and low-density populations. Information relating to the biology and ecology of CoTS in low-density populations is particularly lacking, largely owing to the logistic constraints on gathering sufficient data when starfish are few and far between (e.g., [14]). Conversely, recurrent sampling of outbreak populations [112,128] has revealed marked temporal and spatial variability in the abundance and size structure of CoTS, indicating high rates of mortality among smaller CoTS. Zann et al. [112,128] estimated that 99% of CoTS that recruited to the barrier reef off Suva, Fiji, died within 2 years. However, this high rate of mortality may have been anomalous due to the high incidence (10%) of pathogenesis among small and juvenile CoTS.

Echinoderms are particularly susceptible to disease [163], and disease has been implicated in mass mortalities of numerous species of urchins and starfish (e.g., [164,165]). Accordingly, *A*. cf. *solaris* has been seen to exhibit symptoms of disease, both in captivity [123,166] and in the wild [112,167]. Background levels of pathogenesis among populations of *Acanthaster* spp. appear to be generally very low [168], though the probability of infections arising, as well as rates of transmission among individuals are likely to increase with population density [112,128]. Susceptibility to disease is also likely to increase following prey depletion and declines in the condition of adult starfish, leading to further increases in the incidence and importance of disease after outbreaks are well-established (Section 2.23). Further research is clearly needed to better understand the ultimate fate of individual CoTS, though it seems likely that most succumb to either predation and/or pathogenesis, even if their vulnerability to such processes varies with size, age, prey availability, energetic condition and behavior.

2.22. Question 6 (Adults)—What is the rate of predation on adults on reefs? What are the main predators of adult starfish? Are these predators sufficient to limit adult population levels? Do the densities of these predators fluctuate markedly through time?

There is increasing evidence that adult CoTS are susceptible to predation, but these specific questions are LARGELY UNRESOLVED.

Adult *Acanthaster* spp. were initially thought to be relatively immune to predation due to their elaborate physical and chemical defenses [136]. However, an ever-increasing array of coral reef organisms have been reported to feed on adult CoTS [31,32,169]. For the most part, these nominal predators have been observed feeding on the remains of dead or moribund adult CoTS [31] and it is unknown to what extent these predators actually kill adult CoTS. The main predators that are known to kill and individually consume adult CoTS are the giant triton (*Charonia tritonis*) and the stellar pufferfish (*Arothron stellatus*). The abundance of *C. tritonis* is purported to have been much higher prior to extensive harvesting in the 1950s to 1960s, which coincided with the first reported mass outbreak of *A*. cf. *solaris* on the GBR and prompted concerns that the removal of predators may have caused or contributed to outbreaks (the *predator removal hypothesis*; [170]). *Charonia tritonis* is now universally rare on the GBR and on all other reef systems, potentially contributing to increased instabilities in the abundance of *Acanthaster* spp. Similarly, *A. stellatus* and other large predatory pufferfishes are widespread, but uncommon [31]. Moreover, pufferfishes are not subject to fishing and are unlikely to vary in abundance among reefs that are open versus closed to fishing. On the other hand, lethrinids, such as *Lethrinus nebulosus* or *L. miniatus*, are targeted by many fisheries and may also be important predators of *Acanthaster* spp. [31]. Large polyps of the corallimorpharian *Paracorynactis hoplites* have also been observed fully ingesting and killing adult CoTS [171–173], but the distribution, abundance and rates of CoTS consumption by this highly cryptic predator are unknown. Overall, there is still considerable research needed to identify important predators, let alone establishing their relative preference for CoTS, and respective predation rates at different CoTS densities.

Although there are established methods for measuring relative rates of predation and mortality for echinoderms in the field [174,175], there are no empirical data on rates of predator-induced mortality for *Acanthaster* spp. For now, relative rates of predation are inferred based on the proportion of CoTS with conspicuous injuries [32,169], which are attributed to partial predation. In the Philippines, the incidence of injuries was higher for CoTS populations sampled from inside of marine protected areas (MPAs) where fishing was prohibited, which would be consistent with higher abundance of potential predators. On the GBR, Messmer et al. [32] found marked inter-reef differences in the proportion of CoTS with injuries, but these differences did not correspond with inter-reef differences in fisheries regulations. It is possible that the incidence and severity of injuries has no relation whatsoever to rates of predator-induced mortality, because (i) rates of regeneration and therefore, persistence of injuries vary depending on the physiological condition and energetic reserves of starfish [156], and/or (ii) it is an altogether different suite of predators that cause injuries versus outright mortality of CoTS [32]. High incidence of sub-lethal predation may nonetheless have important effects on the individual fitness and population dynamics of *Acanthaster* spp., diverting energy to tissue repair that would otherwise contribute to increased fecundity.

Predation may contribute to population regulation in several ways, including (i) direct reductions in the local densities of juvenile or adult CoTS; (ii) disrupting spawning aggregations [176]; (iii) reducing individual fecundity through partial predation (V. Messmer, unpublished data); and/or (iv) modifying settlement rates and behavior (Section 2.13). However, the initiation and spread of CoTS outbreaks cannot be definitively attributed to declines in the abundance of reef-based-predators, caused by sustained and on-going fishing (Section 2.22). If local densities of *Acanthaster* spp. are normally regulated by predation then overall declines in the abundance of predators might be expected to result in gradual and sustained increases in baseline abundance of CoTS [55], rather than periodic or recurrent outbreaks. This is why Moran [21] asked whether there are any putative predators that fluctuate markedly in time and potentially account for oscillations in the abundance of *Acanthaster* spp. In reality, reduced predation pressure may contribute to increased instability in the population dynamics of *Acanthaster* spp., thereby contributing to increased incidence or severity of outbreaks on individual reefs subject to increased fishing pressure [133,177]. On this basis, it would be prudent to limit or prohibit fishing in areas known to be important for the initiation of reef-wide outbreaks, though more work is still required to reconcile the specific mechanistic links between fishing and CoTS outbreaks [101].

2.23. Question 9 (Adults)—What causes the rapid disappearance of adult starfish which has been observed in the field? Is it related to density dependent factors (e.g., crowding causing loss of condition)? What happens to the majority of starfish? Do they die (e.g., from disease) or do they move to another reef?

These questions are POORLY RESOLVED and potentially very important for informing management of population outbreaks.

Rapid and pronounced declines in the abundance of CoTS following major outbreaks [2,11,59] are suggested to occur because starfish either die following extensive prey depletion and subsequent starvation, or move en masse to find alternate sources of prey [170]. However, precipitous declines in the local abundance of CoTS may [170] or may not [178] coincide with comprehensive depletion of scleractinian corals. At Lizard Island (northern GBR) in 1999, localized outbreaks ended even though mean coral cover was still >22% [178]. Moreover, CoTS can survive without food for many months [21], and so it seems unlikely that it is a lack of coral per se, that causes elevated mortality or initiates mass exodus from reefs. It is possible, however, that limited access to preferred coral prey (e.g., *Acropora*) leads to compromised health and condition of CoTS [48,123], despite relatively high overall coral cover on reefs subject to moderate CoTS outbreaks.

The extent to which CoTS actually move between reefs is largely unresolved (Section 2.18). Even though starvation is a potential trigger for movement by CoTS [36] and there will be strong motivation to move away from reefs following extensive depletion of coral prey, by the time this happens, the starfish are likely to have already depleted much of their energy reserves thereby constraining the capacity for long-distance migration. Given the high densities and compromised condition of CoTS towards the end of outbreaks, it seems most likely that that rapid transmission of highly virulent and opportunistic pathogens is responsible for localized population collapse [22,112]. Moran [21] argued that mass-mortalities of CoTS are unlikely to have gone completely unnoticed, though it is possible that diseased starfish seek refuge within the reef matrix and are highly cryptic, or that sick individuals are targeted by opportunistic reef predators (e.g., [179]).

There is a potentially important link between food availability, the nutritional condition of CoTS, and their vulnerability to opportunistic pathogens [22], which are highly prevalent on and within the tissues of these starfish [168]. Mills [162] showed that CoTS increasingly invest in prophylaxis at high densities of conspecific, where there is an increased risk of infection. While this would be expected to confer greater resistance to pathogenesis, and therefore, reduced relative rates of mortality, it would be maladaptive to increase investment in prophylaxis unless risks were significantly increased. In addition, there are likely to be significant constraints on energy investment for CoTS towards the end of outbreaks following selective depletion of preferred coral prey (e.g., *Acropora*) if not comprehensive coral loss [178]. If crown-of-thorns starfish continue to invest disproportionately into immune defense even when prey are scarce, energy reserves will be depleted even more rapidly, thereby making individuals even more prone to disease [180,181].

2.24. Question 10 (Adults)—Do the skeletal components of starfish accumulate in the sediments after times of outbreaks? Do more spines tend to accumulate during outbreaks than during times when starfish densities are low?

The specific questions posed here are WELL RESOLVED, but skeletal elements cannot be used to resolve the specific incidence and timing of past outbreaks.

Adult CoTS have about 2000 calcareous skeletal elements; these persist in reef sediments and are readily recognizable. It is expected therefore, that fluctuations in the prevalence of CoTS skeletal elements within distinct layers of reef sediments could be used to test whether outbreaks occurred prior to the first documented outbreaks in the 1950s–1960s, and test whether the frequency of outbreaks has changed through time [182–184]. However, the use of ossicles to reconstruct the history of outbreaks has been controversial. Initial studies sampled the sediments in the lagoons of 44 reefs spread widely across the GBR region and found higher numbers of ossicles in surface sediments at reefs with active outbreaks of CoTS [182]. Sub-surface sampling using an airlift found remains of starfish in sediments that dated from more than 3000 years before present, with some suggestion of higher concentrations at 250–300 year intervals [182]. These results were reassessed by Moran et al. [185] who concluded that, while the occurrence of skeletal elements confirms that CoTS have been present on the GBR for a long time, these data cannot confirm or refute the occurrence of outbreaks prior to the 1960s.

Walbran et al. [183] sampled sediments in multiple sites at two reefs in the north central GBR using a vibro-corer which retained the structure of the sediment cores better than the airlift used by Frankel [182]. Walbran et al. [183] concluded that CoTS have been present on the GBR for at least 8000 years and that the general density of ossicles was noticeably higher in sediments that were 1000–2000 years old than in the more recent deposits. However, individual outbreaks lasting only a few years could not be resolved due to reworking of sediments. A series of subsequent papers identified potential weaknesses in the study which could undermine these broad conclusions: possible taphonomic changes were not considered [186], a number of assumptions that are the basis for the link between recent starfish populations and the incidence of ossicles in surface sediments were untested [187], and differential rates of sedimentation and compaction, both natural and those resulting from the coring process, were not accounted for [188]. While Henderson and Walbran [184] point out that many of these potential problems are unsubstantiated, it is clear that the interpretation of the sediment record is not straightforward. Ultimately, bioturbation and differential compaction mean that the position of ossicles in reef sediments is an unreliable indicator of their relative age. A robust estimate of the timing of past outbreaks awaits the development of a method of aging individual ossicles of CoTS at a cost low enough to allow large sample sizes. The high magnesium calcite content of ossicles makes them unsuitable for established techniques such as U-series dating.

2.25. Question 11 (Adults)—Do adults show a distinct preference for certain types of coral?

This question has been WELL RESOLVED.

Numerous studies have reported that *Acanthaster* spp. feed predominantly on certain coral taxa, mainly *Acropora* and *Montipora*, while rarely feeding on other taxa, such as *Porites* and *Turbinaria* (reviewed by [1,21,22]). Strong selective feeding is expected to cause differential mortality and directional shifts in the structure of coral assemblages, potentially contributing to increased diversity through selective removal of dominant coral taxa [189]. However, differential consumption of coral taxa may not necessarily reflect inherent feeding preferences [190]. Potts [190] argued that field-based patterns of feeding by *Acanthaster* spp. are moderated by accessibility to different prey and that they become conditioned to feed disproportionately on locally abundant corals. Moreover, Moran [21] pointed out that few of the early studies on the feeding habits of CoTS employed methods necessary to document explicit feeding preferences, largely failing to account for the differential abundance or availability of different coral prey. Since that time, there have been several quantitative field studies demonstrating that *Acanthaster* spp. consume and deplete different corals disproportionately to their availability [19,20,144,191,192], reflective of distinct feeding preferences. Compilation of data (forage ratios) from these distinct studies reveal that *Acropora* and *Montipora* are consistently consumed more than expected based on their availability [1]. Conversely, several coral genera (including *Porites*, *Pectinia*, *Galaxea*, and *Echinopora*) were generally consumed less than expected based on their availability, though actual forage ratios and inherent feeding preferences vary with concentration and condition of CoTS, as well as the size, abundance and distribution of prey.

While *Acanthaster* spp. do exhibit distinct feeding preferences, CoTS outbreaks do not necessarily lead to directional shifts in the structure of coral assemblages [6], nor to persistent changes in coral diversity [178]. During major outbreaks, feeding selectivity may be apparent in the sequential depletion of different coral taxa, but even the least-preferred corals (e.g., *Porites*) are consumed and often locally depleted (e.g., [2,6,11]). Selective effects of CoTS feeding on coral assemblages will be most apparent during relatively moderate outbreaks (e.g., [178]). At Lizard Island in the 1990s, outbreaks of *A.* cf. *solaris* caused overall declines in coral cover of <30% [178]. Despite moderate declines in live coral cover, there were marked shifts in the structure of coral assemblages, with disproportionate declines (50–80% depending on species and location) in abundance of formerly dominant *Acropora* corals. Nonetheless, coral diversity declined because CoTS were not sufficiently averse to less common taxa.

2.26. Question 1 (Effects On Communities and Processes)—Do coral communities recover from outbreaks of starfish? How long does this take? Is the pattern of recovery similar for most types of reefs and for different scales of disturbance?

Recovery of coral assemblages (both in terms of total coral cover and community composition) is well studied and LARGELY RESOLVED.

Estimates of the time required for coral assemblages to recover from outbreaks of *Acanthaster* spp. (and other major disturbances) range from <5 years to >100 years [193–196], depending on the spatial extent and magnitude of coral loss, as well as the specific types of corals that are affected. In extreme cases, coral assemblages may never regain their initial structure, even where overall coral cover returns to pre-disturbance levels [197], owing to fundamental shifts in community dynamics. Completely denuded reefs also recover much more slowly than reefs in which at least some corals survive to grow and reproduce [198,199]. Moreover, it is increasingly apparent that major disturbances are occurring too frequently to allow for recovery of coral assemblages in the intervening period [9].

The recovery of coral assemblages following outbreaks of *A*. cf. *solaris* was explicitly studied on the GBR based on temporal dissimilarity in taxonomic composition of benthic assemblages on reefs monitored annually from 1993 to 2005 [200]. While coral assemblages did often exhibit effective recovery from CoTS outbreaks throughout this period, the time taken for community reassembly after outbreaks of *A*. cf. *solaris* was longer than for other major disturbances, such as severe tropical storms and bleaching [200]. Notably, recovery also took longer in reef areas where fishing was permitted, where 8–10 years were necessary for coral communities to return to their pre-outbreak composition, compared to an average of 6–7 years inside no-take areas [200]. This difference was attributed to lower severity of outbreaks and corresponding reductions in the magnitude of coral loss inside no-take areas, potentially linked to greater predation on juvenile or adult starfish where fisheries exploitation was restricted [133].

2.27. Question 2 (Effects on Communities and Processes)—What effect do outbreaks have on other communities (e.g., fish, soft corals)? Is this effect permanent or do these communities recover from such a disturbance?

These questions are WELL RESOLVED, though the effects of coral loss on reef-associated organisms are informed by studies considering a diverse array of different disturbances.

Coral depletion, regardless of whether it is caused by CoTS outbreaks or by other large-scale disturbances (e.g., climate-induced coral bleaching), has adverse effects across a wide range of marine organisms that depend on corals for food, shelter or recruitment [201–204]. Extensive coral depletion also provides increased space for colonisation by soft corals [178,205] or algae [206] which can, in turn, lead to increases in abundance of habitat generalists and herbivores. The specific effects of outbreaks of *Acanthaster* spp. on coral reef fishes will depend on the magnitude (extent and severity) and selectivity of coral loss. However, the corals that are most critical in supporting both coral-dwelling (e.g., *Acropora*, *Stylophora* and *Pocillopora*; [204]) and coral-feeding fishes (*Acropora* and *Pocillopora*; [4]) are the same corals that are preferred by *Acanthaster* spp. (Section 2.25) and are often rapidly and comprehensively depleted during CoTS outbreaks [6]. Not surprisingly, therefore, declines in the abundance of coral-dependent fishes during outbreaks of *Acanthaster* spp., are often disproportionate to overall levels of coral loss [6,207,208].

Approximately 10% of coral reef fishes are directly dependent upon scleractinian coral for food and shelter [202] and mostly small-bodied fishes with limited fisheries importance [204]. However, the effects of extensive coral depletion caused by outbreaks of *Acanthaster* spp. extend well beyond the few fishes that are directly dependent on live corals for food and shelter [209], especially where the effects of coral depletion are compounded by structural collapse of three-dimensional habitats and/or increased dominance of macroalgae. In southern Japan, Sano et al. [209] recorded 65% declines in the abundance and diversity of reef fishes on reefs that were devastated by outbreaks of *A*. cf. *solaris*. Sheltered reef environments that supported extensive growth of tall staghorn *Acropora* corals, providing complex 3-dimensional habitats, were rapidly reduced to flat, homogeneous rubble fields, following the comprehensive consumption of corals by CoTS [209]. Importantly, fundamental shifts in the structure of coral reef habitats, initiated by severe coral loss, can impact on the abundance of many large-bodied and carnivorous fishes [208], directly undermining fisheries productivity [210].

Given their strong reliance on corals and coral-dominated habitats, the abundance and diversity of reef fishes are expected to recover in direct accordance with recovery and reassembly of coral assemblages [211,212]. Inherent lags in recovery may sometimes occur due to vagaries in larval supply and recruitment by fishes, and persistent shifts in community structure may occur due to differential rates of coral recovery [197]. Of greater concern, however, is that the time required for effective recovery relative to projected increases in the incidence of major disturbances [213], whereby the effects of CoTS outbreaks on coral and non-coral communities will be increasingly compounded by other major disturbances, such as mass coral bleaching. Coral recovery and community reassembly may also be constrained by the occupation of reef substrates by non-coral sessile organisms (e.g., macroalgae) that prevent or hamper growth and recruitment of corals [214], highlighting the importance of fisheries management to promote ecosystem resilience on coral reefs. Accordingly, Mellin et al. [200] showed that recovery of both fish and coral communities was faster (<6 years) within no-take areas, relative to reefs that were open to fishing (>9 years).

2.28. Question 3 (Effects on Communities and Processes)—What effect do outbreaks have on reef processes such as calcification, primary production, and reef growth? Question 2 (Overarching)—Whether [CoTS] outbreaks play an important part in reefal processes and the development of reef structure?

The specific effects of widespread coral mortality and corresponding fluctuations in abundance of reef-building corals (regardless of actual cause) on reef processes are WELL RESOLVED.

Net calcification, which is important for rates of reef growth (or accretion) of contemporary reef systems [215,216], is highly sensitive to ecological perturbations and changes in the overall abundance of reef-building corals [217]. Importantly, significant declines in the abundance of corals, and especially fast growing coral species (such as *Acropora*) will significantly affect reef carbonate budgets and net framework production [215,216]. In the Caribbean, for example, sustained and selective coral loss (specifically, functional loss of *Acropora* species) has resulted in >40% declines in rates of carbonate production, such that net carbonate production is below the threshold necessary to sustain positive reef accretion [215]. While the causes of reef degradation in the Caribbean do not include outbreaks of *Acanthaster* spp. (which does not occur in the Atlantic), recurrent and ongoing outbreaks of *Acanthaster* spp. may be undermining critical ecosystem process throughout the Indo-Pacific. If coral assemblages can recover and reassemble reasonably quickly following disturbances (Section 2.26), temporary declines in coral cover and calcification will have negligible effects on carbonate budgets. Periodic mortality and structural collapse of *Acropora* corals may actually contribute to reef accretion through increased production of carbonate sediment [218]. However, outbreaks of *Acanthaster* spp. are a major contributor to sustained declines in the abundance of reef-building corals recorded at key locations throughout the Indo-Pacific (e.g., GBR [7,8]; Japan [16]; Indonesia [20]; Guam [219]; Maldives [17]; French Polynesia [6,9]), and certainly have disproportionate impacts on faster-growing corals (Section 2.25). Functional replacement of reef-building corals by other habitat-forming sessile fauna and flora, may partially offset declines in primary productivity and carbonate production due to widespread coral loss. However, it is now unequivocal that reef ecosystems with very limited (<10%) cover of reef-building corals support a fraction of the species found in coral-rich habitats [220], which has significant consequences for ecosystem function and productivity.

2.29. Question 1 (Overarching)—Why [CoTS] outbreaks occur and whether they are natural or unnatural phenomena?

This question has preoccupied much of the discussion around *Acanthaster* spp., but remains LARGELY UNRESOLVED.

The extent to which outbreaks of *Acanthaster* spp. are caused or exacerbated by anthropogenic activities (e.g., inputs of nutrients and pollutants, and overfishing of key predators) is widely disputed [1,71]. Fundamentally, the unique and extreme life-history characteristics of *Acanthaster* spp., such as exceptional fecundity recorded for *A.* cf. *solaris* [101] predispose them to major population fluctuations [103]. However, when outbreaks of *Acanthaster* spp. were documented in the late 1960s [2,11], it was immediately assumed by many scientists that these were new and unprecedented phenomena, such that scientists were compelled to link the sudden occurrence of outbreaks to sustained and ongoing degradation of coastal environments, due to coastal development [2], land-based run-off of nutrients and pollutants [221], and/or exploitation of marine species [222]. Several scientists (e.g., [137,223–225]) did argue that outbreaks of *Acanthaster* spp. were probably a natural phenomenon that had occurred across the Indo-Pacific well before the 1960s.

While there are no rigorous quantitative estimates of CoTS densities prior to the 1960s, historical accounts, mostly from anecdotal information, support the view that outbreaks have occurred in the past [223,224,226]. In particular, there is information suggesting that CoTS were unusually common at certain locations in the 1940s [113], though some of these reports may refer to normal spawning aggregations of CoTS [227], rather than actual outbreaks. Birkeland [228] and Flanigan and Lamberts [229] also proposed that the significance of CoTS in Micronesian and Samoan cultures could be indicative of previous instances of high abundance. However, Moran [21] argued that the cultural importance of *Acanthaster* spp. could be a result of its sinister appearance and toxicity, rather than periodic abundance. On the GBR, it appears likely that outbreaks of *A.* cf. *solaris* were underway in 1913, given the relative ease with which Clark [230] collected CoTS in the Torres Strait. It is unknown however, whether there were progressive waves of outbreaks that propagated the length of the GBR, and there is insufficient information to establish the timing and spread of outbreaks that occurred prior to the 1960s.

Even if outbreaks did occur on the GBR prior to the 1960s, it has been suggested that the frequency and/or intensity of outbreaks is now much higher than it was in the past [45,113,231]. Fabricius et al. [45] refer to a model of coupled oscillations between coral cover and CoTS densities, suggesting that the incidence of outbreaks has increased from one outbreak in 50–80 years to one in 15 years over the last 200 years, attributing this increased incidence to higher nutrient loads from river discharge. The principal basis of these assertions is that long-term development of the GBR would not have been possible given the current frequency and severity of CoTS outbreaks (see also [221]). However, the increased susceptibility of reef ecosystems to outbreaks of *Acanthaster* spp. and protracted periods required for subsequent recovery and reassembly of contemporary coral assemblages is not in itself evidence that outbreaks are unnatural [22]. Rather, other factors, such as chronic threats posed by increasing anthropogenic activities (e.g., fishing and harvesting, sedimentation, eutrophication, and pollutants) may have undermined the capacity of reef ecosystems to withstand natural disturbances [232], eroding ecosystem resilience and altering population and community responses to persistent and ongoing disturbances. The best (albeit inferential) evidence that there have been temporal shifts in the incidence and/or intensity of CoTS outbreaks comes from cores of massive *Porites* and *Diploastrea* colonies, which purportedly record the incidence of CoTS feeding scars throughout their lifespan [231]. DeVantier and Done [231] concede that the ability to discern previous outbreaks declines as you extend further back, even when comparing among known outbreaks events in the 1960s and 1970s. Nonetheless, there is evidence of outbreaks occurring prior to the 1960s, and mainly in the 1930s. Devantier and Done [231] conclude that outbreaks of CoTS have been a persistent feature of the GBR for centuries, but may have gone from isolated and sporadic events in the 1930s to frequent and widespread events since the 1960s.

Establishing whether outbreaks of CoTS are caused (or exacerbated) by anthropogenic activities was considered fundamental in addressing the dilemma of whether to regulate population densities of *Acanthaster* spp. and prevent future outbreaks [233]. In reality, the decision to aggressively defend coral reef ecosystems against the devastation caused by CoTS outbreaks has already been made, as evidenced by intensive and extensive culling programs in operation throughout the Indo-Pacific, wherein over 17 million CoTS have so far been killed and/or removed from reefs across the Indo-Pacific [1]. The more important questions that must be addressed are whether increased actions to improve water quality (specifically, addressing land-use practices to reduce nutrient inputs) within reef environments and/or increased fisheries restrictions can reduce the frequency or intensity of future CoTS outbreaks. We also need to consider whether these indirect management actions are ultimately going to be more effective than direct intervention (e.g., culling programs) in minimizing the incidence of outbreaks and reducing coral loss at various timeframes. Given persistent uncertainties surrounding the proximal cause(s) of outbreaks and the likelihood that multiple factors will be involved in the initiation and subsequent spread of outbreaks [57,71], it seems prudent to maintain a multipronged approach to managing outbreaks of *Acanthaster* spp. ([18]; Section 2.33). Nevertheless, intensifying efforts to improve water quality, as well as redressing over-fishing on coral reefs, are important regardless of whether they ameliorate the threat posed by outbreaks of *Acanthaster* spp. Meanwhile, evaluating the relative roles of these factors in initiating outbreaks of CoTS will require more integrated and intensive monitoring to explicitly resolve spatial and temporal gradients in biological communities and environmental conditions (Section 2.2) relative to the larval abundance, settlement rates, and post-settlement survival of CoTS [71].

2.30. Question 3 (Overarching)—Why some reefs are more susceptible to [CoTS] outbreaks than others?

This question refers to differences in the occurrence and severity of outbreaks among reefs within the same reef system (e.g., the GBR), which is LARGELY UNRESOLVED

On the GBR, outbreaks of *A.* cf. *solaris* are only ever recorded on a specific subset of reefs [21]. Outbreaks are not, for example, recorded on most nearshore reefs and are only rarely recorded on outermost reefs [21]. The reasons for these patterns are often discussed, but generally unknown. The specific factors that influence susceptibility of reefs to population outbreaks of *Acanthaster* spp. will vary depending on whether outbreaks arise independently (primary outbreaks) or result from extensive delivery of larvae spawned by high-density populations on nearby or upstream reefs (secondary outbreaks). The independent initiation of outbreaks on individual reefs is likely to result from the progressive accumulation of starfish over multiple cohorts [59], which will be conditional upon entrainment of larvae and sustained levels of self-recruitment [57]. The initiation of primary outbreaks may also be promoted by local depletion of putative predators (through fisheries exploitation and trophic cascades [133,177]) leading to increased survival of larvae, settlers, juveniles or adult CoTS within reef environments, and/or localized nutrient enrichment, due to river discharge [45] or upwelling of nutrient rich waters [234], which overcomes normal food-limitation and facilitates increased development and survivorship of CoTS larvae [55]. Outbreaks of *A.* cf. *solaris* on the GBR are initiated on mid-shelf reefs between Cooktown and Cairns, and mainly in the northern portion of this area (commonly referred to as the "initiation box"). This region is characterized by high densities of individual reefs and highly restricted water movement [58], which will promote steady and sustained increases in local densities of *A.* cf. *solaris* during successive years of spawning. However, limited spatial and temporal resolution in monitoring means that it is unclear where exactly outbreaks originate. This prohibits meaningful comparisons of putative predator densities or nutrient concentrations among reefs that do and do not sustain primary outbreaks. There is also very limited monitoring of biological communities and environmental conditions at spatial and temporal scales relevant to explain the initiation of CoTS outbreaks (Sections 2.1 and 2.2).

For secondary outbreaks, the predominant factor that will influence (or at least fundamentally constrain) when and where outbreaks arise is the extent of larval delivery via hydrodynamic connectivity [58,83]. That said, the delivery of high densities of CoTS larvae to individual reefs would not necessarily result in the establishment of population outbreaks if there were local constraints on larval survival and development or high rates post-settlement mortality (Figure 3). Previous discussions on the role of river discharge and nutrient pulses on the GBR have focused largely on the initiation of primary outbreaks [45]. If, however, larval development and survival are severely constrained by very low nutrient concentrations and limited food availability (Section 2.1), it seems logical that this would limit both the initiation and spread of CoTS outbreaks [1,29]. Pratchett et al. [1] suggested that primary outbreaks may propagate over extended periods independent of any major flood events, whereas it is the subsequent spread of outbreaks that might benefit from major flood events, due to enhanced food availability and elevated larval survival across large areas of the reef [29]. Even if there are large numbers of well-fed and competent larvae arriving at a reef, settlement and post-settlement survival might be constrained by habitat structure, availability of coralline algae or coral prey, as well as the local abundance putative predators (Figure 3). Specific habitat requirements and settlement preferences are yet to be established for *Acanthaster* spp. (Section 2.9), though it is likely that there will be inter-reef variation in settlement and survival rates corresponding with differences in reef topography, habitat availability, and also the recent disturbance history. Importantly, there are expected to by interactions between CoTS outbreaks and other disturbances that cause coral loss and habitat degradation [71]. In general, outbreaks might be expected to be less likely to arise on reefs where coral cover has been supressed by recent distrubances, but these disturbances might also increase local availability of critical settlment habitat (dead, but intact coral skeletons, and/or coral rubble). Outbreaks of *A.* cf. *solaris* have certainly occurred on reefs with low coral cover in the aftermath of major disturbances (e.g., GBR [59], Guam [219], Okinawa [235]), and contribute to further coral loss.

2.31. Question 4 (Overarching)—Why some [CoTS] outbreaks cause extensive coral mortality while others do not?

This question is LARGELY RESOLVED, though more research is needed to explore species–specific differences in biology and behavior (Section 3.1).

Although outbreaks of *Acanthaster* spp. are capable of causing extensive coral depletion over vast areas (e.g., [2,11,236,237]), outbreaks vary greatly, not only in the size and density of starfish, but also in their effects [21]. Most notably, high densities of *A.* cf. *solaris* occurred for more than 18 months (1969–1970) at Molokai Island, Hawaii, but had negligble effect on local coral assemblages [238]. Within the Pacific, devastating effects of CoTS outbreaks on coral assemblages are mainly restricted to the central and western Pacific, including French Polynesia [6,9], Australia's GBR [7,11], Micronesia [2,237], and southern Japan [16,239,240]. In contrast, CoTS outbreaks cause minimal coral loss in the eastern Pacific [138,241]. These patterns might be explained by the relative dominance of *Acropora* spp. [22]; *Acropora* spp. tend to dominate coral assemblages in the central and western Pacific, and are consistently among the corals that are first and worst affected during outbreaks (e.g., [19,178]). In the north and east Pacific (e.g., the main Hawaiian islands and Panama) *Acropora* is relatively scarce and coral assemblages tend to be dominated by *Pocillopora*, which is much less susceptible to CoTS attack [138,241,242], owing to the defensive behavior of infauna, especially *Trapezia* crabs.

Geographic variation in the effects of *Acanthaster* spp. may also result from differences in the population dynamics and behavior among the four nominal sister species distributed in different parts of the Indo-Pacific [38,39]. *Acanthaster* spp. from throughout the Indo-Pacific ostensibly look and behave the same way, but devastating impacts of crown-of-thorns starfish appear to be confined to the Pacific, which is the geographical range of *A.* cf. *solaris* [39,243]. This warrants explicit comparisons of reproductive and larval ecology, demographic rates, feeding rates and feeding preferences among *Acanthaster* spp. from each of the four distinct sub-populations identified by Vogler et al. [38], extending the studies in the Pacific and the Red Sea to both southern and northern Indian Ocean regions.

2.32. Question 5 (Overarching)—How [CoTS] outbreaks are propagated over large distances?

This question is ambiguous, but is assumed to refer to the apparent coincidence of outbreaks and population connections across widely separated locations, which has Been Largely Resolved.

The propagation of CoTS outbreaks among adjacent reefs and within reef systems is variously ascribed to larval dispersal (Sections 2.5 and 2.6) versus inter-reef movement by adult starfish (Section 2.18), depending on relevant inter-reef distances. At smaller spatial scales (100 s to 1000 s of meters) it is conceivable that connectivity among reefs is achieved through movement of adults [113], whereas large-scale dispersal is largely, if not exclusively, achieved through dispersal of planktonic larvae. Planktonic larvae may also be dispersed on oceanic currents to provide connections among widely dispersed locations. Indeed, the pan-Pacific range of *A.* cf. *solaris* is a potential indicator of its broad dispersal capability [88]. In the 1960s, and again in the late 1970s, outbreaks of *Acanthaster* spp. occurred more or less synchronously across multiple locations throughout the Indo-Pacific [21] suggesting that these outbreaks were inter-connected. The only viable mechanism that would enable connections among widely separated locations, leading to simultaneous or successive outbreaks in discrete locations, is large-scale dispersal of very large number of larvae. Indeed, outbreaks of *A.* cf. *solaris* that occurred in Central Province, Papua New Guinea in the early 2000s likely represented the ultimate and inevitable conclusion of the northerly progression of outbreaks on the GBR throughout the 1990s [19]. However, Timmers et al. [88] explicitly tested for larval dispersal among distinct geographic regions in the Pacific by examining genetic structure of the highly variable mitochondrial control region (mtDNA). While there was evidence of very occasional larval exchange among geographic regions, there was no possibility that outbreaks were propagated from one region to another through mass larval dispersal. Strong genetic structure, indicative of limited larval exchange, was particularly apparent at scales of >1000 km [88], suggesting that the simultaneous occurrence of outbreaks across a broad range of locations are triggered by large-scale climatic features, such as ENSO [244] or increased temperature [113]. Moreover, the increasing incidence of ENSO anomalies since the 1960s may account for increased incidence of CoTS outbreaks since that time [231]. However, searches for environmental triggers of CoTS outbreaks, based on spatiotemporal correlation have been hampered by imprecise accounts of when outbreaks actually started across different locations [231]. The apparent coincidence in the occurrence of outbreaks may simply result from increased research and reporting following reports of renewed outbreaks at key locations. Even Moran [21] conceded it is unlikely that discrete and disparate outbreaks of *Acanthaster* spp. originate as a result of the same single process.

2.33. Question 6 (Overarching)—Whether special management policies need to be formulated in order to prepare for the occurrence of future [CoTS] outbreaks?

There is an unequivocal need for increased proactive (cf. reactive) management of CoTS outbreaks, but the specific management policies and strategies are LARGELY UNRESOLVED and yet to be implemented.

Contemporary management of CoTS outbreaks is largely focused on culling or removing adult starfish, with the intention of minimizing ongoing coral loss through reductions in the number and size of adult starfish and/or containing the spread of outbreaks. To address the current outbreak on the GBR, for example, a targeted control program is in place that kills upwards of 100,000 *A.* cf. *solaris* each year. Currently, the most efficient and accepted method for culling is to individually inject starfish with specific chemicals [37,179,245] that cause immediate and comprehensive mortality. Manual controls are very labor intensive, though the recent development of a single-injection method [246] has increased the efficiency of in-water culling programs by at least 250%. Single-injection methods also eliminate the need to manually handle adult CoTS, which is purported to result in spontaneous spawning, though this seems unlikely given the delayed responses of CoTS to specific spawning inducers during experimental studies [52]. Single-injection methods are now so efficient that major constraints on effective population control relate to detectability of outbreak populations [247], which is contingent upon both timely surveillance and improved understanding of the spatial dynamics (within and among reefs) of CoTS populations. Even with the current suite of tools, there are significant opportunities to further improve the efficiency and therefore, effectiveness of CoTS control programs (e.g., "CoTSBot" [248]). The concern however, is that effective control of CoTS outbreaks across the entire extent of the GBR will be prohibitively costly. Moreover, ineffective control (e.g., incomplete eradication at specific locations) may simply prolong the outbreak and fail to actually protect local coral assemblages. In reality, outbreaks do not affect all reefs (Section 2.30) and never occur simultaneously across the entire expanse of the GBR [56]. Moreover, outbreaks on most reefs can be traced back to initial outbreaks on a few discrete reefs within the initiation box [56], such that timely investment and focused management activity in these areas could contain the spread of outbreaks and minimize reef-wide coral loss.

Despite the consistent and recurring patterns of CoTS outbreaks on the GBR [1,56] there have been very few management policies that have been specifically formulated to prepare for the occurrence of future outbreaks. This lack of planning, combined with inevitable diversion of research and management focus during non-outbreak periods [249] and limited capacity to detect the early onset of outbreaks, results in inevitable delays in responding to new outbreaks [250]. The issue is even more complex and more pronounced outside of the GBR where there is no apparent pattern to the timing and location of outbreaks. A proactive management policy with dedicated funding that can be immediately accessed when initiation of future outbreaks are imminent or actually detected is sorely needed and currently lacking [249]. Ongoing control programs across the Indo-Pacific are estimated to have cost up to US $44 million [1], and have been largely ineffective in protecting reef systems from outbreaks of *Acanthaster* spp. once they become established [247]. Timely intervention and the containment of outbreaks before they can spread may, therefore, save greatly on management costs and increase management effectiveness (e.g., [148,251]). Established disaster funds have proven successful in the Australian agriculture sector to support pro-active monitoring, early detection and early prevention of locust plagues [252]. Under this scenario, farmers locally monitor and pre-emptively control locusts at the initiation of an outbreak (and are reimbursed by the fund) and the government targets high-risk locations. A similar commission should be established for *Acanthaster* spp. to coordinate early detection and rapid response for future outbreaks. Such a fund needs to have well established trigger levels for action and to be well resourced given the significant cost and efficiency benefits associated with rapid response.

Timely intervention to manage new and renewed outbreaks of *Acanthaster* spp. is partly constrained by the reliance on detection of elevated densities of relatively large starfish to signify the onset of outbreaks [56,59,253,254]. The development and refinement of early-warning systems, focused on measuring larval densities, settlement rates, and/or local abundance of newly settled juveniles is, therefore, a priority not only for improved understanding of population dynamics (Section 2.7), but also for management. Moreover, it is important to maintain research and management traction throughout non-outbreak periods (Figure 4), both to address significant knowledge gaps pertaining to the dynamics of non-outbreak populations and to consider additional preventative actions (e.g., sustained culling of low density populations) that may further increase management effectiveness.

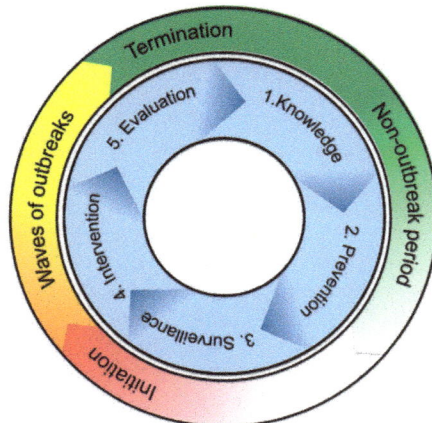

Figure 4. Key phases in the cycle of outbreaks of *Acanthaster* spp. relative to established phases in adaptive management cycles. To effectively manage and prevent future outbreaks, intensive surveillance and intervention are required at/or before the initiation of outbreaks. It is also critically important, that research and management continue throughout the non-outbreak period.

3. New Directions

While many of the questions posed by Moran [21] regarding the biology of CoTS and management of outbreaks remain pertinent, there are several emerging issues and research methods that have deflected much of the recent research attention. For example, when Moran's [21] review was published, taxonomists concurred that crown-of-thorns starfish found on coral reefs throughout the Indo-Pacific (including the Red Sea) were just one species, *Acanthaster planci* (Linnaeus 1758). However, molecular sampling throughout the last 30 years has suggested that there are multiple species of *Acanthaster* that inhabit coral reefs [38,87,255,256]. Most notably, Vogler et al. [38] sampled CoTS from the Red Sea to the eastern Pacific, and revealed at least four strongly differentiated clades, which has stimulated necessary research into the systematics and nomenclature [39] and raises many important questions about the biogeography of CoTS. Moreover, genomic data are providing unprecedented insights into the intrinsic mechanistic basis of CoTS behavior [176], providing new options for population control and management. It is these fields, along with advances in population modelling that are expected to advance understanding and management of CoTS outbreaks in coming years.

3.1. Systematics and Biogeography

It is now unequivocal that coral reef CoTS, which were formerly referred to as *A. planci* throughout the Indo-Pacific (including the Red Sea) comprise at least four distinct species [39]. Most notably, *A. planci* occurs throughout the northeastern Indian Ocean, from the Sea of Oman to Christmas and Cocos Keeling Islands and is both genetically and phenotypically different from *A.* cf. *solaris* [38,256], which occurs throughout the Pacific and the Indo-Pacific archipelago. These two species co-occur and potentially hybridize at Christmas and Cocos (Keeling) Islands [257]. The third major species (*A. mauritiensis*) is generally restricted to the southwestern Indian Ocean, but does co-occur with *A. planci* in the Oman Sea [38]. There is also a fourth distinct species restricted to the Red Sea, which is yet to be formally described and named [39]. The relatively high (8.8–10.6%) genetic distance among the four-aforementioned species, based on the COI marker used for 'barcoding', suggests that these species diverged 1.95 to 3.65 million years ago [38]. Within the four nominal species, genetic divergence was greatest for the Pacific population [39], and other studies have revealed conspicuous genetic structure when comparing among CoTS from different regions across the Pacific [88,99]. Moreover, distinctive phenotypes of *Acanthaster* spp. have been recorded from the southern (Lord Howe Island) and northeastern Pacific (Hawai'i) suggesting that there may actually be more than one species in the Pacific [39]. A recent report by Yuasa et al. [258] also found *A.* cf. *solaris* and its short-spined sibling species, *A. brevispinus* on the same reef in Kushimoto, Japan, which is in contrast with the previous assertion that natural interbreeding between these genetically compatible species is blocked by ecological isolation (different habitats) in regions of sympatry [259]. This suggests that other possible blocks against interbreeding between these two distinct species may be present.

Confirmation that there are multiple CoTS species, which are geographically separated, raises important questions about generality of prior research, whereby much of the research on CoTS has been conducted in the western Pacific on *A.* cf. *solaris* (see Sections 2.1–2.3, 2.5, 2.10, 2.11 and 2.16). Interestingly, devastating outbreaks of *Acanthaster* spp. are also reported mainly from the western Pacific [21], suggesting that species-specific differences in feeding behavior and biology (e.g., reproductive potential) may account for geographic variation in the occurrence of outbreaks as well as their impacts on reef ecosystems [1]. If there are significant interspecific differences in the biology of *Acanthaster* spp., this may also provide significant insights into the intrinsic factors responsible for population outbreaks. Immediate priorities for comparative demographic studies among species would include: (i) estimates of size-specific fecundity, (ii) larval development and competency periods, and (iii) growth functions. Importantly, these demographic processes have already been quantified for *A.* cf. *solaris* [27,34,35,101,129] and are key to understanding population dynamics.

3.2. Genetic and Genomic Sampling

High throughput sequencing technologies (HTS) are providing increasing opportunities to use genomic data to address ecological questions. As such, HTS has been used to describe molecular ecology and evolution, molecular mechanisms of development in animals, and how environmental factors such as those related to climate change influence animal life history. Methods such as RNA-Seq using Illumina technology have been widely used in many marine invertebrates, including corals [260] and echinoderms to detect changes in gene expression [261], or describe genetic variability among populations to assess future adaptation to global change [262]. The era of genomics on CoTS using HTS technologies has just started to emerge. By sequencing the transcriptome of the male gonad of CoTS, using an RNA-Seq approach, key candidate genes involved in reproduction were identified [263]. This study found that CoTS contain proteins, molecules, signalling pathways and key developmental genes that are known to have a role in sperm motility and signalling in other echinoderms [263]. A recent study using genome sequencing and proteomics in combination with behavioral experiments, allowed the identification of key species-specific pheromones involved in their aggregation [176]. Although this result provides a potential solution to control adult populations of CoTS in the GBR, little is known about the molecular basis underlying CoTS life history stages in response to ecological factors. Therefore, future studies using HTS approaches in combination with ecological experiments are necessary. A complete developmental transcriptome will be essential for the identification of (1) key genes and signalling pathways involved in CoTS developmental traits; (2) response to environmental factors addressed in this review; and (3) for understanding the molecular mechanism of calcification in CoTS. Moreover, genome sequencing of closely related species could provide a comparative genomic approach for population genomics and evolution within this group of animals. Other approaches such as eDNA could also serve as a tool to trace the distribution of early life stages of CoTS in the field [53,107,264]. These data will not only contribute to a better understating of CoTS genomics, but will also be required for effective conservation and management.

3.3. Advances in Modelling

There have been significant recent advances in modelling CoTS populations (relevant to demographic and/or dispersal models), building on a long history of theoretical treatment of outbreak dynamics and the spread of outbreaks [265]. Previous demographic models were mostly developed at a reef-scale and typically incorporated demographic processes and predator-prey relationships, but not dispersal [266,267]. For example, Morello et al. [268] developed a model of intermediate complexity for ecosystem assessment (MICE) based on trophic interactions between CoTS and slow versus fast growing corals for Lizard Island. Although this model fit with observed population dynamics based on AIMS LTMP data [133], it implicitly assumed low immigration and self-recruitment levels and failed to capture major peaks in CoTS abundance. The semi-individual based model developed by Chen et al. [269] has further refined those predictions and was able to reconstruct historical abundances on a set of three reefs (including Lizard Island), however such a model remains too data demanding and computationally intensive to be transferred to the entire Great Barrier Reef. Furthermore, this model does not account for connectivity via larval dispersal, though it could be added.

At a regional scale, the development of large-scale hydrodynamic models for the GBR gave researchers the opportunity to recreate the initiation and spread of outbreaks [82–84,93]. However, Wolanski [270] suggested that these predictions were built upon unrealistic assumptions and parameterization of the dispersal process and should therefore not be used for management. Despite this, these hydrodynamic models provided the foundation for Scandol's [271] interactive metapopulation models for CoTS management. Another hydrodynamic model was proposed by Bode and Mason [272] and was used to define self-recruitment and connectivity for 321 reefs on the GBR [273], and in turn model the effect of increasing nutrients on CoTS outbreaks and coral cover over a 150 year time period [45]. This showed that moderate increases in nutrient availability can drastically increase larval survival and reduce coral cover. Importantly, these models were designed to predict population dynamics under a specific scenario (i.e., varying chl-*a* levels) while keeping other parameters constant; however, such parameters were not calibrated against empirical data. Furthermore, such models did not account for uncertainty and how it propagated through the different steps, limiting their usefulness in terms of management and in developing decision-making strategies.

Recent developments in the hydrodynamic modelling for the GBR (e.g., eReefs) have reinvigorated attempts to understand CoTS population dynamics and provide management solutions. A connectivity network model for CoTS on the GBR based on hydrodynamic models generated from eReefs [58] aimed to identify the most important areas for initiating and spreading CoTS outbreaks via their relative importance within the network. This approach has since been used as the basis of an adaptive management strategy, whereby reefs targeted for control are selected based upon their likelihood to spread future outbreaks [274]. While it is a commendable and practical approach to inform management decisions, this model is solely based on dispersal and fails to incorporate demography or important biotic interactions [275]. Furthermore, if models such as these are to be used for management, it is essential for them to be first validated against empirical data, and to account for uncertainty and its propagation throughout the model.

This recent progress in the field of CoTS modelling highlights important directions for future research: (i) despite the fact that there is no one-size-fits-all model, CoTS models should be better integrated across multiple scales so that those developed at reef scales can inform regional models (e.g., role of CoTS behavior, such as aggregation, on spatial distribution patterns). In turn, regional models provide a framework for defining conditions at the boundaries of reef-scale models (e.g., immigration rates from neighbouring reefs); (ii) demographic models should combine multiple and diverse sources of empirical data and specifically highlight where there is limited relevant information [275]. This will provide an opportunity for better interactions between CoTS modellers and biologists to prioritize biological research; (iii) for CoTS models to be useful in terms of management, there needs to be a better appraisal of uncertainty (both data- and model-driven) and how it propagates though the different steps of the model. This implies the need to steer away from purely deterministic models and, instead, account for both environmental and demographic stochasticity. By documenting the uncertainty stemming from each step of the model and each scenario, more transparent CoTS models will allow managers to assess the impacts of different management strategies while accounting for the full range of their possible outcomes.

3.4. CoTS and Climate Change

Acanthaster spp., as well as their coral prey, are increasingly subject to environmental change caused by anthropogenic forcing in global climate systems. This issue was brought to the fore, during the recent mass bleaching on the GBR [276], which is expected to have resulted in levels of coral mortality to rival the cumulative impact of entire cycles and reef-wide outbreaks of *A*. cf. *solaris*. Widespread and significant bleaching [276] and subsequent coral mortality throughout the initiation box may ultimately constrain the initiation of future CoTS outbreaks, which are expected to occur in the early 2020s. However, projected changes in ocean temperatures and seawater chemistry are also expected to have direct effects on *Acanthaster* spp. [23,24,42,74–76,78] and juveniles [74,124,277] especially during early life-history stages.

The optimal temperatures for embryonic and larval development of *A*. cf. *solaris* (28–29 °C), reflects ambient temperatures currently experienced during the reproductive season in the northern GBR [24,76]. Above these temperatures embryonic abnormality and mortality increase [76,78] and larval growth is impaired, as evident in the reduced size of the brachiolaria [75]. Without acclimation to changing climate, even moderate ocean warming (1–2 °C) is therefore, likely to impose significant constraints on reproduction and settlement rates. When ocean acidification is also considered, warming (+2 °C) and reduced pH (pH 7.6–7.8) have additive negative effects, reducing larval size [75]. As a single stressor acidification increases developmental abnormality in CoTS and reduces larval growth [74]. Settlement by CoTS was also negatively affected when coralline algae were grown in low pH conditions [74]. These negative effects on early development may be the bottleneck for population maintenance of CoTS in a changing ocean.

In contrast to larvae, juvenile CoTS are highly tolerant to increased temperature (+2 °C above ambient) and resilient to acidification (pH 7.6) [124]. Growth and feeding rates of the algal-feeding juveniles were highest at 30 °C and pH 7.6. While growth increase at higher temperature is likely due to the direct effects of warming on physiology, faster growth in ocean acidification conditions was not expected. This was largely attributable to indirect effects of acidification on coralline algae [277], which was more palatable (less calcified) and had a higher nutritional value (C:N) when grown at low pH. The cumulative effects of environmental change on individual fitness and population viability of *Acanthaster* spp. still needs further consideration, along with explicit research into the vulnerability of coral feeding juveniles and adults to ocean warming and ocean acidification. However, any such effects may be largely irrelevant, given that climate-change poses a real and considerable threat to the availability of coral prey [276], and has disproportionate impacts on *Acropora* and other corals that are favored by *Acanthaster* spp. Importantly, the loss of preferred coral prey will lead to declines in the quality and quantity of progeny [25,48], with significant consequences for population replenishment regardless of any direct effects of environmental change.

4. Discussion

Despite persistent controversy surrounding the specific cause(s) of outbreaks of *Acanthaster* spp. [1,29], there has been substantial research (>950 publications) on *Acanthaster* spp. over the past 30 years, leading to major advances in knowledge of their biology and ecology, as well as increased understanding of the effects of CoTS outbreaks within reef ecosystems (Table 1). In all, we suggest that 59% (24 out of 41) of the questions posed by Moran [21] have been effectively addressed and largely resolved in the last 30 years (Table 1). Many of the questions that are still largely unresolved relate to ecological processes (e.g., food-limitation and predation) that pertain to population regulation. While these questions are critically important in establishing the fundamental cause(s) of outbreaks of *Acanthaster* spp., a large part of the reason why these questions have not been effectively addressed is that the required research will be logistically very challenging. For example, measuring survivorship of CoTS larvae in the field is not really tractable, and it is clear that experimental studies conducted under laboratory conditions provide very limited insights into natural rates and causes of larval mortality (Section 2.12). There are however, some unresolved questions that are not only tractable, but will contribute directly to increased management efficiency and effectiveness (Table 1), and it is these research topics that should be given immediate priority. Notably, new technologies and opportunities to quantify the temporal and spatial patterns in the abundance of CoTS larvae [53,54], along with other established methods for measuring settlement rates [115] should be incorporated into intensive and ongoing monitoring programs to provide an early warning system for new and renewed outbreaks of *Acanthaster* spp. (Section 2.33).

Table 1. Research progress against each of the specific questions posed by Moran [21]. Numbers reflect the original numbering of questions, as referred to in heading throughout Section 2. "*" indicates those questions that are still unresolved, that are nonetheless very tractable problems which will have significant benefit for understanding and managing outbreaks of *Acanthaster* spp. on relatively short time-frames.

Topic	Addressed/Resolved	Unresolved	Ambiguous
Larvae	Q4–8, Q14	Q1–3, Q9 *, Q10	
Settlement and juveniles	Q12, Q20–21	Q11 *, Q13 *, Q15–18	Q19
Adults	Q2–5, Q8, Q10–11	Q6–7, Q9 *	Q1
Impacts	Q1–3		
Overarching	Q2, Q4–5	Q1, Q3, Q6 *	
Total	21	18	2

A large portion of recent studies on *Acanthaster* spp. are confirmatory in nature, often refining or validating the exceptional insights of early researchers, such as Endean [113] and Moran [21]. For example, Pratchett et al. [127] quantified short-term movement rates for adult *A.* cf. *solaris* over different substrate types. Despite using detailed video analyses to document movement under laboratory conditions, Pratchett et al. [127] reported rates of movement that were broadly similar to those estimated based on field observations in the 1960s [113]. Moreover, the more critical question to establish the capability of CoTS to move between reefs, which was also raised by Endean [113], about how long adult CoTS can actually sustain near maximum rates of movement, has never been addressed. While specific and detailed experimental studies are critical in providing a stronger foundation for future research as well as the application of research findings, major advances in understanding and managing outbreaks of *Acanthaster* spp. are reliant on (i) synthesis of prior research, both through qualitative reviews (e.g., [1,31,129]; this review) and empirical-based models (Section 3.3) to combine diverse data sources and identify critical knowledge gaps; (ii) coordinated and collaborative research efforts to effectively address logistically challenging questions at relevant scales of time and space, and in field settings; and (iii) improved integration of science and management, not only to expedite the update of new and important research findings, but to moderate research objectives to explicitly consider specific management tools and levers, and the relevant time-frames for considering different management actions.

Moran [21] provided a thorough and comprehensive review of the state of knowledge for *Acanthaster* spp. in the mid 1980s, during the height of the second documented outbreak on the GBR. At the time, most of the research and scientific discussion centered on the cause(s) of population outbreaks, and specifically, whether outbreaks are influenced by anthropogenic activities. Moran [21] predicted that the staunch support for individual hypotheses and specific viewpoints would significantly impede scientific advances, such that research should not aim to confirm or refute individual hypotheses, especially not in isolation. Currently, in the midst of the fourth documented outbreak wave on the GBR, there is still ongoing debate and specifically targeted research to address individual causes of outbreaks. Some of this research is entirely pragmatic, focusing on the potential benefits of specific management actions [133], rather than the specific mechanistic causes of outbreaks. However, focus needs to be given to specific, well-defined and tractable issues (Table 2). Outbreaks of *Acanthaster* spp. also need to be considered against the backdrop of other disturbances and threat to coral reef ecosystems. Worryingly for the health of the GBR is that the same areas that have so far been impacted by outbreaks of *A.* cf. *solaris* (northern and central sections; [34,35], have recently (2016 and 2017) been subject to the most widespread and most severe mass coral bleaching ever recorded [276].

In conclusion, outbreaks of *Acanthaster* spp. remain one of the most significant disturbances and major causes of coral loss across the Indo-Pacific [6–9,17,19,20,219,235,278]. While previous efforts to eradicate outbreaks have only been successful in relatively small and isolated areas [22,148], questioning whether it would ever be feasible or practical to prevent outbreaks. However, outbreaks are more amenable to direct intervention than most of the other major causes of widespread and significant coral loss (e.g., climate-induced coral bleaching), and therefore, sustained and ongoing research to improve understanding and management of outbreaks of *Acanthaster* spp. is certainly warranted. Fundamentally, there remain considerable practical and logistical challenges to undertaking necessary research to better understand the population dynamics of *Acanthaster* spp., though emerging technologies continually provide new opportunities and increased efficiencies to tackle research questions that were previously intractable or unfeasible (e.g., [53,54,107]). The purpose of this review was to highlight research questions posed more than 30 years ago, that were considered fundamental in understanding and managing CoTS outbreaks [21]. Despite significant research in the intervening period, a relatively large number (18 out of 41) of these questions remain largely unresolved (Table 1). It was not that scientists completely neglected or disregarded these questions, and there has been some progress made to address many (21 out of 41) of the pertinent questions

(Section 2). There have also been new and emerging issues (Section 3) that have partly deflected research attention. However, ongoing debate regarding the specific cause(s) of CoTS outbreaks is potentially detracting from effective research on this issue. Moving forward, the focus has to be on questions that will improve the effectiveness of management to reduce the frequency and likelihood of outbreaks arising, as well as detecting and containing outbreaks as soon as they start.

Table 2. New questions about crown-of-thorns starfish, intended to stimulate future research. These are additional to unresolved questions that were posed by Moran [21] (Table 1).

Biogeography and systematics
(1)   Is it possible to distinguish distinct species of *Acanthaster* spp. in the field? Do these species hybridize in areas of geographical overlap?
(2)   Is there interspecific variation in demography (e.g., growth and fecundity) and behavior (feeding rates and diets) that might account for geographic variation in incidence and severity of population outbreaks?
Larval biology
(3)   How does the nutritional status of wild larvae vary spatial and temporally? Does the condition, survival and settlement of larvae increase following exposure to nutrient pulses and phytoplankton blooms?
(4)   What is the most critical bottleneck in larval development and survival? Are nutrient (and food) pulses more important for early or late developmental phases?
Adult behavior
(5)   Is interannual variation in spawning intensity and periodicity related intrinsic (adult condition) or extrinsic (e.g., temperature) factors? What are the proximal and ultimate spawning cues?
(6)   Does increased abundance and diversity of reef fishes (and/or invertebrate predators) constrain the reproductive success of *Acanthaster* spp. (either through sub-lethal effects on fecundity or disruption of aggregations and spawning)?
Control strategies
(7)   Can intensive culling contain or prevent, rather than eliminate outbreaks? What are the detection limits and culling efficiencies for immature starfish? What are the longer-term versus short-term benefits of direct control?
(8)   Is it possible to disperse aggregations and/or prevent spawning by CoTS using chemical deterrents? Is it possible to aggregate dispersed populations to increase effectiveness of culling?

Acknowledgments: This study was motivated by initial discussions and planning meetings for the Australian National Environmental Science Program (NESP)—Tropical Water Quality Hub. The authors are grateful to Wei Zhang, whose encouragement and support was central to the realization of this manuscript.

Author Contributions: Morgan S. Pratchett and Sven Uthicke conceived this review; Lauren E. Nadler extracted and analyzed data from Web of Science; Ciemon F. Caballes and Jessica Hoey prepared figures. All authors wrote the paper, addressing questions that were relevant to their specific fields of study.

Conflicts of Interest: The authors declare no conflict of interest.

References

1. Pratchett, M.S.; Caballes, C.F.; Rivera-Posada, J.A.; Sweatman, H.P.A. Limits to understanding and managing outbreaks of crown-of-thorns starfish (*Acanthaster* spp.). *Oceanogr. Mar. Biol. Ann. Rev.* **2014**, *52*, 133–200.

2. Chesher, R.H. Destruction of Pacific corals by the sea star *Acanthaster planci*. *Science* **1969**, *165*, 280–283. [CrossRef] [PubMed]

3. Birkeland, C. The Faustian traits of the crown-of-thorns starfish. *Am. Sci.* **1989**, *77*, 154–163.

4. Cole, A.J.; Pratchett, M.S.; Jones, G.P. Diversity and functional importance of coral-feeding fishes on tropical coral reefs. *Fish Fish.* **2008**, *9*, 286–307. [CrossRef]

5. Rotjan, R.D.; Lewis, S.M. Impact of coral predators on tropical reefs. *Mar. Ecol. Prog. Ser.* **2008**, *367*, 73–91. [CrossRef]

6. Kayal, M.; Vercelloni, J.; Lison de Loma, T.; Bosserelle, P.; Chancerelle, Y.; Geoffroy, S.; Stievenart, C.; Michonneau, F.; Penin, L.; Planes, S.; et al. Predator crown-of-thorns starfish (*Acanthaster planci*) outbreak, mass mortality of corals, and cascading effects on reef fish and benthic communities. *PLoS ONE* **2012**, *7*, e47363. [CrossRef] [PubMed]

7. De'ath, G.; Fabricius, K.E.; Sweatman, H.P.A.; Puotinen, M. The 27-year decline of coral cover on the Great Barrier Reef and its causes. *Proc. Natl. Acad. Sci. USA* **2012**, *109*, 17995–17999. [CrossRef] [PubMed]
8. Osborne, K.; Dolman, A.M.; Burgess, S.C.; Johns, K.A. Disturbance and the dynamics of coral cover on the Great Barrier Reef (1995–2009). *PLoS ONE* **2011**, *6*, e17516. [CrossRef] [PubMed]
9. Trapon, M.L.; Pratchett, M.S.; Penin, L. Comparative effects of different disturbances in coral reef habitats in Moorea, French Polynesia. *J. Mar. Biol.* **2011**, *2011*, 807625. [CrossRef]
10. Goreau, T.F. On the predation of coral by the spiny starfish *Acanthaster planci* (L.) in the southern Red Sea. In *Bulletin of the Sea Fisheries Research Station of Israel, 35*; Sea Fisheries Research Station: Haifa, Israel, 1964; pp. 23–26.
11. Pearson, R.G.; Endean, R.G. A preliminary study of the coral predator *Acanthaster planci* (L.) (Asteroidea) on the Great Barrier Reef. *Queensl. Dep. Harb. Mar. Fish. Notes* **1969**, *3*, 27–55.
12. Wilkinson, C.R.; Macintyre, I.G. Preface: Special Issue on the *Acanthaster* debate. *Coral Reefs* **1992**, *11*, 51–52. [CrossRef]
13. Johnson, C.R. Reproduction, recruitment and hydrodynamics in the crown-of-thorns phenomenon on the Great Barrier Reef: Introduction and synthesis. *Mar. Freshw. Res.* **1992**, *43*, 517–523. [CrossRef]
14. Benzie, J.A.H.; Stoddart, J.A. Genetic structure of outbreaking and non-outbreaking crown-of-thorns starfish (*Acanthaster planci*) populations on the Great Barrier Reef. *Mar. Biol.* **1992**, *112*, 119–130. [CrossRef]
15. Moran, P.J.; De'ath, G. Estimates of the abundance of the crown-of-thorns starfish *Acanthaster planci* in outbreaking and non-outbreaking populations on reefs within the Great Barrier Reef. *Mar. Biol.* **1992**, *113*, 509–515. [CrossRef]
16. Nakamura, M.; Higa, Y.; Kumagai, N.H.; Okaji, K. Using long-term removal data to manage a crown-of-thorns starfish population. *Diversity* **2016**, *8*, 24. [CrossRef]
17. Pisapia, C.; Burn, D.; Yoosuf, R.; Najeeb, A.; Anderson, K.D.; Pratchett, M.S. Coral recovery in the central Maldives archipelago since the last major mass-bleaching, in 1998. *Sci. Rep.* **2016**, *6*, 34720. [CrossRef] [PubMed]
18. Anthony, K.R.N. Coral reefs under climate change and ocean acidification: Challenges and opportunities for management and policy. *Annu. Rev. Environ. Resour.* **2016**, *41*, 59–81. [CrossRef]
19. Pratchett, M.S.; Schenk, T.J.; Baine, M.; Syms, C.; Baird, A.H. Selective coral mortality associated with outbreaks of *Acanthaster planci* L. in Bootless Bay, Papua New Guinea. *Mar. Environ. Res.* **2009**, *67*, 230–236. [CrossRef] [PubMed]
20. Baird, A.H.; Pratchett, M.S.; Hoey, A.S.; Herdiana, Y.; Campbell, S.J. *Acanthaster planci* is a major cause of coral mortality in Indonesia. *Coral Reefs* **2013**, *32*, 803–812. [CrossRef]
21. Moran, P.J. The *Acanthaster* Phenomenon. *Oceanogr. Mar. Biol. Ann. Rev.* **1986**, *24*, 379–480.
22. Birkeland, C.; Lucas, J.S. *Acanthaster planci: Major Management Problem of Coral Reefs*; CRC Press: Boca Raton, FL, USA, 1990; p. 257.
23. Allen, J.D.; Schrage, K.; Foo, S.A.; Watson, S.-A.; Byrne, M. The Effects of salinity and pH on fertilization, early development, and hatching in the crown-of-thorns seastar. *Diversity* **2017**, *9*, 13. [CrossRef]
24. Caballes, C.F.; Pratchett, M.S.; Raymundo, M.L.; Rivera-Posada, J.A. Environmental tipping points for sperm motility, fertilization, and embryonic development in the crown-of-thorns starfish. *Diversity* **2017**, *9*, 10. [CrossRef]
25. Caballes, C.F.; Pratchett, M.S.; Buck, A.C.E. Interactive effects of endogenous and exogenous nutrition on larval development for crown-of-thorns starfish. *Diversity* **2017**, *9*, 15. [CrossRef]
26. Mellin, C.; Lugrin, C.; Okaji, K.; Francis, D.S.; Uthicke, S. Selective feeding and microalgal consumption rates by crown-of-thorns seastar (*Acanthaster* cf. *solaris*) larvae. *Diversity* **2017**, *9*, 8. [CrossRef]
27. Pratchett, M.S.; Dworjanyn, S.A.; Mos, B.; Caballes, C.F.; Thompson, C.; Blowes, S. Larval survivorship and settlement of crown-of-thorns starfish (*Acanthaster* cf. *solaris*) at varying algal cell densities. *Diversity* **2017**, *9*, 2.
28. Nakajima, R.; Nakatomi, N.; Kurihara, H.; Fox, M.; Smith, J.; Okaji, K. Crown-of-thorns starfish larvae can feed on organic matter released from corals. *Diversity* **2016**, *8*, 18. [CrossRef]
29. Brodie, J.E.; Devlin, M.J.; Lewis, S. Potential enhanced survivorship of crown-of-thorns starfish larvae due to near-annual nutrient enrichment during secondary outbreaks on the central mid-shelf of the Great Barrier Reef, Australia. *Diversity* **2017**, *9*, 17. [CrossRef]

30. Cowan, Z.-L.; Dworjanyn, S.A.; Caballes, C.F.; Pratchett, M.S. Benthic predators influence microhabitat preferences and settlement success of crown-of-thorns starfish (*Acanthaster* cf. *solaris*). *Diversity* **2016**, *8*, 27. [CrossRef]

31. Cowan, Z.-L.; Pratchett, M.S.; Messmer, V.; Ling, S.D. Known predators of crown-of-thorns starfish (*Acanthaster* spp.) and their role in mitigating, if not preventing, population outbreaks. *Diversity* **2017**, *9*, 7. [CrossRef]

32. Messmer, V.; Pratchett, M.S.; Chong-seng, K.M. Variation in incidence and severity of injuries among crown-of-thorns starfish (*Acanthaster* cf. *solaris*) on Australia's Great Barrier Reef. *Diversity* **2017**, *9*, 12. [CrossRef]

33. Harrison, H.B.; Pratchett, M.S.; Messmer, V.; Saenz-Agudelo, P.; Berumen, M.L. Microsatellites reveal genetic homogeneity among outbreak populations of crown-of-thorns starfish (*Acanthaster* cf. *solaris*) on Australia's Great Barrier Reef. *Diversity* **2017**, *9*, 16. [CrossRef]

34. Wilmes, J.; Matthews, S.; Schultz, D.; Messmer, V.; Hoey, A.; Pratchett, M.S. Modelling growth of juvenile crown-of-thorns starfish on the Northern Great Barrier Reef. *Diversity* **2016**, *9*, 1. [CrossRef]

35. MacNeil, M.A.; Chong-seng, K.M.; Pratchett, D.; Thompson, C.; Messmer, V.; Pratchett, M.S. Age and growth of an outbreaking *Acanthaster* cf. *solaris* population within the Great Barrier Reef. *Diversity* **2017**, *9*, 18. [CrossRef]

36. Sigl, R.; Laforsch, C. The influence of water currents on movement patterns on sand in the crown-of-thorns seastar (*Acanthaster* cf. *solaris*). *Diversity* **2016**, *8*, 25. [CrossRef]

37. Buck, A.C.E.; Gardiner, N.M.; Boström-Einarsson, L. Citric acid injections: An accessible and efficient method for controlling outbreaks of the crown-of-thorns starfish *Acanthaster* cf. *solaris*. *Diversity* **2016**, *8*, 28. [CrossRef]

38. Vogler, C.; Benzie, J.A.H.; Lessios, H.A.; Barber, P.H.; Wörheide, G. A threat to coral reefs multiplied? Four species of crown-of-thorns starfish. *Biol. Lett.* **2008**, *4*, 696–699. [CrossRef] [PubMed]

39. Haszprunar, G.; Vogler, C.; Wörheide, G. Persistent gaps of knowledge for naming and distinguishing multiple species of crown-of-thorns seastar in the *Acanthaster planci* species complex. *Diversity* **2017**, *9*, 22. [CrossRef]

40. Lucas, J.S. Quantitative studies of feeding and nutrition during larval development of the coral reef asteroid *Acanthaster planci* (L.). *J. Exp. Mar. Biol. Ecol.* **1982**, *65*, 173–193. [CrossRef]

41. Okaji, K. Feeding Ecology in the Early Life Stages of the Crown-of-Thorns Starfish, *Acanthaster planci* (L.). Ph.D. Thesis, James Cook University, Townsville, Queensland, Australia, 1996.

42. Uthicke, S.; Logan, M.; Liddy, M.; Francis, D.S.; Hardy, N.; Lamare, M.D. Climate change as an unexpected co-factor promoting coral eating seastar (*Acanthaster planci*) outbreaks. *Sci. Rep.* **2015**, *5*, 8402. [CrossRef] [PubMed]

43. Wolfe, K.; Graba-Landry, A.; Dworjanyn, S.A.; Byrne, M. Larval starvation to satiation: Influence of nutrient regime on the success of *Acanthaster planci*. *PLoS ONE* **2015**, *10*, e0122010. [CrossRef] [PubMed]

44. Brodie, J.E.; Fabricius, K.E.; De'ath, G.; Okaji, K. Are increased nutrient inputs responsible for more outbreaks of crown-of-thorns starfish? An appraisal of the evidence. *Mar. Pollut. Bull.* **2005**, *51*, 266–278. [CrossRef] [PubMed]

45. Fabricius, K.E.; Okaji, K.; De'ath, G. Three lines of evidence to link outbreaks of the crown-of-thorns seastar *Acanthaster planci* to the release of larval food limitation. *Coral Reefs* **2010**, *29*, 593–605. [CrossRef]

46. Olson, R.R. In situ culturing as a test of the larval starvation hypothesis for the crown-of-thorns starfish, *Acanthaster planci*. *Limnol. Oceanogr.* **1987**, *32*, 895–904. [CrossRef]

47. Wolfe, K.; Graba-Landry, A.; Dworjanyn, S.A.; Byrne, M. Superstars: Assessing nutrient thresholds for enhanced larval success of *Acanthaster planci*, a review of the evidence. *Mar. Pollut. Bull.* **2017**, *116*, 307–314. [CrossRef] [PubMed]

48. Caballes, C.F.; Pratchett, M.S.; Kerr, A.M.; Rivera-Posada, J.A. The role of maternal nutrition on oocyte size and quality, with respect to early larval development in the coral-eating starfish, *Acanthaster planci*. *PLoS ONE* **2016**, *11*, e0158007. [CrossRef] [PubMed]

49. Wolfe, K.; Graba-Landry, A.; Dworjanyn, S.A.; Byrne, M. Larval phenotypic plasticity in the boom-and-bust crown-of-thorns seastar, *Acanthaster planci*. *Mar. Ecol. Prog. Ser.* **2015**, *539*, 179–189. [CrossRef]

50. Devlin, M.J.; DeBose, J.L.; Ajani, P.; Teixeira da Silva, E.; Petus, C.; Brodie, J.E. Phytoplankton in the Great Barrier Reef: Microscopy analysis of community structure in high flow events. In *Report to the National Environmental Research Program*; Reef and Rainforest Research Centre Limited: Cairns, Australia, 2013; p. 68.

51. King, E.; Schroeder, T.; Brando, V.E.; Suber, K. *A Pre-Operational System for Satellite Monitoring of Great Barrier Reef Marine Water Quality*; CSIRO Wealth from Oceans Flagship Report; CSIRO Wealth from Oceans Flagship: Hobart, Australia, 2013; pp. 1–56.

52. Caballes, C.F.; Pratchett, M.S. Environmental and biological cues for spawning in the crown-of-thorns starfish. *PLoS ONE* **2017**, *12*, e0173964. [CrossRef] [PubMed]

53. Uthicke, S.; Doyle, J.R.; Duggan, S.; Yasuda, N.; McKinnon, A.D. Outbreak of coral-eating Crown-of-Thorns creates continuous cloud of larvae over 320 km of the Great Barrier Reef. *Sci. Rep.* **2015**, *5*, 16885. [CrossRef] [PubMed]

54. Doyle, J.R.; McKinnon, A.D.; Uthicke, S. Quantifying larvae of the coralivorous seastar *Acanthaster* cf. *solaris* on the Great Barrier Reef using qPCR. *Mar. Biol.* **2017**, *164*, 176. [CrossRef]

55. Birkeland, C. Terrestrial runoff as a cause of outbreaks of *Acanthaster planci* (Echinodermata: Asteroidea). *Mar. Biol.* **1982**, *69*, 175–185. [CrossRef]

56. Vanhatalo, J.; Hosack, G.R.; Sweatman, H.P.A. Spatio-temporal modelling of crown-of-thorns starfish outbreaks on the Great Barrier Reef to inform control strategies. *J. Appl. Ecol.* **2017**, *54*, 188–197. [CrossRef]

57. Wooldridge, S.A.; Brodie, J.E. Environmental triggers for primary outbreaks of crown-of-thorns starfish on the Great Barrier Reef, Australia. *Mar. Pollut. Bull.* **2015**, *101*, 805–815. [CrossRef] [PubMed]

58. Hock, K.; Wolff, N.H.; Condie, S.A.; Anthony, K.R.N.; Mumby, P.J. Connectivity networks reveal the risks of crown-of-thorns starfish outbreaks on the Great Barrier Reef. *J. Appl. Ecol.* **2014**, *51*, 1188–1196. [CrossRef]

59. Pratchett, M.S. Dynamics of an outbreak population of *Acanthaster planci* at Lizard Island, northern Great Barrier Reef (1995–1999). *Coral Reefs* **2005**, *24*, 453–462. [CrossRef]

60. Okaji, K.; Ayukai, T.; Lucas, J.S. Selective uptake of dissolved free amino acids by larvae of the crown-of-thorns starfish, *Acanthaster planci* (L.). In Proceedings of the 8th International Coral Reef Symposium, Bali, Indonesia, 24–29 June 1997; Volume 1, pp. 613–615.

61. Yamaguchi, M. Early life histories of coral reef asteroids, with special reference to *Acanthaster planci* (L.). In *Biology and Geology of Coral Reefs*; Jones, O.A., Endean, R.G., Eds.; Academic Press, Inc.: New York, NY, USA, 1973; Volume 2, pp. 369–387.

62. Uchida, H.; Nomura, K. On the efficiency of natural plankton for culture of pelagic larvae of the crown-of-thorns starfish. *Bull. Mar. Sci.* **1987**, *41*, 643.

63. Furnas, M.J.; Mitchell, A.W. Phytoplankton dynamics in the central Great Barrier Reef-II. Primary production. *Cont. Shelf Res.* **1987**, *7*, 1049–1062. [CrossRef]

64. Tada, K.; Sakai, K.; Nakano, Y.; Takemura, A.; Montani, S. Size-fractionated phytoplankton biomass in coral reef waters off Sesoko Island, Okinawa, Japan. *J. Plankton Res.* **2003**, *25*, 991–997. [CrossRef]

65. Acevedo-Trejos, E.; Brandt, G.; Bruggeman, J.; Merico, A. Mechanisms shaping size structure and functional diversity of phytoplankton communities in the ocean. *Sci. Rep.* **2015**, *5*, 8918. [CrossRef] [PubMed]

66. Furnas, M.J.; Mitchell, A.W. Phytoplankton dynamics in the central Great Barrier Reef—I. Seasonal changes in biomass and community structure and their relation to intrusive activity. *Cont. Shelf Res.* **1986**, *6*, 363–384. [CrossRef]

67. Ayukai, T. Picoplankton dynamics in Davies Reef lagoon, the Great Barrier Reef, Australia. *J. Plankton Res.* **1992**, *14*, 1593–1606. [CrossRef]

68. Okaji, K.; Ayukai, T.; Lucas, J.S. Selective feeding by larvae of the crown-of-thorns starfish, *Acanthaster planci* (L.). *Coral Reefs* **1997**, *16*, 47–50. [CrossRef]

69. Ayukai, T. Ingestion of ultraplankton by the planktonic larvae of the crown-of-thorns starfish, *Acanthaster planci*. *Biol. Bull.* **1994**, *186*, 90–100. [CrossRef]

70. Hoegh-Guldberg, O. Uptake of dissolved organic matter by larval stage of the crown-of-thorns starfish *Acanthaster planci*. *Mar. Biol.* **1994**, *120*, 55–63.

71. Babcock, R.C.; Dambacher, J.M.; Morello, E.B.; Plagányi, É.E.; Hayes, K.R.; Sweatman, H.P.A.; Pratchett, M.S. Assessing different causes of crown-of-thorns starfish outbreaks and appropriate responses for management on the Great Barrier Reef. *PLoS ONE* **2016**, *11*, e0169048. [CrossRef] [PubMed]

72. Cowan, Z.-L.; Dworjanyn, S.A.; Caballes, C.F.; Pratchett, M.S. Predation on crown-of-thorns starfish larvae by damselfishes. *Coral Reefs* **2016**, *35*, 1253–1262. [CrossRef]

73. Olson, R.R.; Olson, M.H. Food limitation of planktotrophic marine invertebrate larvae: Does it control recruitment success? *Annu. Rev. Ecol. Syst.* **1989**, *20*, 225–247. [CrossRef]

74. Uthicke, S.; Pecorino, D.; Albright, R.; Negri, A.P.; Cantin, N.E.; Liddy, M.; Dworjanyn, S.A.; Kamya, P.Z.; Byrne, M.; Lamare, M.D. Impacts of ocean acidification on early life-history stages and settlement of the coral-eating sea star *Acanthaster planci*. *PLoS ONE* **2013**, *8*, e82938. [CrossRef] [PubMed]

75. Kamya, P.Z.; Dworjanyn, S.A.; Hardy, N.; Mos, B.; Uthicke, S.; Byrne, M. Larvae of the coral eating crown-of-thorns starfish, *Acanthaster planci* in a warmer-high CO_2 ocean. *Glob. Chang. Biol.* **2014**, *20*, 3365–3376. [CrossRef] [PubMed]

76. Lamare, M.D.; Pecorino, D.; Hardy, N.; Liddy, M.; Byrne, M.; Uthicke, S. The thermal tolerance of crown-of-thorns (*Acanthaster planci*) embryos and bipinnaria larvae: Implications for spatial and temporal variation in adult populations. *Coral Reefs* **2014**, *33*, 207–219. [CrossRef]

77. Byrne, M. Impact of ocean warming and ocean acidification on marine invertebrate life history stages: Vulnerabilities and potential for persistence in a changing ocean. *Oceanogr. Mar. Biol. Ann. Rev.* **2011**, *49*, 1–42.

78. Sparks, K.M.; Foo, S.A.; Uthicke, S.; Byrne, M.; Lamare, M.D. Paternal identity influences response of *Acanthaster planci* embryos to ocean acidification and warming. *Coral Reefs* **2017**, *36*, 325–338. [CrossRef]

79. Lucas, J.S. Reproductive and larval biology of *Acanthaster planci* (L.) in Great Barrier Reef Waters. *Micronesica* **1973**, *9*, 197–203.

80. Lucas, J.S. Environmental influences on the early development of *Acanthaster planci* (L.). In *Crown-of-Thorns Starfish Seminar Proceedings*; Australian Government Publishing Service: Canberra, Australia, 1974; pp. 109–121.

81. Rumrill, S.S. Natural mortality of marine invertebrate larvae. *Ophelia* **1990**, *32*, 163–198. [CrossRef]

82. Dight, I.J.; Bode, L.; James, M.K. Modelling the larval dispersal of *Acanthaster planci*. I. Large scale hydrodynamics, Cairns section, Great Barrier Reef Marine Park. *Coral Reefs* **1990**, *9*, 115–123. [CrossRef]

83. Black, K.P.; Moran, P.J. Influence of hydrodynamics on the passive dispersal and initial recruitment of larvae of *Acanthaster planci* (Echinodermata: Asteroidea) on the Great Barrier Reef. *Mar. Ecol. Prog. Ser.* **1991**, *69*, 55–65. [CrossRef]

84. Black, K.P. The relative importance of local retention and inter-reef dispersal of neutrally buoyant material on coral reefs. *Coral Reefs* **1993**, *12*, 43–53. [CrossRef]

85. Timmers, M.A.; Andrews, K.R.; Bird, C.E.; DeMaintenton, M.J.; Brainard, R.E.; Toonen, R.J. Widespread dispersal of the crown-of-thorns sea star, *Acanthaster planci*, across the Hawaiian Archipelago and Johnston Atoll. *J. Mar. Biol.* **2011**, *2011*, 934269. [CrossRef]

86. Yasuda, N.; Taquet, C.; Nagai, S.; Yoshida, T.; Adjeroud, M. Genetic connectivity of the coral-eating sea star *Acanthaster planci* during the severe outbreak of 2006–2009 in the Society Islands, French Polynesia. *Mar. Ecol.* **2015**, *36*, 668–678. [CrossRef]

87. Yasuda, N.; Nagai, S.; Hamaguchi, M.; Okaji, K.; Gérard, K.; Nadaoka, K. Gene flow of *Acanthaster planci* (L.) in relation to ocean currents revealed by microsatellite analysis. *Mol. Ecol.* **2009**, *18*, 1574–1590. [CrossRef] [PubMed]

88. Timmers, M.A.; Bird, C.E.; Skillings, D.J.; Smouse, P.E.; Toonen, R.J. There's no place like home: Crown-of-thorns outbreaks in the central pacific are regionally derived and independent events. *PLoS ONE* **2012**, *7*, e31159. [CrossRef] [PubMed]

89. Benzie, J.A.H.; Black, K.P.; Moran, P.J.; Dixon, P. Small-scale dispersion of eggs and sperm of the starfish (*Acanthaster planci*) in a shallow coral reef habitat. *Biol. Bull.* **1994**, *186*, 153–167. [CrossRef]

90. Ormond, R.F.G.; Campbell, A.C.; Head, S.H.; Moore, R.J.; Rainbow, P.R.; Saunders, A.P. Formation and breakdown of aggregations of crown-of-thorns starfish, *Acanthaster planci* (L.). *Nature* **1973**, *246*, 167–169. [CrossRef]

91. Moran, P.J.; De'ath, G.; Baker, V.J.; Bass, D.K.; Christie, C.A.; Miller, I.R.; Miller-Smith, B.A.; Thompson, A.A. Pattern of outbreaks of crown-of-thorns starfish (*Acanthaster planci* L.) along the Great Barrier Reef since 1966. *Mar. Freshw. Res.* **1992**, *43*, 555–568. [CrossRef]

92. Miller, I.R. Historical patterns and current trends in the broadscale distribution of crown-of-thorns starfish in the northern and central sections of the Great Barrier Reef. In Proceedings of the 9th International Coral Reef Symposium, Bali, Indonesia, 23–27 October 2000; pp. 1478–1484.

93. Dight, I.J.; James, M.K.; Bode, L. Modelling the larval dispersal of *Acanthaster planci*. II. Patterns of reef connectivity. *Coral Reefs* **1990**, *9*, 125–134. [CrossRef]

94. Reichelt, R.E.; Bradbury, R.H.; Moran, P.J. Distribution of *Acanthaster planci* outbreaks on the Great Barrier Reef between 1966 and 1989. *Coral Reefs* **1990**, *9*, 97–103. [CrossRef]

95. Black, K.P.; Moran, P.J.; Burrage, D.; De'ath, G. Association of low-frequency currents and crown-of-thorns starfish outbreaks. *Mar. Ecol. Prog. Ser.* **1995**, *125*, 185–194. [CrossRef]

96. Hock, K.; Mumby, P.J. Quantifying the reliability of dispersal paths in connectivity networks. *J. R. Soc. Interface* **2015**, *12*, 284–292. [CrossRef] [PubMed]

97. Benzie, J.A.H. Review of the genetics, dispersal and recruitment of crown-of-thorns starfish (*Acanthaster planci*). *Mar. Freshw. Res.* **1992**, *43*, 597–610. [CrossRef]

98. Benzie, J.A.H.; Wakeford, M. Genetic structure of crown-of-thorns starfish (*Acanthaster planci*) on the Great Barrier Reef, Australia: Comparison of two sets of outbreak populations occurring ten years apart. *Mar. Biol.* **1997**, *129*, 149–157. [CrossRef]

99. Vogler, C.; Benzie, J.A.H.; Tenggardjaja, K.; Barber, P.H.; Wörheide, G. Phylogeography of the crown-of-thorns starfish: Genetic structure within the Pacific species. *Coral Reefs* **2013**, *32*, 515–525. [CrossRef]

100. Tusso, S.; Morcinek, K.; Vogler, C.; Schupp, P.J.; Caballes, C.F.; Vargas, S.; Wörheide, G. Genetic structure of the crown-of-thorns seastar in the Pacific Ocean, with focus on Guam. *PeerJ* **2016**, *4*, e1970. [CrossRef] [PubMed]

101. Babcock, R.C.; Milton, D.A.; Pratchett, M.S. Relationships between size and reproductive output in the crown-of-thorns starfish. *Mar. Biol.* **2016**, *163*, 234. [CrossRef]

102. Ebert, T.A. Recruitment in echinoderms. In *Echinoderm Studies*; CRC Press: Boca Raton, FL, USA, 1983; Volume 1, pp. 169–203.

103. Uthicke, S.; Schaffelke, B.; Byrne, M. A boom–bust phylum? Ecological and evolutionary consequences of density variations in echinoderms. *Ecol. Monogr.* **2009**, *79*, 3–24. [CrossRef]

104. Babcock, R.C.; Mundy, C.N. Reproductive biology, spawning and field fertilization rates of *Acanthaster planci*. *Mar. Freshw. Res.* **1992**, *43*, 525–534. [CrossRef]

105. Rogers, J.G.D.; Pláganyi, É.E.; Babcock, R.C. Aggregation, Allee effects and critical thresholds for the management of the crown-of-thorns starfish *Acanthaster planci*. *Mar. Ecol. Prog. Ser.* **2017**, *578*, 99–114. [CrossRef]

106. Babcock, R.C.; Mundy, C.N.; Whitehead, D. Sperm diffusion models and in situ confirmation of long-distance fertilization in the free-spawning asteroid *Acanthaster planci*. *Biol. Bull.* **1994**, *186*, 17–28. [CrossRef]

107. Suzuki, G.; Yasuda, N.; Ikehara, K.; Fukuoka, K.; Kameda, T.; Kai, S.; Nagai, S.; Watanabe, A.; Nakamura, T.; Kitazawa, S.; et al. Detection of a high-density brachiolaria-stage larval population of crown-of-thorns sea star (*Acanthaster planci*) in Sekisei Lagoon (Okinawa, Japan). *Diversity* **2016**, *8*, 9. [CrossRef]

108. Hill, A.E. Diel vertical migration in stratified tidal flows: Implications for plankton dispersal. *J. Mar. Res.* **1998**, *56*, 1069–1096. [CrossRef]

109. Metaxas, A. Behaviour in flow: Perspectives on the distribution and dispersion of meroplanktonic larvae in the water column. *Can. J. Fish. Aquat. Sci.* **2001**, *58*, 86–98. [CrossRef]

110. Daigle, R.M.; Metaxas, A. Vertical distribution of marine invertebrate larvae in response to thermal stratification in the laboratory. *J. Exp. Mar. Biol. Ecol.* **2011**, *409*, 89–98. [CrossRef]

111. Sameoto, J.A.; Metaxas, A. Interactive effects of haloclines and food patches on the vertical distribution of 3 species of temperate invertebrate larvae. *J. Exp. Mar. Biol. Ecol.* **2008**, *367*, 131–141. [CrossRef]

112. Zann, L.P.; Brodie, J.E.; Berryman, C.; Naqasima, M. Recruitment, ecology, growth, and behavior of juvenile *Acanthaster planci* (L.) (Echinodermata: Asteroidea). *Bull. Mar. Sci.* **1987**, *41*, 561–575.

113. Endean, R.G. Population explosions of *Acanthaster planci* and associated destruction of hermatypic corals in the Indo-West Pacific region. In *Biology and Geology of Coral Reefs*; Jones, O.A., Endean, R.G., Eds.; Academic Press, Inc.: New York, NY, USA, 1973; pp. 389–438.

114. Johnson, C.R. Settlement and recruitment of *Acanthaster planci* on the Great Barrier Reef: Questions of process and scale. *Mar. Freshw. Res.* **1992**, *43*, 611–627. [CrossRef]

115. Keesing, J.K.; Cartwright, C.M.; Hall, K.C. Measuring settlement intensity of echinoderms on coral reefs. *Mar. Biol.* **1993**, *117*, 399–407.

116. Doherty, P.J.; Davidson, J. Monitoring the distribution and abundance of juvenile *Acanthaster planci* on the central Great Barrier Reef. In Proceedings of the 6th International Coral Reef Symposium, Bali, Indonesia, 8–12 August 1988; Volume 2, pp. 131–136.

117. Habe, T.; Sawamoto, S.; Ueno, S.; Kosaka, M.; Ogura, M. *Studies on the Conservation and Management of Coral Reefs and the Control of Acanthaster planci Juveniles*; Report of Grant-in-Aid for Scientific Research; Ministry of Education, Science and Culture: Tokyo, Japan, 1989; pp. 158–186.

118. Johnson, D.B.; Moran, P.J.; Baker, V.J.; Christie, C.A.; Miller, I.R.; Thompson, A.A. An attempt to locate high density populations of juvenile crown-of-thorns starfish (*Acanthaster planci*) on the central Great Barrier Reef. *Coral Reefs* **1992**, *11*, 122. [CrossRef]

119. Ormond, R.F.G.; Campbell, A.C. Formation and breakdown of *Acanthaster planci* aggregations in the Red Sea. In Proceedings of the 2nd International Coral Reef Symposium, Bali, Indonesia, 22 June–2 July 1974; Volume 1, pp. 595–619.

120. Henderson, J.A.; Lucas, J.S. Larval development and metamorphosis of *Acanthaster planci* (Asteroidea). *Nature* **1971**, *232*, 655–657. [CrossRef] [PubMed]

121. Johnson, C.R.; Sutton, D.C.; Olson, R.R.; Giddins, R. Settlement of crown-of-thorns starfish: Role of bacteria on surfaces of coralline algae and a hypothesis for deepwater recruitment. *Mar. Ecol. Prog. Ser.* **1991**, *71*, 143–162. [CrossRef]

122. Yokochi, H.; Ogura, M. Spawning period and discovery of juvenile *Acanthaster planci* (L.) (Echinodermata: Asteroidea) at northwestern Iriomote-jima, Ryukyu Islands. *Bull. Mar. Sci.* **1987**, *41*, 611–616.

123. Lucas, J.S. Growth, maturation and effects of diet in *Acanthaster planci* (L.) (Asteroidea) and hybrids reared in the laboratory. *J. Exp. Mar. Biol. Ecol.* **1984**, *79*, 129–147. [CrossRef]

124. Kamya, P.Z.; Byrne, M.; Graba-Landry, A.; Dworjanyn, S.A. Near-future ocean acidification enhances the feeding rate and development of the herbivorous juveniles of the crown-of-thorns starfish, *Acanthaster planci*. *Coral Reefs* **2016**, *35*, 1241–1251. [CrossRef]

125. Johnson, C.R.; Sutton, D.C. Bacteria on the surface of crustose coralline algae induce metamorphosis of the crown-of-thorns starfish *Acanthaster planci*. *Mar. Biol.* **1994**, *120*, 305–310. [CrossRef]

126. Keesing, J.K.; Halford, A.R. Importance of postsettlement processes for the population dynamics of *Acanthaster planci* (L.). *Mar. Freshw. Res.* **1992**, *43*, 635–651. [CrossRef]

127. Pratchett, M.S.; Cowan, Z.-L.; Nadler, L.E.; Caballes, C.F.; Hoey, A.S.; Messmer, V.; Fletcher, C.S.; Westcott, D.A.; Ling, S.D. Body size and substrate type modulate movement by the western Pacific crown-of-thorns starfish, *Acanthaster solaris*. *PLoS ONE* **2017**, *12*, e0180805. [CrossRef] [PubMed]

128. Zann, L.P.; Brodie, J.E.; Vuki, V. History and dynamics of the crown-of-thorns starfish *Acanthaster planci* (L.) in the Suva area, Fiji. *Coral Reefs* **1990**, *9*, 135–144. [CrossRef]

129. Caballes, C.F.; Pratchett, M.S. Reproductive biology and early life history of the crown-of-thorns starfish. In *Echinoderms: Ecology, Habitats and Reproductive Biology*; Whitmore, E., Ed.; Nova Science Publishers, Inc.: New York, NY, USA, 2014; pp. 101–146.

130. Keesing, J.K.; Halford, A.R. Field measurement of survival rates of juvenile *Acanthaster planci*: Techniques and preliminary results. *Mar. Ecol. Prog. Ser.* **1992**, *85*, 107–114. [CrossRef]

131. Sweatman, H.P.A. A field study of fish predation on juvenile crown-of-thorns starfish. *Coral Reefs* **1995**, *14*, 47–53. [CrossRef]

132. McCallum, H.I. Effects of predation on organisms with pelagic larval stages: Models of metapopulations. In Proceedings of the 6th International Coral Reef Symposium, Bali, Indonesia, 8–12 August 1988; Volume 2, pp. 101–106.

133. Sweatman, H.P.A. No-take reserves protect coral reefs from predatory starfish. *Curr. Biol.* **2008**, *18*, R598–R599. [CrossRef] [PubMed]

134. Engelhardt, U.; Lassig, B. *The Possible Causes and Consequences of Outbreaks of the Crown-of-Thorns Starfish*; Great Barrier Reef Marine Park Authority: Townsville, Australia, 1993.

135. Yamaguchi, M. Coral-Reef Asteroids of Guam. *Biotropica* **1975**, *7*, 12–23. [CrossRef]

136. Lucas, J.S.; Hart, R.J.; Howden, M.E.; Salathe, R. Saponins in eggs and larvae of *Acanthaster planci* (L.) (Asteroidea) as chemical defences against planktivorous fish. *J. Exp. Mar. Biol. Ecol.* **1979**, *40*, 155–165. [CrossRef]

137. Vine, P.J. Field and laboratory observations of the crown-of-thorns starfish, *Acanthaster planci*. *Nature* **1970**, *228*, 341–342. [CrossRef] [PubMed]

138. Glynn, P.W. Some physical and biological determinants of coral community structure in the eastern Pacific. *Ecol. Monogr.* **1976**, *46*, 431–456. [CrossRef]

139. Littler, M.M. The population and community structure of Hawaiian fringing-reef crustose Corallinaceae (Rhodophyta, Cryptonemiales). *J. Exp. Mar. Biol. Ecol.* **1973**, *11*, 103–120. [CrossRef]

140. Klumpp, D.W.; McKinnon, A.D. Community structure, biomass and productivity of epilithic algal communities on the Great Barrier Reef: Dynamics at different spatial scales. *Mar. Ecol. Prog. Ser.* **1992**, *86*, 77–89. [CrossRef]

141. Fabricius, K.E.; De'ath, G. Environmental factors associated with the spatial distribution of crustose coralline algae on the Great Barrier Reef. *Coral Reefs* **2001**, *19*, 303–309. [CrossRef]

142. Nakamura, M.; Kumagai, N.H.; Sakai, K.; Okaji, K.; Ogasawara, K.; Mitarai, S. Spatial variability in recruitment of acroporid corals and predatory starfish along the Onna coast, Okinawa, Japan. *Mar. Ecol. Prog. Ser.* **2015**, *540*, 1–12. [CrossRef]

143. Johansson, C.L.; Francis, D.S.; Uthicke, S. Food preferences of juvenile corallivorous crown-of-thorns (*Acanthaster planci*) sea stars. *Mar. Biol.* **2016**, *163*, 49. [CrossRef]

144. De'ath, G.; Moran, P.J. Factors affecting the behaviour of crown-of-thorns starfish (*Acanthaster planci* L.) on the Great Barrier Reef. 2: Feeding preferences. *J. Exp. Mar. Biol. Ecol.* **1998**, *220*, 107–126. [CrossRef]

145. Yamaguchi, M. Growth of juvenile *Acanthaster planci* (L.) in the laboratory. *Pac. Sci.* **1974**, *28*, 123–138.

146. Mueller, B.; Bos, A.R.; Graf, G.; Gumanao, G.S. Size-specific locomotion rate and movement pattern of four common Indo-Pacific sea stars (Echinodermata; Asteroidea). *Aquat. Biol.* **2011**, *12*, 157–164. [CrossRef]

147. Keesing, J.K.; Lucas, J.S. Field measurement of feeding and movement rates of the crown-of thorns starfish *Acanthaster planci* (L.). *J. Exp. Mar. Biol. Ecol.* **1992**, *156*, 89–104. [CrossRef]

148. Bos, A.R.; Gumanao, G.S.; Mueller, B.; Saceda-Cardoza, M.M.E. Management of crown-of-thorns sea star (*Acanthaster planci* L.) outbreaks: Removal success depends on reef topography and timing within the reproduction cycle. *Ocean. Coast. Manag.* **2013**, *71*, 116–122. [CrossRef]

149. Suzuki, G.; Kai, S.; Yamashita, H. Mass stranding of crown-of-thorns starfish. *Coral Reefs* **2012**, *31*, 821. [CrossRef]

150. Sigl, R.; Steibl, S.; Laforsch, C. The role of vision for navigation in the crown-of-thorns seastar, *Acanthaster planci*. *Sci. Rep.* **2016**, *6*, 30834. [CrossRef] [PubMed]

151. Beer, S.; Wentzel, C.; Petie, R.; Garm, A. Active control of the visual field in the starfish *Acanthaster planci*. *Vis. Res.* **2016**, *127*, 28–34. [CrossRef] [PubMed]

152. Petie, R.; Hall, M.R.; Hyldahl, M.; Garm, A. Visual orientation by the crown-of-thorns starfish (*Acanthaster planci*). *Coral Reefs* **2016**, *35*, 1139–1150. [CrossRef]

153. Clements, C.S.; Hay, M.E. Size matters: Predator outbreaks threaten foundation species in small Marine Protected Areas. *PLoS ONE* **2017**, *12*, e0171569. [CrossRef] [PubMed]

154. Kenchington, R.A. Growth and recruitment of *Acanthaster planci* (L.) on the Great Barrier Reef. *Biol. Conserv.* **1977**, *11*, 103–118. [CrossRef]

155. Moore, R.J. Persistent and transient populations of the crown-of-thorns starfish. In *Acanthaster and the Coral Reef: A Theoretical Perspective*; Bradbury, R.H., Ed.; Springer: Berlin, Germany, 1990; pp. 236–277.

156. Kettle, B.T. Variations in Biometric and Physiological Parameters of *Acanthaster planci* (L.) (Echinodermata; Asteroidea) during the Course of a High-Density Outbreak. Ph.D. Thesis, James Cook University, Townsville, Queensland, Australia, 1990.

157. Stump, R.J.W. *An Investigation of the Methods to Describe the Population Dynamics of Acanthaster planci (L.) around Lizard Island, Northern Cairns Section, GBR*; CRC Reef Research Tech. Rep. No. 10; Great Barrier Reef Marine Park Authority: Townsville, Australia, 1996.

158. Stump, R.J.W.; Lucas, J.S. Linear growth in spines from *Acanthaster planci* (L.) involving growth lines and periodic pigment bands. *Coral Reefs* **1990**, *9*, 149–154. [CrossRef]

159. Souter, D.W.; Cameron, A.M.; Endean, R.G. Implications of sublethal predation, autotomy and regeneration: Pigment bands on their spines can not be used to determine the ages of adult specimens of the corallivore *Acanthaster planci*. *Mar. Freshw. Res.* **1997**, *48*, 321–328. [CrossRef]

160. Stump, R.J.W. Age Determination and Life-History Characteristics of *Acanthaster planci* (L.) (Echinodermata: Asteroidea). Ph.D. Thesis, James Cook University, Townsville, Queensland, Australia, 1994.

161. Pan, M.; Hilomen, V.; Palomares, M.L.D. Size structure of *Acanthaster planci* populations in Tubbataha Reefs Natural Parks, Sulu Sea, Philippines. In *Marine Biodiversity of Southeast Asian and Adjacent Seas*; Palomares, M.L.D., Pauly, D., Eds.; University of British Columbia Fisheries Centre Research Reports; University of British Columbia Fisheries Centre: Vancouver, BC, Canada, 2010; Volume 18, pp. 70–77.

162. Mills, S.C. Density-dependent prophylaxis in the coral-eating crown-of-thorns sea star, *Acanthaster planci*. *Coral Reefs* 2012, *31*, 603–612. [CrossRef]

163. Jangoux, M. Diseases of Echinodermata. I. Agents microorganisms and protistans. *Mar. Ecol. Prog. Ser.* 1987, *2*, 147–162. [CrossRef]

164. Dungan, M.L.; Miller, T.E.; Thomson, D.A. Catastrophic decline of a top carnivore in the Gulf of California rocky intertidal zone. *Science* 1982, *216*, 989–991. [CrossRef] [PubMed]

165. Lessios, H.A.; Robertson, D.R.; Cubit, J.D. Spread of Diadema mass mortality through the Caribbean. *Science* 1984, *226*, 335–337. [CrossRef] [PubMed]

166. Sutton, D.C.; Trott, L.; Reichelt, J.L.; Lucas, J.S. Assessment of bacterial pathogenesis in crown-of-thorns starfish, *Acanthaster planci* (L.). In Proceedings of the 6th International Coral Reef Symposium, Bali, Indonesia, 8–12 August 1988; Volume 2, pp. 171–176.

167. Pratchett, M.S. An infectious disease in crown-of-thorns starfish on the Great Barrier Reef. *Coral Reefs* 1999, *18*, 272. [CrossRef]

168. Rivera-Posada, J.A.; Pratchett, M.S.; Cano-Gómez, A.; Arango-Gómez, J.D.; Owens, L. Refined identification of *Vibrio* bacterial flora from *Acanthasther*. planci based on biochemical profiling and analysis of housekeeping genes. *Dis. Aquat. Organ.* 2011, *96*, 113–123. [CrossRef] [PubMed]

169. Rivera-Posada, J.A.; Caballes, C.F.; Pratchett, M.S. Size-related variation in arm damage frequency in the crown-of-thorns sea star, *Acanthaster planci*. *J. Coast. Life Med.* 2014, *2*, 187–195.

170. Endean, R.G. *Report on Investigations Made into Aspects of the Current Acanthaster planci (Crown-of-Thorns) Infestations of Certain Reefs of the Great Barrier Reef*; Queensland Department of Primary Industries (Fisheries Branch): Brisbane, Australia, 1969.

171. Bos, A.R.; Gumanao, G.S.; Salac, F.N. A newly discovered predator of the crown-of-thorns starfish. *Coral Reefs* 2008, *27*, 581. [CrossRef]

172. Ocaña, V.O.; Hartog, J.C.D.; Hernández, A.B.; Bos, A.R. On *Pseudocorynactis*. species and another related genus from the Indo-Pacific (Anthozoa: Corallimorphidae). *Revista. de la Academia Canaria de Ciencias* 2009, *21*, 9–34.

173. Bos, A.R.; Gumanao, G.S.; Mueller, B. Feeding biology and symbiotic relationships of the corallimorpharian *Paracorynactis*. hoplites (Anthozoa: Hexacorallia). *Raffles Bull. Zool.* 2011, *59*, 245–250.

174. McClanahan, T.R.; Muthiga, N.A. Patterns of predation on a sea urchin, *Echinometra mathaei* (de Blainville), on Kenyan coral reefs. *J. Exp. Mar. Biol. Ecol.* 1989, *126*, 77–94. [CrossRef]

175. Ling, S.D.; Johnson, C.R. Marine reserves reduce risk of climate-driven phase shift by reinstating size- and habitat-specific trophic interactions. *Ecol. Appl.* 2012, *22*, 1232–1245. [CrossRef] [PubMed]

176. Hall, M.R.; Kocot, K.M.; Baughman, K.W.; Fernandez-Valverde, S.L.; Gauthier, M.E.A.; Hatleberg, W.L.; Krishnan, A.; McDougall, C.; Motti, C.A.; Shoguchi, E.; et al. The crown-of-thorns starfish genome as a guide for biocontrol of this coral reef pest. *Nature* 2017, *544*, 231–234. [CrossRef] [PubMed]

177. Dulvy, N.K.; Freckleton, R.P.; Polunin, N.V.C. Coral reef cascades and the indirect effects of predator removal by exploitation. *Ecol. Lett.* 2004, *7*, 410–416. [CrossRef]

178. Pratchett, M.S. Changes in coral assemblages during an outbreak of *Acanthaster planci* at Lizard Island, northern Great Barrier Reef (1995–1999). *Coral Reefs* 2010, *29*, 717–725. [CrossRef]

179. Rivera-Posada, J.A.; Pratchett, M.S.; Aguilar, C.; Grand, A.; Caballes, C.F. Bile salts and the single-shot lethal injection method for killing crown-of-thorns sea stars (*Acanthaster planci*). *Ocean. Coast. Manag.* 2014, *102*, 383–390. [CrossRef]

180. Rivera-Posada, J.A.; Pratchett, M.S.; Owens, L. Injection of *Acanthaster planci* with thiosulfate-citrate-bile-sucrose agar (TCBS). II. Histopathological changes. *Dis. Aquat. Organ.* 2011, *97*, 95–102. [CrossRef] [PubMed]

181. Rivera-Posada, J.A.; Pratchett, M.S.; Cano-Gómez, A.; Arango-Gómez, J.D.; Owens, L. Injection of *Acanthaster planci* with thiosulfate-citrate-bile-sucrose agar (TCBS). I. Disease induction. *Dis. Aquat. Organ.* 2011, *97*, 85–94. [CrossRef] [PubMed]

182. Frankel, E. Evidence from the Great Barrier Reef of ancient *Acanthaster* aggregations. *Atoll Res. Bull.* **1978**, *220*, 75–93. [CrossRef]

183. Walbran, P.D.; Henderson, R.A.; Jull, A.J.T.; Head, M.J. Evidence from sediments of long-term *Acanthaster planci* predation on corals of the Great Barrier Reef. *Science* **1989**, *245*, 847–850. [CrossRef] [PubMed]

184. Henderson, R.A.; Walbran, P.D. Interpretation of the fossil record of *Acanthaster planci* from the Great Barrier Reef: A reply to criticism. *Coral Reefs* **1992**, *11*, 95–101. [CrossRef]

185. Moran, P.J.; Reichelt, R.E.; Bradbury, R.H. An assessment of the geological evidence for previous *Acanthaster* outbreaks. *Coral Reefs* **1986**, *4*, 235–238. [CrossRef]

186. Pandolfi, J.M. A palaeobiological examination of the geological evidence for recurring outbreaks of the crown-of-thorns starfish, *Acanthaster planci* (L.). *Coral Reefs* **1992**, *11*, 87–93. [CrossRef]

187. Keesing, J.K.; Bradbury, R.H.; DeVantier, L.M.; Riddle, M.J.; De'ath, G. Geological evidence for recurring outbreaks of the crown-of-thorns starfish: A reassessment from an ecological perspective. *Coral Reefs* **1992**, *11*, 79–85. [CrossRef]

188. Fabricius, K.E.; Fabricius, F.H. Re-assessment of ossicle frequency patterns in sediment cores: Rate of sedimentation related to *Acanthaster planci*. *Coral Reefs* **1992**, *11*, 109–114. [CrossRef]

189. Porter, J.W. Predation by *Acanthaster* and its effect on coral species diversity. *Am. Nat.* **1972**, *106*, 487–492. [CrossRef]

190. Potts, D.C. Crown-of-thorns starfish—Man-induced pest or natural phenomenon. In *The Ecology of Pests: Some Australian Case Histories*; Kitching, R.L., Jones, R.E., Eds.; CSIRO: Melbourne, Australia, 1981; pp. 54–86.

191. Chess, J.R.; Hobson, E.S.; Howard, D.F. Interactions between *Acanthaster planci* (Echinodermata, Asteroidea) and scleractinian corals at Kona, Hawai'i. *Pac. Sci.* **1997**, *51*, 121–133.

192. Bouchon, C. Quantitative study of Scleractinian coral communities of Tiahura Reef (Moorea Island, French Polynesia). In Proceedings of the 5th International Coral Reef Symposium, Bali, Indonesia, 27 May–1 June 1985; Volume 1, pp. 279–284.

193. Pearson, R.G. Recovery and recolonization of coral reefs. *Mar. Ecol. Prog. Ser.* **1981**, *4*, 105–122. [CrossRef]

194. Done, T.J.; DeVantier, L.M. Fundamental change in coral community structure at Green Island. *Coral Reefs* **1990**, *9*, 166. [CrossRef]

195. Lourey, M.J.; Ryan, D.A.J.; Miller, I.R. Rates of decline and recovery of coral cover on reefs impacted by, recovering from and unaffected by crown-of-thorns starfish *Acanthaster planci*: A regional perspective of the Great Barrier Reef. *Mar. Ecol. Prog. Ser.* **2000**, *196*, 179–186. [CrossRef]

196. Wakeford, M.; Done, T.J.; Johnson, C.R. Decadal trends in a coral community and evidence of changed disturbance regime. *Coral Reefs* **2008**, *27*, 1–13. [CrossRef]

197. Berumen, M.L.; Pratchett, M.S. Recovery without resilience: Persistent disturbance and long-term shifts in the structure of fish and coral communities at Tiahura Reef, Moorea. *Coral Reefs* **2006**, *25*, 647–653. [CrossRef]

198. Connell, J.H.; Hughes, T.P.; Wallace, C.C. A 30-year study of coral abundance, recruitment, and disturbance at several scales in space and time. *Ecol. Monogr.* **1997**, *67*, 461–488. [CrossRef]

199. Gilmour, J.P.; Smith, L.D.; Heyward, A.J.; Baird, A.H.; Pratchett, M.S. Recovery of an isolated coral reef system following severe disturbance. *Science* **2013**, *340*, 69–71. [CrossRef] [PubMed]

200. Mellin, C.; MacNeil, M.A.; Cheal, A.J.; Emslie, M.J.; Caley, M.J. Marine protected areas increase resilience among coral reef communities. *Ecol. Lett.* **2016**, *19*, 629–637. [CrossRef] [PubMed]

201. Jones, G.P.; McCormick, M.I.; Srinivasan, M.; Eagle, J.V. Coral decline threatens fish biodiversity in marine reserves. *Proc. Natl. Acad. Sci. USA* **2004**, *101*, 8251–8253. [CrossRef] [PubMed]

202. Pratchett, M.S.; Munday, P.L.; Wilson, S.K.; Graham, N.A.J.; Cinner, J.E.; Bellwood, D.R.; Jones, G.P.; Polunin, N.V.C.; McClanahan, T.R. Effects of climate-induced coral bleaching on coral-reef fishes: Ecological and economic consequences. *Oceanogr. Mar. Biol. Ann. Rev.* **2008**, *46*, 251–296.

203. Stella, J.S.; Pratchett, M.S.; Hutchings, P.A.; Jones, G.P. Coral-associated invertebrates: Diversity, ecological importance and vulnerability to disturbance. *Oceanogr. Mar. Biol. Ann. Rev.* **2011**, *49*, 43–104.

204. Coker, D.J.; Wilson, S.K.; Pratchett, M.S. Importance of live coral habitat for reef fishes. *Rev. Fish Biol. Fish.* **2014**, *24*, 89–126. [CrossRef]

205. Chou, L.M.; Yamazato, K. Community structure of coral reefs within the vicinity of Motobu and Sesoko, Okinawa, and the effects of human and natural influences. *Galaxea* **1990**, *9*, 9–75.

206. Larkum, A.W.D. High rates of nitrogen fixation on coral skeletons after predation by the crown of thorns starfish *Acanthaster planci*. *Mar. Biol.* **1988**, *97*, 503–506. [CrossRef]

207. Bouchon-Navaro, Y.; Bouchon, C.; Harmelin-Vivien, M.L. Impact of coral degradation on a chaetodontid fish population (Moorea, French Polynesia). In Proceedings of the 5th International Coral Reef Symposium, Bali, Indonesia, 27 May–1 June 1985; Volume 5, pp. 427–432.

208. Pratchett, M.S.; Coker, D.J.; Jones, G.P.; Munday, P.L. Specialization in habitat use by coral reef damselfishes and their susceptibility to habitat loss. *Ecol. Evol.* **2012**, *2*, 2168–2180. [CrossRef] [PubMed]

209. Sano, M.; Shimizu, M.; Nose, Y. Long-term effects of destruction of hermatypic corals by *Acanthaster planci* infestation on reef fish communities at Iriomote Island, Japan. *Mar. Ecol. Prog. Ser.* **1987**, *37*, 191–199. [CrossRef]

210. Ainsworth, C.H.; Mumby, P.J. Coral-algal phase shifts alter fish communities and reduce fisheries production. *Glob. Chang. Biol.* **2015**, *21*, 165–172. [CrossRef] [PubMed]

211. Sano, M. Stability of reef fish assemblages: Responses to coral recovery after catastrophic predation by *Acanthaster planci*. *Mar. Ecol. Prog. Ser.* **2000**, *198*, 121–130. [CrossRef]

212. Halford, A.R.; Cheal, A.J.; Ryan, D.; Williams, D.M. Resilience to large-scale disturbance in coral and fish assemblages on the Great Barrier Reef. *Ecology* **2004**, *85*, 1892–1905. [CrossRef]

213. Wilson, S.K.; Graham, N.A.J.; Pratchett, M.S.; Jones, G.P.; Polunin, N.V.C. Multiple disturbances and the global degradation of coral reefs: Are reef fishes at risk or resilient? *Glob. Chang. Biol.* **2006**, *12*, 2220–2234. [CrossRef]

214. Hughes, T.P.; Rodrigues, M.J.; Bellwood, D.R.; Ceccarelli, D.; Hoegh-Guldberg, O.; McCook, L.; Moltschaniwskyj, N.A.; Pratchett, M.S.; Steneck, R.S.; Willis, B.L. Phase shifts, herbivory and the resilience of coral reefs to climate change. *Curr. Biol.* **2007**, *17*, 360–365. [CrossRef] [PubMed]

215. Kennedy, E.V.; Perry, C.T.; Halloran, P.R.; Iglesias-Prieto, R.; Schönberg, C.H.L.; Wisshak, M.; Form, A.U.; Carricart-Ganivet, J.P.; Fine, M.; Eakin, C.M.; et al. Avoiding coral reef functional collapse requires local and global action. *Curr. Biol.* **2013**, *23*, 912–918. [CrossRef] [PubMed]

216. Perry, C.T.; Murphy, G.N.; Kench, P.S.; Smithers, S.G.; Edinger, E.N.; Steneck, R.S.; Mumby, P.J. Caribbean-wide decline in carbonate production threatens coral reef growth. *Nat. Commun.* **2013**, *4*, 1402. [CrossRef] [PubMed]

217. Gattuso, J.-P.; Pichon, M.; Delesalle, B.; Canon, C.; Frankignoulle, M. Carbon fluxes in coral reefs. I. Lagrangian measurement of community metabolism and resulting air-sea CO_2 disequilibrium. *Mar. Ecol. Prog. Ser.* **1996**, *145*, 109–121. [CrossRef]

218. Perry, C.T.; Spencer, T.; Kench, P.S. Carbonate budgets and reef production states: A geomorphic perspective on the ecological phase-shift concept. *Coral Reefs* **2008**, *27*, 853–866. [CrossRef]

219. Caballes, C.F. The Role of Chemical Signals on the Feeding Behavior of the Crown-of-Thorns Seastar, *Acanthaster planci* (Linnaeus, 1758). Master Thesis, University of Guam, Mangilao, Guam, USA, 2009.

220. Pratchett, M.S.; Hoey, A.S.; Wilson, S.K. Reef degradation and the loss of critical ecosystem goods and services provided by coral reef fishes. *Curr. Opin. Environ. Sustain.* **2014**, *7*, 37–43. [CrossRef]

221. Randall, J.E. Chemical pollution in the sea and the crown-of-thorns starfish (*Acanthaster planci*). *Biotropica* **1972**, *4*, 132–144. [CrossRef]

222. Endean, R.G. *Acanthaster planci* infestations of reefs of the Great Barrier Reef. In Proceedings of the 3th International Coral Reef Symposium, Bali, Indonesia, 23 May 1977; Volume 1, pp. 185–191.

223. Dana, T.F. *Acanthaster*: A rarity in the past? *Science* **1970**, *169*, 894. [CrossRef] [PubMed]

224. Vine, P.J. Crown of thorns (*Acanthaster planci*) plagues: The natural causes theory. *Atoll Res. Bull.* **1973**, *166*, 1–10. [CrossRef]

225. Weber, J.N.; Woodhead, P.M.J. Ecological studies of the coral predator *Acanthaster planci* in the South Pacific. *Mar. Biol.* **1970**, *6*, 12–17. [CrossRef]

226. Newman, W.A. *Acanthaster*: A disaster? *Science* **1970**, *167*, 1274. [CrossRef] [PubMed]

227. Branham, J.M. The crown of thorns on coral reefs. *Bioscience* **1973**, *23*, 219–226. [CrossRef]

228. Birkeland, C. *Acanthaster* in the culture of high islands. *Atoll Res. Bull.* **1981**, *255*, 55–58.

229. Flanigan, J.M.; Lamberts, A.E. *Acanthaster* as a recurring phenomenon in Samoan history. *Atoll Res. Bull.* **1981**, *255*, 59–62.

230. Clark, H.L. *The Echinoderm Fauna of Torres Strait: Its Composition and Its Origin*; Carnegie Institution Washington: Washington, DC, USA, 1921; Volume 10, pp. 1–224.

231. DeVantier, L.M.; Done, T.J. Inferring past outbreaks of the crown-of-thorns seastar from scar patterns on coral heads. In *Geological Approaches to Coral Reef Ecology*; Aronson, R.B., Ed.; Springer: New York, NY, USA, 2007; pp. 85–125.

232. Seymour, R.M.; Bradbury, R.H. Lengthening reef recovery times from crown-of-thorns outbreaks signal systemic degradation of the Great Barrier Reef. *Mar. Ecol. Prog. Ser.* **1999**, *176*, 1–10. [CrossRef]

233. Kenchington, R.A.; Kelleher, G. Crown-of-thorns starfish management conundrums. *Coral Reefs* **1992**, *11*, 53–56. [CrossRef]

234. Houk, P. The transition zone chlorophyll front can trigger *Acanthaster planci* outbreaks in the Pacific Ocean: Historical confirmation. *J. Oceanogr.* **2007**, *63*, 149–154. [CrossRef]

235. Nakamura, M.; Okaji, K.; Higa, Y.; Yamakawa, E.; Mitarai, S. Spatial and temporal population dynamics of the crown-of-thorns starfish, *Acanthaster planci*, over a 24-year period along the central west coast of Okinawa Island, Japan. *Mar. Biol.* **2014**, *161*, 2521–2530. [CrossRef]

236. Randall, R.H. Distribution of corals after *Acanthaster planci* (L.) infestation at Tanguisson Point, Gŭam. *Micronesica.* **1973**, *3*, 213–222.

237. Colgan, M.W. Coral reef recovery on Guam (Micronesia) after catastrophic predation by *Acanthaster planci*. *Ecology* **1987**, *68*, 1592–1605. [CrossRef]

238. Branham, J.M.; Reed, S.A.; Bailey, J.H.; Caperon, J. Coral-eating sea stars *Acanthaster planci* in Hawaii. *Science* **1971**, *172*, 1155–1157. [CrossRef] [PubMed]

239. Nishihira, M.; Yamazato, K. Human interference with the coral reef community and *Acanthaster* infestation of Okinawa. In Proceedings of the 2th International Coral Reef Symposium, Bali, Indonesia, 22 June–2 July 1973; Volume 5, pp. 577–590.

240. Yamaguchi, M. *Acanthaster planci* infestations of reefs and coral assemblages in Japan: A retrospective analysis of control efforts. *Coral Reefs* **1986**, *5*, 23–30. [CrossRef]

241. Glynn, P.W. The impact of *Acanthaster* on corals and coral reefs in the Eastern Pacific. *Environ. Conserv.* **1974**, *1*, 295–304. [CrossRef]

242. Pratchett, M.S. Influence of coral symbionts on feeding preferences of crown-of thorns starfish *Acanthaster planci* in the western Pacific. *Mar. Ecol. Prog. Ser.* **2001**, *214*, 111–119. [CrossRef]

243. Haszprunar, G.; Spies, M. An integrative approach to the taxonomy of the crown-of-thorns starfish species group (Asteroidea: *Acanthaster*): A review of names and comparison to recent molecular data. *Zootaxa* **2014**, *3841*, 271–284. [CrossRef] [PubMed]

244. Zann, L.P. Status of crown-of-thorns starfish in the Indian Ocean. In *Coral Reefs of the Indian Ocean*; McClanahan, T.R., Sheppard, C.R.C., Obura, D.A., Eds.; Oxford University Press: New York, NY, USA, 2000.

245. Boström-Einarsson, L.; Rivera-Posada, J.A. Controlling outbreaks of the coral-eating crown-of-thorns starfish using a single injection of common household vinegar. *Coral Reefs* **2015**, *35*, 223–228. [CrossRef]

246. Rivera-Posada, J.A.; Caballes, C.F.; Pratchett, M.S. Lethal doses of oxbile, peptones and thiosulfate-citrate-bile-sucrose agar (TCBS) for *Acanthaster planci*; exploring alternative population control options. *Mar. Pollut. Bull.* **2013**, *75*, 133–139. [CrossRef] [PubMed]

247. Westcott, D.A.; Fletcher, C.S.; Babcock, R.C.; Plaganyi-Lloyd, E. *A Strategy to Link Research and Management of Crown-of-Thorns Starfish on the Great Barrier Reef: An Integrated Pest Management Approach*; Report to the National Environmental Science Programme; Reef and Rainforest Research Centre Ltd.: Cairns, Australia, 2016; 80p.

248. Dayoub, F.; Dunbabin, M.; Corke, P. Robotic detection and tracking of Crown-of-Thorns starfish. *IEEE Int. Conf. Intell. Robot. Syst.* **2015**, 1921–1928. [CrossRef]

249. Hoey, J.; Campbell, M.; Hewitt, C.; Gould, B.; Bird, R. *Acanthaster planci* invasions: Applying biosecurity practices to manage a native boom and bust coral pest in Australia. *Manag. Biol. Invasions* **2016**, *7*, 213–220. [CrossRef]

250. Lassig, B.; Gladstone, W.; Moran, P.J.; Engelhardt, U. A crown-of-thorns starfish contingency plan. In Proceedings of the 2nd International Coral Reef Symposium, Bali, Indonesia, 22 June–2 July 1973; Volume 2, pp. 780–788.

251. Dumas, P.; Moutardier, G.; Ham, J.; Kaku, R.; Gereva, S.; Lefèvre, J.; Adjeroud, M. Timing within the reproduction cycle modulates the efficiency of village-based crown-of-thorns starfish removal. *Biol. Conserv.* **2016**, *204*, 237–246. [CrossRef]

252. Hunter, D.M. Advances in the control of locusts (Orthoptera: Acrididae) in eastern Australia: From crop protection to preventive control. *Aust. J. Entomol.* **2004**, *43*, 293–303. [CrossRef]
253. MacNeil, M.A.; Mellin, C.; Pratchett, M.S.; Hoey, J.; Anthony, K.R.N.; Cheal, A.J.; Miller, I.R.; Sweatman, H.P.A.; Cowan, Z.-L.; Taylor, S.; et al. Joint estimation of crown of thorns (*Acanthaster planci*) densities on the Great Barrier Reef. *PeerJ* **2016**, *4*, e2310. [CrossRef] [PubMed]
254. Kayal, M.; Bosserelle, P.; Adjeroud, M. Bias associated with the detectability of the coral-eating pest crown-of-thorns seastar and implications for reef management. *R. Soc. Open Sci.* **2017**, *4*, 170396. [CrossRef] [PubMed]
255. Nishida, M.; Lucas, J.S. Genetic differences between geographic populations of the crown-of-thorns starfish throughout the Pacific region. *Mar. Biol.* **1988**, *98*, 359–368. [CrossRef]
256. Benzie, J.A.H. Major genetic differences between crown-of-thorns starfish (*Acanthaster planci*) populations in the Indian & Pacific Oceans. *Evolution* **1999**, *53*, 1782–1795. [PubMed]
257. Hobbs, J.-P.A.; Salmond, J.K. Cohabitation of Indian and Pacific Ocean species at Christmas and Cocos (Keeling) Islands. *Coral Reefs* **2008**, *27*, 933. [CrossRef]
258. Yuasa, H.; Higashimura, Y.; Nomura, K.; Yasuda, N. Diet of *Acanthaster brevispinus*, sibling species of the coral-eating crown-of-thorns startfish, *Acanthaster planci sensu lato. Bull. Mar. Sci.* **2017**, *93*. [CrossRef]
259. Lucas, J.S.; Jones, M.M. Hybrid crown-of-thorns starfish (*Acanthaster planci* x *A. brevispinus*) reared to maturity in the laboratory. *Nature* **1976**, *263*, 409–412. [CrossRef] [PubMed]
260. Moya, A.; Huisman, L.; Ball, E.E.; Hayward, D.C.; Grasso, L.C.; Chua, C.M.; Woo, H.N.; Gattuso, J.-P.; Forêt, S.; Miller, D.J. Whole transcriptome analysis of the coral *Acropora. millepora* reveals complex responses to CO$_2$-driven acidification during the initiation of calcification. *Mol. Ecol.* **2012**, *21*, 2440–2454. [CrossRef] [PubMed]
261. Evans, T.G.; Pespeni, M.H.; Hofmann, G.E.; Palumbi, S.R.; Sanford, E. Transcriptomic responses to seawater acidification among sea urchin populations inhabiting a natural pH mosaic. *Mol. Ecol.* **2017**, *26*, 2257–2275. [CrossRef] [PubMed]
262. Pespeni, M.H.; Sanford, E.; Gaylord, B.; Hill, T.M.; Hosfelt, J.D.; Jaris, H.K.; LaVigne, M.; Lenz, E.A.; Russell, A.D.; Young, M.K.; et al. Evolutionary change during experimental ocean acidification. *Proc. Natl. Acad. Sci. USA* **2013**, *110*, 6937–6942. [CrossRef] [PubMed]
263. Stewart, M.J.; Stewart, P.; Rivera-Posada, J.A. De novo assembly of the transcriptome of *Acanthaster planci* testes. *Mol. Ecol. Resour.* **2015**, *15*, 953–966. [CrossRef] [PubMed]
264. Yasuda, N.; Kajiwara, K.; Nagai, S.; Ikehara, K.; Nadaoka, K. First report of field sampling and identification of crown-of-thorns starfish larvae. *Galaxea.* **2015**, *17*, 15–16. [CrossRef]
265. Reichelt, R.E. Dispersal and control models of *Acanthaster planci* populations on the Great Barrier Reef. In *Acanthaster and the Coral Reef: A Theoretical Perspective*; Bradbury, R.H., Ed.; Springer: Berlin, Germany, 1990; pp. 6–16.
266. McCallum, H.I. *Are crown-of-thorns starfish populations chaotic? In The Possible Causes and Consequences of Outbreaks of the Crown-of-Thorns Starfish*; Engelhardt, U., Lassig, B., Eds.; Great Barrier Reef Marine Park Authority: Townsville, Australia, 1993; pp. 83–93.
267. Antonelli, P.; Auger, P.; Bradbury, R.H. Corals and starfish waves on the Great Barrier Reef: Analytical trophodynamics and 2-patch aggregation methods. *Math. Comput. Model.* **1998**, *27*, 121–135. [CrossRef]
268. Morello, E.B.; Plagányi, É.E.; Babcock, R.C.; Sweatman, H.P.A.; Hillary, R.; Punt, A.E. Model to manage and reduce crown-of-thorns starfish outbreaks. *Mar. Ecol. Prog. Ser.* **2014**, *512*, 167–183. [CrossRef]
269. Chen, C.; Drovandi, C.; Keith, J.; Anthony, K.R.N.; Caley, M.J.; Mengersen, K. Bayesian semi-individual based model with ABC for parameter calibration: Modelling Crown-of-Thorns populations on the Great Barrier Reef. *Methods Ecol. Evol.* **2017**. under review.
270. Wolanski, E. Facts and numerical artefacts in modelling the dispersal of crown-of-thorns starfish larvae in the Great Barrier Reef. *Mar. Freshw. Res.* **1993**, *44*, 427–436. [CrossRef]
271. Scandol, J.P. CotSim—An interactive *Acanthaster planci* metapopulation model for the central Great Barrier Reef. *Mar. Model.* **1999**, *1*, 39–81. [CrossRef]
272. Bode, L.; Mason, L.B. Application of an implicit hydrodynamic model over a range of spatial scales. In *Computational Techniques and Applications: CTAC93*; World Scientific: Singapore, 1994; pp. 112–121.
273. James, M.K.; Armsworth, P.R.; Mason, L.B.; Bode, L. The structure of reef fish metapopulations: Modelling larval dispersal and retention patterns. *Proc. R. Soc. B Biol. Sci.* **2002**, *269*, 2079–2086. [CrossRef] [PubMed]

274. Hock, K.; Wolff, N.H.; Beeden, R.; Hoey, J.; Condie, S.A.; Anthony, K.R.N.; Possingham, H.P.; Mumby, P.J. Controlling range expansion in habitat networks by adaptively targeting source populations. *Conserv. Biol.* **2016**, *30*, 856–866. [CrossRef] [PubMed]

275. Mellin, C.; Lurgi, M.; Matthews, S.; MacNeil, M.A.; Caley, M.J.; Bax, N.; Przeslawski, R.; Fordham, D.A. Forecasting marine invasions under climate change: Biotic interactions and demographic processes matter. *Biol. Conserv.* **2016**, *204*, 459–467. [CrossRef]

276. Hughes, T.P.; Kerry, J.T.; Álvarez-Noriega, M.; Álvarez-Romero, J.G.; Anderson, K.D.; Baird, A.H.; Babcock, R.C.; Beger, M.; Bellwood, D.R.; Berkelmans, R.; et al. Global warming and recurrent mass bleaching of corals. *Nature* **2017**, *543*, 373–377. [CrossRef] [PubMed]

277. Kamya, P.Z.; Byrne, M.; Mos, B.; Hall, L.; Dworjanyn, S.A. Indirect effects of ocean acidification drive feeding and growth of juvenile crown-of-thorns starfish, *Acanthaster planci. Proc. R. Soc. B Biol. Sci.* **2017**, *284*, 20170778. [CrossRef] [PubMed]

278. Roche, R.C.; Pratchett, M.S.; Carr, P.; Turner, J.R.; Wagner, D.; Head, C.; Sheppard, C.R.C. Localized outbreaks of *Acanthaster planci* at an isolated and unpopulated reef atoll in the Chagos Archipelago. *Mar. Biol.* **2015**, *162*, 1695–1704. [CrossRef]

diversity

MDPI

Opinion

Persistent Gaps of Knowledge for Naming and Distinguishing Multiple Species of Crown-of-Thorns-Seastar in the *Acanthaster planci* Species Complex

Gerhard Haszprunar [1,2,*], Catherine Vogler [3] and Gert Wörheide [3,4]

1   Staatliche Naturwissenschaftliche Sammlungen Bayerns (SNSB)-Zoological State Collection,
    Münchhausenstraße 21, D-81247 Munich, Germany
2   Department Biology II and GeoBio-Center, Ludwig-Maximilians-Universität München,
    D-80539 Munich, Germany
3   Department of Earth and Environmental Sciences, Paleontology & Geobiology, and GeoBio-Center,
    Ludwig-Maximilians-Universität München, Richard-Wagner-Straße 10, D-80333 Munich, Germany;
    catherine.vogler@gmail.com (C.V.); woerheide@lmu.de (G.W.)
4   SNSB-Bavarian State Collections of Palaeontology and Geology, D-81827 Munich, Germany
*   Correspondence: haszi@zsm.mwn.de; Tel.: +49-89-8107-104

Academic Editors: Morgan Pratchett and Sven Uthicke
Received: 19 September 2016; Accepted: 9 May 2017; Published: 12 May 2017

Abstract: Nearly a decade ago, DNA barcoding (partial mitochondrial COI gene sequences) showed that there are at least four species in the Indo-Pacific within what was previously conceived to be a single Crown-of-Thorns-Seastar (COTS) species, *Acanthaster planci*. Two of these species—*A. planci* Linnaeus, 1758, distributed in the North Indian Ocean, and *A. mauritiensis* de Loriol, 1885, distributed in the South Indian Ocean—have been already unequivocally named. In contrast, the Pacific COTS (proposed name: *A. solaris* (Schreber, 1795) and the COTS from the Red Sea (still to be named) require further taxonomic work. COI barcoding sequences and Barcode Identification Numbers (BINs) are available for all four COTS species in the global Barcode of Life Database (BOLD). We recommend depositing voucher specimens or tissue samples suitable for DNA analyses when studying any aspect of COTS, and use BINs to identify species, to ensure that no information is lost on species allocation until unequivocal Linnean names are available for the Pacific and Red Sea species as well. We also review the differences between COTS species with respect to morphology, ecology, and toxicity. Future studies should widen the current biogeographic coverage of the different COTS species by strategically sampling neglected areas, especially at the geographic distribution limits of each species, to enhance our understanding of the diversity of this reef coral predator.

Keywords: COTS; taxonomy; Linnean names; DNA-barcodes; Barcode Index Numbers; biogeography

1. Introduction

There is little doubt that the Crown-Of-Thorns-Seastar (COTS), usually referred to as *Acanthaster planci* (Linnaeus, 1758), with its corallivorous lifestyle and strong tendency to mass outbreaks, is one of the most serious threats to coral reefs throughout the Indo-Pacific marine biome, and numerous studies over the last decades have addressed all aspects concerning this point (reviewed in [1]). These circumstances have led to *Acanthaster planci*'s being the most researched and cited of all echinoderm species.

For more than 250 years, researchers have noticed morphological differences among specimens of COTS from different geographical areas. Indeed, a number of taxa have been proposed, and these have subsequently been merged again or split further (for a review, see [2]). Currently, only one other species aside from *Acanthaster planci* is uniformly accepted as valid within the genus, *Acanthaster brevispinus* Fisher, 1917 [3]. This species does not feed on coral and thus does not threaten coral reefs. Yet during the last 30 years, various molecular datasets have suggested that *Acanthaster planci* might be more than a single species [4–8]. The most thorough molecular and phylogeographic study that included *Acanthaster planci* specimens across its entire Indo-Pacific distribution range (from the Red Sea to Mauritius and the Eastern Pacific) revealed no less than four deeply divergent clades [9]. The "barcoding fragment" of the mitochondrial COI gene showed high inter-clade divergence (8.8–10.6%) compared to <0.7% intra-clade divergence, strongly suggesting that *Acanthaster planci* represents a species complex with four different species instead of a single one [9]. These species (and we use the term "species" henceforth) show distinct geographical distribution patterns across the Indo-Pacific: one species might be restricted to the Red Sea, one occurs in the Northern Indian Ocean, another one mainly in the Southern Indian Ocean (with the exception of Northern Oman), and the fourth species shows a pan-Pacific distribution (Table 1; Figure 1, references [9–11]). Moreover, the Pacific species shows some internal phylogeographic structure [12–14], accordingly it cannot be excluded that additional (sub-)species be discovered here in the future using higher-resolution molecular markers, such as SNPs.

Figure 1. Geographic distribution of COI-barcoded species and of type localities of names (combined and modified from [2] and [9–11]: red—Red Sea (RS) species; blue—South Indian Ocean (SIO) species (*A. mauritiensis*); yellow—North Indian Ocean (NIO) species (*A. planci*); green—Pacific Ocean (PO) species (*A. solaris*). Location of type localities of nominal (sub-)species: asterisk—*A. planci*; cross—*A. echinites*; triangle—*A. solaris*, square—*A. mauritiensis*; circle—*A. ellisii pseudoplanci*; "?"—the type locality of *A. ellisii* was not specified: in South American waters of the East Pacific.

Table 1. Summary of Linnean nomenclature, Biological Index Number (BIN) and distribution of currently recognized Organismic Taxonomic Units (OTUs) of the *Acanthaster planci* species complex based on Vogler et al. [9–11] and Haszprunar & Spies [2]. See text for details.

OTU (Vogler et al., 2008 [9])	NIO Species	SIO Species	Pacific Species	Red Sea Species
Name [proposed]	*A. planci* (Linnaeus, 1758)	*A. mauritiensis* de Loriol, 1885	[*A. solaris* (Schreber, 1795)]	NN (not yet named)
Type locality	off Goa, West Indian coast	off Mauritius	Magellan Street, Philippines	-
BIN (BOLD)	AAA1633	AAA1631	AAA1630	AAA1632
Synonyms [proposed]	[*A. echinites* (Ellis & Solander in Watt, 1786)]	-	*A. ellisii* (Gray, 1840) *A. e. pseudoplanci* Caso, 1962	-
Distribution	North Indian Ocean	South Indian Ocean (except West Australian coast)	Pacific Ocean and West Australian coast	Red Sea
Outbreak ability	Yes	yes	Yes	yes
number of arms	max. 23	max. 23	max. 23	max. 13-14
Specific color	"electric-blue"	light blue to rusty	gray-green to gray-purple, bulls-eye appearance	gray-green to gray-purple, bulls-eye appearance
Harmfulness to humans	harmful	harmful	very harmful	less harmful

2. Nomenclatorical Status and Problems

When established in 2008, Vogler et al. [9–11] did not correlate the four species with any Linnean taxon names. This could explain why their results have not been considered in most studies on COTS biology and ecology published since then, despite the potential implications of these findings for COTS research and conservation.

This unfortunate situation was only partially overcome by a recent thorough nomenclatorial study [2], which linked available names of two of the four species—(1) *Acanthaster planci* (Linnaeus, 1758) (with type locality Goa, West Indian coast) and *Acanthaster mauritiensis* (de Loriol, 1885) (with type locality Mauritius Island, [15]) with the Northern Indian Ocean (NIO) species and the Southern Indian Ocean (SIO) species, respectively. The Red Sea species still needs formal description, which is currently in progress by Gerhard Haszprunar and Gert Wörheide.

In contrast, the Pacific species, the most frequently studied one, is still somewhat equivocal in its nomenclature. The type localities of two of the previously described nominal species fall into the actual distribution area (Figure 1): (a) *Acanthaster solaris* (Schreber, 1795) has been described as being likely from the area around Cebu in the Philippines (see [2] for the specifics of the problems concerning the exact localization of the type locality). (b) *Acanthaster echinites* (Ellis & Solander in Watt, 1786) has its type locality "from Batavia", i.e., off Jakarta, Indonesia. According to Vogler et al. [9], the latter region shows sympatry between the North Indian Ocean (*A. planci*) and the Pacific species. There is no type material of *A. echinites* left, and the original description can be applied to both species. Since Schreber [16] expressively distinguished his specimen from the description by Ellis & Solander (in [17]), we prefer the interpretation made by the authors of [2] that *A. echinites* is a synonym of *A. planci* and that *A. solaris* is the valid name for the Pacific species. For formal clarification, the designation of a neotype from off Jakarta is necessary to make the species name *Acanthaster solaris* (Schreber, 1793) [16] unequivocally available for this species (currently in progress by Gerhard Haszprunar and Gert Wörheide).

Acanthaster ellisii (Gray, 1850) and *Acanthaster ellisii pseudoplanci* (Caso, 1962) have both been described as from East Pacific regions (Figure 1). However, all molecular analyses to date have united these populations within the Pacific species (same Barcoding Identification Number (BIN), see below), although the Pacific clade shows a significant substructure [12–14]. The Eastern Pacific COTS populations certainly deserve more in-depth future studies to clarify their taxonomic relationships.

3. Why Taxonomy of COTS Does Matter—Differences between *A. planci* Species

The fact that the existence of these four species has largely been ignored by the COTS research community over the last decade may also be due to a certain reluctance to accept the identification of species based on molecular data rather than on more traditional methods, such as morphological characters or biological traits. However, the most recent publications on COTSs increasingly reflect the reception of the species complex and its implications (e.g., [18–23]).

Yet, although no thorough study was ever conducted to directly compare the biology of any of these four putative species, when digging through the COTS literature of the past 50 years, there are some indications pointing towards differences between them. The comparisons are limited by the fact that the overwhelming majority of COTS research was conducted on the Pacific species (*Acanthaster solaris*), nevertheless the few studies conducted in the Red Sea and the Indian Ocean indicated some differences. Here, we present a brief review of these differences (and in some cases lack of differences) for various aspects.

3.1. Morphology

There are indications of species-specific differences in the number of arms between the various species of the *A. planci* species complex. Reports and photos available through various websites reveal a maximum of 23 arms for the Pacific species, whereas the Red Sea species has a maximum of only

13–14 arms (e.g., [24]). However, since the number of arms increase with age and size, this character appears to be of minor importance. The same appears to be true for the length of arms which seem to be shorter in East Pacific populations (often called *Acanthaster ellisii*), although not statistically significant [25].

Differences between *A. brevispinus* and all COTS species are currently restricted to the shape of pedicellariae and spines [23], but thorough morphological comparisons by Scanning Electron Microscopy (SEM) or Micro-Tomography (μCT) may uncover new differences between these taxa. A focus of future morphological comparisons using these techniques may especially be the East Pacific populations of the Pacific species, which were formally described as *Acanthaster ellisii pleudoplanci* (Caso, 1962) [26]. Finally, there are significant differences in spine length between the population of COTSs from Ambon Islands (Molluken Archipelago, Indonesia) and that from Papua Islands [27], which may reflect a difference between the true *A. planci* (Northern Indian Ocean species) and the Pacific species, *A. solaris*. We cannot rule out that further significant morphological differences between species will be detected in more comprehensive morphometric studies.

The most obvious aspects of morphological variation identified so far are the differences in color detected between the Pacific species and those of the Indian Ocean (*A. planci* and *A. mauritiensis*). COTSs in the Pacific Ocean are variable in color [27]: they are usually gray-green to gray-purple, often with reddish papulae that typically take on a "bulls-eye" appearance due to two rings of darker papulae (Figure 2a; [27]). Specimens of the Red Sea species are of a similar type of color (Figure 2b; [27]). However, specimens of the true *A. planci* tend to have a very different color, sometimes referred to as "electric blue" (Figure 2c,d; [27]; personal observations by all authors). This "electric blue" is also not found in *A. mauritiensis*. The latter species appears to range from a light blue to a rusty color (Figure 2e,f; Catherine Vogler and Gerhard Haszprunar pers. obs.). Whether the "electric blue" color exactly matches the genetic affiliation of the true *A. planci*, especially in the contact zone with other species, has yet to be fully confirmed.

Figure 2. "Typical" color morphs found in the sister-species: (**a**) Pacific species, Fiji (tentatively *Acanthaster solaris* (Schreber, 1793)) (credit: Nina Yasuda), (**b**) yet unnamed Red Sea species (credit: Jessica Bouwmeester), (**c**) and (**d**) Northern Indian Ocean species, i.e., the true *Acanthaster planci* (Linnaeus, 1758), (**c**) UAE (credit: Maral Shuriqi); (**d**) Oman (credit: David Mothershaw), (**e**, **f**) Southern Indian Ocean species, i.e., *Acanthaster mauritiensis* de Loriol, 1885, (**e**) Kenya (credit: Kevin Ransom), and (**f**) Chagos Archipelago (credit: Anne Sheppard).

Summing up, it is not unlikely that distinct morphologies can be applied to distinguish the various species of COTS, but up to now an integrative approach to resolve COTS taxonomy is still missing.

3.2. Toxicity

The toxicity of Red Sea and Pacific COTSs also appears to be different. There is no evidence for a strong venom in Red Sea COTS, based on the reaction of divers from the Cambridge Coral Starfish Research Group to the penetration of spines: "although most members of the expedition were often pricked by their sharp spines, only one animal gave any real discomfort to both of the people who handled it" [28]. Pacific COTS, on the other hand, were found to inflict a range of different symptoms, from severe pain for several hours to persistent nausea and fever for several weeks or even permanent abscesses, apoptosis, hemolysis, and bone-destroying processes (reviewed in [27]). This depended on the number of spines penetrating, whether they broke off in the wounds, possible variability between individual sea-stars and different sensitivities of the victims [27]. Because it is improbable that all members of the Cambridge Coral Starfish Research Group were all particularly insensitive to COTS-inflicted wounds, Red Sea COTSs probably have significantly lower toxicity than the Pacific COTSs (no clear data are available on the other two species from the Indian Ocean). As toxicity is likely to evolve as a defense against predation, this could even be part of the reason why Red Sea COTSs are predominantely nocturnal (Gert Wörheide pers.obs.), as their defense system may be less efficient than that of Pacific COTSs. Since overexploitation of predators has been proposed as one possible cause of COTS outbreaks [27,29], these differences in cryptic behavior and toxicity are far from trivial. Indeed, outbreaks are much more intensive in the Pacific Ocean compared to the Red Sea, and investigations of whether higher predator pressure, higher toxicity, or different behavior in the latter is the main cause of this difference may shed new light on this question.

3.3. Outbreaks

All species of COTS tend to produce outbreaks and mass aggregation, although this appears more frequently in the Pacific species [30]. The causes of outbreaks are still debated (see recent review [1]), and two main hypotheses have gained the highest support to date: (1) the "larval starvation hypothesis", which argues that increased nutrient levels favor the survival of COTS larvae [27,30–32], and (2) the "predator removal hypothesis", which suggests that the overexploitation of predators allows more larvae and juveniles to survive to maturity [27,29]. Outbreaks in the different COTS species could have different causes, and the different species could, for example, react differently to increased nutrient levels or the removal of predators. As such, obtaining a more thorough understanding of the biology of the Red Sea and the two Indian Ocean species and the causes of outbreaks in these species could also help understand more about the Pacific species as well, e.g., by giving new clues and opening new avenues of research. Therefore, comparative studies between the four species are needed to gain more insight into this major management issue of coral reefs.

3.4. Conclusions

To conclude, despite the limited amount of information available on COTSs from the Red Sea and the Indian Ocean, several differences suggest divergent biology and ecology of the four COTS species. Further research is definitely needed to establish whether these rather anecdotal observations can be substantiated to significantly increase our appreciation of the differences in the biology of the individual species in this species complex.

It is likely that there are differences in color patterns between the four species, and in-depth morphological (SEM, μCT) and morphometric analyses may uncover further differences that may characterize each species. In particular, the larval and juvenile stages are in urgent need of more comparative data outside the Pacific species, *Acanthaster solaris*, since these probably play a critical role in the causes of outbreaks [27,29–32]. Finally, to understand differences between the sister-species and the degree to which they have diverged, interbreeding experiments are also important. Sympatric occurrence of different COTS species has been observed in two areas: the Gulf of Oman, where *A. planci* and *A. mauritiensis* co-occur and at the Indian/Pacific break (Indonesian waters), where individuals of *A. planci* and of the Pacific sister-species *Acanthaster solaris* co-occur [2,9–11]. Due to its maternal inheritance, the available data from mitochondrial DNA [9–12] prevent unequivocal inference of reproductive barriers. Data from biparentally inherited allozymes suggest there was little gene exchange between the Northern Indian (*Acanthaster planci*) and Pacific Ocean (*Acanthaster solaris*) species, but that there may have been introgression in the mixed population of Pulau Seribu [14]. However, the data are somewhat inconclusive due to small sample sizes, and the potential incomplete segregation of ancestral genotypes [5]. Further research with high-resolution nuclear molecular markers will thus be necessary to determine whether there are introgression events or even hybridization zones in the contact areas. Lucas et al. [33] crossed specimens from the Pacific species *A. solaris* with *A. brevispinus*, the short-spined sister-species of the COTS species complex. The authors found that first-generation hybrids were viable but that both second-generation hybrids and back-crossed offsprings were of poor viability. Some showed morphological abnormalities, which suggested there were barriers to reproduction between *A. brevispinus* and COTSs. However, crossing experiments are very time- and resource-intensive, so indirect genetic methods are preferable.

4. The Usage of BINs for the Species of the *Acanthaster planci* Complex

Since the seminal paper by Hebert et al. [34], partial mitochondrial COI gene sequences have been used as so called "DNA barcodes" with high success in nearly all metazoan taxa. Problems are mainly restricted to taxa with a net-like genealogy (hybrids, introgressions, etc.) and non-bilaterians, which frequently show highly conserved COI sequences [35,36]. However, available data strongly suggest that DNA barcoding is an excellent tool in discriminating echinoderm species [37–39]. As a general advantage, DNA barcode sequences do not change over an organism's lifetime, and can thus equally be applied to eggs, larva, adults, or the remains of animals.

Ratnasingham and Hebert [40,41] subsequently established the Barcoding of Life Data System (BOLD: [42]) as a global and open-access database for DNA barcodes. Already somewhat foreshadowed by certain authors (e.g., [43]), they [44] also introduced the Barcode Identification Number (BIN) system as a third step. A website with a distinct BIN number is provided for clusters of sequences, where sophisticated statistical analyses (see [45] for a comparative evaluation) have provided evidence for *intra*specific continuity versus *inter*specific discontinuity (the so-called barcoding gap, [46]). Meanwhile, the BIN-framework is well established and has already become an international standard for re-identification of still unnamed or undetermined species (e.g., [47]). The BIN-site of the BOLD handbook [48] lists several points offered by the BIN-framework concerning all reliable data of specimens and species for the user, all these data and options are open access: (1) Details of BIN. (2) Taxonomy. (3) Tags & Comments to be added. (4) Distance distribution. (5) Publications. (6) Tree reconstructions of BIN and nearest neighbor. (7) Haplotype network. (8) Collection location and data managers. For the BIN-numbers of the four currently recognized species of COTS, see Table 1.

Whereas there are as yet no clear diagnostic morphological characters available to clearly distinguish COTS species, species distinction based on DNA barcoding sequences are so far unequivocal and clear. Moreover, an openly accessible database with the Barcode Index Numbers for all currently recognized species of COTS is available, where sequences can be easily checked, compared, and added. Moreover, fast, cheap, and reliable routine protocols for mtDNA COI DNA barcode sequencing have been established to unequivocally determine the species identity of each COTS sample using the BIN framework [9–11].

5. Recommendations

In order to unequivocally allocate any kind of data on COTSs to the respective species, we make a number of recommendations:

(1) Provide and deposit a whole- or partial specimen voucher or tissue sample to your local public museum or collection with the exact sampling locality and inventory data, and request a voucher number. Add voucher number and metadata to any publication that arises. These vouchers, or at least parts thereof, should be suitable for molecular analyses, for example, stored in high percentage (>95%) alcohol, but may be freeze-dried or frozen. The issue of voucher deposition is of particular importance for molecular studies [49,50] and genome sequencing project such that future studies do not mismatch different species and one can go back to the actual specimens for (morphological) verification. Regrettably, no voucher specimens were kept from the two *A. solaris* specimens from which draft genomes were recently sequenced [42], so their morphology unfortunately cannot be examined.

(2) If possible, provide a COI barcode of your specimens, larvae or tissues, and use the BIN-code for unequivocal classification until the nomenclatorial prerequisites for valid Linnean binomens of all species are fulfilled (see [10] and above). Meanwhile, a good number of institutions (and the labs of Gerhard Haszprunar and Gert Wörheide) offer DNA barcoding services for a small fee, e.g., to cover consumables and sequencing.

(3) Extend the knowledge about the biogeographic distribution of the at least four species of COTS. Of particular interest are potential areas of sympatry and areas that have not been covered so far (see [2,9–11]). Such areas include

   a. areas of putative species sympatry/clade overlap (e.g., the southern coast of Oman, Yemen, and Somalia; Gulf of Aden; Western Indonesia);
   b. sites close to the southern distribution limits of both Indian Ocean species;
   c. sites close to the northern and southern distribution limits of *A. solaris*;
   d. sites in the eastern Pacific to resolve the status of the putative species *Acanthaster ellisii* and *Acanthaster ellisii pseudoplanci*.

We here call for, and propose, an open biogeographic study within the framework of the recently established "Diversity of the Indo-Pacific Network" [51] to enhance our knowledge of the diversity of this coral predator. If collectors are unable to produce DNA barcode samples themselves, they are invited to send specimens or tissue samples suitable for molecular work to the lab of Gert Wörheide at LMU Munich.

Acknowledgments: Gert Wörheide acknowledges funding through "Institutional Strategy LMU excellent" by the Ludwig-Maximilians-Universität München within the framework of the German Excellence Initiative. We thank three anonymous reviewers and the editor for their valuable comments and recommendations.

Author Contributions: Gerhard Haszprunar and Gert Wörheide conceived the typescript, Catherine Vogler provided the photos and the basic parts of the text from a chapter of her Ph.D. thesis. Gerhard Haszprunar wrote the initial draft of the typescript, which was subsequently edited by Gert Wörheide and Catherine Vogler. All authors finally approved the typescript.

Conflicts of Interest: The authors declare no conflict of interest.

References

1. Pratchett, M.S.; Caballes, C.F.; Rivera-Posada, J.A.; Sweatman, P.A. Limits to understanding and managing outbreaks of crown-of-thorns starfish (*Acanthaster* spp.). *Oceanogr. Mar. Biol. Ann. Rev.* **2014**, *52*, 133–200.
2. Haszprunar, G.; Spies, M. An integrative approach of the crown-of-thorns starfish species group (Asteroidea: *Acanthaster*): A review of names and comparison to recent molecular data. *Zootaxa* **2014**, *3841*, 271–284. [CrossRef] [PubMed]
3. Fisher, W.K. Starfishes of the Philippine seas and adjacent waters. *Bull. US Natl. Mus.* **1919**, *3*, 100.
4. Nishida, M.; Lucas, J.S. Genetic differences between geographic populations of the crown-of-thorns starfish throughout the Pacific Ocean. *Mar. Biol.* **1990**, *98*, 359–368. [CrossRef]
5. Benzie, J.A.H. Major genetic differences between crown-of-thorns starfish (*Acanthaster planci*) populations in the Indian and Pacific Oceans. *Evolution* **1999**, *53*, 1782–1795. [CrossRef]

6. Benzie, J.A.H. The detection of spatial variation in widespread marine species: methods and bias in the analysis of population structure in the crown of thorns starfish (Echinodermata: Asteroidea). *Hydrobiologia* **2000**, *420*, 1–14. [CrossRef]

7. Gérard, K.; Roby, C.; Chevalier, N.; Thomassin, B.; Chenuil, A.; Féral, J.-P. Assessment of three mitochondrial loci variability for the crown-of-thorns starfish: A first insight into *Acanthaster* phylogeography. *C. R. Biol.* **2008**, *331*, 137–143. [CrossRef] [PubMed]

8. Yasuda, N.; Nagai, S.; Hamaguchi, M.; Okaji, K.; Gérard, K.; Nadaoka, K. Gene flow of *Acanthaster planci* (L.) in relation to ocean currents revealed by microsatellite analysis. *Molec. Ecol.* **2009**, *18*, 1574–1590. [CrossRef] [PubMed]

9. Vogler, C.; Benzie, J.; Lessios, H.; Barber, P.; Wörheide, G. A threat to coral reefs multiplied? Four species of crown-of-thorns starfish. *Biol. Lett.* **2008**, *4*, 696–699. [CrossRef] [PubMed]

10. Vogler, C.; Benzie, J.; Barber, P.H.; Erdmann, M.V.; Ambariyanto, S.C.; Tenggardjaja, K.; Gérard, K.; Wörheide, G. Phylogeography of the crown-of-thorns starfish in the Indian Ocean. *PLoS ONE* **2012**, *7*, e43499. [CrossRef] [PubMed]

11. Vogler, C.; Benzie, J.A.H.; Tenggardjaja, K.; Ambariyanto, S.C.; Barber, P.H.; Wörheide, G. Phylogeography of the crown-of-thorns starfish: Genetic structure within the Pacific species. *Coral Reefs* **2013**, *32*, 515–525. [CrossRef]

12. Yasuda, N.; Hamaguchi, M.; Sasaki, M.; Nagai, S.; Saba, M.; Nadaoka, K. Complete mitochondrial genome sequences for crown-of-thorns starfish *Acanthaster planci* and *Acanthaster brevispinus*. *BMC Genom.* **2006**, *7*, 17. [CrossRef] [PubMed]

13. Timmers, M.A.; Bird, C.E.; Skillings, D.J.; Smouse, P.E.; Toonen, R.J. There's no place like home: Crown-of-thorns outbreaks in the central Pacific are regionally derived and independent events. *PLoS ONE* **2012**, *7*, e31159. [CrossRef] [PubMed]

14. Tusso, S.; Morcinek, K.; Vogler, C.; Schupp, P.J.; Caballes, C.F.; Vargas, S.; Wörheide, G. Genetic structure of the crown-of-thorns sea star in the Pacific. *Peer J.* **2016**, *5*, e1970. [CrossRef] [PubMed]

15. De Loriol, P. Catalogue raisonné des Echinodermes recuellis par M.V. de Robillard à l'Ile Maurice. II. Stellérides. *Mem. Soc. Phys. Hist. Nat. Genéve* **1885**, *29*, 1–84. (In French)

16. Schreber, J.C.D. von Beschreibung der Seesonne, einer Art Seesterne, mit 21 Strahlen. *Der Naturforscher Halle a/d Saale* **1793**, *27*, 1–6. (In German)

17. Ellis, J. *The Natural History of Many Curious and Uncommon Zoophytes, Collected from Various Parts of The Globe*; Benjamin White and Son: London, UK, 1786; pp. 1–12.

18. Nakajima, R.; Nakatomi, N.; Kurihara, H.; Fox, M.D.; Smith, J.E.; Okaji, K. Crown-of-Thorns starfish larvae can feed on organic matter released from corals. *Diversity* **2016**, *8*, 18. [CrossRef]

19. Nakamura, M.; Higa, Y.; Kumagai, N.H.; Okaji, K. Using long-term removal data to manage a Crown-of-Thorns Starfish population. *Diversity* **2016**, *8*, 24. [CrossRef]

20. Sigl, R.; Laforsch, C. The influence of water currents on movement patterns on sand in the Crown-of-Thorns Seastar (*Acanthaster* cf. *solaris*). *Diversity* **2016**, *8*, 25. [CrossRef]

21. Cowan, Z.-L.; Dworjanyn, S.A.; Caballes, C.F.; Pratchett, M. Benthic predators influence microhabitat preferences and settlement success of Crown-of-Thorns Starfish (*Acanthaster* cf. *solaris*). *Diversity* **2016**, *8*, 27. [CrossRef]

22. Buck, A.C.E.; Gardiner, N.M.; Bostrom-Einarsson, L. Citric acid injections: An accessible and efficient method for controlling outbreaks of the Crown-of-Thorns Starfish *Acanthaster* cf. *solaris*. *Diversity* **2016**, *8*, 28. [CrossRef]

23. Hall, M.R.; Kocot, K.M.; Baughman, K.W.; Fernandez-Valverde, S.L.; Gauthier, M.E.A.; Hatleberg, W.L.; Krishnan, A.; McDougall, C.; Motti, C.A.; Shoguchi, E.; et al. The crown-of-thorns starfish genome as a guide for biocontrol of this coral reef pest. *Nature* **2017**, *544*, 231–234. [CrossRef] [PubMed]

24. Bruckner, A.W.; Dempsey, A.C. The status, threats, and resilience of reef-building corals of the Saudi Arabian Red Sea. In *The Red Sea. The Formation, Morphology, Oceanography and Environment of a Young Ocean Basin*; Rasul, N.M.A., Steward, C.F., Eds.; Springer: Berlin/Heidelberg, Germany, 2015; pp. 470–486.

25. Madsen, F.J. A note on the seastar genus *Acanthaster*. *Vidensk. Medd. Dansk Naturh. Foren.* **1955**, *117*, 179–192.

26. Caso, M.E. Estudios sobre Astéridos de México. Observaciones sobre especies pacíficas del género *Acanthaster* y descripción de una subespecie nueva, *Acanthaster ellisii pseudoplanci*. *Anales Inst. Biol. Univ. Nacl. Autón. México* **1962**, *32*, 313–331. (In Spanish)

27. Birkeland, C.; Lucas, J.S. *Acanthaster Planci: Major Management Problem of Coral Reefs*; CRC Press: Boca Raton, FL, USA, 1990; p. 267.
28. Campbell, A.; Ormond, R.F.G. The threat of the "crown-of-thorns" starfish (*Acanthaster planci*) to coral reefs in the Indo-Pacific area: Observations on a normal population in the Red Sea. *Biol. Conserv.* **1970**, *2*, 246–251. [CrossRef]
29. Sweatman, H. No-take reserves protect coral reefs from predatory starfish. *Curr. Biol.* **2008**, *18*, R598–R599. [CrossRef] [PubMed]
30. Houk, P.; Raubani, J. *Acanthaster planci* outbreaks in Vanuatu coincide with ocean productivity, furthering trends throughout the Pacific Ocean. *J. Oceanogr.* **2011**, *66*, 435–438. [CrossRef]
31. Houk, P.; Bograd, S.; van Woesik, R. The transition zone chlorophyll front can trigger *Acanthaster planci* outbreaks in the Pacific Ocean: Historical confirmation. *J. Oceanogr.* **2007**, *63*, 149–154. [CrossRef]
32. Fabricius, K.E.; Okaji, K.; De'ath, G. Three lines of evidence to link outbreaks of the crown-of-thorns seastar *Acanthaster planci* to the release of larval food limitation. *Coral Reefs* **2010**, *29*, 593–605. [CrossRef]
33. Lucas, J.S.; Jones, M.M. Hybrid crown-of-thorns starfish (*Acanthaster planci* X *A. brevispinus*) reared to maturity in the laboratory. *Nature* **1976**, *263*, 409–412. [CrossRef] [PubMed]
34. Hebert, P.D.N.; Cywinska, A.; Ball, S.L.; de Waard, J.R. Biological identifications through DNA barcodes. *Proc. R. Soc. Lond.* **2003**, *B270*, 313–321. [CrossRef] [PubMed]
35. Shearer, T.L.; van Oppen, M.J.H.; Romano, S.L.; Wörheide, G. Slow mitochondrial DNA sequence evolution in the Anthozoa (Cnidaria). *Molec. Ecol.* **2002**, *11*, 2475–2487. [CrossRef]
36. Wörheide, G.; Erpenbeck, D. DNA taxonomy of sponges—Progress and perspectives. *J. Mar. Biol. Ass. UK* **2007**, *87*, 1629–1633. [CrossRef]
37. Ward, R.; Holmes, B.H.; O'Hara, T.D. DNA barcoding discriminates echinoderm species. *Molec. Ecol. Res.* **2008**, *8*, 1202–1211. [CrossRef] [PubMed]
38. Janosik, A.M.; Mahon, A.R.; Halanych, K.M. Evolutionary history of Southern Ocean *Odontaster* sea star species (Odontasteridae; Asteroidea). *Polar Biol.* **2011**, *34*, 575–586. [CrossRef]
39. Bucklin, A.; Steinke, D.; Blanco-Bercial, L. DNA barcoding of marine Metazoa. *Ann. Rev. Mar. Sci.* **2011**, *3*, 471–508. [CrossRef] [PubMed]
40. Ratnasingham, S.; Hebert, P.D.N. Bold: The barcode of life data system (www.barcodinglife.org). *Molec. Ecol. Notes* **2007**, *7*, 355–364. [CrossRef] [PubMed]
41. Ratsasingham, S.; Hebert, P.D.N. BOLD's role in barcode data management and analysis: A response. *Molec. Ecol. Res.* **2011**, *11*, 941–942. [CrossRef]
42. BOLD Systems. Available online: http://www.boldsystems.org/ (accessed on 10 May 2017).
43. Jones, M.; Ghoorah, A.; Blaxter, M. jMOTU and Taxonerator: Turning DNA barcode sequences into annotated operational taxonomic units. *PLoS ONE* **2011**, *6*, e19259. [CrossRef] [PubMed]
44. Ratsasingham, S.; Hebert, P.D.N. A DNA-based registry for all animal species: The Barcode Index Number (BIN) System. *PLoS ONE* **2013**, *8*, e66213.
45. Jörger, K.M.; Norenburg, J.L.; Wilson, N.G.; Schrödl, M. Barcoding against a paradox? Combined molecular species delineations reveal multiple cryptic lineages in elusive meiofaunal sea slugs. *BMC Evol. Biol.* **2012**, *12*, 245. [CrossRef] [PubMed]
46. Meyer, C.P.; Paulay, G. DNA barcoding: Error rates based on comprehensive sampling. *Publ. Lib. Sci. Biol.* **2005**, *3*, 2229–2238. [CrossRef] [PubMed]
47. Hausmann, A.; Godfray, H.C.J.; Huemer, P.; Mutanen, M.; Rougerie, R.; van Nieukerken, E.J.; Ratnasingham, S.; Hebert, P.D.N. Genetic patterns in European geometrid moths revealed by the Barcode Index Number (BIN) System. *PLoS ONE* **2013**, *8*, e84518. [CrossRef]
48. BOLD-Handbook—Barcode Index Numbers. Available online: www.boldsystems.org/index.php/resources/handbook?chapter=2_databases.html&section=bins (accessed on 10 May 2017).
49. Pleijel, F.; Jondelius, U.; Norlinder, E.; Nygren, A.; Oxelman, B.; Schander, C.; Sundberg, P.; Thollesson, M. Phylogenies without roots? A plea for the use of vouchers in molecular phylogenetic studies. *Mol. Phylogenet. Evol.* **2008**, *48*, 369–371. [CrossRef] [PubMed]

50. Astrin, J.J.; Zhou, X.; Misof, B. The importance of biobanking in molecular taxonomy, with proposed definitions for vouchers in a molecular context. *ZooKeys* **2013**, *365*, 67–70. [CrossRef] [PubMed]
51. Diversity of the Indopacific Network. Available online: http://diversityindopacific.net (accessed on 10 May 2017).

diversity

MDPI

Article

Environmental Tipping Points for Sperm Motility, Fertilization, and Embryonic Development in the Crown-of-Thorns Starfish

Ciemon Frank Caballes [1,*], Morgan S. Pratchett [1], Maia L. Raymundo [2] and Jairo A. Rivera-Posada [3]

[1]   ARC Centre of Excellence for Coral Reef Studies, James Cook University, Townsville, QLD 4811, Australia; morgan.pratchett@jcu.edu.au

[2]   School of Biological Sciences, University of Queensland, Brisbane, QLD 4072, Australia; maia.raymundo@uqconnect.edu.au

[3]   Academic Environmental Corporation, Universidad de Antioquia, Medellín, Antioquia 050010, Colombia; jriveraposada@yahoo.com.au

*   Correspondence: ciemon.caballes@my.jcu.edu.au; Tel.: +61-4-4781-5747

Academic Editors: Sven Uthicke and Michael Wink
Received: 30 November 2016; Accepted: 10 February 2017; Published: 15 February 2017

Abstract: For broadcast spawning invertebrates such as the crown-of-thorns starfish, early life history stages (from spawning to settlement) may be exposed to a wide range of environmental conditions, and could have a major bearing on reproductive success and population replenishment. Arrested development in response to multiple environmental stressors at the earliest stages can be used to define lower and upper limits for normal development. Here, we compared sperm swimming speeds and proportion of motile sperm and rates of fertilization and early development under a range of environmental variables (temperature: 20–36 °C, salinity: 20–34 psu, and pH: 7.4–8.2) to identify environmental tipping points and thresholds for reproductive success. We also tested the effects of water-soluble compounds, derived from eggs, on sperm activity. Our results demonstrate that gametes, fertilization, and embryonic development are robust to a wide range of temperature, salinity, and pH levels that are outside the range found at the geographical limits of adult distribution and can tolerate environmental conditions that exceed expected anomalies as a result of climate change. Water-soluble compounds derived from eggs also enhanced sperm activity, particularly in environmental conditions where sperm motility was initially limited. These findings suggest that fertilization and embryonic development of crown-of-thorns starfish are tolerant to a wide range of environmental conditions, though environmental constraints on recruitment success may occur at later ontogenic stages.

Keywords: cleavage; gastrulation; sperm activity; temperature; salinity; pH; *Acanthaster* outbreaks

1. Introduction

Outbreaks of the coral-eating crown-of-thorns starfish (CoTS), *Acanthaster* spp., are one of the most significant biological threats to coral reefs and account for a substantial proportion of coral mortality in the Indo-Pacific region [1–3]. CoTS are predisposed to major population fluctuations, whereby local densities may vary by several orders of magnitude [4], due to inherent features of their reproductive biology and behavior [5,6]. Reproductive success is central to explaining periodic increases in local densities [7]. Understanding the critical events in the early life history of CoTS is key to identifying population bottlenecks that could be strategically targeted to improve control programs and mitigate coral mortality [8]. Despite this, environmental drivers of variation in reproductive success for *Acanthaster* spp. remain poorly understood.

Achieving high fertilization rates is vital in ensuring reproductive success [9]. Fertilization had initially been thought to be non-limiting, given that broadcast spawners, such as CoTS, release copious amounts of gametes during spawning [6,10]. Population replenishment in CoTS is believed to be largely regulated by larval provisioning, larval delivery, post-settlement competition and predation [11–14]. However, a host of factors, at the gamete, individual, and population levels, as well as prevailing environmental conditions, can influence fertilization success [9]. For example, changes in sperm swimming speeds and the proportion of motile sperm affect fertilization success in CoTS [15] and other echinoderms [16,17]. Previous studies have shown that the number and distribution of individuals and the prevailing flow conditions during spawning dictate the local concentration of gametes [18–20]. Fertilization rates of CoTS have been reported to reach up to 83% at the peak of a major spawning event [21]. In induced spawning experiments in the field, fertilization rates can be as high as 95% when male and female starfish are in very close proximity [20]. As expected, fertilization rates drop significantly as the distance between spawning individuals increases. Nevertheless, 70% fertilization success was still achieved at distances of up to 8 m between spawning individuals and more than 20% at a distance of 60 m [20]. Fertilization success per unit distance in CoTS is higher compared to other asteroid species and significantly greater than those reported for other marine invertebrates [7]. Despite achieving high fertilization rates at given sperm concentrations, at greater distances, and at longer durations from the point of gamete release [22], very little is known on the tolerance of gametes, fertilization, and early development of CoTS to a wide range of environmental conditions.

For broadcast spawning invertebrates such as CoTS, early life history stages occur in the water column where environmental factors could disrupt the initial phases in the process of population replenishment. The persistence and success of populations require that all developmental stages be completed successfully and the variable sensitivity of planktonic stages (i.e., gametes, fertilization, and early development) to environmental stressors (e.g., temperature, salinity, and pH) may be a potential population bottleneck [23,24]. Evaluating the effects of environmental stress on gametes and early life history stages is important as this can result in detrimental flow-on effects where physiological performance and cellular responses of subsequent ontogeny depend on the success of preceding stages [23]. In addition, marine organisms are exposed not only to natural environmental stressors, but also the compounding effects of anthropogenic stressors, notably increasing global temperatures, pulses of decreased salinity brought about by higher frequency of cyclones and freshwater runoff, and reduced pH [24]. Climate change causes changes in baseline environmental conditions, such that inherent fluctuations of temperature, salinity, and pH, particularly in nearshore waters, may increasingly exceed tolerance thresholds, especially for populations currently living at physiological limits [25].

Recent studies on the response of early life history stages of marine invertebrates to ocean warming and acidification, have improved our knowledge on environmental thresholds of several species [24]. Generally, temperature affects everything an organism does through its pervasive physiological impact on all biological functions [26]. Ocean acidification has negative impacts on development due to direct pH effects and hypercapnic suppression of metabolism, and is a major threat to marine calcifiers because acidification decreases carbonate saturation with a negative impact on skeleton formation [23,27]. Pulses of reduced salinity brought by heavy rainfall or freshwater lenses of river plumes have been reported to result in decreased growth and reproduction rates in some invertebrates [28] and affect the cellular osmoregulation in gametes and embryos [29]. The responses of echinoderms to these environmental stressors are stage- and species-specific, but gametes and fertilization appear to be robust to a wide range of temperature, salinity, and pH levels [30–32]. Environmental tolerances of echinoderm embryos are generally narrower than for gametes and fertilization [16,31,33].

Spermatozoa of free-spawning marine organisms remain immobile at the time of gamete release but become motile spontaneously upon dilution in seawater. Evaluating the response of spermatozoa to environmental factors is important since activation is influenced by seawater temperature, osmotic pressure, extracellular pH, ultraviolet radiation, and the concentration of specific ions relative to that in the seminal plasma in echinoderms [34–37]. Sperm swimming speeds in the polychaete *Galeolaria caespitosa* have been reported to be enhanced under increased water temperatures [38], but comparable research is yet to be undertaken for most echinoderms [23,27]. Decreased motility and inactivation of sperm at low salinities has also been reported in sea urchins [29,39]. Previous studies on echinoids also show reductions in the percentage of motile sperm at decreased pH and ultimately reproductive success [18–20]. There is also evidence that oocytes from conspecifics release attractants that induce chemotaxis toward the egg [40–42]. In some marine invertebrates, chemoattractants may not only change the direction of sperm swimming but also increase sperm swimming speeds and the proportion of motile sperm [43–45]. The interactive or additive effects of environmental stressors and egg-derived chemoattractants warrant further attention, especially given potential impacts of climate change on fertilization success.

The purpose of this study is to compare sperm behavior and rates of fertilization, as well as early development under a range of environmental variables to identify environmental tipping points and thresholds for reproductive success. As fertilization immediately follows spawning, existing environmental conditions during gamete release could potentially limit fertilization rates and early development even when sperm-to-egg ratios are optimal, which is expected when spawning individuals are aggregated, gamete release is synchronized, and flow conditions are low to moderate [7]. Here, we examine temperature, salinity, and pH thresholds of sperm motility, fertilization, cleavage, and gastrulation. Reproductive failure in echinoderms has been reported at different levels of these environmental parameters, but few have looked whether this is due to the sensitivity of gametes, failure of fertilization, or failure of fertilized eggs to cleave or hatch [46,47]. We also tested the excitatory effect of water-soluble egg extracts on sperm behavior to add a maternal dimension to the characterization of sperm motility. Sperm swimming speeds and proportion of motile sperm are discussed in relation to fertilization rates. Previous studies on the impacts of these environmental variables on marine invertebrates have mostly set experimental conditions with respect to projections by the Intergovernmental Panel on Climate Change [48] for temperature rise (2 °C to 4 °C above ambient), pulses of decreased salinity (regionally variable), and ocean acidification (0.2 to 0.4 pH units below ambient) [49]. Here, we included extreme environmental stressor treatments to determine how far gametes, fertilization, and early development can be pushed to identify tipping points and thresholds for deleterious effects. Developmental arrest in response to multiple environmental stressors at the earliest stages can be used to define lower and upper limits for normal development. Quantifying environmental regulation of initial elements of reproductive success is important in understanding the spatial and temporal dynamics of populations of *Acanthaster* spp., as well as understanding vulnerability to environmental changes.

2. Materials and Methods

2.1. Collection and Maintenance of Animals for Experiments

Adult individuals of the Pacific species of crown-of-thorns starfish (*Acanthaster* cf. *solaris*) were collected from aggregations in reefs around Puntan Dos Amantes (13° 32.346′ N, 144° 48.200′ E) on the northwest coast of the island of Guam, Micronesia in October 2013. Starfish were immediately transported to the University of Guam Marine Laboratory and allowed to acclimatize to ambient conditions for 48 h (28.79 ± 0.23 °C; 34.19 ± 0.04 psu; pH 8.23 ± 0.02) in 1000-L concrete tanks with flow-through seawater. Individuals were sexed by drawing contents from gonads along the arm junction using a syringe with a large-bore biopsy needle [7]. Male and female CoTS were placed in separate tanks prior to experiments. Gametes from gravid individuals were examined under a compound microscope to generally assess reproductive maturity of oocytes and sperm motility.

2.2. Preparation of Experimental Seawater

2.2.1. Water-Soluble Egg Extracts

Water-soluble egg extracts and seawater solutions (ESW) were prepared by incubation of unfertilized eggs from five females (standardized to 100 egg·mL^{-1}) for 60–90 min [43,45] under different levels of temperature, salinity, or pH as described below. Eggs were filtered through a 0.22-μm syringe filter (Millipore, Darmstadt, Germany) and immediately used in experiments. Filtered seawater (0.2-μm) was used as controls and incubated under different levels of environmental treatments. Experimental seawaters were kept in sealed Nalgene® glass containers prior to experiments.

2.2.2. Temperature

Preliminary pilot studies have shown that temperature below 20 °C resulted in zero fertilization and cleavage. Temperatures ranging from 20 °C to 36 °C, at 2 °C intervals, were tested in this study. This experiment was done inside a temperature-controlled room set at 16 °C. Parafilm®-sealed beakers with 0.2-μm filtered seawater were placed in water baths with aquarium heaters (Eheim Jäger, Deizisau, Germany) connected to digital controllers (Aqua Logic Inc., San Diego, CA, USA) to maintain set temperatures. Pre-calibrated digital thermometers were placed in each water bath to monitor and stabilize set temperatures.

2.2.3. Salinity

Initial rangefinder experiments showed zero fertilization at 18 psu. Eight salinity levels were tested in this study: 20, 22, 24, 26, 28, 30, 32, and 34 psu. Salinity treatments below ambient conditions (<34 psu) were prepared by adding distilled freshwater to 0.2-μm filtered seawater until set levels were reached. This experiment was done in an incubator (VWR International, Radnor, PA, USA) set at 28 °C. Beakers were fitted with plastic lids that had a 12-rpm synchronous motor attached to a plastic stirrer to maintain set conditions and prevent the formation of artificial haloclines within beakers. Salinity of seawater samples from experimental beakers was also measured before and after experiments using HI 96822 Seawater Refractometer (Hanna Instruments, Woonsocket, RI, USA) with automatic temperature compensation.

2.2.4. pH

This experiment was conducted to test the tolerance of fertilization and embryonic development in CoTS to different pH$_{NIST}$ levels: 7.4, 7.6, 7.8, 8.0, and 8.2. Experimental seawater pH levels (below ambient pH 8.2) was achieved by gently bubbling CO_2 into reservoir overhead tanks, using a pH computer (Aqua-Medic of North America, CO, USA) connected to a solenoid valve, until programmed levels were reached. Experimental 0.2-μm filtered seawater was gravity-fed to containers with 45-mm mesh windows enclosed by a plastic jacket placed in water baths set at 28 °C. Seawater pH in experimental containers were measured before and after experiments using Orion 3-Star benchtop pH meter (Thermo Scientific, MA, USA), which was triple calibrated with NIST-certified buffers (pH 4.01, 7.00, 10.01).

2.3. Sperm Speed and Motility

Sperm speed (sperm point-to-point velocity = total distance travelled per second) and sperm motility (percentage of motile sperm) were measured from five male starfish, using techniques described for crown-of-thorns starfish [15] and sea urchins [17]. Experimental seawater treatments were prepared as described in the previous section. For each dilution, 2 μL of dry sperm were diluted with 4 mL of experimental seawater. One drop (~100 μL) of this sperm suspension was placed on an albumin-coated microscope slide and a coverslip, which were separated by a 0.75 mm thick O-ring and focus set midplane to minimize wall effects on sperm swimming speed [16]. Sperm behavior was captured using a Canon EOS 60D single lens reflex camera coupled with a Zeiss Axio Scope A1 (A-Plan ph1 10×/0.25 objective). The video camera was remotely controlled using Canon EOS Utility and set to take 25 frames per second over a two second period. All recordings were made within 10 s of the sperm suspension being placed on the slide. For each male, three replicate observations (slides) were made for three independent sperm dilutions under each temperature, salinity, or pH level and water-soluble egg extract treatment combination. Video recordings were post-processed with Sony Vegas Movie Studio HD (Sony Creative Software Inc., Middleton, WI, USA), and 1-s video clips from each slide (replicate) were analyzed using computer-assisted sperm analysis (CASA) plugin in Image J [50]. From an average of 200 sperm tracks analyzed per slide, mean sperm speed and percentage of motile sperm was determined for each replicate (slide) and standard deviation (SD) was calculated.

2.4. Bioassays for Fertilization and Embryonic Development

Three sets of experiments were conducted to quantify fertilization, cleavage, and gastrulation rates in response to different levels of (1) temperature, (2) salinity, and (3) pH. Ripe ovary lobes were dissected from two female starfish and gently placed in glass dishes with 0.2-μm filtered seawater (FSW) at 28 °C, to which, 1-methyladenine (1-MA) as added at a final concentration of 1×10^{-4} M. Eggs were spawned after 60 min and pooled by transferring to a large glass beaker with FSW. For each experiment, eggs were split into triplicate containers (with 150-mL experimental seawater) for each treatment level. Approximately 300 eggs were rinsed with experimental seawater and transferred to beakers so that final density was ~2 eggs mL^{-1}. Testes lobes were dissected from three male starfish and sperm that were shed after ~3 min were pooled together and placed in experimental seawater for ~10 s at a concentration of 1×10^4 sperm mL^{-1} to ensure appropriate treatment conditions when added to containers with eggs. There was no water movement in the beaker at this point to minimize immotile sperm from artificially coming in contact with eggs. After 30 min, eggs were rinsed three times in experimental FSW to remove excess sperm and resuspended in experimental FSW. Gametes were pooled to reflect a population of spawners, as might occur in nature, and to record the mean response of the system under investigation. Gamete concentrations used in this study resulted in high fertilization rates (>95%) during procedural control experiments and none of the eggs showed fertilization envelopes without the addition of sperm, demonstrating that there was no contamination during the preparation and handling of gametes. After two hours, ~100 eggs from each replicate were placed in a scintillation vial and 7% formalin was added to prevent further development. Fertilization (presence of fertilization envelope, Figure 1a) and/or holoblastic radial cleavage (cell division, Figure 1b) were assessed in the first 50 eggs seen across a gridded slide viewed under a compound microscope at low power. Beakers containing the remaining embryos were then resealed and maintained in experimental temperature, salinity, or pH conditions. After 24 h, 50 embryos were scored as either "gastrula" if they had developed archenteron, or "non-gastrula", where invagination had not occurred (Figure 1c). Five independent runs using different sets of gamete sources were undertaken with full replication for each treatment. Mean values from three containers within runs were used as replicates in each experiment ($n = 5$) and SD calculated. Temperature, salinity, and temperature-compensated pH measurements of seawater in experimental beakers were monitored using a HI 9828 multiparameter handheld probe (Hanna Instruments, RI, USA), with only minimal fluctuation from set values (<0.1).

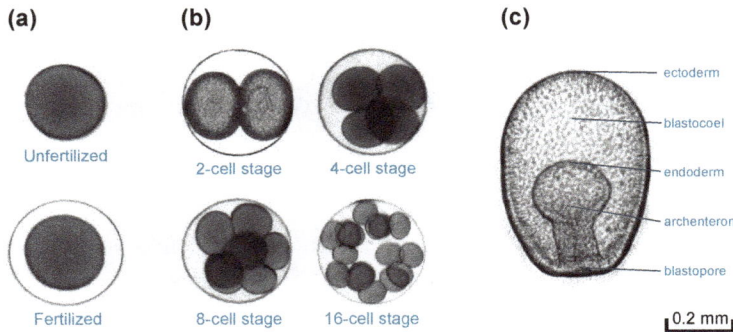

Figure 1. Early life history processes or stages assessed in this study: (**a**) fertilization; (**b**) early cleavage; and (**c**) gastrulation.

2.5. Statistical Analyses

Statistical comparisons of sperm speed between combinations of environmental treatment (temperature, salinity, or pH) and water-soluble egg extracts was performed using a two-factor analysis of variance (ANOVA) followed by post hoc pairwise comparisons using the "lsmeans" function in R with Tukey's adjustment [51]. No significant departures from normality and homogeneity of variance were detected for all data. A generalized linear model (GLM) with binomial errors and logit link function was used to analyse the effect of each environmental treatment and water-soluble egg extracts (fixed categorical predictors) on the proportion of motile sperm. Significant overall tests were followed by post hoc pairwise comparisons between different levels of temperature, salinity, or pH with corrected p-values [52] using the "glht" function from the "multcomp" package in R [53]. Mean sperm speed and motility for each male ($n = 5$) across 3 replicate dilutions (slides) were used in these analyses.

A generalized linear model (GLM) with binomial errors and logit link function was used to analyse the effect of temperature, salinity, or pH (categorical predictors) on fertilization, cleavage, or gastrulation rates (binomial response variables). Quasibinomial error distributions were used in place of binomial errors to correct for overdispersion when detected [54]. This was followed by post hoc multiple comparisons with corrected p-values [52] using the "glht" function from the "multcomp" package in R [53]. Data from within treatments that had zero variance were excluded in the analyses.

3. Results

3.1. Temperature

Seawater temperature ($F_{8, 72} = 85.96$, $p < 0.0001$) and exposure to water-soluble egg extracts ($F_{1, 72} = 13.16$, $p = 0.0005$) had a significant effect on sperm swimming speeds in CoTS (Table A1). Sperm velocity was lowest at the minimum temperature tested, 20 °C (FSW: 100.75 μm·s^{-1} $\pm$ 9.48 SD, here and in all instances hereafter; ESW: 127.05 $\pm$ 14.40 μm·s^{-1}), and peaked at a temperature range of 28 °C to 34 °C (FSW: >221 μm·s^{-1}; ESW: >225 μm·s^{-1}) before slightly dropping back to 219 $\pm$ 96 μm·s^{-1}(FSW) and 228.01 $\pm$ 25.59 μm·s^{-1} (ESW) at 36 °C (Figure 2a). Sperm exposed to water-soluble egg extracts had consistently faster swimming speeds compared to controls, but this difference was most prominent between 20 °C and 26 °C where sperms swimming speeds were relatively slow in controls (Figure 2a). We found a significant variation in sperm motility between temperature treatment levels ($\chi^2 = 1233.07$, $df = 8$, $p < 0.0001$) and between control and water-soluble egg extract treatments ($\chi^2 = 31.34$, $df = 1$, $p = 0.0008$). The proportion of motile sperm was steadily increasing from a minimum of 8.80% $\pm$ 3.02% (FSW) and 21.47% $\pm$ 7.49% (ESW) at 20 °C then peaking at >65% (FSW) and >70% (ESW) for temperatures between 28 °C and 34 °C (Figure 2b).

Water temperature, ranging from 20 to 36 °C, had a significant effect on fertilization, cleavage and gastrulation for *A* cf. *solaris*, whereby reproductive performance would be maximized at intermediate temperatures (26–30 °C). For fertilization, there was significant variation across the full range of temperatures tested (Table A1; $\chi^2 = 1316.20$, $df = 8$, $p < 0.0001$), mainly due to low fertilization under low and high temperature extremes. Fertilization rates were >89% between 24 °C to 32 °C (Figure 2c). For cleavage, there was significant variation with temperature (Table A1; $\chi^2 = 521.09$, $df = 7$, $p < 0.0001$). Cleavage was >75% for 26 °C to 32 °C (Figure 2d), but greatly reduced at lower and higher temperatures. Temperature also had a significant effect on gastrulation rates (Table A1; $\chi^2 = 822.66$, $df = 7$, $p < 0.0001$). The proportion of embryos undergoing gastrulation was maximized between 26 °C and 32 °C (Figure 2e).

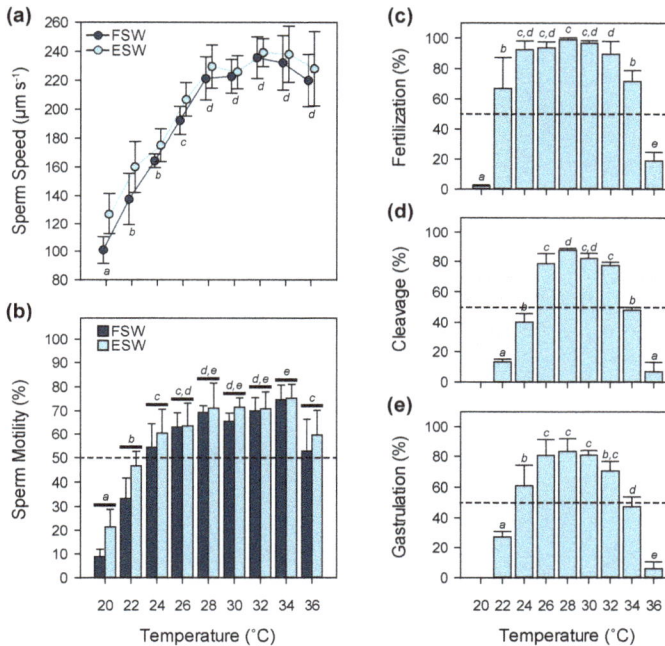

Figure 2. Thermal tolerance of sperm, fertilization, and embryonic development: (**a**) sperm speed (points slightly displaced for clarity); (**b**) sperm motility; (**c**) fertilization; (**d**) cleavage; and (**e**) gastrulation (*n* = 5). Letters next to error bar caps (± SD) indicate significant differences based on post hoc pairwise comparisons with corrected *p*-values. FSW = 0.2-μm filtered seawater (control); ESW = solution with water-soluble egg extract.

3.2. Salinity

Salinity ($F_{7, 64} = 5.83$, $p < 0.0001$) had a significant effect on sperm swimming speeds in CoTS (Table A1). The disparity in sperm velocity between treatments exposed to water-soluble egg extracts and controls was progressively wider from high to low salinity, but differences were not statistically significant ($F_{1, 64} = 2.93$, $p = 0.0918$) (Figure 3a). Variation between salinity treatments was mainly driven by differences between three groups: low sperm swimming speeds for treatments ranging from 20 to 22 psu, intermediate velocity at 24 and 26 psu, and significantly higher sperm velocity from 28 to 34 psu (Figure 3a). Sperm swimming speeds were relatively high across all treatments, with mean sperm velocity all above 170 μm·s^{-1}. Salinity ($\chi^2 = 523.43$, $df = 7$, $p < 0.0001$) also had a significant effect on sperm motility, but not water-soluble egg extracts ($\chi^2 = 16.42$, $df = 1$, $p = 0.0682$) (Table A1). The proportion of motile sperm was above 40% for salinities ranging from 24 to 34 psu.

Salinity had a significant effect on overall fertilization rates ($\chi^2 = 597.86$, $df = 7$, $p < 0.0001$). Fertilization envelopes did not form at salinities <20 psu in preliminary experiments, while 31.43% ± 13.89% and 36.67% ± 12.41% of eggs were fertilized in 20 psu and 22 psu treatments, respectively. Highest fertilization rates were achieved at 30 psu (89.33% ± 8.29%), 32 psu (97.60% ± 2.34%), and 34 psu (96.40% ± 3.35%). The proportion of embryos undergoing cleavage was significantly different between salinity treatments (Table A1; $\chi^2 = 369.59$, $df = 5$, $p < 0.0001$). Fertilized eggs did not cleave at 20 and 22 psu, while only 15.03% ± 8.76% cleaved under the 24-psu treatment. Percentage of normal cleavage in CoTS was optimal (>85%) when exposed to salinities ranging from 30–34 psu. Cleavage rates at 26 psu (57.04% ± 14.64%) and 28 psu (65.80% ± 11.50%) treatments were significantly lower than those under 30–34 psu (Figure 3d). There was also a significant variation in the proportion of embryos undergoing gastrulation after 24 h between salinity treatments (Table A1; $\chi^2 = 504.40$, $df = 5$, $p < 0.0001$). As with cleavage rates, no gastrulation occurred at 20 and 22 psu, and the proportion of embryos at gastrula stage was significantly higher at salinities between 30 and 34 psu compared to 26 psu (56.80% ± 8.81%) and 28 psu (65.20% ± 10.07%) treatments, which were also significantly higher than 24 psu treatment (8.80% ± 5.55%) (Figure 3e).

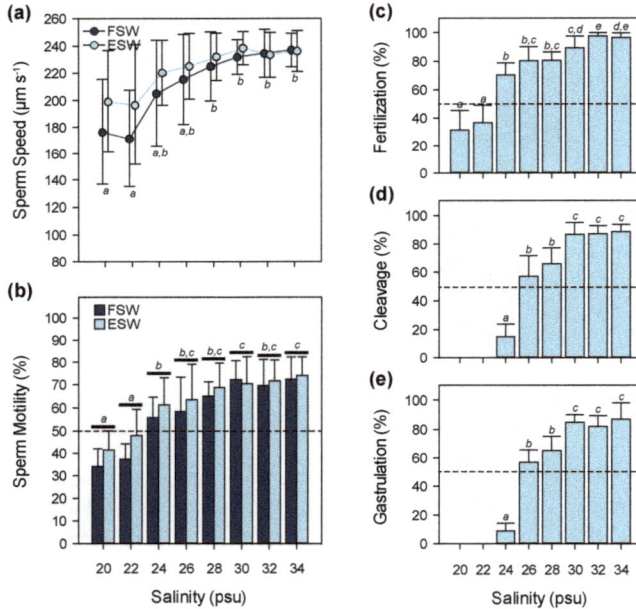

Figure 3. Effect of salinity on sperm behavior, fertilization, and early development: (**a**) sperm speed (points slightly displaced for clarity); (**b**) proportion of motile sperm, and proportion of eggs undergoing (**c**) fertilization; (**d**) cleavage; and (**e**) gastrulation ($n = 5$). Letters above error bars (± SD) indicate significant differences based on post hoc pairwise comparisons with corrected *p*-values. FSW = 0.2-μm filtered seawater (control); ESW = solution with water-soluble egg extract.

3.3. pH

Mean sperm swimming speeds differed significantly (Table A1) between pH treatments ($F_{4, 40} = 28.57$, $p < 0.0001$), but not between egg-derived extracts and controls ($F_{1, 40} = 3.85$, $p = 0.0568$). For this experiment, sperm velocity was highest at pH 8.2 (FSW: 228.89 ± 17.89 μm·s^{-1};

ESW: $235.40 \pm 15.44\ \mu m \cdot s^{-1}$) and pH 8.0 treatments (FSW: $224.23 \pm 24.05\ \mu m \cdot s^{-1}$; ESW: $222.23 \pm 27.65\ \mu m \cdot s^{-1}$). Apart from pH 7.4 treatments, where sperm velocity was lowest (FSW: $118.69 \pm 31.73\ \mu m \cdot s^{-1}$; ESW: $147.52 \pm 30.22\ \mu m \cdot s^{-1}$), sperm swimming speeds were relatively high ($>180\ \mu m \cdot s^{-1}$) for pH levels ranging from 7.6 to 8.2 (Figure 4a). We also found significant variations in the proportion of motile sperm under different pH ($\chi^2 = 669.24$, $df = 4$, $p < 0.0001$) and egg extract ($\chi^2 = 38.11$, $df = 1$, $p = 0.0033$) treatments (Table A1). The proportion of motile sperm was consistently higher for treatments exposed to water-soluble egg extracts (Figure 4b). For sperm under pH levels ranging from 7.6 to 8.2, motility was over 50%, while the proportion of motile sperm was relatively low at pH 7.4 (FSW: $14.93\% \pm 6.74\%$; ESW: $29.87\% \pm 13.50\%$).

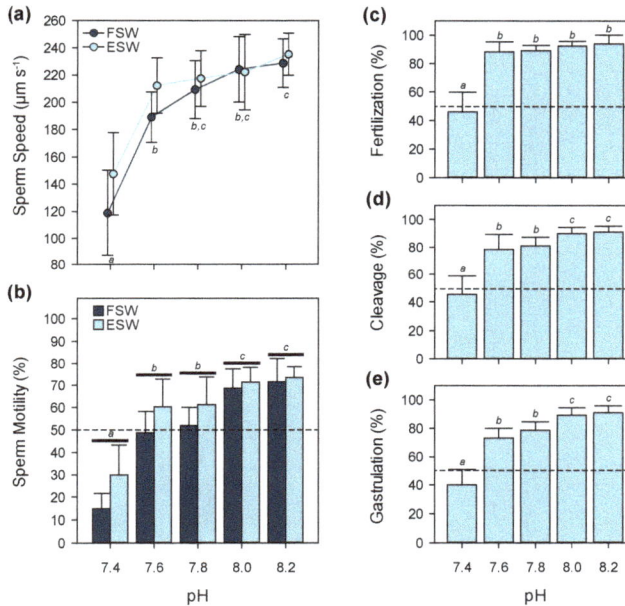

Figure 4. Influence of pH on sperm behavior, fertilization, and early development: (**a**) sperm speed (points slightly displaced for clarity); (**b**) proportion of motile sperm, and proportion of eggs undergoing (**c**) fertilization; (**d**) cleavage, and (**e**) gastrulation. Letters above error bars ($\pm$SD) indicate significant differences based on post hoc pairwise comparisons with corrected p-values. FSW = 0.2-μm filtered seawater (control); ESW = solution with water-soluble egg extract.

Percentage of fertilization was high across all pH levels tested (Figure 4c), except for eggs in pH 7.4 ($46.09\% \pm 13.73\%$), which was significantly lower than fertilization success at pH 7.6 to pH 8.2 ($>88\%$). The effect of low pH levels was more evident when looking at the frequency of normal cleavage (Figure 4d) and gastrulation (Figure 4e). Cleavage ($45.48\% \pm 13.17\%$) and gastrulation rate ($40.13\% \pm 10.75\%$) at pH 7.4 was lowest among all the pH levels tested. The range of pH levels for optimum normal cleavage and gastrulation ($>89\%$) was between pH 8.0 and pH 8.2. Proportion of embryos undergoing cleavage and gastrulation was significantly higher at optimum pH levels (8.0–8.2) compared to pH 7.6 and pH 7.8 (Table A1).

4. Discussion

This study shows that CoTS gametes, fertilization, and embryonic development are robust to a wide range of environmental conditions. Notably, these early life-stages could tolerate temperature, salinity, and pH conditions well beyond those experienced by *Acanthaster* spp. across their normal geographic range, even accounting for extreme anomalies in contemporary environmental conditions and predicted climate change impacts that are likely to occur at the end of this century [48]. If general to all populations, these findings have important implications for the reproductive success and dispersal of CoTS. A common pattern observed in this study was that sperm motility, fertilization, cleavage, and gastrulation were maximized at local summer temperature, salinity, and pH conditions, which generally coincides with periods of peak reproduction for CoTS [3,55] (Figure 5). This suggests that spawning in CoTS occurs at an optimal time when environmental conditions favor enhanced fertilization and early development. Our results also show that chemoattractants (water-soluble egg extracts) play some role in sperm activity across all environmental parameters tested.

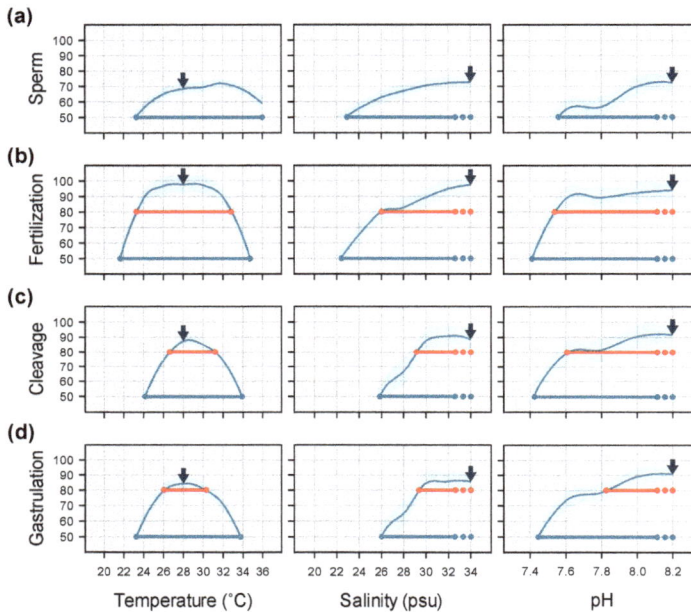

Figure 5. Environmental tipping points for: (**a**) sperm motility; (**b**) fertilization; (**c**) cleavage; and (**d**) gastrulation. Arrows signify mean ambient levels during spawning. Curves are loess smoothers fitted to dataset with proportions >50%; bold lines cover the range where: (**a**) proportion of motile sperm; (**b**) fertilization; (**c**) cleavage; and (**d**) gastrulation rates were >50% (dark blue) or >80% (red). Ellipses indicate that upper limits (above ambient) were not examined.

4.1. Temperature

Fertilization rates for CoTS were high (>80%) over a wide temperature range (24–32 °C), but do appear to be adversely affected by even higher temperatures (34–36 °C), as shown for many other tropical echinoderms [30]. Thermal enhancement of fertilization as a result of increased motility and respiratory rates of spermatozoa, with concomitant decrease in ATP concentration, has been previously demonstrated in other echinoderms [56]. Thermal robustness of fertilization may be also due to the loading of protective maternal factors (e.g., heat shock proteins) during oogenesis [57,58]. This protection may be enhanced in species with large eggs—CoTS, for example, have larger eggs compared to other planktotrophic tropical asteroids and maternal provisioning to the egg influences early larval development [5]. Increased temperature and associated decrease in viscosity increases fertilization success due to increased sperm swimming speeds [38]. This was evident for high fertilization rates achieved at temperatures above ambient levels (28 °C) and low fertilization at lower temperature extremes (zero fertilization at 18 °C and below). However, reduced sperm activity at 22 °C and 24 °C still resulted in relatively high fertilization rates, while heightened sperm activity at 34 °C and 36 °C did not correspond with significant reductions in fertilization rates. The limiting factor appears to be the restriction placed on the viability of sperm subjected to temperature extremes [56]. At temperature extremes above normal, reduced fertilization was associated with increases in the incidence of polyspermy and granular fertilization membranes that adhere to the egg [59]. Increased sperm activity due to elevated temperature, as observed in this study, could also result in mechanical damage to the sperm and incur metabolic costs and exhaustion of energy reserves [29]. Physiological and viscosity-based aspects of high temperatures can influence sperm longevity, and hence fertilization success, by directly affecting sperm velocity [38].

Temperatures that do not restrict fertilization may nonetheless be detrimental for embryonic development [60]. Embryos of the temperate sea urchin, *Strongylocentrotus purpuratus*, subjected to seawater 8 °C above ambient showed normal fertilization, but subsequently resulted in abnormal cleavage [61]. This was also consistent with earlier work by Rupp [30] where fertilization rates of CoTS decreased by 20% while cleavage fell by 60% at 34 °C. Similarly, more recent work by Sparks et al. [62] showed that the proportion of cleaved embryos was significantly lower at 31 °C compared to 27 °C and 29 °C treatments. Our study revealed that cleavage and gastrulation for CoTS were maximized over a relatively narrow temperature range (26–32 °C) and closely reflects the range of temperatures to which CoTS are likely to be exposed throughout their geographic range [3]. Conversely, Habe et al. [63] showed that gastrulation was possible at a wider temperature range (13–34 °C) than cleavage. This suggests that if post-gastrula embryos are swept into cooler waters, normal development can proceed during transport and will have important implications for long-range dispersal. However, the proportion of embryos that successfully cleave limits the proportion of embryos undergoing gastrulation. In this study, embryonic development in CoTS ceased at 20 °C and below, which was slightly higher than the lower thermal limit for embryonic development reported for CoTS from the GBR, which was between 18 and 19 °C [33,64]. This might reflect the less variable thermal environment of adult CoTS from Guam used in this study compared to CoTS from the GBR [3]. Thermal acclimatization of adults, particularly during gametogenesis, has been found to shift the thermotolerance of echinoderm embryos [33,65].

4.2. Salinity

Out of the three pervasive environmental stressors investigated in this study, response to salinity is perhaps the least studied for CoTS. Here, we found that the lower salinity limit for successful fertilization (>50%) in *Acanthaster* spp. was about 24 psu (Figure 5). No fertilization occurred after 2 h at salinities below 20 psu. At 20 and 22 psu, less than 10% of eggs produced fertilization envelopes. This range and lower salinity limit appears to be common in asteroids (24 to 32 psu in *Asterias amurensis* [31]; 22 to 34 psu in *Asterina pectinifera* [66]), echinoids (26 to 36 psu in *Echinocardium cordatum* [67]; 24 to 32 psu in *Echinarachnius parma* [47]), and holothuroids (24 to 32 psu in *Eupentacta fraudatrix* [68]). Fertilization was highest at mean ambient salinity conditions experienced by adults in their natural habitat throughout most of the year. Dinnel et al. [39] found that fertilization of gametes of the sea urchin, *Strongylocentrotus purpuratus*, was best at the salinity at which the adults were held. Contrary to these observations, Roller and Stickle [69] found no evidence of acclimation of echinoid gametes when *Lytechinus variegatus* were exposed to different salinities prior to spawning.

Developmental failure at low salinity is often thought to reflect limited fertilization, possibly due to substantial reductions in sperm motility. There is a paucity of work on the response of echinoderm spermatozoa to salinity fluctuations and most examples come from research on sperm activity in commercially valuable teleost fishes [44,70,71]. Sperm swimming speeds and sperm motility were relatively high between 24 and 34 psu and decreased slightly at 20 and 22 psu, which partly mirrored the range observed for fertilization. Minor improvements in sperm activation when exposed to water-soluble egg extracts were observed, but were not significant. The influence of egg extracts on sperm velocity and motility was greater at lower salinities.

Although there was some fertilization at 20–22 psu, eggs failed to cleave at these salinities and less than 20% cleaved at 24 psu. The failure of eggs to develop at low salinity largely reflects an inability of fertilized eggs to complete meiosis and cleave, rather than simply an inability of eggs to become fertilized at these low salinities. Salinity changes appear to have most detrimental effects for ova, which are unable to control water flow in and out of the cell. Osmotic shock experiments on the spermatozoa and ova of the echinoid, *Parechinus angulosus*, prior to fertilization under optimal temperature and salinity conditions indicated that temperature gradients exerted a greater effect on spermatozoa while low salinity was more deleterious to ova—at salinities below 15 psu, water was imbibed by the ova, which swelled and lysed. Salinity tolerance of gastrulation mirrored that of cleavage. Salinity levels as low as 10 psu have been observed to persist in nearshore and mid-shelf waters in the GBR after flood events [72]. Since embryos were not as tolerant to low salinities as previously expected [63], the timing of reduced salinity events would be critical in predicting the population response.

4.3. pH

Our results show that fertilization in *Acanthaster* spp. was robust to reduced pH. Patterns of fertilization success in relation to pH were coincident with relatively high sperm swimming speeds and proportion of motile sperm down to 0.6 pH units below ambient (pH 8.2) and significant reductions at pH 7.4 (Figure 5). In looking at the potential effects of near-future ocean acidification on CoTS recruitment, Uthicke et al. [15] found that low pH reduced sperm motility and velocity, which resulted in reduction of fertilization rates by 0.7% at pH 7.9 and 25% at pH 7.7 across a wide range of sperm concentrations. It was not clear whether impaired sperm motility, resulting in reduced fertilization at low pH, may be due to acidosis or the narcotic effect of hypercapnia on sperm [73]. For the sea urchins *Hemicentrotus pulcherrimus* and *Echinometra mathaei*, seawater acidified by CO_2 had a more severe effect on fertilization compared to HCl-acidified seawater, suggesting that hypercapnia may be more influential to fertilization. However, cross-factorial experiments showed no significant difference in fertilization rates between different combinations of temperature and pH (7.6 to 8.1) treatments [74]. This is consistent with our findings wherein no significant differences in fertilizations rates were found for pH ranging from 7.6 to 8.2. The mechanism of hypercapnic stress on sperm involves the control intracellular pH; although these effects may be overcome through respiratory dilution effects when sperm is released into the water column [75]. Coelomic fluid surrounding CoTS gonads has a mean pH of 7.49 [15], which is relatively low, hence may be activated when seawater pH levels are above this. This could explain the robustness of sperm motility and fertilization in CoTS even at relatively low pH. In addition, our results also demonstrated that water-soluble compounds derived from eggs also promoted sperm motility at low pH. Activation of nonmotile sperm by egg-derived compounds may provide a mechanism by which the energy reserves of sperm can be conserved in the absence of eggs, thereby maintaining sperm viability for extended periods [38,43]. This response has been reported for many species of corals, molluscs, echinoderms and ascidians [49].

The pH tolerance range for cleavage and gastrula embryos coincided with that of fertilization, albeit with slight reductions in frequency. Similarly, Kamya et al. [74] reported that pH had no significant effect on gastrulation in CoTS. Marine invertebrates that do not calcify during early developmental stages are generally robust to reduced pH [23,32]. Later stages (bipinnaria and brachiolaria) in the life history of CoTS are more sensitive to reduced pH and have been shown to suffer high rates of larval abnormality and mortality at low pH [15,74].

4.4. Interactive Effects and Implications for Subsequent Larval Development

Our results show that absolute sperm velocity (221–237 $\mu m \cdot s^{-1}$), at ambient temperature (28 °C), salinity (34 psu), and pH (8.2) levels, was slightly higher compared to previous estimates on CoTS sperm swimming speeds (210 μm s^{-1} [15]). These values are generally higher compared to estimates of sperm swimming speeds in other marine invertebrates, e.g., echinoids (*Heliocidaris erythrogramma*, 26–38 μm·s^{-1} [16,17]; *L. variegatus*, 153–275 μm·s^{-1} [76]), bivalves (*Macoma calcarea*, ~60 μm·s^{-1} [77]; *Mytilus galloprovincialis*, ~50 μm·s^{-1} [77]; *Crassostrea gigas*, 94 μm·s^{-1} [78]), and polychaetes (*G. caespitosa*, 45–114 μm·s^{-1} [45,79]). High sperm velocity over a wide range of temperature, salinity, and pH levels partly explains high fertilization rates of CoTS in the field [21]. However, there is a possible trade-off between sperm velocity and sperm longevity, which also influences fertilization success [76]. Sperm longevity was not quantified in this study, but previous studies have shown that CoTS sperm can also remain competent for longer periods relative to other echinoderm species, resulting in relatively higher fertilization rates at greater distances [22].

The response of gametes and early life history stages to multiple environmental stressors may have significant flow-on effects on the survival and development of subsequent larval stages, and thus, on successful recruitment. In the GBR, spawning of CoTS have usually coincided with peak summer temperatures, as well as high precipitation. Although fertilization and embryonic development may be robust to high temperatures (up to 34 °C), survival may be low when salinities drop (below 25 psu) during heavy rainfall events that result in high freshwater discharge from rivers. Disregarding the influence of other variables (i.e., predation, dispersal), the proportion of embryos progressing to subsequent larval stages will be substantially reduced. Tolerance of CoTS larvae has also been shown to be stage-specific and may constrain successful recruitment further [7]. Bipinnaria larvae of CoTS can to tolerate temperatures between 14.5 and 32 °C for up to 48 h, while the brachiolaria stage is more sensitive to temperature variation [63]. In terms of tolerance to salinity, bipinnaria larvae can tolerate abrupt salinity changes down to 21 psu [63,80], while brachiolaria larvae rupture even with a decrease in salinity of 2 psu [81]. High flow events have also been associated with elevated nutrient levels and phytoplankton densities, which have been shown to improve larval survival and development [12,82], even more so when modulated by increased temperatures up to 30 °C [83].

Here we showed that CO_2-acidified seawater (down to pH 7.6) did not have a significant effect on fertilization and early embryonic development. The detrimental effects of ocean acidification have been shown to be more apparent in subsequent larval stages. Uthicke et al. [15] found that normal development and settlement in CoTS larvae kept at pH 7.6 was significantly reduced compared to pH 8.1 treatments. Low pH (7.6) coupled with elevated temperatures (30 °C) also had an additive negative effect on larval size and development [74]. However, the positive effects of increased temperature on larval growth [83] may ameliorate the detrimental effects of low pH.

5. Conclusions

Taken together, our results show that CoTS gametes, fertilization, and embryonic development are robust to a wide range of temperature, salinity, and pH levels, well beyond environmental conditions found within the current geographical distribution of *Acanthaster* spp. Majority of sperm are motile at temperatures between 24 and 36 °C, salinities between 24 and 34 psu, and pH between 7.6 and 8.2. Over 50% of eggs are fertilized at wide range of temperature (22–34), salinity (24–34), and pH (7.6–8.2) levels. The robustness of fertilization to these pervasive environmental stressors may be attributed to the molecular predisposition of CoTS sperm [84], which possesses an enhanced capacity for high fertilization rates, compared to other echinoderms [22]. Compared to fertilization, tolerance range for cleavage was mostly narrower for temperature (26–32 °C), salinity (26–34 psu), and pH (7.6–8.2). Gastrulation under salinity and pH levels tested coincided with cleavage rates, while thermotolerance range for gastrulation was slightly wider than cleavage (24–32 °C). In general, the effects of temperature and pH on fertilization and early development mostly corresponded with the sensitivity of sperm to these stressors, while response to salinity was largely due to detrimental effects on osmotic balance in eggs. Water-soluble compounds associated with eggs also enhanced sperm activity, particularly in environmental conditions where sperm motility was initially limited. Although the response to multiple environmental stressors was tested in this study, these pervasive environmental parameters impact marine organisms simultaneously. Future work should include cross-factorial studies to tease out additive, antagonistic, and synergistic interactions between these factors [24]. The tolerance of the earliest stages of development to a wide range of environmental stressors suggests that later ontogenic stages (larvae, juveniles, adults) may be more vulnerable to small fluctuations in environmental conditions.

Acknowledgments: The ARC Centre of Excellence for Coral Reef Studies provided funding for this study through a PhD research allocation to CFC and annual research appropriation to MSP. We are grateful to A.M. Kerr, L.J. Raymundo, R.G. Rowan, M. Kitamura, and M. Byrne for technical advice. We also thank J. Miller, J. Cummings, T. Reynolds, J. Gault and the staff at University of Guam Marine Laboratory for field and laboratory assistance. Comments from two anonymous reviewers have significantly improved this manuscript.

Author Contributions: C.F.C., M.S.P., and J.R.P. conceived and designed the experiments; C.F.C and M.L.R. performed the experiments; C.F.C. analyzed the data; M.S.P. and J.R.P. contributed reagents, materials, and analysis tools; and C.F.C., M.S.P., M.L.R., J.R.P. wrote the paper.

Conflicts of Interest: The authors declare no conflict of interest. The funder had no role in the design of the study; in the collection, analyses, or interpretation of data; in the writing of the manuscript, and in the decision to publish the results.

Appendix A

Table A1. Results of statistical analyses on the effects of temperature, salinity, and pH on sperm behavior, fertilization, and early development.

Source	DF	Statistic (F, χ^2)	p
Temperature			
Sperm Speed [1]			
temperature	8	85.96	<0.0001
egg extract	1	13.16	0.0005
temperature × egg extract	8	0.76	0.6353
Sperm Motility [2]			
temperature	8	1233.07	<0.0001
egg extract	1	31.34	0.0008
temperature × egg extract	8	32.79	0.1612
Fertilization [2]	8	1316.20	<0.0001
Cleavage [2]	7	521.09	<0.0001
Gastrulation [2]	7	632.82	<0.0001
Salinity			
Sperm Speed [1]			
salinity	7	5.83	<0.0001
egg extract	1	2.93	0.0918
salinity × egg extract	7	0.31	0.9449
Sperm Motility [2]			
salinity	7	525.43	<0.0001
egg extract	1	16.42	0.0682
salinity × egg extract	7	9.62	0.9626
Fertilization [2]	7	597.86	< 0.0001
Cleavage [2]	5	369.59	< 0.0001
Gastrulation [2]	5	504.40	< 0.0001
pH			
Sperm Speed [1]			
pH	4	28.57	<0.0001
egg extract	1	3.85	0.0568
pH × egg extract	4	0.74	0.5706
Sperm Motility [2]			
pH	4	669.24	<0.0001
egg extract	1	38.11	0.0033
pH × egg extract	4	18.05	0.3943
Fertilization [2]	4	234.28	<0.0001
Cleavage [2]	4	95.37	<0.0001
Gastrulation [2]	4	213.24	<0.0001

[1] Two-way Analysis of Variance (ANOVA): *F* value. [2] Analysis of Deviance for generalized linear models (GLM): χ^2 value.

References

1. De'ath, G.; Fabricius, K.E.; Sweatman, H.P.A.; Puotinen, M. The 27-year decline of coral cover on the Great Barrier Reef and its causes. *Proc. Natl. Acad. Sci. USA* **2012**, *109*, 17995–17999. [CrossRef] [PubMed]
2. Baird, A.H.; Pratchett, M.S.; Hoey, A.S.; Herdiana, Y.; Campbell, S.J. *Acanthaster planci* is a major cause of coral mortality in Indonesia. *Coral Reefs* **2013**, *32*, 803–812. [CrossRef]
3. Pratchett, M.S.; Caballes, C.F.; Rivera-Posada, J.A.; Sweatman, H.P. Limits to understanding and managing outbreaks of crown-of-thorns starfish (*Acanthaster* spp.). *Oceanogr. Mar. Biol. An Annu. Rev.* **2014**, *52*, 133–200.
4. Uthicke, S.; Schaffelke, B.; Byrne, M. A boom-bust phylum? Ecological and evolutionary consequences of density variations in echinoderms. *Ecol. Monogr.* **2009**, *79*, 3–24. [CrossRef]
5. Caballes, C.F.; Pratchett, M.S.; Kerr, A.M.; Rivera-Posada, J.A. The role of maternal nutrition on oocyte size and quality, with respect to early larval development in the coral-eating starfish, *Acanthaster planci*. *PLoS ONE* **2016**, *11*, e0158007. [CrossRef] [PubMed]
6. Babcock, R.C.; Milton, D.A.; Pratchett, M.S. Relationships between size and reproductive output in the crown-of-thorns starfish. *Mar. Biol.* **2016**, *163*, 234. [CrossRef]
7. Caballes, C.F.; Pratchett, M.S. Reproductive biology and early life history of the crown-of-thorns starfish. In *Echinoderms: Ecology, Habitats and Reproductive Biology*; Whitmore, E., Ed.; Nova Science Publishers, Inc.: New York, NY, USA, 2014; pp. 101–146.
8. Hoey, J.; Campbell, M.; Hewitt, C.; Gould, B.; Bird, R. *Acanthaster planci* invasions: Applying biosecurity practices to manage a native boom and bust coral pest in Australia. *Manag. Biol. Invasions* **2016**, *7*, 213–220. [CrossRef]
9. Levitan, D.R. The ecology of fertilization in free-spawning invertebrates. In *Ecology of Marine Invertebrate Larvae*; McEdward, L.R., Ed.; CRC Press: Boca Raton, FL, USA, 1995; pp. 123–156.
10. Conand, C. Distribution, reproductive cycle and morphometric relationships of *Acanthaster planci* (Echinodermata: Asteroidea) in New Caledonia, western tropical Pacific. In Proceedings of the 5th International Echinoderm Conference, Galway, Ireland, 24–29 September 1985; pp. 499–506.
11. Yamaguchi, M. Recruitment of coral reef asteroids, with emphasis on *Acanthaster planci* (L.). *Micronesica* **1973**, *9*, 207–212.
12. Fabricius, K.E.; Okaji, K.; De'ath, G. Three lines of evidence to link outbreaks of the crown-of-thorns seastar *Acanthaster planci* to the release of larval food limitation. *Coral Reefs* **2010**, *29*, 593–605. [CrossRef]
13. Hock, K.; Wolff, N.H.; Condie, S.A.; Anthony, K.R.N.; Mumby, P.J. Connectivity networks reveal the risks of crown-of-thorns starfish outbreaks on the Great Barrier Reef. *J. Appl. Ecol.* **2014**, *51*, 1188–1196. [CrossRef]
14. Cowan, Z.L.; Dworjanyn, S.A.; Caballes, C.F.; Pratchett, M.S. Predation on crown-of-thorns starfish larvae by damselfishes. *Coral Reefs* **2016**, *35*, 1253–1262. [CrossRef]
15. Uthicke, S.; Pecorino, D.; Albright, R.; Negri, A.P.; Cantin, N.; Liddy, M.; Dworjanyn, S.A.; Kamya, P.Z.; Byrne, M.; Lamare, M.D. Impacts of ocean acidification on early life-history stages and settlement of the coral-eating sea star *Acanthaster planci*. *PLoS ONE* **2013**, *8*, e82938. [CrossRef] [PubMed]
16. Havenhand, J.N.; Buttler, F.R.; Thorndyke, M.C.; Williamson, J.E. Near-future levels of ocean acidification reduce fertilization success in a sea urchin. *Curr. Biol.* **2008**, *18*, 651–652. [CrossRef] [PubMed]
17. Schlegel, P.; Havenhand, J.N.; Gillings, M.R.; Williamson, J.E. Individual variability in reproductive success determines winners and losers under ocean acidification: A case study with sea urchins. *PLoS ONE* **2012**, *7*, e53118. [CrossRef] [PubMed]
18. Denny, M.W.; Shibata, M.F. Consequences of surf-zone turbulence for settlement and external fertilization. *Am. Nat.* **1989**, *134*, 859–889. [CrossRef]
19. Levitan, D.R.; Sewell, M.A.; Chia, F.-S. How distribution and abundance influence fertilization success in the sea urchin *Strongylocentotus franciscanus*. *Ecology* **1992**, *73*, 248–254. [CrossRef]
20. Babcock, R.C.; Mundy, C.N.; Whitehead, D. Sperm diffusion models and in situ confirmation of long-distance fertilization in the free-spawning asteroid *Acanthaster planci*. *Biol. Bull.* **1994**, *186*, 17–28. [CrossRef]
21. Babcock, R.C.; Mundy, C.N. Reproductive biology, spawning and field fertilization rates of *Acanthaster planci*. *Mar. Freshw. Res.* **1992**, *43*, 525–534. [CrossRef]
22. Benzie, J.A.H.; Dixon, P. The effects of sperm concentration, sperm: Egg ratio, and gamete age on fertilization success in crown-of-thorns starfish (*Acanthaster planci*) in the laboratory. *Biol. Bull.* **1994**, *186*, 139–152. [CrossRef]

23. Byrne, M. Global change ecotoxicology: Identification of early life history bottlenecks in marine invertebrates, variable species responses and variable experimental approaches. *Mar. Environ. Res.* **2012**, *76*, 3–15. [CrossRef] [PubMed]

24. Przeslawski, R.; Byrne, M.; Mellin, C. A review and meta-analysis of the effects of multiple abiotic stressors on marine embryos and larvae. *Glob. Chang. Biol.* **2015**, *21*, 2122–2140. [CrossRef] [PubMed]

25. Helmuth, B.; Mieszkowska, N.; Moore, P.; Hawkins, S.J.; Hawkins, S.J. Living on the edge of two changing worlds: Forecasting the responses of rocky intertidal ecosystems to climate change. *Annu. Rev. Ecol. Evol. Syst.* **2006**, *37*. [CrossRef]

26. Hoegh-Guldberg, O.; Pearse, J.S. Temperature, food availability, and the development of marine invertebrate larvae. *Am. Zool.* **1995**, *35*, 415–425. [CrossRef]

27. Przeslawski, R.; Ahyong, S.; Byrne, M.; Wörheide, G.; Hutchings, P.A. Beyond corals and fish: The effects of climate change on noncoral benthic invertebrates of tropical reefs. *Glob. Chang. Biol.* **2008**, *14*, 2773–2795. [CrossRef]

28. Roberts, D.E.; Davis, A.R.; Cummins, S.P. Experimental manipulation of shade, silt, nutrients and salinity on the temperate reef sponge *Cymbastela concentrica. Mar. Ecol. Prog. Ser.* **2006**, *307*, 143–154. [CrossRef]

29. Greenwood, P.J.; Bennett, T. Some effects of temperature-salinity combinations on the early development of the sea urchin *Parechinus angulosus* (Leske). Fertilization. *J. Exp. Mar. Biol. Ecol.* **1981**, *51*, 119–131. [CrossRef]

30. Rupp, J.H. Effects of temperature on fertilization and early cleavage of some tropical echinoderms, with emphasis on *Echinometra mathaei. Mar. Biol.* **1973**, *23*, 183–189. [CrossRef]

31. Kashenko, S.D. Responses of embryos and larvae of the starfish *Asterias amurensis* to changes in temperature and salinity. *Russ. J. Mar. Biol.* **2005**, *31*, 294–302. [CrossRef]

32. Dupont, S.; Ortega-Martínez, O.; Thorndyke, M.C. Impact of near-future ocean acidification on echinoderms. *Ecotoxicology* **2010**, *19*, 449–462. [CrossRef] [PubMed]

33. Johnson, L.G.; Babcock, R.C. Temperature and the larval ecology of the crown-of-thorns starfish, *Acanthaster planci. Biol. Bull.* **1994**, *187*, 304–308. [CrossRef]

34. Mita, M.; Nakamura, M. Energy metabolism of sea urchin spermatozoa: An approach based on echinoid phylogeny. *Zool. Sci.* **1998**, *15*, 1–10. [CrossRef] [PubMed]

35. Shirai, H.; Ikegami, S.; Kanatani, H.; Mohri, H. Regulation of sperm motility in starfish. I. Initiation of movement. *Dev. Growth Differ.* **1982**, *24*, 419–428. [CrossRef]

36. Lu, X.Y.; Wu, R.S.S. Ultraviolet damages sperm mitochondrial function and membrane integrity in the sea urchin *Anthocidaris crassispina. Ecotoxicol. Environ. Saf.* **2005**, *61*, 53–59. [CrossRef] [PubMed]

37. Lu, X.Y.; Wu, R.S.S. UV induces reactive oxygen species, damages sperm, and impairs fertilisation in the sea urchin Anthocidaris crassispina. *Mar. Biol.* **2005**, *148*, 51–57. [CrossRef]

38. Kupriyanova, E.; Havenhand, J.N. Effects of temperature on sperm swimming behaviour, respiration and fertilization success in the serpulid polychaete, *Galeolaria caespitosa* (Annelida: Serpulidae). *Invertebr. Reprod. Dev.* **2005**, *48*, 7–17. [CrossRef]

39. Dinnel, P.A.; Link, J.M.; Stober, Q.J. Improved methodology for a sea urchin sperm cell bioassay for marine waters. *Arch. Environ. Contam. Toxicol.* **1987**, *16*, 23–32. [CrossRef] [PubMed]

40. Miller, R.L. Demonstration of sperm chemotaxis in Echinodermata: Asteroidea, Holothuroidea, Ophiuroidea. *J. Exp. Zool.* **1985**, *234*, 383–414.

41. Cook, S.P.; Brokaw, C.J.; Muller, C.H.; Babcock, D.F. Sperm chemotaxis: Egg peptides control cytosolic calcium to regulate flagellar responses. *Dev. Biol.* **1994**, *165*, 10–19. [CrossRef] [PubMed]

42. Nishigaki, T.; Chiba, K.; Miki, W.; Hoshi, M. Structure and function of asterosaps, sperm-activating peptides from the jelly coat of starfish eggs. *Zygote* **1996**, *4*, 237–245. [CrossRef] [PubMed]

43. Bolton, T.F.; Havenhand, J.N. Chemical mediation of sperm activity and longevity in the solitary ascidians *Ciona intestinalis* and *Ascidiella aspersa. Biol. Bull.* **1996**, *190*, 329–335. [CrossRef]

44. Litvak, M.K.; Trippel, E.A. Sperm motility patterns of Atlantic cod (*Gadus morhua*) in relation to salinity: Effects of ovarian fluid and egg presence. *Can. J. Fish. Aquat. Sci.* **1998**, *55*, 1871–1877. [CrossRef]

45. Kupriyanova, E.; Havenhand, J.N. Variation in sperm swimming behaviour and its effect on fertilization success in the serpulid polychaete Galeolaria caespitosa. *Invertebr. Reprod. Dev.* **2002**, *41*, 21–26. [CrossRef]

46. Byrne, M.; Ho, M.A.; Selvakumaraswamy, P.; Nguyen, H.D.; Dworjanyn, S.A.; Davis, A.R. Temperature, but not pH, compromises sea urchin fertilization and early development under near-future climate change scenarios. *Proc. R. Soc. B Biol. Sci.* **2009**, *276*, 1883–1888. [CrossRef] [PubMed]

47. Allen, J.D.; Pechenik, J.A. Understanding the effects of low salinity on fertilization success and early development in the sand dollar *Echinarachnius parma. Biol. Bull.* **2010**, *218*, 189–199. [CrossRef] [PubMed]

48. Intergovernmental Panel on Climate Change (IPCC). *Climate Change 2014: Synthesis Report. Contribution of Working Groups I, II and III to the Fifth Assessment Report of the Intergovernmental Panel on Climate Change*; IPCC: Geneva, Switzerland, 2014.

49. Byrne, M. Impact of ocean warming and ocean acidification on marine invertebrate life history stages: Vulnerabilities and potential for persistence in a changing ocean. *Oceanogr. Mar. Biol. Annu. Rev.* **2011**, *49*, 1–42.

50. Wilson-Leedy, J.G.; Ingermann, R.L. Development of a novel CASA system based on open source software for characterization of zebrafish sperm motility parameters. *Theriogenology* **2007**, *67*, 661–672. [CrossRef] [PubMed]

51. R Development Core Team. R: A Language and Environment for Statistical Computing. R Foundation for Statistical Computing, Vienna, Austria. Available online: http://www.R-project.org/ (accessed on 31 May 2016).

52. Benjamini, Y.; Hochberg, Y. Controlling the false discovery rate: A practical and powerful approach to multiple testing. *J. R. Stat. Soc. Ser. B* **1995**, *57*, 289–300.

53. Hothorn, T.; Bretz, F.; Westfall, P. Simultaneous Inference in General Parametric Models. *Biometrical J.* **2008**, *50*, 346–363. [CrossRef] [PubMed]

54. Crawley, M.J. *The R Book*; John Wiley & Sons Ltd.: Sussex, UK, 2013.

55. Cheney, D.P. Spawning and aggregation of Acanthaster planci in Micronesia. In *Proceedings of the Second International Coral Reef Symposium*; Great Barrier Reef Committee: Brisbane, Australia, 1974; Volume 1, pp. 591–594.

56. Mita, M.; Hino, A.; Yasumasu, I. Effect of temperature on interaction between eggs and spermatozoa of sea urchin. *Biol. Bull.* **1984**, *166*, 68–77. [CrossRef]

57. Yamada, K.; Mihashi, K. Temperature-independent period immediately after fertilization in sea urchin eggs. *Biol. Bull.* **1998**, *195*, 107–111. [CrossRef]

58. Hamdoun, A.; Epel, D. Embryo stability and vulnerability in an always changing world. *Proc. Natl. Acad. Sci. USA* **2007**, *104*, 1745–1750. [CrossRef] [PubMed]

59. Hagström, B.E.; Hagström, B. The effect of decreased and increased temperatures on fertilization. *Exp. Cell Res.* **1959**, *16*, 174–183. [CrossRef]

60. Andronikov, V.B. Heat resistance of gametes of marine invertebrates in relation to temperature conditions under which the species exist. *Mar. Biol.* **1975**, *30*. [CrossRef]

61. Farmanfarmaian, A.; Giese, A.C. Thermal tolerance and acclimation in the western purple sea urchin, *Strongylocentrotus purpuratus. Physiol. Zool.* **1963**, *36*, 237–243. [CrossRef]

62. Sparks, K.M.; Foo, S.A.; Uthicke, S.; Byrne, M.; Lamare, M.D. Paternal identity influences response of *Acanthaster planci* embryos to ocean acidification and warming. *Coral Reefs* **2017**, *36*, 325–338. [CrossRef]

63. Habe, T.; Sawamoto, S.; Ueno, S.; Kosaka, M.; Ogura, M. *Studies on the Conservation and Management of Coral Reefs and the Control of Acanthaster Planci Juveniles*; Report of Grant-in-Aid for Scientific Research; Ministry of Education, Science and Culture: Tokyo, Japan, 1989; pp. 158–186. (In Japanese)

64. Lamare, M.D.; Pecorino, D.; Hardy, N.; Liddy, M.; Byrne, M.; Uthicke, S. The thermal tolerance of crown-of-thorns (*Acanthaster planci*) embryos and bipinnaria larvae: Implications for spatial and temporal variation in adult populations. *Coral Reefs* **2014**, *33*, 207–219. [CrossRef]

65. Johnson, L.G.; Chenoweth, J.E.; Bingham, B.L. Population differences and thermal acclimation in temperature responses of developing sea urchin embryos. *Proc. S. D. Acad. Sci.* **1990**, *69*, 99–108.

66. Kashenko, S.D. The combined effect of temperature and salinity on development of the sea star *Asterina pectinifera. Russ. J. Mar. Biol.* **2006**, *32*, 37–44. [CrossRef]

67. Kashenko, S.D. Adaptive responses of embryos and larvae of the heart-shaped sea urchin *Echinocardium cordatum* to temperature and salinity changes. *Russ. J. Mar. Biol.* **2007**, *33*, 381–390. [CrossRef]

68. Kashenko, S.D. Combined effect of temperature and salinity on the development of the holothurian *Eupentacta fraudatrix. Russ. J. Mar. Biol.* **2000**, *26*, 188–193. [CrossRef]

69. Roller, R.A.; Stickle, W.B. Effects of temperature and salinity acclimation of adults on larval survival, physiology, and early development of *Lytechinus variegatus* (Echinodermata: Echinoidea). *Mar. Biol.* **1993**, *116*, 583–591. [CrossRef]

70. Griffin, F.J.; Pillai, M.C.; Vines, C.A.; Kääriä, J.; Hibbard-Robbins, T.; Yanagimachi, R.; Cherr, G.N. Effects of salinity on sperm motility, fertilization, and development in the Pacific herring, *Clupea pallasi*. *Biol. Bull.* **1998**, *194*, 25–35. [CrossRef]

71. Elofsson, H.; Van Look, K.; Borg, B.; Mayer, I. Influence of salinity and ovarian fluid on sperm motility in the fifteen-spined stickleback. *J. Fish Biol.* **2003**, *63*, 1429–1438. [CrossRef]

72. Devlin, M.J.; DeBose, J.L.; Ajani, P.; Teixeira da Silva, E.; Petus, C.; Brodie, J.E. Phytoplankton in the Great Barrier Reef: Microscopy analysis of community structure in high flow events. In *Report to the National Environmental Research Program*; Reef and Rainforest Research Centre Limited: Cairns, Australia, 2013; p. 68.

73. Johnson, C.H.; Clapper, D.L.; Winkler, M.M.; Lee, H.C.; Epel, D. A volatile inhibitor immobilizes sea urchin sperm in semen by depressing the intracellular pH. *Dev. Biol.* **1983**, *98*, 493–501. [CrossRef]

74. Kamya, P.Z.; Dworjanyn, S.A.; Hardy, N.; Mos, B.; Uthicke, S.; Byrne, M. Larvae of the coral eating crown-of-thorns starfish, *Acanthaster planci* in a warmer-high CO_2 ocean. *Glob. Chang. Biol.* **2014**, *20*, 3365–3376. [CrossRef] [PubMed]

75. Chia, F.-S.; Bickell, L.R. Echinodermata. In *Reproductive Biology of Invertebrates, Volume 2*; Adiyodi, K.G., Adiyodi, R.G., Eds.; New York, NY, USA, 1983; pp. 545–620.

76. Levitan, D.R. Sperm velocity and longevity trade off each other and infuence fertilization in the sea urchin *Lytechinus variegatus*. *Proc. R. Soc. B Biol. Sci.* **2000**, *267*, 531–534. [CrossRef] [PubMed]

77. Vihtakari, M.; Havenhand, J.N.; Renaud, P.E.; Hendriks, I.E. Variable individual- and population-level responses to ocean acidification. *Front. Mar. Sci.* **2016**, *3*, 51. [CrossRef]

78. Havenhand, J.N.; Schlegel, P. Near-future levels of ocean acidification do not affect sperm motility and fertilization kinetics in the oyster Crassostrea gigas. *Biogeosciences* **2009**, *6*, 3009–3015. [CrossRef]

79. Schlegel, P.; Havenhand, J.N.; Obadia, N.; Williamson, J.E. Sperm swimming in the polychaete *Galeolaria caespitosa* shows substantial inter-individual variability in response to future ocean acidification. *Mar. Pollut. Bull.* **2014**, *78*, 213–217. [CrossRef] [PubMed]

80. Henderson, J.A. Preliminary observations on the rearing and development of *Acanthaster planci* (L.) (Asteroidea) larvae. *Fish. Notes Queensl. Dep. Harb. Mar.* **1969**, *3*, 69–79.

81. Henderson, J.A.; Lucas, J.S. Larval development and metamorphosis of *Acanthaster planci* (Asteroidea). *Nature* **1971**, *232*, 655–657. [CrossRef] [PubMed]

82. Wolfe, K.; Graba-Landry, A.; Dworjanyn, S.A.; Byrne, M. Larval starvation to satiation: Influence of nutrient regime on the success of *Acanthaster planci*. *PLoS ONE* **2015**, *10*, e0122010. [CrossRef] [PubMed]

83. Uthicke, S.; Logan, M.; Liddy, M.; Francis, D.S.; Hardy, N.; Lamare, M.D. Climate change as an unexpected co-factor promoting coral eating seastar (*Acanthaster planci*) outbreaks. *Sci. Rep.* **2015**, *5*, 8402. [CrossRef] [PubMed]

84. Stewart, M.J.; Stewart, P.; Rivera-Posada, J.A. De novo assembly of the transcriptome of *Acanthaster planci* testes. *Mol. Ecol. Resour.* **2015**, *15*, 953–966. [CrossRef] [PubMed]

Communication

Crown-of-Thorns Starfish Larvae Can Feed on Organic Matter Released from Corals

Ryota Nakajima [1,*], Nobuyuki Nakatomi [2], Haruko Kurihara [3], Michael D. Fox [1], Jennifer E. Smith [1] and Ken Okaji [4]

1   Scripps Institution of Oceanography, University of California San Diego, 9500 Gilman Drive, La Jolla, CA 92083-0202, USA; fox@ucsd.edu (M.D.F.); smithj@ucsd.edu (J.E.S.)
2   Faculty of Science and Engineering, Soka University, Hachioji, Tokyo 192-8577, Japan; nnakatomi@gmail.com
3   Faculty of Science, University of the Ryukyus, Nishihara, Okinawa 903-0213, Japan; harukoku@sci.u-ryukyu.ac.jp
4   Coralquest Inc., Atsugi, Kanagawa 243-0014, Japan; cab67820@pop06.odn.ne.jp
*   Correspondence: rnakajima@ucsd.edu; Tel.: +1-858-822-3271

Academic Editor: Morgan Pratchett
Received: 10 August 2016; Accepted: 28 September 2016; Published: 6 October 2016

Abstract: Previous studies have suggested that Crown-of-Thorns starfish (COTS) larvae may be able to survive in the absence of abundant phytoplankton resources suggesting that they may be able to utilize alternative food sources. Here, we tested the hypothesis that COTS larvae are able to feed on coral-derived organic matter using labeled stable isotope tracers (^{13}C and ^{15}N). Our results show that coral-derived organic matter (coral mucus and associated microorganisms) can be assimilated by COTS larvae and may be an important alternative or additional food resource for COTS larvae through periods of low phytoplankton biomass. This additional food resource could potentially facilitate COTS outbreaks by reducing resource limitation.

Keywords: *Acanthaster*; COTS; coral reefs; coral mucus; food limitation; isotope analysis; Japan

1. Introduction

Outbreaks of Crown-of-Thorns starfish (COTS), *Acanthaster* spp., can have devastating effects on coral reefs throughout the Pacific and Indian Oceans [1,2]. Determining the causes and spatial variability of COTS outbreaks has proven to be a major challenge for coral reef managers [2,3]. Efforts to determine the causes of COTS outbreaks on the Great Barrier Reef have identified anthropogenic eutrophication as an important correlate [3–6]. The increased phytoplankton concentrations that result from high inorganic nutrient concentrations in the water are thought to promote abnormally high survival rates of COTS larvae by acting as an important food resource [4–6]. However, most coral reefs are often considered oligotrophic systems with low phytoplankton biomass, which may help keep COTS populations stable in unperturbed conditions [4]. Maximal survival of COTS larvae has been directly linked to specific concentrations of food resources [6–9]. In the studies using natural phytoplankton assemblages as food sources, COTS larvae have been found to benefit from increasing food availability from 0.25 to 0.8 µg chl-*a* L^{-1}, with the highest growth rates observed at concentrations of >0.8 µg chl-*a* L^{-1} and resource limited mortality occurring below 0.25 µg chl-*a* L^{-1} [6,8]. Furthermore, Wolfe et al. [7] recently reported using a single microalgae species that phytoplankton levels of 1 µg chl-*a* L^{-1} were optimal for COTS larval development success.

On many coral reefs, typical concentrations of chl-*a* are approximately 0.2–0.6 µg chl-*a* L^{-1} (Table 1), which may be limiting to the survival of COTS larvae [6,8]. In Okinawa, chl-*a* concentrations often fall below 0.25 µg chl-*a* L^{-1}, the critical concentration for COTS larval survival [10–14].

However, local abundance of COTS suggests their larvae may be able to survive under low chl-*a* conditions, possibly by utilizing additional food sources [15,16]. Recent studies have reported that naturally-occurring COTS larvae were in the advanced developmental stages in the vicinity of coral communities [17–19], suggesting the utilization of organic matter derived from coral reefs by COTS larvae.

Table 1. Chlorophyll-*a* (Chl-*a*) concentrations in various coral reef waters. Some values were visually interpreted from figures. The values in parenthesis are the mean of several data in the study. GBR, Great Barrier Reef.

Site	Chl-*a* ($\mu g \cdot L^{-1}$)	Reference
Miyako Island (Okinawa, Japan)	0.10–0.15	[10]
Miyako Island (Okinawa, Japan)	0.1–0.4	[11]
Sesoko Island (Okinawa, Japan)	0.11–0.77 (0.45)	[12]
West coast of Okinawa Island (Japan)	<0.05–1.79 (0.17)	[13]
Ishigaki Island (Okinawa, Japan)	0.09–0.55	[14]
Princess Charlotte Bay (GBR, Australia)	0.06–0.28 (0.16)	[20]
Princess Charlotte Bay (GBR, Australia)	0.40	[21]
Cairns-Innisfail sector (GBR, Australia)	0.03–0.64 (0.25)	[20]
Wet Tropics (GBR, Australia)	0.70	[21]
Central GBR (Australia)	0.19–0.72 (0.38)	[22]
Whitsunday Islands (GBR, Australia)	0.31–1.21 (0.79)	[23]
Uvea Atoll (New Caledonia)	0.23	[24]
The Southwest lagoon (New Caledonia)	0.25–2.14 (0.60)	[25]
Maître Island (New Caledonia)	0.26–0.42 (0.30)	[26]
Tikehau Atoll (Tuamotu, French Polynesia)	0.17	[27]
Takapoto Atoll (Tuamotu, French Polynesia)	0.23	[27]
Takapoto Atoll (Tuamotu, French Polynesia)	0.21–0.23 (0.22)	[28]
Fakarava/Rangiroa Atolls (French Polynesia)	0.008–0.25	[29]
Ahe Atoll (French Polynesia)	0.08–0.85 (0.34)	[30]
Tioman Island (Malaysia)	0.20–0.24 (0.22)	[31]
Bidong Island (Malaysia)	0.28–0.30 (0.29)	[32]

Scleractinian corals release a large amount of organic matter in the form of mucus [33] as a result of basic metabolic activities and as a protective mechanism against various stresses [34]. Coral mucus particles contain carbohydrates, proteins, lipids, and numerous microorganisms [33,35]. When coral mucus particles become suspended in the water column it aggregates various organic particles (such as microbes and phytoplankton) and becomes more enriched in organic matter over time [33,36,37]. These coral mucus aggregates are considered to be one of the major contributors to the origin of particulate organic matter (POM) in reef waters [36,38]. These mucus aggregates have also been reported as an important food source for various reef animals, such as fish, zooplankton, and several benthic taxa, such as coral crabs and brittle stars [39]. Experimental evidence has shown that copepods and mysids, two common zooplankton taxa, directly utilize coral mucus aggregates as a food resource [40,41]. The release of mucus by corals increases with increasing ambient light and water temperature [42] as it originates from by-products of the photosynthesis of zooxanthellae [43]. Thus, mucus production is likely to be maximized in summer (July–August in Okinawa) which corresponds with the peak spawning of COTS (July in Okinawa). Coral mucus could, therefore, be an important food source for COTS larvae that develop on or near coral reefs during periods of low phytoplankton biomass.

Here, we examined whether COTS larvae are able to feed on coral-derived organic matter. Since understanding the key food sources of COTS larvae is critical for determining future recruitment of adults [7], it is important to investigate all possible food sources in the natural environment. If COTS larvae feed on organic matter released by corals, this would represent an additional food source not previously considered. This additional food resource may enhance the survival of COTS larvae and subsequent recruitment of adults in areas that are naturally low in phytoplankton abundance and/or during times of reduced phytoplankton biomass.

2. Materials and Methods

2.1. Collection of COTS Larvae

Adult COTS (ca. 25 cm diameter) were collected on shallow reefs around Onna Village (N26.508496; E127.854283), Okinawa, Japan, in June 2015. Individuals were immediately transferred to the flow-through indoor aquaria at the Sesoko Station, Tropical Biosphere Research Center (University of the Ryukyus). We refer to the individuals from the Pacific Ocean clade COTS used in this study as *Acanthaster* cf. *solaris* [44,45].

COTS larvae were obtained using the method of Birkeland and Lucas [1]. Several mL of 1-methyladenine solution (1×10^{-3} M) were added to a crystallization dish with mature ovary lobes from the collected adults to release the oocytes. The oocytes released were pipetted into a separate glass dish filled with filtered seawater. Two drops of dense spermatozoa suspension were then added and gently stirred to fertilize oocytes. Fertilized oocytes were repeatedly rinsed with filtered seawater and introduced into a 9 L cylindrical plastic container with filtered seawater. After approximately 24 h, actively swimming gastrula larvae were transferred to another container where they were reared in filtered seawater at 28 °C with enriched phytoplankton using *Dunaliella tertiolecta* at a cell density of 500 cells mL^{-1} (= 0.9–1.0 µg chl-a L^{-1}), until they reached the advanced bipinnaria stage. The algae *D. tertiolecta* (NRIA-0109) was provided by GeneBank of the Japan Fisheries Research and Education Agency (Yokohama, Japan). Daily seawater changes were carried out by gentle reverse filtration using a 60 µm mesh screen.

2.2. Collection of Coral Mucus

Colonies of the branching corals, *Acropora muricata* and *A. intermedia* were collected from the reefs of Sesoko Island, and raised in the outdoor aquaria facility at the Sesoko Station for more than a year. Two weeks before the experiment, the coral colonies were transferred to a 30 L flow-through outdoor aquarium tank. Fresh seawater was directly pumped from the reef to the aquaria at the flow rate of 2.5 L·min^{-1}.

The corals were labeled with enriched concentrations of stable carbon and nitrogen isotopes by incubating corals in isotopically-enriched seawater following the methods of Naumann et al. [46]. The water inflow was stopped at 1000 h, and the aquarium was treated with $NaH^{13}CO_3$ and $Na^{15}NO_3$ (Cambridge Isotope Laboratories, MA, USA, 98 atom %) to a final concentration of 20 mg·L^{-1} and 1 mg·L^{-1}, respectively. After a 5 h (at 1500 h) incubation, the flow-through seawater was resumed. The tank was aerated with a powerful air-pump for sufficient seawater agitation during each of the incubation periods. The labeling procedure was repeated for three days. On the final incubation day, labeled mucus was collected at 1600 h, one hour after the incubation period. To collect the mucus, corals were removed from the tank and washed with unlabeled filtered seawater [46]. The corals were then exposed to air and hung in direct sunlight to trigger mucus production [33]. The released mucus was collected in sterilized 50 mL Corning tubes. The collection of the labeled mucus was conducted twice, on July 18 and 25. The mucus was used immediately after each collection for the following feeding experiments. Triplicate subsamples of the labeled mucus from each collection period were placed in pre-weighed tin capsules and dried for 24 h at 60 °C, then stored in a desiccator until isotopic analysis was performed.

2.3. Feeding Experiments

Feeding experiments were conducted following each of the mucus collections using bipinnaria larvae. During the first feeding experiment, five-day-old (fertilized on 13 July) and 14-day-old (fertilized on 4 July) larvae were used (Experiment 1). For the second feeding experiment only five-day-old larvae (fertilized on 20 July) were used (Experiment 2). Prior to the experiments, actively-swimming larvae were gently siphoned out, concentrated, and rinsed thoroughly with GF/F (Whatman, NJ, USA) filtered seawater. These larvae were kept in GF/F filtered seawater for 24 h to empty their guts. Feeding incubations were conducted using 12 polycarbonate bottles filled with 1 L of GF/F filtered seawater. In Experiment 1, six bottles were used for five-day-old larvae and the other six for 14-day-old larvae. In each set of six bottles, three bottles treated with 2 mL of labeled mucus and the other three bottles were used as controls (without mucus). The feeding experiments contained approximately 300 starved larvae in each bottle. All bottles were incubated in the laboratory for 24 h keeping the water temperature at 28 °C. The bottles were rotated on a plankton wheel (Model II, Wheaton Instruments, NJ, USA) during the incubation to keep larvae and coral mucus particles in suspension. In Experiment 2, five-day-old larvae were introduced to all 12 bottles (300 individuals per bottle) and filled with 1 L of GF/F filtered seawater. Of these 12 bottles, six were used for the feeding treatment (with mucus) and the other six were used for the control treatment (without mucus). This second incubation was conducted for 48 h, and three bottles were collected from each treatment at 24 and 48 h.

After the incubations, the COTS larvae were collected on a 60 μm mesh screen and washed with unlabeled Milli-Q water (Whatman, NJ, USA) 10 times to remove remaining labeled external mucus material. The larvae were then transferred into 15 mL Corning tube full of Milli-Q water. These tubes were refrigerated until the larvae settled to the bottom. The settled larvae were gently pipetted out (this procedure allowed for the collection of all larvae within a sample in a small volume of water), then placed into a tin capsule, dried (40 °C, 24 h), and stored in a desiccator until isotopic analysis was performed. Due to low biomass, COTS larvae collected from triplicate bottles for each feeding treatment were pooled to ensure a good signal. Therefore, only four samples for each feeding treatment were analyzed. Samples were not acidified with HCl to remove possible inorganic carbon because the acid could have damaged the larvae and altered their isotopic signature.

2.4. Analysis

Carbon- and nitrogen-stable isotope ratios of the larvae were determined by elemental analysis/isotope ratio mass spectrometry (EA/IRMS) using a Flash EA1112-DELTA V PLUS ConFlo III System (Thermo Fisher Scientific, MA, USA) at SI Science Co., Ltd. (Saitama, Japan). The carbon and nitrogen isotopic ratios are expressed in δ notation (Vienna-PeeDee Belemnite limestone for carbon and atmospheric nitrogen for nitrogen) as the deviation from standards in parts per mill (‰) using the following equation: δX (‰) = $[(R_{sample}/R_{standard}) - 1] \times 1000$, where X is ^{13}C or ^{15}N and R is the ratio of ^{13}C/^{12}C or ^{15}N/^{14}N. The analytical error was less than $\pm 0.13‰$ for carbon and $\pm 0.61‰$ for nitrogen. Significant differences (at the $p < 0.05$ level) of δ^{13}C and δ^{15}N values between control and mucus were determined using Student's *t*-test.

3. Results and Discussion

The average value ($\pm$ SE) of δ^{13}C and δ^{15}N of labeled coral mucus was 358.5‰ $\pm$ 92.7‰ and 1683.4 $\pm$ 112.1‰ for the first experiment and 707.4‰ $\pm$ 31.2‰ and 1983.8‰ $\pm$ 111.1‰ for the second experiment, respectively (Table 2). The δ^{13}C and δ^{15}N ratios of the labeled mucus showed higher values in the second experiment, likely because of label accumulation in coral tissue and/or the biological community in the mucous on the coral surface (e.g., bacteria and zooxanthellae) [46]. We measured the δ^{13}C and δ^{15}N values of raw coral mucus, which could contain bacteria and zooxanthellae. Therefore, we are not able to offer direct evidence of mucus labeling via coral tissue but rather the labeling of the aggregation of organic matter within the mucus particles. However, Naumann et al. [46] measured the δ^{15}N of coral mucus that was filtered through 0.2 μm filter to remove microorganisms, and showed coral mucus had been labeled successfully. Considering the δ^{15}N value in our labeled raw coral mucus (1.68‰–1.98‰) are similar to values reported in other studies ([46] ca. 2.00‰), we consider that the coral mucus in our experiment was successfully labeled.

Table 2. δ^{13}C and δ^{15}N signatures of labeled mucus released by *Acropora* corals. C/N ratios were determined as %C/%N of the samples. Bold values indicate mean ($\pm$ SE) of triplicate measurements.

Collection Day (mm/dd/yy)	δ^{13}C (‰)	δ^{15}N (‰)	C/N
07/18/2015	288.9	1562.5	6.7
	463.7	1783.9	7.7
	322.9	1703.7	7.2
	358.5 ± 53.5	**1683.4 ± 64.7**	
07/24/2015	690.6	2091.3	7.8
	688.1	1869.4	7.2
	743.3	1990.6	7.3
	707.4 ± 18.0	**1983.8 ± 64.1**	

After both the 24 and 48 h incubations, the COTS larvae in the mucus addition treatments exhibited highly-enriched δ^{13}C and δ^{15}N values compared to those in control treatments (Figure 1). The average value of δ^{13}C (40.7‰ ± 34.9‰) and δ^{15}N (133.4‰ ± 55.9‰) of larvae in the mucus treatments (n = 4) was significantly (P = 0.017 for δ^{13}C and 0.0040 for δ^{15}N) higher than those in controls (δ^{13}C, −16.8‰ ± 0.3‰; δ^{15}N, 7.1‰ ± 0.2‰, n = 4) (Figure 1). Since δ^{13}C and δ^{15}N of the labeled mucus in the second experiment showed higher values, the following δ^{13}C and δ^{15}N values of COTS larvae also showed higher values compared to those in the first experiment. These results indicate that δ^{13}C- and δ^{15}N-labeled mucus are transferred into COTS larvae, providing further evidence for the potential use of organic matter derived from corals as a food source.

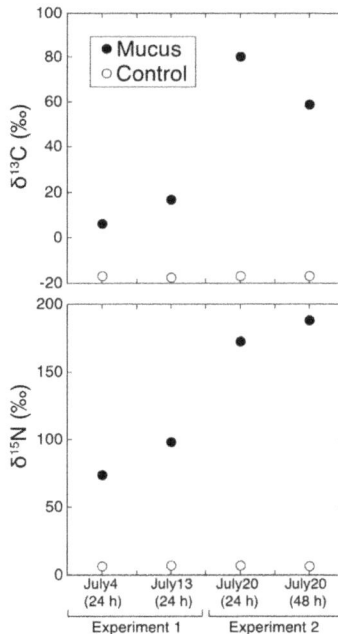

Figure 1. δ^{13}C and δ^{15}N signatures of COTS larvae after 24 h or 48 h of incubation with δ^{13}C- and δ^{15}N-labeled coral mucus and under control conditions. Closed and opened circles indicate fed larvae (with mucus) and unfed larvae (without mucus), respectively. Dates indicate the day of fertilization. Each plot comes from a single analysis.

This is the first study to show that COTS larvae can take up organic matter derived from corals. However, the raw coral mucus we provided to the COTS larvae may contain zooxanthellae and other microorganisms as mentioned above. Therefore, whether COTS larvae fed on whole coral mucus aggregates or selectively fed on the associated microorganisms remains unclear. Regardless, it is evident that the COTS larvae, in our feeding experiments, assimilated organic matter derived from corals, either directly (i.e., pure mucus), or indirectly via the associated microorganisms in/on the coral mucus aggregates (i.e., bacteria and zooxanthellae). It is also unclear whether COTS larvae fed on the particulate and/or dissolved fraction of the coral mucus, as we did not size-fractionate the coral mucus. Previous reports have shown that COTS larvae can utilize both POM and dissolved organic matter (DOM) [8,47,48]. Naumann et al. [46] also reported that coral associated epizoic acoelomorph *Wamioa* worms utilized both the particulate and dissolved fraction of mucus released by corals. Thus, it is possible that COTS larvae are capable of feeding on both POM (mucus particles) and DOM released by corals.

Previous studies that have observed high survival rates of COTS larvae even when exposed to low concentrations of phytoplankton, suggesting that they may utilize other food sources (e.g., [8,9]). For COTS larvae in near-reef waters, organic matter derived from corals may provide an important and previously underappreciated source of nutrition. Peak spawning for COTS tends to occur in summer (July in Japan). During this period, elevated water temperature and light intensity may also cause higher organic matter release by corals [42]. In Okinawa, numerous advanced COTS brachiolaria larvae were found near coral communities [19] and it appears that some populations of COTS larvae are retained in near-shore waters due to prevailing along-shore currents in the archipelago in summer [17]. If COTS larvae are entrained near-shore they are more likely to encounter coral-derived organic matter. When the reef water residence time is short and the net transport of water is offshore, some fraction of coral mucus may be physically transported offshore before this material settles or is consumed within the reef environment [39,49]. Thus, coral-derived organic matter may also provide an important food source to COTS larvae in pelagic habitats adjacent to coral reefs. Consequently, coral mucus and its aggregates may provide one of the energy pathways for the conversion of coral primary production to COTS larvae.

Coral-derived organic matter may be a particularly important resource for COTS larval under oligotrophic conditions typical of many coral reefs where nutrients and phytoplankton biomass are low. While phytoplankton can be ephemerally high in areas with internal tidal influence or with episodic upwelling events, coral-derived organic matter may be a more consistent source of nutrition to COTS larvae on most reefs. Further, these larvae may be able to extend their planktonic durations during times of low phytoplankton biomass using endogenous nutrients, or by increasing their capacity for food capture by extending ciliary bands [50]. Although it is not likely that coral mucus, alone, is a direct causal factor of COTS outbreaks, it does seem likely that coral-derived organic matter could be an important food source, especially during periods of low phytoplankton levels. In sum, while more data are needed to understand the quantitative importance of coral-derived organic matter to COTS population dynamics, our results suggest that the feeding ecology of this species is more complex than previously thought. Given the nutritional quality of coral mucus and associated material, it is not surprising that COTS larvae, as well as many other species, take advantage of this resource in an ecosystem where nutrients are often limiting.

Acknowledgments: We are grateful to three anonymous reviewers for their time and valuable suggestions to improve this article. We thank Sesoko Station staffs, M. Kitamura (Incorporated Foundation Okinawa Prefecture Environment Science Center) for his support on collection of COTS larvae, M. Okauchi and M. Tokuda (Japan Fisheries Research and Education Agency) for providing algal culture, and Y. Tadokoro for his help in sample collection. This study was made possible by a research project titled "Comprehensive management program of the Crown-of-Thorns Starfish outbreaks in Okinawa" supported by Nature Conservation Division, Okinawa Prefectural Government, by JSPS Fellowships for Research Abroad to the first author, and by Grant-in-Aid for JSPS Fellows (14J00884) to the second author.

Author Contributions: R.N. and N.N. conceived and designed the experiments; R.N., N.N. and K.O. performed the experiments; N.N. analyzed the data; H.K., N.N. and K.O. contributed reagents/materials/analysis tools; R.N., N.N., H.K., M.D.F., J.E.S. and K.O. wrote the paper.

Conflicts of Interest: The authors declare no conflict of interest.

References

1. Birkeland, C.; Lucas, J. *Acanthaster Planci: Major Management Problem of Coral Reefs*; CRC Press: Florida, FL, USA, 1990.
2. Pratchett, M.; Caballes, C.; Rivera Posada, J.; Sweatman, H. Limits to understanding and managing outbreaks of crown-of-thorns starfish (*Acanthaster* spp.). *Oceanogr. Mar. Biol. Annu. Rev.* **2014**, *52*, 133–200.
3. De'ath, G.; Fabricius, K.E.; Sweatman, H.; Puotinen, M. The 27-year decline of coral cover on the Great Barrier Reef and its causes. *Proc. Natl. Acad. Sci. USA* **2012**, *109*, 17995–17999. [CrossRef] [PubMed]
4. Lucas, J. Quantitative studies of feeding and nutrition during larval stages of the coral reef asteroid *Acanthaster planci* (L.). *J. Exp. Mar. Biol. Ecol.* **1982**, *65*, 173–193. [CrossRef]
5. Brodie, J.; Fabricius, K.; De'ath, G.; Okaji, K. Are increased nutrient inputs responsible for more outbreaks of crown-of-thorns starfish? An appraisal of the evidence. *Mar. Pollut. Bull.* **2005**, *51*, 266–278. [CrossRef] [PubMed]
6. Fabricius, K.E.; Okaji, K.; De'ath, G. Three lines of evidence to link outbreaks of the crown-of-thorns seastar *Acanthaster planci* to the release of larval food limitation. *Coral Reefs* **2010**, *29*, 593–605. [CrossRef]
7. Wolfe, K.; Graba-Landry, A.; Dworjanyn, S.A.; Byrne, M. Larval starvation to satiation: Influence of nutrient regime on the success of *Acanthaster planci*. *PLoS ONE* **2015**, *10*, 1–17. [CrossRef] [PubMed]
8. Okaji, K.; Ayukai, T.; Lucas, J.S. Selective feeding by larvae of the crown-of-thorns starfish, *Acanthaster planci* (L.). *Coral Reefs* **1997**, *16*, 47–50. [CrossRef]
9. Okaji, K. Feeding Ecology in the Early Life Stages of the Crown-of-Thorns Starfish, *Acanthaster planci* (L.). Ph.D. Thesis, James Cook University, Brisbane, Australia, June 1996.
10. Ferrier-Pagès, C.; Gattuso, J.P. Biomass, production and grazing rates of pico- and nanoplankton in coral reef waters (Miyako Island, Japan). *Microb. Ecol.* **1998**, *35*, 46–57. [CrossRef] [PubMed]
11. Casareto, B.E.; Charpy, L.; Blanchot, J.; Suzuki, Y.; Kurosawa, K.; Ihikawa, Y. Photorophic prokaryotes in Bora Bay, Miyako Island, Okinawa, Japan. *Proc. 10th Int. Coral Reef Symp.* **2006**, *31*, 844–853.
12. Tada, K.; Sakai, K.; Nakano, Y.; Takemura, A.; Montani, S. Size-fractionated phytoplankton biomass in coral reef waters off Sesoko Island, Okinawa, Japan. *J. Plankton Res.* **2001**, *25*, 991–997. [CrossRef]
13. Kinjyo, K.; Yamakawa, E. Survey of chlorophyll distribution. *A Report on the Comprehensive Management Program of the Crown-of-Thorns Starfish Outbreaks in Okinawa, 2014*; Okinawa Prefectural Government, Department of Environmental Affairs, Nature Conservation and Afforestation Promotion Division: Okinawa, Japan, 2014; pp. 99–112.
14. Fukuoka, K.; Shimoda, T.; Abe, K. Community structure and abundance of copepods in summer on a fringing coral reef off Ishigaki Island, Ryukyu Islands, Japan. *Plankt. Benthos Res.* **2015**, *10*, 225–232. [CrossRef]
15. Olson, R. In situ culturing of larvae of the crown-of-thorns starfish *Acanthaster planci. Mar. Ecol. Prog. Ser.* **1985**, *25*, 207–210. [CrossRef]
16. Olson, R. In situ culturing as a test of the larval starvation hypothesis for the crown-of-throns starfish, Acanthaster planci. *Limnol. Oceanogr.* **1987**, *32*, 895–904. [CrossRef]
17. Nakamura, M.; Kumagai, N.H.; Sakai, K.; Okaji, K.; Ogasawara, K.; Mitarai, S. Spatial variability in recruitment of acroporid corals and predatory starfish along the Onna coast, Okinawa, Japan. *Mar. Ecol. Prog. Ser.* **2015**, *540*, 1–12. [CrossRef]
18. Uthicke, S.; Doyle, J.; Duggan, S.; Yasuda, N.; McKinnon, A.D. Outbreak of coral-eating Crown-of-Thorns creates continuous cloud of larvae over 320 km of the Great Barrier Reef. *Sci. Rep.* **2015**, *5*, 16885. [CrossRef] [PubMed]
19. Suzuki, G.; Yasuda, N.; Ikehara, K.; Fukuoka, K.; Kameda, T.; Kai, S.; Nagai, S.; Watanabe, A.; Nakamura, T.; Kitazawa, S.; et al. Detection of a high-density brachiolaria-stage larval population of crown-of-thorns sea star (*Acanthaster planci*) in Sekisei Lagoon (Okinawa, Japan). *Diversity* **2016**, *8*, 9. [CrossRef]
20. McKinnon, A.; Duggan, S.; De'ath, G. Mesozooplankton dynamics in nearshore waters of the Great Barrier Reef. *Estuar. Coast. Shelf Sci.* **2005**, *63*, 497–511. [CrossRef]

21. Fabricius, K.; De'ath, G.; McCook, L.; Turak, E.; Williams, D.M. Changes in algal, coral and fish assemblages along water quality gradients on the inshore Great Barrier Reef. *Mar. Pollut. Bull.* **2005**, *51*, 384–398. [CrossRef] [PubMed]

22. Furnas, M.J.; Mitchell, A.W.; Gilmartin, M.; Revelante, N. Phytoplankton biomass and primary production in semi-enclosed reef lagoons of the central Great Barrier Reel Australia. *Coral Reefs* **1990**, *9*, 1–10. [CrossRef]

23. Van Woesik, R.; Tomascik, T.; Blake, S. Coral assemblages and physico-chemical characteristics of the Whitsunday Islands: Evidence of recent community changes. *Mar. Freshw. Res.* **1999**, *50*, 427–440. [CrossRef]

24. Le Borgne, R.; Rodier, M.; Le Bouteiller, A.; Kulbicki, M. Plankton biomass and production in an open atoll lagoon: Uvea, New Caledonia. *J. Exp. Mar. Biol. Ecol.* **1997**, *212*, 187–210. [CrossRef]

25. Rochelle-Newall, E.J.; Torréton, J.P.; Mari, X.; Pringault, O. Phytoplankton-bacterioplankton coupling in a subtropical South Pacific coral reef lagoon. *Aquat. Microb. Ecol.* **2008**, *50*, 221–229. [CrossRef]

26. Torréton, J.P.; Rochelle-Newall, E.; Pringault, O.; Jacquet, S.; Faure, V.; Briand, E. Variability of primary and bacterial production in a coral reef lagoon (New Caledonia). *Mar. Pollut. Bull.* **2010**, *61*, 335–348. [CrossRef] [PubMed]

27. Charpy, L.; Dufour, P.; Garcia, N. Particulate organic matter in sixteen Tuamotu atoll lagoons (French Polynesia). *Mar. Ecol. Ser.* **1997**, *151*, 55–65. [CrossRef]

28. Sakka, A.; Legendre, L. Carbon budget of the planktonic food web in an atoll lagoon (Takapoto, French Polynesia). *J. Plankton Res.* **2002**, *24*, 301–320. [CrossRef]

29. Ferrier-Pagès, C.; Furla, P. Pico- and nanoplankton biomass and production in the two largest atoll lagoons of French Polynesia. *Mar. Ecol. Prog. Ser.* **2001**, *211*, 63–76. [CrossRef]

30. Charpy, L.; Rodier, M.; Fournier, J.; Langlade, M.J.; Gaertner-Mazouni, N. Physical and chemical control of the phytoplankton of Ahe lagoon, French Polynesia. *Mar. Pollut. Bull.* **2012**, *65*, 471–477. [CrossRef] [PubMed]

31. Nakajima, R.; Yoshida, T.; Othman, B.; Toda, T. Biomass and estimated production rates of metazoan zooplankton community in a tropical coral reef of Malaysia. *Mar. Ecol.* **2014**, *35*, 112–131. [CrossRef]

32. Nakajima, R.; Tsuchiya, K.; Nakatomi, N.; Yoshida, T.; Tada, Y.; Konno, F.; Toda, T.; Kuwahara, V.S.; Hamasaki, K.; Othman, B.H.R.; et al. Enrichment of microbial abundance in the sea-surface microlayer over a coral reef: Implications for biogeochemical cycles in reef ecosystems. *Mar. Ecol. Prog. Ser.* **2013**, *490*, 11–22. [CrossRef]

33. Wild, C.; Huettel, M.; Klueter, A.; Kremb, S.G. Coral mucus functions as an energy carrier and particle trap in the reef ecosystem. *Nature* **2004**, *428*, 66–70. [CrossRef] [PubMed]

34. Brown, B.E.; Bythell, J.C. Perspectives on mucus secretion in reef corals. *Mar. Ecol. Prog. Ser.* **2005**, *296*, 291–309. [CrossRef]

35. Wild, C.; Naumann, M.; Niggl, W.; Haas, A. Carbohydrate composition of mucus released by scleractinian warm- and cold-water reef corals. *Aquat. Biol.* **2010**, *10*, 41–45. [CrossRef]

36. Huettel, M.; Wild, C.; Gonelli, S. Mucus trap in coral reefs: Formation and temporal evolution of particle aggregates caused by coral mucus. *Mar. Ecol. Prog. Ser.* **2006**, *307*, 69–84. [CrossRef]

37. Naumann, M.S.; Richter, C.; El-Zibdah, M.; Wild, C. Coral mucus as an efficient trap for picoplanktonic cyanobacteria: Implications for pelagic-benthic coupling in the reef ecosystem. *Mar. Ecol. Prog. Ser.* **2009**, *385*, 65–76. [CrossRef]

38. Hata, H.; Kudo, S.; Yamano, H.; Kurano, N.; Kayanne, H. Organic carbon flux in Shiraho coral reef (Ishigaki Island, Japan). *Mar. Ecol. Prog. Ser.* **2002**, *232*, 129–140. [CrossRef]

39. Nakajima, R.; Tanaka, Y. The role of coral mucus in the material cycle in reef ecosystems: Biogeochemical and ecological perspectives. *J. Jpn. Coral Reef Soc.* **2014**, *16*, 3–27. [CrossRef]

40. Richman, S.; Loya, Y.; Slobodkin, L. The rate of mucus production by corals and its assimilation by the coral reef copepod Acartia negligens. *Limnol. Oceanogr.* **1975**, *20*, 918–923. [CrossRef]

41. Gottfried, M.; Roman, M.R. Ingestion and incorporation of coral-mucus detritus by reef zooplankton. *Mar. Biol.* **1983**, *72*, 211–218. [CrossRef]

42. Naumann, M.S.; Haas, A.; Struck, U.; Mayr, C.; el-Zibdah, M.; Wild, C. Organic matter release by dominant hermatypic corals of the Northern Red Sea. *Coral Reefs* **2010**, *29*, 649–659. [CrossRef]

43. Crossland, C. In situ release of mucus and DOC-lipid from the corals Acropora variabilis and Stylophora pistillata in different light regimes. *Coral Reefs* **1987**, *6*, 35–42. [CrossRef]

44. Vogler, C.; Benzie, J.; Lessios, H.; Barber, P.; Wörheide, G. A threat to coral reefs multiplied? Four species of crown-of-thorns starfish. *Biol. Lett.* **2008**, *4*, 696–699. [CrossRef] [PubMed]

45. Haszprunar, G.; Spies, M. An integrative approach to the taxonomy of the crown-of-thorns starfish species group (Asteroidea: *Acanthaster*): A review of names and comparison to recent molecular data. *Zootaxa* **2014**, *3841*, 271–284. [CrossRef] [PubMed]

46. Naumann, M.S.; Mayr, C.; Struck, U.; Wild, C. Coral mucus stable isotope composition and labeling: Experimental evidence for mucus uptake by epizoic acoelomorph worms. *Mar. Biol.* **2010**, *157*, 2521–2531. [CrossRef]

47. Hoegh-Guldberg, O.; Ayukai, T. *Assessment of the Role of Dissolved Organic Matter and Bacteria in the Nutrition of Crown-of-Thorns Starfish Larva*; Australian Institute of Marine Science: Townsville, Australia, 1992.

48. Hoegh-Gulberg, O. Is *Acanthaster planci* able to utilise dissolved organic matter (DOM) to satisfy the energetic requirements of larval development? In *The Possible Causes and Consequences of Outbreaks of the Crown-of-Thorns Starfish*; Great Barrier Reef Marine Park Authority Workshop: Townsville, Australia, 1992; pp. 37–54.

49. Wyatt, A.S.J.; Lowe, R.J.; Humphries, S.; Waite, A.M. Particulate nutrient fluxes over a fringing coral reef: Source—Sink dynamics inferred from carbon to nitrogen ratios and stable isotopes. *Limnol. Oceanogr.* **2013**, *58*, 409–427. [CrossRef]

50. Wolfe, K.; Graba-Landry, A.; Dworjanyn, S.A.; Byrne, M. Larval phenotypic plasticity in the boom-and-bust crown-of-thorns seastar, *Acanthaster planci. Mar. Ecol. Prog. Ser.* **2015**, *539*, 179–189. [CrossRef]

diversity

MDPI

Article

Interactive Effects of Endogenous and Exogenous Nutrition on Larval Development for Crown-Of-Thorns Starfish

Ciemon Frank Caballes [1,*]**, Morgan S. Pratchett** [1] **and Alexander C. E. Buck** [1,2]

[1]   ARC Centre of Excellence for Coral Reef Studies, James Cook University, Townsville, QLD 4811, Australia; morgan.pratchett@jcu.edu.au

[2]   Marine Biology and Aquaculture, College of Science and Engineering, James Cook University, Townsville, QLD 4811, Australia; alexander.buck@my.jcu.edu.au

*   Correspondence: ciemon.caballes@my.jcu.edu.au; Tel.: +61-7-4781-5747

Academic Editors: Sven Uthicke and Michael Wink
Received: 30 November 2016; Accepted: 27 February 2017; Published: 4 March 2017

Abstract: Outbreaks of crown-of-thorns starfish are often attributed to step-changes in larval survivorship following anomalous increases in nutrients and food availability. However, larval growth and development is also influenced by the nutritional condition of spawning females, such that maternal provisioning may offset limitations imposed by limited access to exogenous sources of nutrients during the formative stages of larval development. This study examined the individual, additive, and interactive effects of endogenous (maternal diet: *Acropora*, *Porites*, mixed, and starved) and exogenous (larval diet: high concentration at 10^4 cells·mL^{-1}, low concentration at 10^3 algal cells·mL^{-1}, and starved) nutrition on the survival, growth, morphology, and development of larvae of the crown-of-thorns starfish. Female starfish on *Acropora* and mixed diet produced bigger oocytes compared to *Porites*-fed and starved treatments. Using oocyte size as a proxy for maternal provisioning, endogenous reserves in the oocyte had a strong influence on initial larval survival and development. This suggests that maternal reserves can delay the onset of obligate exogenous food acquisition and allow larvae to endure prolonged periods of poor environmental nutritive conditions or starvation. The influence of exogenous nutrition became more prominent in later stages, whereby none of the starved larvae reached the mid-to-late brachiolaria stage 16 days after the onset of the ability to feed. There was no significant difference in the survival, development, and competency of larvae between high and low food treatments. Under low algal food conditions, larvae compensate by increasing the length of ciliated feeding bands in relation to the maximum length and width, which improve food capture and feeding efficiency. However, the effects of endogenous nutrition persisted in the later developmental stages, as larvae from starved females were unable to develop larger feeding structures in response to food-limiting conditions. Phenotypic plasticity influenced by endogenous provisions and in response to exogenous food availability may be an important strategy in boosting the reproductive success of crown-of-thorns starfish, leading to population outbreaks.

Keywords: *Acanthaster* outbreak; maternal provisioning; larval nutrition; larval development; phenotypic plasticity

---

1. Introduction

Marked and acute increases in the local abundance of the coral-eating crown-of-thorns starfish (CoTS), *Acanthaster planci* s. l., often termed "outbreaks", contribute significantly to global declines in coral cover [1–3] and are a central focus of ongoing research and management to secure the future of coral reef ecosystems [4,5]. Effective long-term management of CoTS outbreaks is fundamentally dependent upon identifying the ultimate cause(s) of changes in key demographic properties that potentially differentiate outbreak and non-outbreak populations [6]. However, given the exceptional fecundity and reproductive potential of CoTS [7], it has been suggested that very subtle changes in recruitment rates could be sufficient to initiate outbreaks [8], especially if primary outbreaks represent the accumulation of individuals over several successive recruitment events (e.g., [9]). Conversely, step-changes in developmental rates and survivorship of CoTS larvae have been reported across relatively moderate gradients in chlorophyll concentrations, such that periodic influxes or concentrations of nutrients (e.g., during major flood events [10,11], upwelling [12], or from oceanographic features such as chlorophyll fronts [13]) may be an important precursor to CoTS outbreaks [14–16].

For planktotrophic larvae, the energy required for survival, growth, and development can be derived from two sources: parental investment in the oocyte and nutrient acquisition by pelagic larvae from the external environment [17–19]. There is a clear dissociation between the adult and larval nutritional environments of CoTS and the factors influencing the abundance of adult food (coral) and larval food (microalgae) may be quite different. The importance of exogenous food, acquired through the filter feeding activity of larvae, to complete development is well established for CoTS [14–16,20–22]. The "larval-starvation hypothesis" is predicated on the notion that normally low levels of nutrients in near-reef environments would generally constrain growth and development of CoTS larvae and the release from starvation during periods of nutrient-induced phytoplankton blooms significantly enhances larval survival and recruitment success leading to population outbreaks [20,23]. Fabricius et al. [14] reported dramatic increases in survival and competency of CoTS larvae with every doubling of chlorophyll concentration (a proxy for phytoplankton abundance) above 0.25 $\mu g \cdot L^{-1}$. In addition, Uthicke et al. [22] also demonstrated that algal food concentration has a strong influence on larval development, with temperature as a modulator. Apart from exogenous food availability, the condition of CoTS larvae (at least during early larval stages) will also be partly influenced by the nutritional condition of females during spawning, whereby well-fed females produce larger and faster developing larvae [24]. The question is whether the effects of maternal provisioning are sufficient to offset potential limitations on larval growth and survivorship when larvae are exposed to low levels of food in the environment?

The nutritional condition, and therefore fitness, of adult CoTS is clearly dependent on availability of coral prey [25], and may also vary with differences in the local availability of different types of corals [24]. It is well known that adult CoTS have very specific prey preferences, generally feeding on *Acropora* and *Montipora* corals to the exclusion of all other coral genera when available [3]. Although CoTS will eat virtually all scleractinian corals and can cause comprehensive depletion of corals during severe population outbreaks, there is often serial depletion of different coral genera (e.g., [26]), whereby less preferred coral prey are generally consumed only after other more preferred corals are locally depleted. While it is yet to be effectively shown, strong feeding preferences by CoTS likely reflect the variation in the nutritional content and/or food value of different coral prey [27]. If so, it can be inferred that the nutritional condition of CoTS would be maximized when feeding on generally preferred corals such *Acropora*, while condition is likely to decline after preferred coral prey have been locally depleted, such that feeding is restricted to less preferred corals (e.g., *Porites*). In Guam, Caballes et al. [24] showed that CoTS maintained for 8-weeks on a diet of *Acropora abrotanoides* gained weight and produced bigger oocytes, compared to conspecifics that were maintained on a diet of *Porites rus* or starved. Moreover, differences in the diet and nutritional condition of females had a significant bearing on the quality and quantity of their progeny [24].

Tropical reef waters are typically oligotrophic in the absence of flood or upwelling events [20,28]. Under conditions of scarce exogenous food, echinoid larvae have been known to respond through adaptive changes in shape and by increasing the size of feeding structures to improve the efficiency of food capture [29]. For example, pluteus larvae of sea urchins use ciliated bands on feeding arm rods to capture food, hence developing longer arm rods can improve clearance rates of particulate food [30]. For asteroid larvae, arms do not elongate until the later stages; in lieu of this, asteroids modify larval shape to maintain high clearance rates at food-limiting conditions. Bipinnaria larvae of the sand starfish, *Luidia foliolata*, maintained at high food concentration had pointed anterodorsal and posterodorsal arms, whereas larvae at low food levels had rounded arms, which was associated with high clearance rates [31]. Wolfe et al. [32] documented phenotypic plasticity in 7-day old CoTS larvae in response to a range of algal food concentrations. However, the effect of adult diet and corresponding maternal investment on how larvae respond to variable conditions of exogenous food availability is unknown. This has important implications in understanding how CoTS larvae thrive even in oligotrophic conditions.

The purpose of this study was to evaluate the individual, additive, and interactive effects of endogenous ("Maternal") and exogenous ("Larval") nutrition on larval vitality and morphology. Previous studies [24] have shown that there are significant maternal effects on the quality of CoTS larvae, but it is unknown whether significant levels of endogenous nutrients could offset potential limitations associated with low availability of planktonic food (e.g., [20]). Maternal nutrition could affect larval planktonic duration by affecting larval growth rates, but effects of exogenous food (phytoplankton) could overwhelm maternal effects [18]. This study aims to determine whether the effects of maternal provisioning disappears though compensation or persists throughout development under different conditions of food availability for larvae. The gonad index and size of oocytes from females fed with *Acropora*, mixed diet (*Acropora, Pocillopora, Porites*), *Porites*, and starved females were compared as a proxy for maternal provisioning and oocyte quality. The effects of endogenous and exogenous nutrition on (1) absolute survival; (2) development and competency; and (3) growth and morphology of larvae were specifically addressed in this paper.

2. Materials and Methods

2.1. Collection and Maintenance of Specimens

Adult individuals of the Pacific crown-of-thorns starfish (*Acanthaster* cf. *solaris*) were collected on 26 October 2015 from Eyrie Reef (14.705660° S, 145.379154° E), located 8 km west of Lizard Island in the northern Great Barrier Reef (GBR), Australia. Starfish were transported to the Lizard Island Research Station and placed in 1000-L oval tanks with continuous flow of fresh seawater (27.15 ± 0.97 °C; 35.46 ± 0.07 psu; pH 8.17 ± 0.01). Sex was determined by examining contents drawn from gonads along the arm junction using a syringe with a large-bore biopsy needle [33]. Twelve female starfish were allowed to acclimatize to ambient aquarium conditions for three days, without food prior to being assigned to one of four different feeding treatments (described below). A fresh batch of male starfish was collected from Eyrie Reef on 29 November 2015 and gravid males were placed in 1000-L oval tanks with flow-through seawater and maintained on a mix of *Acropora intermedia, Porites cylindrica*, and *Pocillopora damicornis* corals for three days. Coral fragments used for experimental feeding treatments were collected from within the Lizard Island lagoon (14.697030° S, 145.451410° E) and allowed to acclimatize in plastic aquaria for 24 h (GBRMPA Permit No. G15/38002.1). Coral infauna (e.g., *Trapezia* crabs) were physically removed from all coral fragments so as not to deter feeding by CoTS [34].

2.2. Maternal Feeding Treatments

Twelve intact female starfish of approximately similar size (diameter = 338 ± 8 mm; wet weight = 1633 ± 91 g) were randomly split into four groups and each group of three starfish placed in 300-L plastic aquaria with flow through seawater. All females were nearing reproductive maturity based on microscopic examination of oocytes drawn from starfish using the biopsy procedure described in the previous section. Oogenesis of CoTS from the GBR is usually most active between September and November, while some have been observed to rapidly complete oogenesis within a month (i.e., between November and December) [35]. Starfish were assigned to one of four different "Maternal" feeding treatments (*n* = 3) for 30 days: (i) *Acropora* (fed with *Acropora intermedia*); (ii) Mixed (fed with *Acropora intermedia*, *Porites cylindrica*, and *Pocillopora damicornis*); (iii) *Porites* (fed with *Porites cylindrica*); and (iv) Starved (no food provided, only dead coral skeletons). Supply of coral food for fed treatments was replenished as soon as the piece of coral provided had been completely consumed. Sample size (*n* = 3) was kept low to limit the amount of coral fed *ad libitum* to starfish. Wet weight of each starfish, prior to and 30 days after feeding treatments, was measured. Gonads and pyloric caeca were also weighed after feeding treatments to calculate the gonad index (GSI) and the pyloric caeca index (PCI) for each individual. The average weight of gonads and pyloric caeca from three arms was multiplied with the total number of arms of each starfish to estimate the total gonad or pyloric caeca weight. GSI and PCI were expressed as the ratio of gonad or pyloric caeca weight to the total weight of the starfish [36].

2.3. Spawning Induction and Oocyte Metrics

Gonad lobes were dissected from the twelve females and ovaries were rinsed in 0.2-μm filtered seawater (FSW) to remove loose oocytes. Ovary lobes were treated in 10^{-5} M 1-methyladenine to induce ovulation. Released oocytes were transferred into containers with filtered seawater and wet mounted on glass slides for microscopic examination. Diameters (D_{oocyte}) of the long and short axes of 100 randomly selected mature oocytes (have undergone germinal vesicle breakdown) from each treatment were measured using image analysis of micrometer-scaled photographs of oocytes in Image J [37]. Oocyte volume (V_{oocyte}) was calculated using the formula for an oblate spheroid:

$$V_{oocyte} = 4/3 \times \pi \times a^2 \times b, \tag{1}$$

where *a* is the radius of the major axis (long axis) and *b* is the radius of the minor axis (short axis).

2.4. Fertilization and Larval Rearing

Oocytes from each female were placed in separate 1-L beakers with FSW kept at 28 °C. Approximately 200 oocytes from each female were transferred into triplicate 250-mL beakers using a glass pipette. Spermatozoa were collected from the testes of five males and checked for motility under a microscope. Roughly equal amounts of spermatozoa from each male were combined and counted using a haemocytometer. Oocytes were fertilized with spermatozoa diluted to achieve a spermatozoa-to-oocyte ratio of 100:1.

Fertilized eggs were pooled for each maternal diet treatment group as variation among females had previously been found to be minimal [24]. Zygotes from each group were separately reared in round acrylic containers equipped with stirrers. After 48 h, 100 actively swimming bipinnaria larvae with fully formed mouth, stomach, and anus were siphoned into separate plastic culture bottles with 150 mL FSW. Algal food was prepared from a mixture of sterile cultures of *Dunaliella tertiolecta* at 30% (strain CS-175) and *Chaetoceros muelleri* at 70% (strain CS-176). CoTS have been previously reared to settlement using *D. tertiolecta* [20] and *C. muelleri* (Caballes et al., unpublished manuscript) individually and as a mixture [22]. Sampling of phytoplankton communities in the GBR has shown *Chaetoceros* to be one of the most dominant microalgae taxa during flood events [38], hence we used a higher proportion of *C. muelleri* in this study. Final cell densities were quantified using a haemocytometer. Each group of larvae was assigned to three exogenous ("Larval") nutrition treatments: (i) "High Food" (fed twice daily at 10^4 cells·mL^{-1}) (ii) "Low Food" (fed twice daily at 10^3 cells·mL^{-1}) and (iii) "Starved" (no algal food, 0.2-μm FSW). There were six replicate culture bottles for each of the 12 combinations of endogenous and exogenous nutrition treatments (total of 72 culture bottles). Each culture bottle was connected to an air hose set at one bubble per second to prevent larvae from settling on the bottom. Water changes with fresh FSW were performed daily. Surviving larvae in each culture bottle were counted every four days for 16 days during water changes. At 4 and 10 days after the start of feeding, 10 normally developing larvae from each bottle were immediately photographed using a camera mounted on a microscope. Maximum length, maximum width, posterior width, ciliated band length, and gut area were measured using ImageJ (Figure 1). At day 4, 8, and 16 after the start of feeding, all surviving larvae were categorized into the following developmental stages: **(1) early bipinnaria**—gut fully formed, preoral and anal lobes present, coelomic pouches below or close to mouth; **(2) advanced bipinnaria**—coelomic pouches fuse as axohydrocoel above the mouth, anterodorsal and posterolateral arms start to form; **(3) early brachiolaria**—brachiolar arms start to appear as stump-like projections from the anteroventral surface of the larvae, anterior extension of axohydrocoel, anterodorsal, posterolateral, and posterodorsal arms start to elongate, preoral arms start to form; **(4) mid-late brachiolaria**—brachiolar arms prominent, starfish rudiment developing in the posterior region of larvae, postoral arms form, and other larval arms more elongated; and **(5) abnormal**—stunted and deformed larvae [24,39,40].

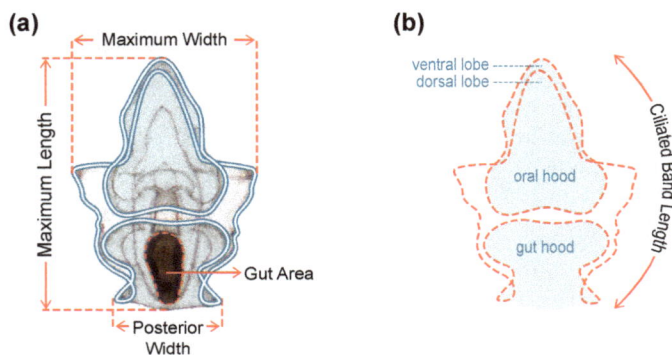

Figure 1. Morphometric measurements of larvae taken four and ten days after commencement of feeding: (**a**) size of morphological features; (**b**) Ciliated band length = sum of the traced perimeter measurements of the oral hood and ventral lobe, gut hood, larval sides, and dorsal lobe (red dashed outline).

2.5. Statistical Analyses

Statistical comparison of oocyte diameter and volume was made using a two-factor mixed model hierarchical analysis of variance (ANOVA) with "Maternal Nutrition" as a fixed effect (four levels) and "Female" (three levels, random) nested within "Maternal Nutrition". No departures from normality and homogeneity of variance were detected for all data. A post hoc Tukey's test was used for pairwise comparisons of fixed factor means. A generalized linear model (GLM) with quasibinomial errors and logit link function was used to analyze the effect of maternal nutrition and larval feeding treatments (fixed categorical predictors) on the proportion of surviving and normally developing larvae and percentage of larvae at the brachiolaria stage (response variables). Treatments with zero variance (e.g., 0% larvae at the mid-late brachiolaria stage across all replicates for treatment with no algal food) were excluded from this analysis. Pairwise post hoc tests were subsequently performed using the Tukey's method in "lsmeans" function in R [41]. The frequency distribution of larvae under different developmental stages was analyzed as a contingency table using log-linear models with log link and Poisson error terms [42] to examine larval progression in relation to "Maternal" and "Larval" nutrition treatments. "Developmental Stage" was considered as a response variable so all models included the interaction between "Maternal" and "Larval" nutrition [43]. Degrees of freedom (*df*) were calculated and deviance statistics (χ^2) were used to compare models in R [41]. Pairwise comparisons were done using G-test of independence with correction for false discovery rate [44]. Data for measurements of morphological traits were analyzed using two-way ANOVA testing for the main and interactive effects of "Maternal" and "Larval" nutrition treatments. Data were log-transformed when assumptions of normality or homogeneity of variance were not met. Significant tests were followed by post hoc Tukey's test for pairwise comparisons within fixed effects. Principal component analysis (PCA),

Diversity *2017*, *9*, 15

implemented using the "vegan" package in R, was used to visualize the effect of "Maternal" and "Larval" nutrition treatments on larval morphology. Morphometric data were log-transformed and the average per replicate culture bottle was used to avoid pseudoreplication. Further morphometric comparisons were performed using permutational multivariate ANOVA (PERMANOVA) on 9999 permutations under a reduced model [45]. Morphological traits were log-transformed and Euclidean distances were used to generate a resemblance matrix in PRIMER v.6 (Primer-E Ltd., Plymouth, UK). Means and standard deviation ($\pm$ SD) were calculated for all data in each treatment. All statistical analyses were performed in R, unless stated otherwise, and p-values (p) below 0.05 were considered statistically significant in all tests.

3. Results

3.1. Maternal and Oocyte Metrics

Initial diameter ($F_{3,8} = 0.19$, $p = 0.9011$) and weight ($F_{3,8} = 0.21$, $p = 0.8898$) of female starfish used under the four maternal nutrition treatments were not significantly different. Weight change after 30 days was also not significantly different between maternal diet treatments ($F_{3,8} = 0.47$, $p = 0.7130$). Maternal diet had a significant effect on pyloric caeca indices (PCI; $F_{3,8} = 4.60$, $p = 0.0374$), mainly due to females under the mixed diet treatment having significantly higher PCI than starved starfish. The gonadosomatic index (GSI) of females given *Acropora* (24.1 $\pm$ 1.7% SD, standard deviation in all instances hereafter) and mixed (24.2 $\pm$ 1.4%) diets were significantly higher than *Porites*-fed (13.2 $\pm$ 1.6%) and starved (11.2 $\pm$ 2.5%) starfish ($F_{3,8} = 14.29$, $p = 0.0014$).

Variation in oocyte diameter (D) and volume (V) was consistent with patterns for GSI between treatments (Figure 2). The diameter and volume of oocytes from starfish placed under *Acropora* (D = 0.22 $\pm$ 0.01 mm; V = 0.0051 $\pm$ 0.0009 mm^3) and mixed (D = 0.22 $\pm$ 0.02 mm; V = 0.0051 $\pm$ 0.0010 mm^3) diet treatments were significantly larger compared to *Porites*-fed (D = 0.19 $\pm$ 0.02 mm; V = 0.0033 $\pm$ 0.0008 mm^3) and starved (D = 0.19 $\pm$ 0.01 mm; V = 0.0031 $\pm$ 0.0006 mm^3) females. Maternal diet treatments accounted for 60% and 63% of the variation in oocyte diameter and volume, respectively (Table S1). There was a significant difference in oocyte size among females within treatments, but this only accounted for 5% of the variation (Table S1).

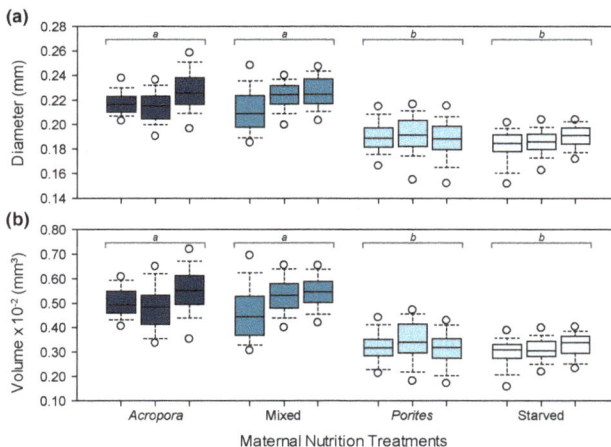

Figure 2. Size of oocytes from female starfish under different maternal nutrition treatments: (**a**) oocyte diameter; (**b**) oocyte volume. Boxplots with different letters above are significantly different.

93

3.2. Larval Survival

Absolute survival of larvae (without taking into account normal development and larval stage) was high (>85%) across all treatments at day 4 (Figure 3) and no significant differences were found between maternal nutrition and between larval diet treatments (GLM, Table 1). At day 8, maternal effects were significant, with a higher proportion (>70%) of surviving larvae from females that were fed (*Acropora*, Mixed, *Porites*) compared to the starved treatment (56.9 ± 13.1%) (Figure 3). At 12 and 16 days after the onset of larval ability to feed, maternal and larval nutrition treatments had a significant additive effect on larval survival. At day 12, survival was >60% for larvae from females on a coral diet, while survival was only 48.4 ± 14.2% for larvae from unfed starfish. Survival was also >60% for larvae provided with exogenous food, while only 52 ± 15.6% of starved larvae survived at day 12. At the end of the experiment (day 16), survival was almost twice as high for larvae from maternally fed treatments (*Acropora* = 50.1 ± 14.6%, Mixed = 49.4 ± 15.7%, *Porites* = 45.3 ± 19.0%) compared to those from the starved treatment (37.7 ± 17.1%). Larvae that were fed with microalgae also had a higher survival rate (High: 55.9 ± 13.1%; Low: 53.8 ± 10.7%) compared to larvae with no food (28.4 ± 10.5%) at day 16 (Figure 3).

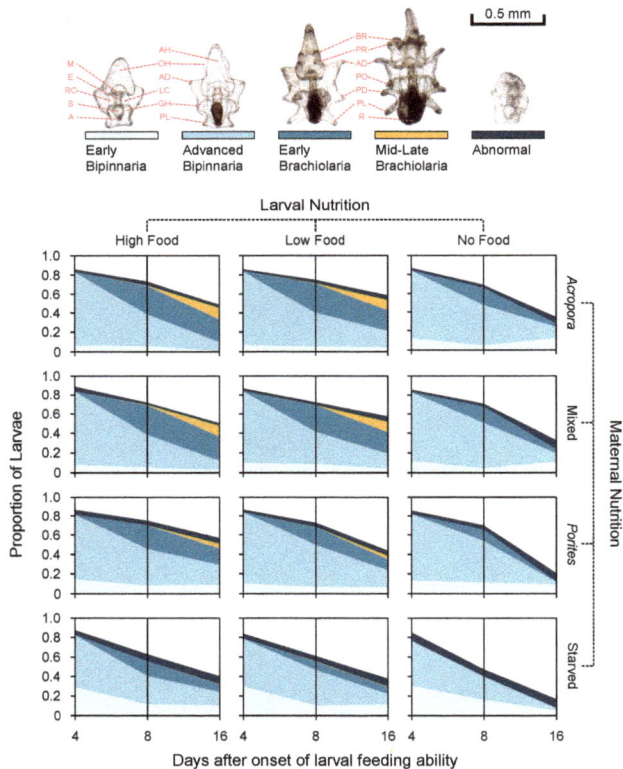

Figure 3. Larval survival and progression of larval development at 4, 8, and 16 days. Morphological traits used in scoring larvae are: mouth (**M**), esophagus (**E**), right coelomic pouch (**RC**), stomach (**S**), anus (**A**), axohydrocoel (**AH**), oral hood (**OH**), anterodorsal arm (**AD**), left coelomic pouch (**LC**), gut hood (**GH**), posterolateral arm (**PL**), brachiolar arm (**BR**), preoral arm (**PR**), anterodorsal arm (**AD**), postoral arm (**PO**), posterodorsal arm (**PD**), posterolateral arm (**PL**), and rudiment (**R**).

Table 1. Analysis of deviance for binomial generalized linear models (GLMs) testing the effects of maternal nutrition and larval feeding treatments on the proportion of surviving larvae at 4, 8, 12, and 16 days after the onset of the ability of larvae to feed. Hereafter, Maternal Diet: **Acr** = *Acropora*, **Mix** = mixed diet, **Por** = *Porites*, **Stv** = starved; Algal Food Concentration (cells·mL^{-1}): **Hi** = 10^4, **Lo** = 10^3, **No** = 0.

SOURCE	df	χ^2	p	Post Hoc
Day 4				
Maternal Nutrition	3	2.58	0.9349	
Larval Nutrition	2	2.12	0.8396	
Maternal Nutrition × Larval Nutrition	6	4.71	0.9927	
Day 8				
Maternal Nutrition	3	140.00	**0.0001**	Acr = Mix = Por > Stv
Larval Nutrition	2	27.92	0.1281	
Maternal Nutrition × Larval Nutrition	6	15.63	0.8901	
Day 12				
Maternal Nutrition	3	98.03	**0.0035**	Acr = Mix = Por > Stv
Larval Nutrition	2	60.24	**0.0152**	Hi = Lo > No
Maternal Nutrition × Larval Nutrition	6	20.05	0.8349	
Day 16				
Maternal Nutrition	3	61.00	**0.0072**	Acr = Mix = Por > Stv
Larval Nutrition	2	435.23	**<0.0001**	Hi = Lo > No
Maternal Nutrition × Larval Nutrition	6	5.81	0.9794	

3.3. Larval Development

At day four after the onset of larval feeding ability, the majority of the larvae across all treatments were at the advanced bipinnaria stage. The distribution of larvae among the different developmental stages was dependent on maternal nutrition (Table 2), with a higher proportion of larvae that were still at early bipinnaria under the starved treatment ($35.7 \pm 18.1\%$) compared to larvae from starfish placed on *Acropora* ($9.9 \pm 4.9\%$), mixed ($11.8 \pm 4.6\%$), and *Porites* ($15.4 \pm 10.5\%$) coral diet (Figure 3). At day 8, larval development was dependent on maternal and larval nutrition treatments (Table 2). The stage of development was bimodal (Figure 3), with the majority of larvae remaining at advanced bipinnaria and the main driver of variation was the proportion of larvae at the brachiolaria stage (Maternal Nutrition: *Acropora* = $36.9 \pm 9.9\%$, Mixed = $35.2 \pm 10.0\%$, *Porites* = $24.1 \pm 10.8\%$, Starved = $13.4 \pm 13.4\%$; Larval Nutrition: High = $35.3 \pm 11.0\%$, Low = $29.9 \pm 13.8\%$, No Food = $16.9 \pm 12.1\%$). At day 16, maternal provisioning and larval diet had a significant additive effect on the developmental progression of larvae (Table 2). The developmental stage frequency distribution of larvae from starfish on *Acropora* and mixed diets were significantly different from *Porites*-fed and starved treatments (Figure 3). Retrogression of larvae from brachiolaria or advanced bipinnaria back to early bipinnaria was evident for larvae from starved females and for starved larvae from *Porites*-fed females (Figure 3).

Results of statistical analyses of normal development and larval competency are summarized in Table S2. At Day 8, maternal diet had a significant effect on the proportion of normally developing larvae (Figure 4a) and the proportion of larvae reaching the brachiolaria stage (Figure 4c). Under maternal treatments that were fed with coral, the proportion of normally developing larvae (*Acropora* = $95.7 \pm 3.2\%$, Mixed = $96.4 \pm 2.7\%$, *Porites* = $94.1 \pm 3.8\%$) was significantly higher compared to those from the starved treatment ($88.9 \pm 11.1\%$). The proportion of larvae from starfish on *Acropora* and mixed diets that reached the brachiolaria stage was 1.5 times higher compared to *Porites*-fed treatments and 2.7 times higher compared to starved treatments. At this point, larval nutrition did not have a significant effect on the proportion of normally developing larvae, but the proportion of larvae reaching the brachiolaria stage was twice as high for treatments provided with algal food compared to starved larvae (Figure 4c). At 16 days after the onset of larval feeding capability, maternal provisioning and larval diet had a significant additive effect on the proportion of larvae that developed normally

(Figure 4b). The proportion of normally developing larvae from CoTS on *Acropora* and mixed diets was 8% higher compared to *Porites*-fed treatments and 18% higher compared to starved treatments. Fed larvae were also 1.4 times more likely to undergo normal development compared to starved larvae (Figure 4b). Maternal condition also had a strong influence on the proportion of larvae that reached the mid-to-late brachiolaria stage after 16 days. Treatments on *Acropora* and mixed diets were 2.3 and 13.7 times more likely to reach competency compared to *Porites*-fed and starved treatments, respectively. None of the unfed larvae reached the mid-to-late brachiolaria stage at 16 days and were excluded in the analysis due to zero variance. There was no significant difference in the proportion of larvae at the mid-to-late brachiolaria stage between high and low algal food treatments.

Table 2. Analysis of deviance for log-linear models testing complete and conditional dependence of larval development on maternal provisioning and larval diet at 4, 8, and 16 days.

Source	df	χ^2	p	Post Hoc
Day 4				
Maternal Nutrition	36	82.80	<0.0001	Acr = Mix = Por $\neq$ Stv
Larval Nutrition	32	7.34	0.9663	
Maternal Nutrition × Larval Nutrition	24	4.94	1.0000	
Day 8				
Maternal Nutrition	36	79.62	<0.0001	Acr = Mix $\neq$ Por = Stv
Larval Nutrition [1]	32	51.55	**0.0157**	
Maternal Nutrition × Larval Nutrition	24	22.05	0.5763	
Day 16				
Maternal Nutrition	36	103.49	<0.0001	Acr = Mix $\neq$ Por = Stv
Larval Nutrition	32	143.85	<0.0001	Hi = Lo $\neq$ No
Maternal Nutrition × Larval Nutrition	24	12.38	0.9753	

[1] Power not sufficient to show significant differences in post hoc pairwise comparisons.

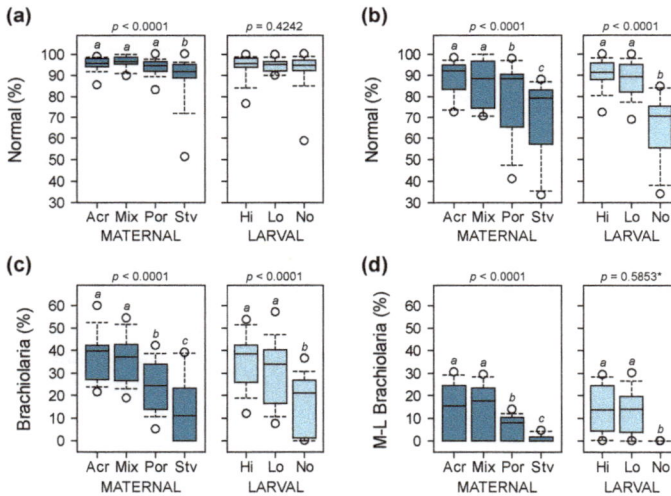

Figure 4. Proportion of normally developing larvae at 8 days (**a**) and 16 days (**b**) after the onset of the ability of larvae to feed, and proportion of larvae at the brachiolaria stage after 8 days (**c**) and larvae at the mid-to-late (M-L) brachiolaria stage after 16 days (**d**). *p*-values are from overall binomial GLMs and different letters are significantly different based on post hoc pairwise comparisons (* *p*-value for comparison between high and low algal food treatments only; starved larvae not included in analysis due to zero variance).

3.4. Larval Growth and Morphometry

Maternal diet had a significant effect on initial size across all morphological traits at four days after the onset of larval feeding ability (Table S3). Larval diet did not influence larval growth at this stage. Variation in maximum larval length, maximum width, posterior width, ciliated band length, and gut area was mainly due to maternal treatments (Figure 5). In particular, variation in larval size was driven by differences in growth rates, which was consistently higher for coral-fed treatments compared to larvae from starved starfish. Among the coral-fed treatments, larvae from starfish on *Acropora* and mixed diets were bigger compared to *Porites*-fed treatments. Patterns of variation in maximum width and posterior width were consistent with differences in ciliated band length and gut area, respectively, suggesting proportional growth of feeding structures and stomach with overall larval size (Figure 5).

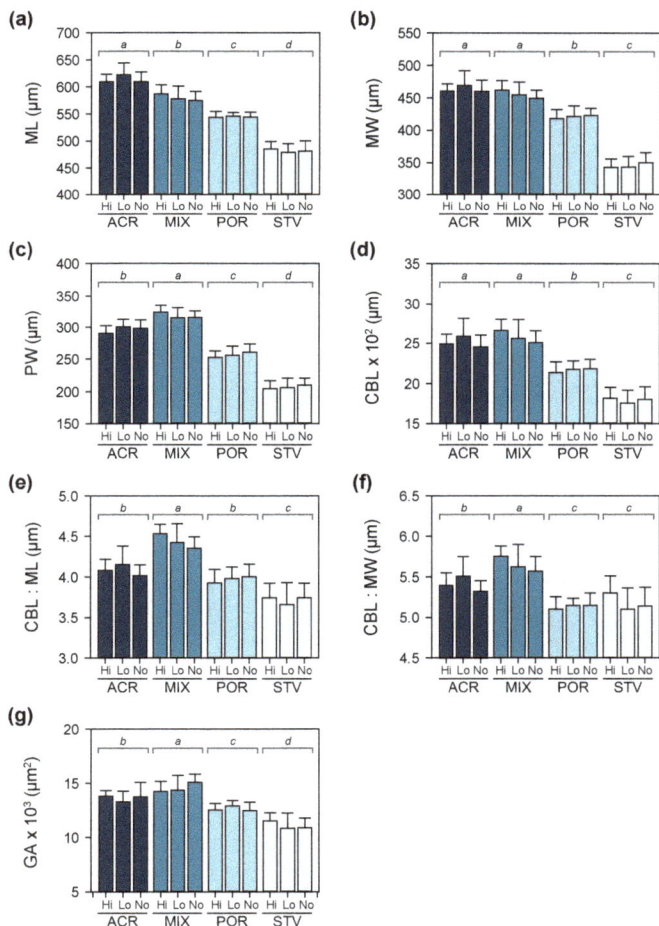

Figure 5. Morphometric measurements of larvae ($\pm$ SD) at Day 4: (**a**) maximum length (**ML**), (**b**) maximum width (**MW**), (**c**) posterior width (**PW**), (**d**) ciliated band length (**CBL**), (**e**) ratio of **CBL** to **ML**, (**f**) ratio of **CBL** to **MW**, and (**g**) gut area (**GA**). Letters above bars denote significant differences as determined by Tukey's post hoc tests following two-way ANOVA.

At 10 days after the onset of larval feeding ability, there was a significant interaction in the effects of maternal and larval diet on larval size (Table S4). Initial differences in larval size due to maternal nutritional condition persisted at this stage, while larval diet also had a significant effect on larval growth, i.e., fed larvae were longer and wider compared to starved larvae (Figure 6a–c). Larvae under low algal food concentration had disproportionately longer ciliated bands (Figure 6d) in relation to maximum length and width (Figure 6e,f). The influence of maternal effects on gut area was reduced, while the effect of larval nutrition was more pronounced at this stage (Figure 6g).

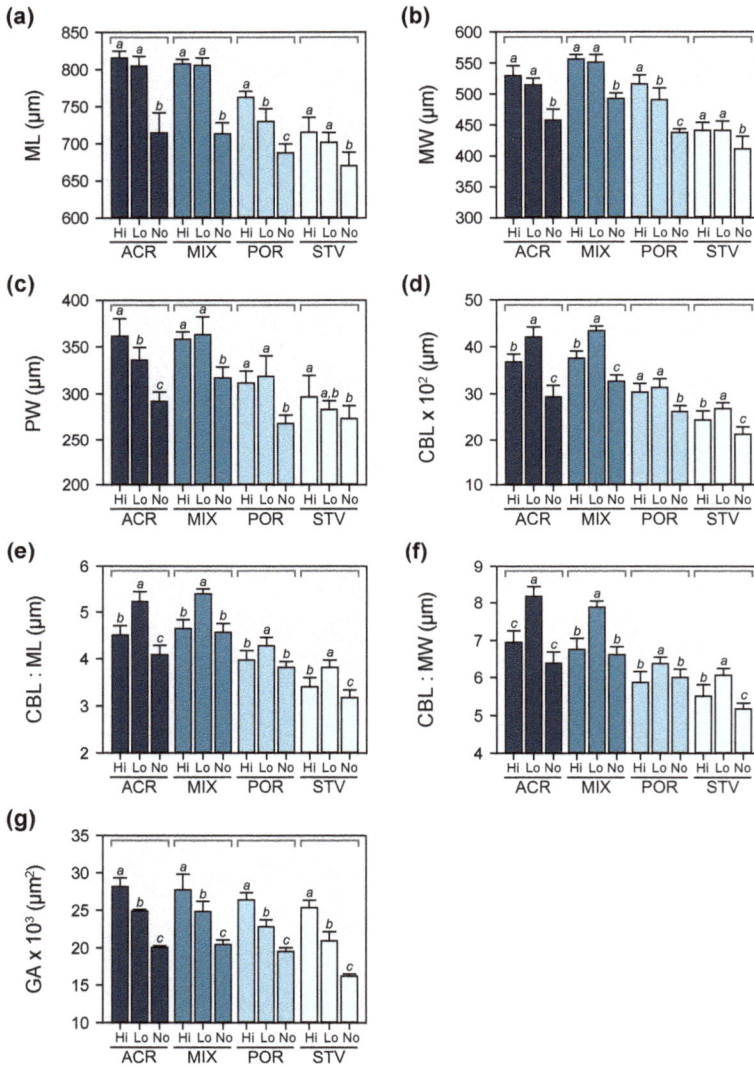

Figure 6. Morphometric measurements of larvae (± SD) at Day 10: (**a**) maximum length (**ML**), (**b**) maximum width (**MW**), (**c**) posterior width (**PW**), (**d**) ciliated band length (**CBL**), (**e**) ratio of **CBL** to **ML**, (**f**) ratio of **CBL** to **MW**, and (**g**) gut area (**GA**). Different letters above bars indicate significant differences based on Tukey's post hoc tests following two-way ANOVA.

Consistent with individual measurements of morphological traits, maternal diet explained 66.2% of the variation in overall larval allometry at four days after the onset of larval feeding ability (Table S5). Larvae from females under *Acropora* and mixed diets were generally longer and wider. At this stage, ciliated band length and gut area coincided with larval growth (i.e., increase in larval length and width), irrespective of changes in larval morphology (Figure 7a).

At day 10, endogenous and exogenous nutrition explained 37.5% and 38.9% of the variation in larval morphology, respectively (Table S5). The effects of these factors were interactive and mostly driven by differences in ciliated band length and gut area (Figure 7b). Ciliated bands were disproportionately longer in relation to maximum length and maximum width, which indicates phenotypic change in response to the concentration of algal food. Gut area also varied with algal food concentration, i.e., larvae under high food concentration treatments had larger stomachs compared to low food and starved treatments.

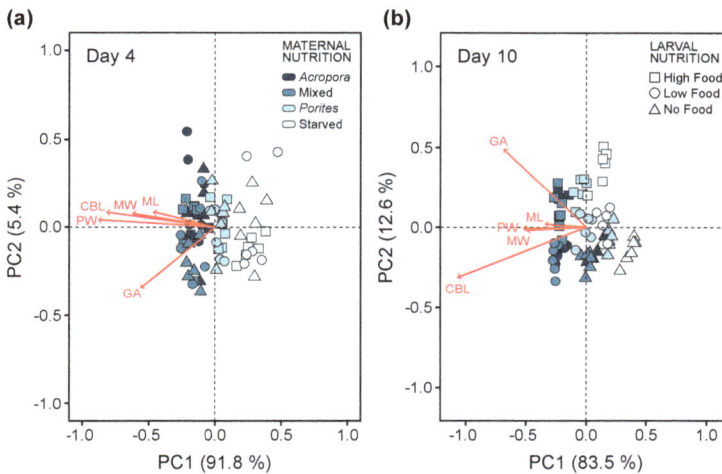

Figure 7. Principal component analysis (PCA) plot of morphological traits measured to analyze similarities in larval morphology at (**a**) four days and (**b**) ten days after the onset of feeding ability. Red vectors are morphometric measurements of maximum length (**ML**), maximum width (**MW**), posterior width (**PW**), ciliated band length (**CBL**), and gut area (**GA**). Maternal nutrition treatments are indicated by different colors and larval feeding treatments are indicated by symbols. Individual data points are average values from 10 replicate larvae per experimental bottle.

4. Discussion

This study shows that the inferred nutritional condition, and thereby the diet, of crown-of-thorns starfish has a significant and lasting effect on larval vitality. Notably, CoTS fed on either an exclusive diet of *Acropora* or a mixed diet in which *Acropora* represented ~33% of available prey produced larger gonads and oocytes, which were correlated with larval growth and morphology, as well as rates of larval survival, development, and competency. The influence of endogenous nutrition was most apparent in the earlier stages of development (i.e., bipinnaria stage) while the significance of exogenous nutrition was manifested in later stages (i.e., advanced bipinnaria to the brachiolaria stage). Our results suggest that the quantity and quality of coral food rations to female starfish differed sufficiently to affect reproductive investment in CoTS, as evidenced by significant variations in oocyte diameter and volume. This is consistent with a previous study by Caballes et al. [24], which showed that CoTS fed *ad libitum* with *Acropora* for two months during peak oogenesis produced significantly larger oocytes compared to oocytes from *Porites*-fed and starved females. Intraspecific variation in oocyte size in many echinoderms is often mediated by differences in food quality and quantity [46]. Within species, larger oocytes are generally associated with specimens collected from field sites with abundant food and with animals under high food treatments in laboratory studies [18,46,47]. Using the gonad index and oocyte size as an index for maternal investment, the present study demonstrates that maternal effects persist through to later stages of development and affect how larvae respond to varying conditions of food availability in the external environment.

Neither endogenous provisioning to the egg or exogenous food availability influenced initial larval survival (all above 85% at day 4), but larvae from fed adults, regardless of diet, generally showed faster growth and development compared to those from starved adults. This suggests that early larval success is significantly influenced by maternal condition. The influence of adult diet and nutritional condition on initial growth and development has been previously shown for CoTS [24] and other echinoderms [19,46,48,49]. After the digestive tract differentiates (usually two days after fertilization), CoTS larvae enter a facultative feeding period where they are capable of feeding, but do not necessarily require food for growth and development due to existing maternal reserves [17]. Starved larvae have been shown to survive for long periods and develop all the way to early brachiolaria [20,24]. The onset of the ability to feed occurs at around the same stage across several species with planktotrophic larvae, but the onset of the need to feed and the duration of the facultative feeding period vary dramatically according to oocyte size and quality [50], and hence maternal diet. In the present study, the influence of maternal effects on survival and development remained strong at eight days after the onset of the ability to feed. This stage is normally the transition phase from bipinnaria to brachiolaria larval form. Continued development at this stage suggests that maternal reserves can buffer the effects of low food availability or starvation. Rate of survival, normal development, and larval progression was higher in larvae from coral-fed adults (*Acropora*, mixed, *Porites* diet) compared to starved treatments. Moreover, the proportion of larvae that progressed to the brachiolaria stage was higher in starfish on *Acropora* and mixed diet compared to *Porites*-fed treatments. Differences in the inferred nutritional condition between adult starfish on *Acropora* and mixed diet versus *Porites*-fed females was most likely influenced by variable consumption rates between coral species. In the present study, estimated coral consumption was higher for *Acropora intermedia* and *Pocillopora damicornis* compared to *Porites cylindrica*. Similarly, when coral food was rationed *ad libitum* to CoTS for 60 days, consumption rates of *Acropora abrotanoides* were significantly higher compared to *Porites rus* [24]. Efficiency in feeding and digestion was significantly higher on acroporid and pocilloporid corals compared to poritids [51]. This is in accordance with CoTS feeding preferences observed in the field and in controlled laboratory assays, where the frequency of predation and predation rates on *Porites* was consistently lower compared to *Acropora* and *Pocillopora* corals (reviewed by Pratchett et al. [3]). General models of optimal diet theory would predict that CoTS would prefer to feed on corals with the highest nutritional value to maximize energetic return [27]. Nutritional analyses of corals showed that the energetic and protein contents of acroporid and pocilloporid corals were marginally higher compared to poritids [51].

Variation in larval survival, development, and competency at the 16 days after onset of feeding ability was mainly driven by exogenous food availability. High mortality rates at day 16 were documented for starved larvae and none of the larvae reached the mid-to-late brachiolaria stage. Maternal effects on initial larval vitality in the earlier stages persisted in the later stages. Even when provided with a high concentration of algal food, very few larvae from starved females progressed to the mid-to-late brachiolaria stage. However, larvae from *Porites*-fed females were able to partially compensate for these initial differences by feeding in the plankton. The facultative feeding period of CoTS larvae appears to exceed 10 days after fertilization, and potentially longer for larvae from well-fed starfish. The assumption that the length of the facultative feeding period is correlated with oocyte size [17] is supported by our results. The oocyte size of CoTS is relatively bigger compared other tropical planktotrophic asteroids [24] and other echinoderms, which may allow storage of surplus nutrients essential for larval growth [19,52]. The success of CoTS in exploiting a lipid-rich food resource (i.e., scleractinian corals [53]) more so than any other reef organism also allows reproductively mature females to allocate energetic and structural resources directly towards the oocyte and indirectly to the juvenile (e.g., in the forcipulate starfish, *Pisaster ochraceous* [47]). Maternal provisioning of surplus nutritional reserves to the oocyte may allow larvae to withstand prolonged periods of starvation [54] and produce larger larvae with structures that improve feeding efficiency in food-limited environments [32].

Larvae from starved females were shorter, narrower, and had smaller stomachs compared to those from fed starfish, while female CoTS on *Acropora* and mixed diets produced bigger larvae with larger gut areas compared to *Porites*-fed treatments. At this stage, ciliated band length was proportional to overall larval size. In a previous study, Caballes et al. [24] reported that four-day old larvae from CoTS fed with *Acropora* or *Porites* were bigger in terms of length, width and stomach area compared to those from starved starfish. Divergence in initial larval size and form was mainly driven by maternal diet treatments. The influence of exogenous nutrients is negligible at this stage (4 days after the onset of the ability of larvae to feed) due to available maternal reserves. While maternal provisions were still present, Byrne et al. [19] did not observe a significant size difference between fed and unfed larvae of the echinoid, *Tripneustus gratilla*.

The onset of phenotypic response of larvae comes in later, influenced by the synergistic effects of endogenous and exogenous nutrition. Initial differences in larval size and development rate influenced by endogenous maternal reserves were carried over to later stages. Continuous supply of high concentrations of algal food for 16 days did not compensate for initial deficiencies of larvae from starved and *Porites*-fed females. Likewise, growth in larvae from starfish on *Acropora* and mixed diets was stunted in the absence of particulate food. Supplemental storage lipids from maternal provisions may be important in allocating resources for larval growth and for building larval feeding structures. Exceptionally high feeding rates on a lipid-rich food source such as hermatypic corals [55–57] uniquely predisposes CoTS to increased maternal reserves. Lipid levels in corals are higher than most marine invertebrates [58]. Maternally derived energetic lipids, particularly triglycerides, fuel early development in echinoderms [52]. Moreover, elevated levels of lipids during egg production in corals [59] coincides with oogenesis in CoTS in the GBR, hence an increase in the amount of lipids in CoTS diet prior to spawning. Although the proportion of triglycerides in *A. intermedia* [60], *P. damicornis* and *P. cylindrica* [57] were almost identical, variable consumption rates on these species could drive differences in maternal provisioning. Byrne et al. [19] suggest that the presence of triglycerides later in development may be a bet-hedging strategy to maintain a buffer against uncertain food supply for larvae. Our results support this proposed strategy, i.e., the degree of allometric elongation of ciliated bands in relation to larval size was more pronounced among larvae from starfish on *Acropora* and mixed diets compared to larvae from starfish under poor maternal nutritive conditions.

Enhancement of feeding capacity is set by the total length of the ciliated band, which requires complementary increases in body size or changes in larval shape to maximize the length of ciliated bands [29,30]. George [31,49] demonstrated that the bipinnaria larvae of asteroids are capable of changing the size of their feeding structures in response to the amount of available algal food. For CoTS larvae, allometric growth of the ciliated band relative to body size can be achieved by increasing the length and width of the larval body coupled with allometric development of bigger oral and gut hoods. Larval CoTS in starved and low food condition had longer ciliated bands relative to body size [32]. Few studies have proposed reliable cues that stimulate these changes in the size of larval feeding structures. Shilling [61] found that echinoderms respond morphologically to organic compounds in the environment that may indicate the availability of dissolved and particulate nutrients. Larvae may also respond morphologically upon detection of chemical and physical cues from algal cells [62].

Given that larval size and shape influence feeding capability, changes in larval morphology will have important functional consequences. In this study, phenotypic plasticity aided by maternal provisions and in response to the environmental nutritive regime may explain the differential success in survival, growth, and development of larvae. Plasticity in larval development has been shown to reduce pelagic larval duration [30], which consequently increases survival by reducing exposure to planktonic predators [63] and by reducing the probability of advective loss from adult habitat [64]. The ability of larvae from well-fed females to modify feeding structures in response to oligotrophic conditions, which is typical for reef waters, may help explain reported outbreaks in locations where the likelihood of elevated phytoplankton levels induced by terrestrial runoff is low [26,65–67].

5. Conclusions

Maternal diet had strong effects on larval survival, development, and growth at the earlier stages. Ciliated band length was proportional to larval growth at this stage. The effect of exogenous diet becomes more pronounced at the later stages, presumably when maternal provisions have been exhausted. Under low algal food conditions, larvae compensate by increasing the length of ciliated feeding bands in relation to larval size, which improves food capture and feeding efficiency. However, the effects of endogenous nutrition persist through to the later stages of larval development, as larvae from starved females did not possess supplemental maternal reserves to develop longer ciliated bands in response to low-food conditions. Resilience of CoTS larvae from starvation and food-limiting conditions is influenced, in part, by the availability of surplus maternal reserves in the earlier stages of development and then later through compensatory morphological plasticity to improve the efficiency of food capture. Although acquisition of particulate food may still be necessary to fuel larval growth for successful metamorphosis, initial advantages or deficiencies in larval survival, growth, and development are carried over in later stages. Phenotypic plasticity influenced by endogenous provisions and in response to exogenous food availability may be an important strategy in boosting the reproductive success of CoTS, leading to population outbreaks.

Supplementary Materials: The following are available online at http://www.mdpi.com/1424-2818/9/1/15/s1, Tables S1–S5. Table S1: Results of mixed model hierarchical ANOVA for diameter and volume of oocytes from female starfish under four maternal diet treatments: Acr = *Acropora*, Mix = mixed diet, Por = *Porites*, Stv = starved. Table S2: Analysis of deviance for binomial generalized linear models (GLMs) testing the effects of maternal nutrition and larval feeding treatments on the proportion of normally developing larvae and larvae at brachiolaria stage after 8 days and normally developing and larvae at mid-to-late brachiolaria after 16 days. Maternal Diet: Acr = *Acropora*, Mix = mixed diet, Por = *Porites*, Stv = starved; Algal Food Concentration (cells·mL^{-1}): Hi = 10^4, Lo = 10^3, No = 0. Table S3: Results of two-way ANOVA testing the main and interactive effects of maternal nutrition and larval feeding treatments taken on different morphometric measurements taken 4 days after the onset of larval feeding. Maternal Diet: Acr = *Acropora*, Mix = mixed diet, Por = *Porites*, Stv = starved; Algal Food Concentration (cells·mL^{-1}): Hi = 10^4, Lo = 10^3, No = 0. Table S4: Results of two-way ANOVA testing the main and interactive effects of maternal nutrition (Acr = *Acropora*, Mix = mixed diet, Por = *Porites*, Stv = starved) and larval feeding (Hi = 10^4, Lo = 10^3, No = 0 cells·mL^{-1}) treatments on different morphometric measurements taken at day 10 after onset of larval feeding ability. Table S5: Results of permutational multivariate ANOVA (PERMANOVA) testing the main and interactive effects of maternal diet and larval feeding treatments on larval morphology. Maternal Diet: Acr = *Acropora*, Mix = mixed diet, Por = *Porites*, Stv = starved; Algal Food Concentration (cells·mL^{-1}): Hi = 10^4, Lo = 10^3, No = 0.

Acknowledgments: Funding for this study was provided by the Ian Potter Foundation 50th Anniversary Commemorative Crown-of-Thorns Starfish Research Grant to M.S.P. and C.F.C., and the Australian Research Council Centre of Excellence for Coral Reef Studies annual research funding to M.S.P. and PhD research allocation to C.F.C. We thank Andrew Hoey, Vanessa Messmer, Arun Oakley-Cogan, Alexia Graba-Landry, and Zara-Louise Cowan for field and laboratory assistance. We also acknowledge Maia Raymundo and Chao-yang Kuo for assistance in preparing the manuscript, Maria Byrne and Symon Dworjanyn for technical advice, and Lyle Vail, Anne Hoggett and the staff at the Australian Museum's Lizard Island Research Station for logistical support. Comments from two anonymous reviewers significantly improved this manuscript.

Author Contributions: C.F.C. and M.S.P. conceived and designed the experiments; C.F.C., M.S.P., A.C.E.B. performed the experiments; C.F.C. and A.C.E.B. analyzed the data; M.S.P. contributed reagents/materials/analysis tools; C.F.C and M.S.P. wrote the paper.

Conflicts of Interest: The authors declare no conflict of interest. The funders had no role in the design of the study; in the collection, analyses, or interpretation of data; in the writing of the manuscript, and in the decision to publish the results.

References

1. Bellwood, D.R.; Hughes, T.P.; Folke, C.; Nyström, M. Confronting the coral reef crisis. *Nature* **2004**, *429*, 827–833. [CrossRef] [PubMed]
2. Hughes, T.P.; Graham, N.A.J.; Jackson, J.B.C.; Mumby, P.J.; Steneck, R.S. Rising to the challenge of sustaining coral reef resilience. *Trends Ecol. Evol.* **2010**, *25*, 633–642. [CrossRef] [PubMed]
3. Pratchett, M.S.; Caballes, C.F.; Rivera-Posada, J.A.; Sweatman, H.P.A. Limits to understanding and managing outbreaks of crown-of-thorns starfish (*Acanthaster* spp.). *Oceanogr. Mar. Biol. Annu. Rev.* **2014**, *52*, 133–200.
4. Westcott, D.A.; Fletcher, C.S.; Babcock, R.C.; Plaganyi-Lloyd, E. *A Strategy to Link Research and Management of Crown-of-Thorns Starfish on the Great Barrier Reef: An Integrated Pest Management Approach*; National Environment Science Programme: Cairns, Australia, 2016.
5. Hoey, J.; Campbell, M.; Hewitt, C.; Gould, B.; Bird, R. *Acanthaster planci* invasions: Applying biosecurity practices to manage a native boom and bust coral pest in Australia. *Manag. Biol. Invasions* **2016**, *7*, 213–220. [CrossRef]
6. Moore, R.J. Persistent and transient populations of the crown-of-thorns starfish, *Acanthaster planci*. *Lect. Notes Biomath.* **1990**, *88*, 236–277.
7. Babcock, R.C.; Milton, D.A.; Pratchett, M.S. Relationships between size and reproductive output in the crown-of-thorns starfish. *Mar. Biol.* **2016**, *163*, 234. [CrossRef]
8. Uthicke, S.; Schaffelke, B.; Byrne, M. A boom–bust phylum? Ecological and evolutionary consequences of density variations in echinoderms. *Ecol. Monogr.* **2009**, *79*, 3–24. [CrossRef]
9. Pratchett, M.S. Dynamics of an outbreak population of *Acanthaster planci* at Lizard Island, northern Great Barrier Reef (1995–1999). *Coral Reefs* **2005**, *24*, 453–462. [CrossRef]
10. Brodie, J.E.; Fabricius, K.E.; De'ath, G.; Okaji, K. Are increased nutrient inputs responsible for more outbreaks of crown-of-thorns starfish? An appraisal of the evidence. *Mar. Pollut. Bull.* **2005**, *51*, 266–278. [CrossRef] [PubMed]
11. Wooldridge, S.A.; Brodie, J.E. Environmental triggers for primary outbreaks of crown-of-thorns starfish on the Great Barrier Reef, Australia. *Mar. Pollut. Bull.* **2015**, *101*, 805–815. [CrossRef] [PubMed]
12. Houk, P.; Raubani, J. *Acanthaster planci* outbreaks in Vanuatu coincide with ocean productivity, furthering trends throughout the Pacific Ocean. *J. Oceanogr.* **2010**, *66*, 435–438. [CrossRef]
13. Houk, P. The transition zone chlorophyll front can trigger *Acanthaster planci* outbreaks in the Pacific Ocean: Historical confirmation. *J. Oceanogr.* **2007**, *63*, 149–154. [CrossRef]
14. Fabricius, K.E.; Okaji, K.; De'ath, G. Three lines of evidence to link outbreaks of the crown-of-thorns seastar *Acanthaster planci* to the release of larval food limitation. *Coral Reefs* **2010**, *29*, 593–605. [CrossRef]
15. Wolfe, K.; Graba-Landry, A.; Dworjanyn, S.A.; Byrne, M. Larval starvation to satiation: Influence of nutrient regime on the success of *Acanthaster planci*. *PLoS ONE* **2015**, *10*, 1–18. [CrossRef] [PubMed]
16. Pratchett, M.S.; Dworjanyn, S.; Mos, B.; Caballes, C.F.; Thompson, C.; Blowes, S. Larval survivorship and settlement of crown-of-thorns starfish (*Acanthaster* cf. *solaris*) at varying algal cell densities. *Diversity* **2017**, *9*, 2. [CrossRef]
17. McEdward, L.R. Reproductive strategies of marine benthic invertebrates revisited: Facultative feeding by planktotrophic larvae. *Am. Nat.* **1997**, *150*, 48–72. [CrossRef] [PubMed]
18. Bertram, D.F.; Strathmann, R.R. Effects of maternal and larval nutrition on growth and form of planktotrophic larve. *Ecology* **1998**, *79*, 315–327. [CrossRef]
19. Byrne, M.; Sewell, M.A.; Prowse, T.A.A. Nutritional ecology of sea urchin larvae: Influence of endogenous and exogenous nutrition on echinopluteal growth and phenotypic plasticity in *Tripneustes gratilla*. *Funct. Ecol.* **2008**, *22*, 643–648. [CrossRef]
20. Lucas, J.S. Quantitative studies of feeding and nutrition during larval development of the coral reef asteroid *Acanthaster planci* (L.). *J. Exp. Mar. Biol. Ecol.* **1982**, *65*, 173–193. [CrossRef]
21. Okaji, K.; Ayukai, T.; Lucas, J.S. Selective feeding by larvae of the crown-of-thorns starfish, *Acanthaster planci* (L.). *Coral Reefs* **1997**, *16*, 47–50. [CrossRef]
22. Uthicke, S.; Logan, M.; Liddy, M.; Francis, D.S.; Hardy, N.; Lamare, M.D. Climate change as an unexpected co-factor promoting coral eating seastar (*Acanthaster planci*) outbreaks. *Sci. Rep.* **2015**, *5*, 8402. [CrossRef] [PubMed]

23. Birkeland, C. Terrestrial runoff as a cause of outbreaks of *Acanthaster planci* (Echinodermata: Asteroidea). *Mar. Biol.* **1982**, *69*, 175–185. [CrossRef]
24. Caballes, C.F.; Pratchett, M.S.; Kerr, A.M.; Rivera-Posada, J.A. The role of maternal nutrition on oocyte size and quality, with respect to early larval development in the coral-eating starfish, *Acanthaster planci*. *PLoS ONE* **2016**, *11*, e0158007. [CrossRef] [PubMed]
25. Lucas, J.S. Growth, maturation and effects of diet in *Acanthaster planci* (L.) (Asteroidea) and hybrids reared in the laboratory. *J. Exp. Mar. Biol. Ecol.* **1984**, *79*, 129–147. [CrossRef]
26. Kayal, M.; Vercelloni, J.; Lison de Loma, T.; Bosserelle, P.; Chancerelle, Y.; Geoffroy, S.; Stievenart, C.; Michonneau, F.; Penin, L.; Planes, S.; et al. Predator crown-of-thorns starfish (*Acanthaster planci*) outbreak, mass mortality of corals, and cascading effects on reef fish and benthic communities. *PLoS ONE* **2012**, *7*, e47363. [CrossRef] [PubMed]
27. Ormond, R.F.G.; Hanscomb, N.J.; Beach, D.H. Food selection and learning in the crown-of-thorns starfish, *Acanthaster planci* (L.). *Mar. Behav. Physiol.* **1976**, *4*, 93–105. [CrossRef]
28. Revelante, N.; Gilmartin, M. Dynamics of phytoplankton in the Great Barrier Reef lagoon. *J. Plankton Res.* **1982**, *4*, 47–76. [CrossRef]
29. McEdward, L.R. Comparative morphometrics of echinoderm larvae. II. Larval size, shape, growth, and the scaling of feeding and metabolism in echinoplutei. *J. Exp. Mar. Biol. Ecol.* **1986**, *96*, 267–286. [CrossRef]
30. Hart, M.W.; Strathmann, R.R. Functional consequences of phenotypic plasticity in echinoid larvae. *Biol. Bull.* **1994**, *186*, 291–299. [CrossRef]
31. George, S.B. Phenotypic plasticity in the larvae of Luidia foliolata (Echinodermata: Asteroidea). In *Echinoderms Through Time*; David, B., Guille, A., Féral, J.-P., Roux, M., Eds.; August Aimé Balkema: Rotterdam, The Netherlands, 1994; p. 20.
32. Wolfe, K.; Graba-Landry, A.; Dworjanyn, S.A.; Byrne, M. Larval phenotypic plasticity in the boom-and-bust crown-of-thorns seastar, *Acanthaster planci*. *Mar. Ecol. Prog. Ser.* **2015**, *539*, 179–189. [CrossRef]
33. Caballes, C.F.; Pratchett, M.S. Reproductive biology and early life history of the crown-of-thorns starfish. In *Echinoderms: Ecology, Habitats and Reproductive Biology*; Whitmore, E., Ed.; Nova Science Publishers, Inc.: New York, NY, USA, 2014; pp. 101–146.
34. Pratchett, M.S. Influence of coral symbionts on feeding preferences of crown-of-thorns starfish *Acanthaster planci* in the western Pacific. *Mar. Ecol. Prog. Ser.* **2001**, *214*, 111–119. [CrossRef]
35. Lucas, J.S. Reproductive and larval biology of *Acanthaster planci* (L.) in Great Barrier Reef Waters. *Micronesica* **1973**, *9*, 197–203.
36. Conand, C. Distribution, reproductive cycle and morphometric relationships of *Acanthaster planci* (Echinodermata: Asteroidea) in New Caledonia, western tropical Pacific. In Proceedings of the 5th International Echinoderm Conference, Galway, Ireland, 24–29 September 1984; pp. 499–506.
37. Schneider, C.A.; Rasband, W.S.; Eliceiri, K.W. NIH Image to ImageJ: 25 years of image analysis. *Nat. Methods* **2012**, *9*, 671–675. [CrossRef] [PubMed]
38. Devlin, M.J.; DeBose, J.L.; Ajani, P.; Teixeira da Silva, E.; Petus, C.; Brodie, J.E. Phytoplankton in the Great Barrier Reef: Microscopy analysis of community structure in high flow events. In *Report to the National Environmental Research Program*; Reef and Rainforest Research Centre Limited: Cairns, Australia, 2013; p. 68.
39. Yamaguchi, M. Early life histories of coral reef asteroids, with special reference to *Acanthaster planci* (L.). In *Biology and Geology of Coral Reefs*; Jones, O.A., Endean, R., Eds.; Academic Press, Inc.: New York, NY, USA, 1973; Volume 2, pp. 369–387.
40. Byrne, M.; Barker, M.F. Embryogenesis and larval development of the asteroid Patiriella regularis viewed by light and scanning electron microscopy. *Biol. Bull.* **1991**, *180*, 332–345. [CrossRef]
41. R Core Team. *R: A Language and Environment for Statistical Computing*; R Foundation for Statistical Computing: Vienna, Austria, 2016; Available online: http://www.R-project.org/ (accessed on 28 February 2017).
42. Agresti, A. *Introduction to Categorical Data Analysis*; John Wiley & Sons Ltd.: New York, NY, USA, 1996.
43. Quinn, G.P.; Keough, M.J. *Experimental Design and Data Analysis for Biologists*; Cambridge University Press: New York, NY, USA, 2002.
44. Benjamini, Y.; Hochberg, Y. Controlling the false discovery rate: A practical and powerful approach to multiple testing. *J. R. Stat. Soc. Ser. B* **1995**, *57*, 289–300.
45. Anderson, M.J. Permutation tests for univariate or multivariate analysis of variance and regression. *Can. J. Fish. Aquat. Sci.* **2001**, *58*, 626–639. [CrossRef]

46. George, S.B. Echinoderm egg and larval quality as a function of adult nutritional state. *Oceanol. Acta* **1996**, *19*, 297–308.

47. George, S.B. Population differences in maternal size and offspring quality for *Leptasterias epichlora* (Brandt) (Echinodermata: Asteroidea). *J. Exp. Mar. Biol. Ecol.* **1994**, *175*, 121–131. [CrossRef]

48. De Jong-Westman, M.; Qian, P.-Y.; March, B.E.; Carefoot, T.H. Artificial diets in sea urchin culture: Effects of dietary protein level and other additives on egg quality, larval morphometrics, and larval survival in the green sea urchin, *Strongylocentrotus droebachiensis*. *Can. J. Zool.* **1995**, *73*, 2080–2090. [CrossRef]

49. George, S.B. Egg quality, larval growth and phenotypic plasticity in a forcipulate seastar. *J. Exp. Mar. Biol. Ecol.* **1999**, *237*, 203–224. [CrossRef]

50. Herrera, J.C.; McWeeney, S.K.; McEdward, L.R. Diversity of energetic strategies among echinoid larvae and the transition from feeding to nonfeeding development. *Oceanol. Acta* **1996**, *19*, 313–321.

51. Keesing, J.K. Feeding Biology of the Crown-of-Thorns Starfish, *Acanthaster planci* (Linnaeus). Ph.D. Thesis, James Cook University, Townsville, Australia, 1990.

52. Prowse, T.A.A.; Sewell, M.A.; Byrne, M. Fuels for development: Evolution of maternal provisioning in asterinid sea stars. *Mar. Biol.* **2008**, *153*, 337–349. [CrossRef]

53. Stimson, J.S. Location, quantity and rate of change in quantity of lipids in tissue of Hawaiian hermatypic corals. *Bull. Mar. Sci.* **1987**, *41*, 889–904.

54. George, S.B.; Cellario, C.; Fenaux, L. Population differences in egg quality of *Arbacia lixula* (Echinodermata: Echinoidea): Proximate composition of eggs and larval development. *J. Exp. Mar. Biol. Ecol.* **1990**, *141*, 107–118. [CrossRef]

55. Patton, J.S.; Battey, J.F.; Rigler, M.W.; Porter, J.W.; Black, C.C.; Burris, J.E. A comparison of the metabolism of bicarbonate ^{14}C and acetate 1-^{14}C and the variability of species lipid compositions in reef corals. *Mar. Biol.* **1983**, *75*, 121–130. [CrossRef]

56. Harland, A.D.; Navarro, J.C.; Spencer Davies, P.; Fixter, L.M. Lipids of some Caribbean and Red Sea corals: total lipid, wax esters, triglycerides and fatty acids. *Mar. Biol.* **1993**, *117*, 113–117. [CrossRef]

57. Yamashiro, H.; Oku, H.; Higa, H.; Chinen, I.; Sakai, K. Composition of lipids, fatty acids and sterols in Okinawan corals. *Comp. Biochem. Physiol. Part B Biochem. Mol. Biol.* **1999**, *122*, 397–407. [CrossRef]

58. Giese, A.C. Lipids in the economy of marine invertebrates. *Physiol. Rev.* **1966**, *46*, 244–298. [PubMed]

59. Arai, I.; Kato, M.; Heyward, A.; Ikeda, Y.; Iizuka, T.; Maruyama, T. Lipid composition of positively buoyant eggs of reef building corals. *Coral Reefs* **1993**, *12*, 71–75. [CrossRef]

60. Imbs, A.B.; Yakovleva, I.M. Dynamics of lipid and fatty acid composition of shallow-water corals under thermal stress: an experimental approach. *Coral Reefs* **2012**, *31*, 41–53. [CrossRef]

61. Shilling, F.M. Morphological and physiological responses of echinoderm larvae to nutritive signals. *Am. Zool.* **1995**, *35*, 399–411. [CrossRef]

62. Miner, B.G. Larval feeding structure plasticity during pre-feeding stages of echinoids: Not all species respond to the same cues. *J. Exp. Mar. Biol. Ecol.* **2007**, *343*, 158–165. [CrossRef]

63. Sinervo, B.; McEdward, L.R. Developmental consequences of an evolutionary change in egg size: An experimental test. *Evolution* **1988**, *42*, 885. [CrossRef]

64. Strathmann, R.R. Length of pelagic period in echinoderms with feeding larvae from the Northeast Pacific. *J. Exp. Mar. Biol. Ecol.* **1978**, *34*, 23–27. [CrossRef]

65. Lane, D.J.W. *Acanthaster planci* impact on coral communities at permanent transect sites on Bruneian reefs, with a regional overview and a critique on outbreak causes. *J. Mar. Biol. Assoc. UK* **2011**, *92*, 803–809. [CrossRef]

66. Roche, R.C.; Pratchett, M.S.; Carr, P.; Turner, J.R.; Wagner, D.; Head, C.; Sheppard, C.R.C. Localized outbreaks of *Acanthaster planci* at an isolated and unpopulated reef atoll in the Chagos Archipelago. *Mar. Biol.* **2015**, *162*, 1695–1704. [CrossRef]

67. Suzuki, G.; Yasuda, N.; Ikehara, K.; Fukuoka, K.; Kameda, T.; Kai, S.; Nagai, S.; Watanabe, A.; Nakamura, T.; Kitazawa, S.; et al. Detection of a high-density brachiolaria-stage larval population of crown-of-thorns sea star (*Acanthaster planci*) in Sekisei Lagoon (Okinawa, Japan). *Diversity* **2016**, *8*, 9. [CrossRef]

diversity

MDPI

Article

The Effects of Salinity and pH on Fertilization, Early Development, and Hatching in the Crown-of-Thorns Seastar

Jonathan D. Allen [1,*], Kharis R. Schrage [1], Shawna A. Foo [2], Sue-Ann Watson [3] and Maria Byrne [2,*]

[1] College of William and Mary, Williamsburg, VA 23187, USA; krschrage@email.wm.edu
[2] School of Medical and Biological Sciences, University of Sydney, Sydney, NSW 2006, Australia; shawna@anatomy.usyd.edu.au
[3] ARC Centre of Excellence for Coral Reef Studies, James Cook University, Townsville, QLD 4811, Australia; sueann.watson@jcu.edu.au
* Correspondence: jdallen@wm.edu (J.D.A.); mbyrne@anatomy.usyd.edu.au (M.B.); Tel.:+1-757-221-7498 (J.D.A.); +61-2-9351-5166 (M.B.)

Academic Editors: Sven Uthicke and Michael Wink
Received: 2 December 2016; Accepted: 10 February 2017; Published: 22 February 2017

Abstract: Understanding the influence of environmental factors on the development and dispersal of crown-of-thorns seastars is critical to predicting when and where outbreaks of these coral-eating seastars will occur. Outbreaks of crown-of-thorns seastars are hypothesized to be driven by terrestrial runoff events that increase nutrients and the phytoplankton food for the larvae. In addition to increasing larval food supply, terrestrial runoff may also reduce salinity in the waters where seastars develop. We investigated the effects of reduced salinity on the fertilization and early development of seastars. We also tested the interactive effects of reduced salinity and reduced pH on the hatching of crown-of-thorns seastars. Overall, we found that reduced salinity has strong negative effects on fertilization and early development, as shown in other echinoderm species. We also found that reduced salinity delays hatching, but that reduced pH, in isolation or in combination with lower salinity, had no detectable effects on this developmental milestone. Models that assess the positive effects of terrestrial runoff on the development of crown-of-thorns seastars should also consider the strong negative effects of lower salinity on early development including lower levels of fertilization, increased frequency of abnormal development, and delayed time to hatching.

Keywords: fertilization; embryonic development; salinity; pH; hatching; crown-of-thorns seastar; *Acanthaster*

1. Introduction

Coral reefs, among the world's most diverse and valuable ecosystems, are under threat from global stressors associated with climate change and local stressors such as overfishing, pollution, and outbreaks of the coral-eating crown-of-thorns seastars, *Acanthaster* cf. *solaris* (COTS) [1,2]. The vulnerability of coral reef ecosystems to global change is seen in the mass bleaching and coral mortality across the tropics caused by the 2016 El Nino-driven ocean warming [3]. Australia's Great Barrier Reef (GBR), and Indo-Pacific reefs in general, are in the midst of a multi-decadal decline in coral cover [4,5]. Outbreaks of COTS cause major damage to coral reefs [5–7], prompting large-scale removal programs of this seastar, albeit with equivocal effectiveness [8]. Analysis of data from the GBR monitoring program estimates that 30–40 percent of the decline in coral cover can be attributed to COTS predation [5,9]. In addition, coral recovery after bleaching and cyclones is greatly reduced if these events are followed by *A.* cf. *solaris* predation [9–11].

It is unclear whether outbreaks of COTS are becoming more common or are simply reported with increasing frequency [8]. However, there is evidence that COTS outbreaks on the GBR occurred historically every 50–80 years, while today they occur approximately every 15 years [12]. Regardless of their frequency, given the far-reaching consequences of COTS outbreaks (e.g., [13]), understanding the factors underlying increases in abundance of this species is critical to predicting when and where loss of coral cover due to COTS will occur. Despite decades of research, the factors behind the boom and bust population cycles of COTS are not understood. The very high fecundity (up to 200 million eggs/female) and resilient larval stage indicate that success in the plankton is a key consideration in determining the causes of recruitment pulses [14–17].

There are two primary hypotheses for the increasing frequency of COTS outbreaks: the predator removal hypothesis (top-down) and the terrestrial runoff (enhanced nutrients) hypothesis (bottom-up). While there is some evidence for both hypotheses (reviewed by [8]), a consensus seems to be emerging that the terrestrial runoff hypothesis best explains outbreak dynamics on the GBR, possibly combined with local hydrodynamic conditions that govern secondary outbreaks ([18], but see [19]). In brief, the terrestrial runoff hypothesis attributes outbreak events to increasing agricultural runoff, which in turn increases the abundance of nutrients in the waters surrounding the GBR. The addition of limiting nutrients enhances the growth of phytoplankton, providing more food for *A.* cf. *solaris* larvae, and thereby increasing larval survival and ultimately recruitment into the reproductive population [20].

In addition to increasing nutrients, freshwater runoff from terrestrial sources simultaneously reduces seawater salinity [20,21]. In large flood events that often occur around the time that COTS larvae are in the plankton, flood plumes can extend up to 100 km offshore in GBR waters causing pulses of low salinity extending to mid-shelf reefs [22–24]. These low salinity events are predicted to increase by recent projections of global climate change [25]. As most echinoderms have limited tolerance for low salinity as adults and larvae [26,27], even short-term exposure to low salinity may be detrimental to COTS development. For example, even small reductions in salinity cause abnormal developmental phenotypes in sand dollars and sea urchins, beginning at fertilization and continuing at least through hatching [25,26]. However, in experiments where larval COTS were transferred to a range of salinity treatments, reduced salinities of 30 ppt actually enhanced survival relative to 32 and 35 ppt treatments [27]. Lucas (1973) also found that development was completed in larval COTS transferred to 26 ppt, but not 22 ppt. Developmental resilience to reduced salinities might be another trait of the life history of COTS contributing to its success during flooding events. However, the tolerance of gametes and early developmental stages (e.g., zygotes, cleavage stage embryos, blastulae) of COTS to lower salinity has not been determined.

To understand the effects of freshwater runoff on COTS reproductive success, we examined the effects of reduced salinity conditions on fertilization, normal development, and hatching in this seastar across a salinity gradient (19 to 34 ppt). Based on previous studies demonstrating the negative effects of low salinity on echinoderm development [26,28,29], we predicted that decreased salinity would lower fertilization success and reduce the percentage of embryos exhibiting normal development. We also predicted that decreased salinity would cause a delay in hatching, as shown in other echinoderms [30,31]. The potential for polyembrony, the phenomenon where low salinity induces fission of early embryos to generate multiple embryos per egg, as described for echinoid embryos [32], was also investigated. Finally, we examined the effect of salinity reductions in combination with shifts in pH to explore how changes in water chemistry more generally could affect development to hatching. We use our data to address the possibility that the resilience of the planktonic phase of *A.* cf. *solaris* to decreased salinity may contribute to its success in runoff conditions.

2. Methods

2.1. Adult Collection and Maintenance

Crown-of-thorns seastars were collected on snorkel in December 2015 as encountered on reefs around Lizard Island (14°40′44.0″ S 145°26′53.7″ E), Northern Great Barrier Reef, Australia. Upon collection, the animals were transported by boat to the Lizard Island Research Station and the gender of each specimen was determined by gonad biopsy with the tissue removed using forceps through a small incision at the base of the arms. The males and females were placed in separate large tanks of ambient flow-through seawater at 28 °C and ~34 ppt salinity. Animals were kept in flow-through tanks and used within a week of collection for experiments.

To obtain gametes for fertilization, a small portion of gonad was removed through an incision. The ovaries were rinsed with 1 μm filtered seawater (FSW) and placed in 10^{-5} M 1-methyl adenine in FSW to induce ovulation. After 30–40 min the eggs were collected and placed in ~100 mL of FSW. Eggs were checked microscopically for quality and to confirm germinal vesicle breakdown. Sperm was collected directly from dissected testes and placed in a small dish at room temperature (~28 °C). Each sperm source was checked microscopically for motility and used promptly. For each fertilization, sperm from a single male was combined to fertilize the eggs of a single female (salinity only experiments) or sperm from two to three males was combined with eggs from two to three females (salinity plus pH experiments) at a sperm to egg ratio of 100:1. Sperm and egg concentrations were estimated using a hemocytometer. Fertilization was checked microscopically and confirmed to be >90% before the eggs were rinsed in FSW to remove excess sperm. For pH experiments, levels of salinity, pH, and DO (dissolved oxygen) were measured using a Hach Hqd portable temperature-compensated multiprobe (Hach Company, Loveland, CO, USA).

2.2. Effects of Salinity on Fertilization, Development, and Hatching

For all experiments where salinity was manipulated, FSW at ambient temperature (~28 °C) and salinity (34 ppt) was mixed with deionized water to create treatment salinities of 19, 23, 25, 27, 29, 31, and 34 ppt. New seawater was mixed each day for a complete water change in experiments, thereby minimizing changes in salinity due to evaporation. To determine the effects of salinity on fertilization success, development, and hatching, embryos from single females were reared in water at a range of salinities. Use of one female and two to three males for each fertilization generated populations of embryos for the salinity experiment. The eggs were pipetted into 250 mL plastic beakers with ~100 mL FSW at salinities of 19, 23, 25, 27, 29, 31, and 34 ppt. After adding eggs to each beaker, a few drops of dilute sperm solution were added to each beaker. After a brief stir, gametes were left for a few minutes before checking for fertilization. A subsample of eggs was photographed using a dissecting microscope to score the number of fertilized eggs as indicated by the presence of a fertilization envelope. All eggs in focus were scored ($n = 39$–97 per picture). Initial trials (two to three per salinity) revealed that development was inhibited at 19, 23, or 25 ppt, and so those salinities were not used in subsequent trials. We then conducted 11 fertilizations and for each of these had one well of embryos for each salinity level. Variable numbers of crosses were examined for measures of fertilization ($n = 3$–11 crosses), normal development ($n = 2$–9 crosses), and hatching ($n = 5$ crosses) for each salinity treatment.

Six-well plates were prepared with four wells in each plate filled with 10 mL of experimental seawater (27, 29, 31, or 34 ppt) and fertilized eggs from respective salinity treatments pipetted into individual wells. Plates were covered and left at ambient temperature (~28 °C). At 14, 16, and 18 h post-fertilization (hpf), a subsample of 30 embryos from each well were examined microscopically to score developmental stage. Scoring categories were as follows: unfertilized (no fertilization envelope), fertilized (one cell), dead, early cleavage (two cells to blastula), abnormal cleavage (blastomeres varying in size and shape), abnormal blastula (irregular shapes with blebbing cells), blastula, gastrula, and hatched. The frequency of normal development (the sum of blastula, gastrula, and hatched categories) was determined in counts of 30 embryos per well. The percentage of normal development was calculated by dividing the number of normally developing embryos by the total number of embryos that were fertilized, in order to avoid confounding the failed development of fertilized eggs with the previously observed failure of eggs to fertilize at low salinities. All cultures were examined closely for the incidence of polyembryony.

2.3. Effects of Salinity and pH on Development to Hatching

The salinity-pH experiment used four salinities (27, 29, 31, 34 ppt) and two pH_{NIST} levels (Mean $\pm$ SE, control 8.07 $\pm$ 0.02, and 7.61 $\pm$ 0.01, n = 12). The pH treatments were within model projections for near-future (2300) conditions [25]. Unmanipulated FSW served as the control. The water was first conditioned to achieve the salinity levels as above and then the pH was adjusted. To achieve experimental pH levels, FSW was bubbled with 100% CO_2 and pH adjustment was tracked using a Hach Hqd portable multiprobe. Probes were calibrated using NIST high precision buffers pH 4.0, 7.0, and 10.0 (ProSciTech, Kirwan, Queensland, Australia). To ground truth pH_{NIST} values, pH on the total scale was checked for adjusted FSW samples across all salinities. The spectrophotometric approach was used with *m*-cresol purple indicator dye (Acros Organics lot AO321770) and a USB4000 spectrophotometer following the procedures outlined in Standard Operating Procedures (SOP) 6b of [33] and the equations of [34]. All values fell in the expected range, confirming the accuracy of the pH_{NIST} values. Water samples (250 mL) collected for each pH level and fixed with 100 µL of saturated HgCl were used to determine total alkalinity (TA) by potentiometric titration. Experimental pCO_2 was determined from pH_{NIST}, TA, temperature, and salinity data using CO2SYS [35] (Table 1) applying the dissociation constants of [36] as refitted by [37].

Table 1. Experimental pH conditions in experiments with *A.* cf. *solaris*. Mean values for pH_{NIST} ($\pm$SE, n = 3) for each salinity across experimental runs is presented, as well as the overall mean pH_{NIST} (n = 12). pCO_2 was calculated in CO2SYS using data on total alkalinity (TA, n = 2–4 per salinity), salinity, and pH_{NIST}.

	pH 8.1				pH 7.6			
	27	29	31	34	27	29	31	34
	8.08 $\pm$ 0.02	8.07 $\pm$ 0.02	8.07 $\pm$ 0.02	8.07 $\pm$ 0.01	7.61 $\pm$ 0.01	7.61 $\pm$ 0.01	7.61 $\pm$ 0.01	7.61 $\pm$ 0.02
pH_{NIST}	8.09 $\pm$ 0.01				7.61 $\pm$ 0.00			
TA (µmol/kg)	1968.3$\pm$12.0	2026.9$\pm$7.1	2166.4$\pm$12.9	2309.4$\pm$2.0	1968.3$\pm$12.0	2026.9$\pm$7.1	2166.4$\pm$12.9	2309.4$\pm$2.0
pCO_2 (ppm)	492.4	493.1	514.3	529.9	1682.1	1696.2	1779.6	1851.0

Using gametes from multiple males and females (n = 2–3 each), males for each fertilization generated populations of embryos for the salinity-pH experiment. Thus, each experimental container was considered to be a replicate. Combined eggs of three females were placed in plastic beakers containing 50 mL of experimental salinity-pH seawater at 460 eggs/mL, as determined in egg counts, and fertilized with dilute sperm to achieve ~90% fertilization. Approximately 1 mL of these eggs was then pipetted into full 40–80 mL containers (~12 eggs/mL) of the same experimental water conditions and sealed. They were left in water baths maintained at ambient temperature (~28 °C).

At 14 and 24 hpf, 30 embryos were pipetted from each container and scored as dead, abnormal blastula, blastula, gastrula, or hatched (as above). They were sampled in order of replicate (all replicate one pots, then two, then three), so that time was not a confounding variable. This experiment was repeated three times with different gamete sources.

2.4. Statistical Analysis

Data describing the effects of salinity on fertilization success were analysed using a nonparametric Kruskal-Wallis test since the distribution of residuals resulting from ANOVA tests were non-normal, even following standard (arc-sine square root) transformations. The percentage of embryos that exhibited normal development at 14 h post-fertilization under different salinity conditions were normally distributed and were analysed using a mixed-model ANOVA where female was modelled as a random factor and salinity was modelled as a fixed factor. Data describing the effects of time and salinity on the percentage of embryos exhibiting normal development were non-normally distributed, preventing analysis in an ANOVA framework. Instead, we used a binomial logistic regression to analyse these data. A Hosmer-Lemeshow test was used to analyse the fit of our binomial regression model to the data, although the Hosmer-Lemeshow test is known to yield significant departures from a perfect fit when observations exceed a few hundred (our $N = 2140$) [38].

For the data on the percentage of hatching in salinity and pH treatments, a mixed-model ANOVA was used with data at two different time points (14 and 24 hpf). In this case, we modelled pH, salinity, time, and their interactions as fixed effects and block (different runs of the experiment) were modelled as a random effect. Each block was conducted on independent days using unique and non-overlapping combinations of male and female gametes. In all cases, normality of residuals was assessed using a Kolmogorov-Smirnov test with an alpha level of 0.05 and in all cases the test yielded a $p \geq 0.200$, meeting this basic assumption of ANOVA. In cases where post-hoc tests were used to assess differences among levels of a significant main effect, we used the Bonferroni adjustment to correct p-values for multiple pairwise comparisons [39]. The one exception to this was analysis of pairwise comparisons following the Kruskal-Wallis nonparametric test where we used Dunn's adjustment [40] to correct for family-wise type I error. All analyses were conducted using IBM SPSS Statistics (version 23, Armonk, NY, USA).

3. Results

3.1. Effects of Salinity on Fertilization, Early Development, and Hatching Time of Crown-of-Thorns Seastars

Salinity had a significant effect on the percentage of eggs that were fertilized (Figure 1), as determined by a Kruskal-Wallis test ($H = 14.636$, d.f. = 5, $p = 0.021$). Multiple pairwise comparisons across all salinity levels revealed that fertilization at 23 ppt was significantly lower than fertilization success at 34 ppt. All other pairwise comparisons were not significantly different from one another after the p-values were corrected for multiple comparisons. While not statistically significant, the decrease in fertilization at 25 ppt in some crosses indicates that this salinity level may approach the threshold for fertilization success.

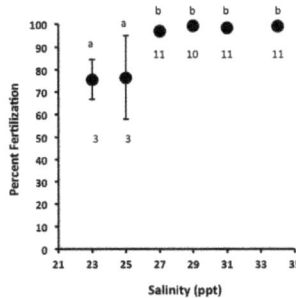

Figure 1. Fertilization percentage of *A.* cf. *solaris* eggs across a range of salinities (ppt). Circles represent the mean (±SE) percent fertilization of 3–11 independent, replicate crosses for each salinity treatment. Numbers beneath each data point indicate the number of replicate females for that point. Letters above each data point indicate significant differences in percent fertilization among salinities based on a Kruskal-Wallis H test and post-hoc tests (see text for details). For symbols where error bars are not visible, the error bars are contained within that symbol.

Salinity had a significant effect on the proportion of embryos that exhibited normal development. Using a mixed-model ANOVA, we found that the proportion of embryos exhibiting normal development was significantly reduced when salinity was reduced ($F_{1,4} = 9.989$; $p < 0.001$; Figure 2). Post-hoc tests with Bonferroni corrections revealed that the frequency of normal development at 25 ppt was significantly lower than at all other salinities ($p < 0.001$) except for 27 ppt ($p = 0.070$). The frequency of normal development at 27 ppt was also lower than the frequency at 29 ppt ($p = 0.028$) and 31 ppt ($p = 0.039$), but was not different from the frequency at 34 ppt ($p = 0.110$; Figure 2). The decrease in the frequency of normal development at 27 ppt indicates that this salinity level approximates the embryo tolerance levels for reduced salinity.

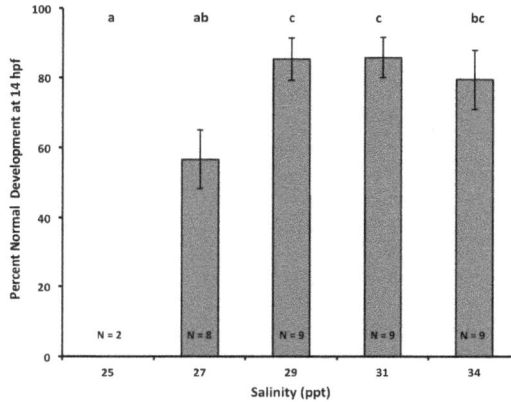

Figure 2. The percentage of embryos exhibiting normal development 14 h post fertilization (hpf) across a range of salinities (ppt). Bars represent the means ($\pm$SE) for $N = 2$–9 independent, replicate crosses. The number of crosses tested for a given salinity is recorded within each bar. At low salinity (25) only two crosses were conducted as no normal development was seen in either of the first two replicates. Letters above each bar denote statistically significant differences in the percentage of embryos completing normal development based on a mixed model ANOVA and post-hoc tests (see text for details).

We found that salinity, time, and male/female cross all had significant effects on the percentage of embryos that hatched between 12 and 16 hpf (Figure 3). We initially tested for the main effects of salinity, time, and male/female cross using a full model with all two-way and three-way interactions included, but found that none of the interactions were significant and all exceeded a threshold p-value of 0.250 for removal from the model to create a reduced model focused on testing our main effects of interest [41]. The results of the reduced model are shown in Table 2 and show that all three main effects were highly significant predictors of hatching status in our experiments. We tested whether the fit of our model departed significantly from the data using a Hosmer-Lemeshow test, and found that it did ($p < 0.002$); however, our model correctly predicted hatching status in 82.3% of cases.

Table 2. Binomial logistic regression of hatching probability in *A.* cf. *solaris*. A full model found that all possible interactions yielded $p > 0.250$, and so the results of a reduced model using only main effects are presented. Significant effects are in bold.

Variable	β	*p*-Value	Exp(β)
Salinity	0.295	**>0.001**	1.343
Male/Female Cross	−0.146	**>0.001**	0.864
Time	1.025	**>0.001**	2.788

Plotting the mean success of fertilization and normal development by female shows different effects of salinity treatments across the male/female crosses (Figure 4). At fertilization, performance across all 11 females was reduced at all experimental salinity levels. At 14 hpf, however, it can be seen that developmental success was dependent on female identity where females 7 and 8 showed enhanced development with respect to the control salinity (Figure 4B).

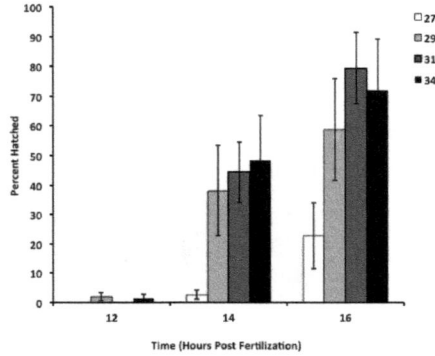

Figure 3. The relationship between percent hatching, salinity, and time in *A.* cf. *solaris*. Percent hatching increases with both time and salinity and varied significantly with male/female cross (see Table 1). Each bar represents the mean (±SE) for each of five replicate crosses, measured at three time points (12, 14, and 16 h post-fertilization).

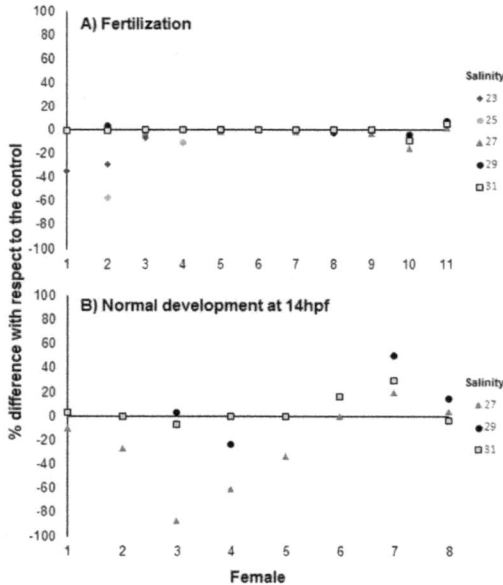

Figure 4. The difference in fertilization and early developmental success with respect to the control treatment (salinity of 34 ppt) grouped by female across experimental salinity treatments. Mean success per female is displayed for the different salinity levels across fertilization (**A**) and normal development 14 h post fertilization (**B**). Symbols above the line display higher success than the control, while success was lower than the control for those symbols below the line.

3.2. Effects of Salinity and pH on Early Development of Crown of Thorns Seastars

Salinity and time, but not pH, had a significant effect on the percentage of hatched embryos (Figure 5). We tested for the main fixed effects of salinity, pH, time, and for all possible interactions of these variables, along with the random effect of block. After running the full model described above, we removed all interactions where $p > 0.250$ and re-ran the reduced model presented in Table 3. The main effects of salinity and time both significantly affected the percent of embryos hatched ($p < 0.001$), but pH did not (Table 3). The random effect of block was also a significant factor affecting the percent of embryos hatched ($F_{2,136} = 5.550$; $p = 0.005$). For the main effect of salinity, 27 ppt yielded significantly lower percentages of hatching ($p < 0.004$) from all other levels, while 29 ppt was significantly lower ($p < 0.005$) than 31 ppt and the highest salinities of 31 and 34 ppt were not different from one another ($p > 0.9$).

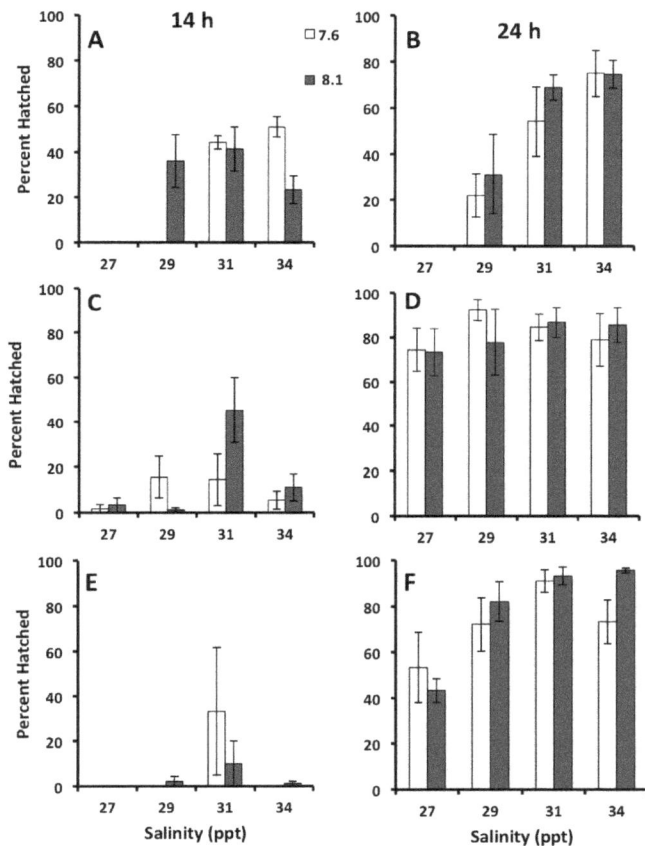

Figure 5. The effect of salinity (ppt) and pH on the percent of hatched *A.* cf. *solaris* embryos at 14 and 24 h post fertilization. Each bar represents the mean (±SE) of three replicate blocks. Panels (**A,B**) represent the first block. Panels (**C,D**) represent the second block and panels (**E,F**) represent the third block. The colour of the bars represents the pH treatments. Dark grey = pH 8.1 and white = pH 7.6.

Table 3. ANOVA for the effects of salinity, pH, time, block (male/female cross), and their interactions on the percent of embryos hatched. Data were arcsine-square root transformed prior to analysis. Significant effects ($p < 0.05$) are in bold.

Dependent Variable	Fixed Effects	df	F	p
Percent embryos hatched	Salinity	3136	18.244	**<0.001**
	Time	1136	196.418	**<0.001**
	pH	2136	1.094	0.298

4. Discussion

We found that salinity has a strong negative effect on the early development of COTS, beginning with fertilization and continuing through to hatching of embryos from the fertilization envelope. Fertilization was relatively resilient to reductions in salinity, remaining above 90% until salinity dropped to 25 ppt or below. However, by 14 hpf embryos began to exhibit sensitivity to salinity, as seen in the presence of high numbers of abnormal phenotypes at salinities <27 ppt. This pattern of increasing sensitivity to lower salinities continued into hatching. We detected negative effects of reduced salinity on the proportion of embryos hatching at <29 ppt, relative to embryos at 31 and 34 ppt. Overall, this suggests that salinity is an important factor to include in models of reproductive success in COTS, especially as it is likely to covary with other environmental conditions such as nutrient runoff and subsequent phytoplankton blooms that are currently associated with COTS outbreaks on the GBR [20].

As seen here for COTS development, echinoderms generally exhibit a narrow tolerance for decreased salinity during both planktonic and benthic life stages, although there is considerable variation among species [26,27]. Previous studies of the impact of lower salinity on fertilization (where gamete union was conducted in treatment water) in asteroids indicate that 24–26 ppt approximates the salinity when the percentage of fertilization decreases, and this is also influenced by temperature (e.g., 23% at 5 °C, 69% at 10 °C in *Asterias amurensis*) [42,43]. For COTS at 23–25 ppt, fertilization was fairly high (~75%), although it does appear that this level of decreased salinity approximates a threshold tolerance level. Similar results were obtained for fertilization in *Acanthaster* cf. *solaris* from Guam, where fertilization drops below 50% between 24 and 22 ppt [44]. By comparison, it appears that echinoid fertilization is more sensitive to decreased salinity, with significant deleterious effects at salinities below 28–29 ppt for sea urchins [45–47] and at 25 ppt for a sand dollar [28]. For *Lytechinus variegatus*, fertilization was reduced to ~10% at 25 ppt [42].

The percentage of normal development for COTS was significantly lower at 27 ppt and it appears that this level of reduced salinity may be a tipping point for deleterious effects. Similarly, increased embryo abnormality was observed in the asteroids *Pisaster ochraceus* and *Asterias amurensis* at ~25–26 ppt [42–45] and in the echinoid *Evechinus chloroticus* at 29 ppt [44]. While reduced salinity has been reported to induce polyembryony in echinoid species [32], the phenomenon of polyembryony was not observed in COTS under any experimental treatment. There was variation in the performance, with the progeny of some females showing enhanced development with respect to the control salinity. As seen in studies of other stressors, the variable performance of progeny with respect to parental source indicates that there may be standing genetic variation in tolerance to salinity and potential for adaptation [48].

In general, decreased salinity is well known to delay or retard development in echinoderms [26,28,29,45,47,49–52]. For COTS, hatching was significantly delayed in embryos reared from fertilization at 27 ppt. Similarly, for *Lytechinus variegatus* time to hatching time was delayed by ~5 days at 28 ppt (six days vs. one day) [45]. Although we did not rear our cultures to the larval stage, delayed hatching in COTS due to low salinity is likely to be associated with larval abnormality, as shown for *Pisaster ochraceus* and *Echinaster* sp. [49,53]. Despite our finding of delayed hatching of COTS, no signs of changes in stage at hatching were observed, as has been found in some echinoids [29].

Regardless, any delay of hatching would be expected to increase the time spent in the risky planktonic phase [54–56], as it delays the onset of larval feeding. However, the negative developmental effects of salinity reductions may be offset by positive developmental effects associated with runoff events, especially phytoplankton blooms, that are known to covary with outbreaks of COTS [20].

Our results suggest that decreased salinity has a greater deleterious effect on hatching than does decreased pH, as also found for *Acanthaster* cf. *solaris* in Guam [44]. Thus, it appears that the hatching stage is relatively resilient to lower pH while other time points in development are more vulnerable in both earlier (e.g., gametes) and later (e.g., larvae) life stages of COTS, creating potential life history bottlenecks [44,57–59], while there are potential benefits of near future ocean acidification for the early juvenile [60].

The degree to which COTS eggs, embryos, and larvae encounter lower salinity water in nature remains an open question. The eggs and sperm of COTS are neutrally buoyant (sperm) or nearly so (eggs), and thus their dispersion is largely dependent on hydrodynamic conditions at the time of spawning [61]. The vertical distribution of COTS larvae is unknown, but larvae are widely distributed horizontally on the GBR, and recently have been shown to be dispersed across most of the GBR during the spawning season, even at sites distant from known outbreaks [62]. This wide geographic coverage suggests that at least some embryos and larvae are likely to encounter low salinity waters from either river plumes or intense rain events, both of which have been demonstrated to reduce salinity over the GBR during the wet season [23,63,64]. Wet season flood events can create flood plumes that extend up to 100 km offshore in GBR waters, causing pulses of low salinity onto mid-shelf reefs [21–23]. Salinity exposure in the upper 50 m of the water column calculated using a hydrodynamic model (eReefs http://www.bom.gov.au/environment/activities/coastal-info.shtml) during the 2010–2016 wet seasons showed salinity minima of 27 and 29 extended up to ~65 and 80 km, respectively, off the north Queensland coast onto some mid-shelf reefs. In addition, freshwater impacts on the GBR have become more frequent since the time of European settlement, with high flow events occurring in one out of every six years since 1948 [24]. This leaves open the possibility that low salinity events on the GBR may continue to increase in frequency and/or intensity.

While we show that the embryonic stages of COTS are sensitive to low salinity, Lucas (1973) reported that larvae are less so [27]. Larvae transferred to 26 ppt, a salinity level deleterious to embryos, were capable of completing development. This indicates that COTS larvae that encounter plumes of lower salinity water in nature might be robust. Similarly, the bipinnaria larvae of *Asterias amurensis* tolerated being transferred from ambient salinity seawater to 20–32 ppt [42]. After initial osmotic shock these larvae were able to restore swimming activity, although they swam more slowly at decreased salinity. Larval tolerance to transfer to low salinity increased with larval age [42]. This indicates asteroid larvae, perhaps especially those of resilient boom and bust species such as *Asterias amurensis* and *Acanthaster* cf. *solaris* [14], can tolerate low salinity perturbations. An extreme example of salinity tolerance among asteroids is found in populations of the seastar *Asteria rubens* in brackish waters of the Baltic Sea, where it occurs at salinities from 15 to 35 ppt [65]. In some populations of this species, survival of fertilized eggs to embryos is actually highest at 24 ppt [65] and development to metamorphosis can be completed at salinities as low as 15 ppt [66].

Across all asteroids, even the extreme cases described above, there is a limit to low salinity tolerance in larvae that is likely to differ not only taxonomically, but also with respect to the duration of the perturbation and across larval ages. *Pisaster ochraceus* gastrulae and larvae exposed to low salinity (20 ppt) for 20 days, as occurs during precipitation events, developed into shorter and wider larvae, while those exposed to shorter pulses of lower salinity (three days) developed into longer and more slender larvae [67]. This salinity-induced morphological change in seastar larvae likely has consequences for swimming and feeding, as both of these functions are strongly influenced by larval size and shape [68]. The ability of asteroid larvae to change shape with respect to salinity treatments may indicate an ability to phenotypically adjust their body profile to maximize feeding and swimming efficiency with regard to salinity conditions, albeit with a lower limit [67]. For COTS and other asteroid and echinoid larvae, phenotypic plasticity with respect to their food environment is a key mechanism of resilience [15,69,70]. That larvae may also be able to adjust their phenotype in response to the salinity environment warrants further investigation. In addition, asteroid and echinoid larvae exhibit avoidance behaviour swimming away from, or not swimming into, low salinity water [71,72], and this ability has recently been shown to be affected by prior exposure to low salinity during early development [73]. This behavioural plasticity may increase larval survival and, given the plasticity of echinoderm larval growth, may also be associated with differing body profiles. Empirical data on the influence of the timing of exposure (with respect to larval stage/age) on low salinity tolerance of COTS larvae, the life stage most likely to encounter pulses of low salinity conditions during flooding periods, and the potential for salinity-induced phenotypic plasticity is needed to more fully understand the resilience of this species with respect to the influence of freshwater incursions.

Acknowledgments: The research was supported by an Ian Potter Foundation Grant from Lizard Island Research Station, a facility of the Australian Museum, a PhD scholarship from the University of Sydney (SF), a Yulgilbar Foundation Fellowship (S.-A.W.), a Reves Center International Grant and a Mary E. Ferguson Research Grant from William and Mary (K.R.S.). Thanks to Symon Dworjanyn, Kennedy Wolfe, Mailie Galle, and Juselyn Tupik for assistance. We gratefully acknowledge the assistance of the team of Morgan Pratchett in obtaining specimens, particularly Ciemon Caballes, Morgan Pratchett, and Vanessa Messmer.

Author Contributions: J.D.A., M.B., K.R.S., S.A.F., and S.-A.W. conceived and designed the experiments; K.R.S., S.A.F., M.B., and S.-A.W. performed the experiments; J.D.A. and S.A.F. analysed the data; M.B. and S.-A.W. contributed reagents/materials/analysis tools; J.D.A., M.B., K.R.S., S.A.F., and S.-A.W. wrote the paper.

Conflicts of Interest: The authors declare no conflict of interest.

References

1. Anthony, K.R.N. Coral reefs under climate change and ocean acidification: Challenges and opportunities for management and policy. *Ann. Rev. Environ. Resour.* **2016**, *41*, 59–81. [CrossRef]
2. Lough, J.M. Coral reefs: Turning back time. *Nature* **2016**, *531*, 314–315. [CrossRef] [PubMed]
3. Normile, D. El Niño's warmth devastating reefs worldwide. *Science* **2016**, *352*, 15–16. [CrossRef] [PubMed]
4. Bruno, J.F.; Selig, E.R. Regional Decline of Coral Cover in the Indo-Pacific: Timing, Extent, and Subregional Comparisons. *PLoS ONE* **2007**, *2*, e711. [CrossRef] [PubMed]
5. De'ath, G.; Fabricius, K.E.; Sweatman, H.; Puotinen, M. The 27—Year decline of coral cover on the Great Barrier Reef and its causes. *Proc. Nat. Acad. Sci. USA* **2012**, *109*, 17995–17999. [CrossRef] [PubMed]
6. Baird, A.H.; Pratchett, M.S.; Hoey, A.S.; Herdiana, Y.; Campbell, S.J. *Acanthaster planci* is a major cause of coral mortality in Indonesia. *Coral Reefs* **2013**, *32*, 803–812. [CrossRef]
7. Plass-Johnson, J.G.; Schwieder, H.; Heiden, J.; Weiand, L.; Wild, C.; Jompa, J.; Ferse, S.C.A.; Teichberg, M. A recent outbreak of crown-of-thorns starfish (*Acanthaster planci*) in the Spermonde Archipelago, Indonesia. *Reg. Environ. Chang.* **2015**, *15*, 1157–1162. [CrossRef]
8. Pratchett, M.S.; Caballes, C.F.; Rivera-Posada, J.A.; Sweatman, H.P.A. Limits to Understanding and Managing Outbreaks of Crown-of-Thorns Starfish (*Acanthaster* spp.). *Oceanogr. Mar. Biol. Annu. Rev.* **2014**, *52*, 133–200.
9. Osborne, K.; Dolman, A.M.; Burgess, S.C.; Kerryn, J.A. Disturbance and dynamics of coral cover on the Great Barrier Reef (1995–2009). *PLoS ONE* **2011**, *6*, e17516. [CrossRef] [PubMed]
10. Lourey, M.J.; Ryan, D.A.J.; Miller, I.R. Rates of decline and recovery of coral cover on reefs impacted by, recovering from and unaffected by crown-of-thorns starfish *Acanthaster planci*: A regional perspective of the Great Barrier Reef. *Mar. Ecol. Prog. Ser.* **2000**, *196*, 179–186. [CrossRef]
11. Halford, A.; Cheal, A.J.; Ryan, D.; Williams, D.M. Resilience to large-scale disturbance in coral and fish assemblages on the Great Barrier Reef. *Ecology* **2004**, *85*, 1892–1905. [CrossRef]
12. Fabricius, K.E.; Okaji, K.; De'ath, G. Three lines of evidence to link outbreaks of the crown-of-thorns seastar *Acanthaster planci* to the release of larval food limitation. *Coral Reefs* **2010**, *29*, 593–605. [CrossRef]
13. Kayal, M.; Vercelloni, J.; Lison de Loma, T.; Bosserelle, P.; Chancerelle, Y.; Geoffroy, S.; Stievenart, C.; Michonneau, F.; Penin, L.; Planes, S.; et al. Predator Crown-of-Thorns Starfish (*Acanthaster planci*) Outbreak, Mass Mortality of Corals, and Cascading Effects on Reef Fish and Benthic Communities. *PLoS ONE* **2012**, *7*. [CrossRef] [PubMed]

14. Uthicke, S.; Schaffelke, B.; Byrne, M. A boom-bust phylum? Ecological and evolutionary consequences of density variations in echinoderms. *Ecol. Monogr.* **2009**, *79*, 3–24. [CrossRef]
15. Wolfe, K.; Graba-Landry, A.; Dworjanyn, S.A.; Byrne, M. Larval phenotypic plasticity in the boom-and-bust crown-of-thorns seastar, *Acanthaster planci*. *Mar. Ecol. Prog. Ser.* **2015**, *539*, 179–189. [CrossRef]
16. Wolfe, K.; Graba-Landry, A.; Dworjanyn, S.A.; Byrne, M. Larval starvation to satiation: influence of nutrient regime on the success of *Acanthaster planci*. *PLoS ONE* **2015**, *10*, e0122010. [CrossRef] [PubMed]
17. Babcock, R.C.; Milton, D.A.; Pratchett, M.S. Relationships between size and reproductive output in the crown-of-thorns starfish. *Mar. Biol.* **2016**, *163*, 234. [CrossRef]
18. Wooldridge, S.A.; Brodie, J.E. Environmental triggers for primary outbreaks of crown-of-thorns star fish on the Great Barrier Reef, Australia. *Mar. Pollut. Bull.* **2015**, *101*, 805–815. [CrossRef] [PubMed]
19. Wolfe, K.; Graba-Landry, A.; Dworjanyn, S.A.; Byrne, M. Superstars: Assessing nutrient thresholds for enhance larval success of *Acanthaster planci*, a review of the evidence. *Mar. Pollut. Bull.* **2017**. [CrossRef] [PubMed]
20. Brodie, J.; Fabricius, K.; De'ath, G.; Okaji, K. Are increased nutrient inputs responsible for more outbreaks of crown-of-thorns starfish? An appraisal of the evidence. *Mar. Pollut. Bull.* **2005**, *51*, 266–278. [CrossRef] [PubMed]
21. Wolanski, E.; Jones, M. Physical properties of Great Barrier Reef lagoon waters near Townsville. I. Effects of Burdekin River floods. *Mar. Freshw. Res.* **1981**, *32*, 305–319. [CrossRef]
22. King, B.A.; McAllister, F.A.; Wolanski, E.J.; Done, T.J.; Spagnol, S.B. River plume dynamics in the central Great Barrier Reef. In *Oceanographic Processes of Coral Reefs: Physical and Biological Links in the Great Barrier Reef*; Wolanski, E.J., Ed.; CRC Press: Boca Raton, FL, USA, 2001; Chapter 10; pp. 145–160, 356.
23. Brodie, J.E.; Kroon, F.J.; Schaffelke, B.; Wolanski, E.C.; Lewis, S.E.; Devlin, M.J.; Bohnet, I.C.; Bainbridge, Z.T.; Waterhouse, J.; Davis, A.M. Terrestrial pollutant runoff to the Great Barrier Reer: An update of issues, priorities and management responses. *Mar. Pollut. Bull.* **2012**, *65*, 81–100. [CrossRef] [PubMed]
24. Lough, J.M.; Lewis, S.E.; Cantin, N.E. Freshwater impacts in the central Great Barrier Reef: 1648–2011. *Coral Reefs* **2015**, *34*, 739–751. [CrossRef]
25. Intergovernmental Panel on Climate Change (IPCC). *Climate Change 2014: Synthesis Report. Contribution of Working Groups I, II and III to the Fifth Assessment Report of the Intergovernmental Panel on Climate Change*; IPCC: Geneva, Switzerland, 2014.
26. Stickle, W.B.; Diehl, W.J. Effects of salinity on echinoderms. In *Echinoderm Studies*; Jangoux, M., Lawrence, J.M., Eds.; Balkema: Rotterdam, The Netherlands, 1987; Volume 2, pp. 235–285.
27. Lucas, J. Reproductive and larval biology of *Acanthaster planci* (L.) in Great Barrier Reef waters. *Micronesica* **1973**, *9*, 197–203.
28. Russell, M.P. Echinoderm responses to variation in salinity. *Adv. Mar. Biol.* **2013**, *66*, 171–212. [PubMed]
29. Allen, J.D.; Pechenik, J.A. Understanding the effects of low salinity on fertilization success and early development in the sand dollar *Echinarachnius parma*. *Biol. Bull.* **2010**, *218*, 189–199. [CrossRef] [PubMed]
30. Armstrong, A.F.; Blackburn, H.N.; Allen, J.D. A novel report of hatching plasticity in the phylum Echinodermata. *Am. Nat.* **2013**, *181*, 264–272. [CrossRef] [PubMed]
31. Yu, P.C.; Sewell, M.A.; Matson, P.G.; Rivest, E.B.; Kapsenberg, L.; Hofmann, G.E. Growth attenuation with developmental schedule progression in embryos and early larvae of *Sterechinus neumayeri* raised under elevated CO_2. *PLoS ONE* **2013**, *8*, e52448. [CrossRef] [PubMed]
32. Allen, J.D.; Armstrong, A.F.; Ziegler, S.L. Environmental induction of polyembryony in echinoid echinoderms. *Biol. Bull.* **2015**, *229*, 221–231. [CrossRef] [PubMed]
33. Dickson, A.G.; Sabine, C.L.; Christian, J.R. *Guide to Best Practices for Ocean O_2 Measurements*; PICES Special Publication: Sidney, Canada, 2007; Volume 3, p. 119.
34. Liu, X.; Patsavas, M.C.; Byrne, R.H. Purification and characterization of meta-cresol purple for spectrophotometric seawater pH measurements. *Environ. Sci. Technol.* **2011**, *45*, 4862–4868. [CrossRef] [PubMed]
35. Pierrot, D.; Lewis, E.; Wallace, D.W.R. *MS Excel Program Developed for CO_2 System Calculations. ORNL/CDIAC-105a*; Carbon Dioxide Information Analysis Center, Oak Ridge National Laboratory, US Department of Energy: Oak Ridge, TE, USA, 2006.

36. Mehrbach, C.; Culberson, C.H.; Hawley, J.E.; Pytkowicz, R.M. Measurement of the apparent dissociation constants of carbonic acid in seawater at atmospheric pressure. *Limnol. Oceanogr.* **1973**, *18*, 897–907. [CrossRef]

37. Dickson, A.G.; Millero, F.J. A comparison of the equilibrium constants for the dissociation of carbonic acid in seawater media. *Deep Sea Res.* **1987**, *34*, 1733–1743. [CrossRef]

38. Kramer, A.A.; Zimmerman, J.E. Assessing the calibration of mortality benchmarks in critical care: The Hosmer-Lemeshow test revisited. *Crit. Care Med.* **2007**, *35*, 2052–2056. [CrossRef] [PubMed]

39. Dunn, O.J. Multiple Comparisons among Means. *J. Am. Stat. Assoc.* **1961**, *56*, 52–64. [CrossRef]

40. Dunn, O.J. Multiple Comparisons Using Rank Sums. *Technometrics* **1964**, *6*, 241–252. [CrossRef]

41. Quinn, G.P.; Keough, M.J. *Experimental Design and Data Analysis for Biologists*; Cambridge University Press: Cambridge, UK, 2002.

42. Kashenko, S.D. Responses of embryos and larvae of the starfish *Asterias amurensis* to changes in temperature and salinity. *Russ. J. Mar. Biol.* **2005**, *31*, 294–302. [CrossRef]

43. Kashenko, S.D. The combined effect of temperature and salinity on development of the sea star *Asterina pectinifera. Russ. J. Mar. Biol.* **2006**, *32*, 37–44. [CrossRef]

44. Caballes, C.F.; Pratchett, M.S.; Raymundo, M.L.; Rivera-Posada, J.A. Environmental tipping points for sperm motility, fertilization, and embryonic development in the Crown-of-Thorns starfish, *Acanthaster* cf. *solaris. Diversity* **2017**, *9*, 10.

45. Roller, R.A.; Stickle, W.B. Effects of temperature and salinity acclimation of adults on larval survival, physiology, and early development of *Lytechinus variegatus. Mar. Biol.* **1993**, *116*, 583–591. [CrossRef]

46. Carballeira, C.; Martin-Diaz, L.; DelValls, T.A. Influence of salinity on fertilization and larval development toxicity tests with two species of sea urchin. *Mar. Environ. Res.* **2011**, *72*, 196–203. [CrossRef] [PubMed]

47. Delorme, N.J.; Sewell, M.A. Temperature and salinity: Two climate change stressors affecting early development of the New Zealand sea urchin *Evechinus chloroticus. Mar. Biol.* **2014**, *161*, 1999. [CrossRef]

48. Foo, S.A.; Byrne, M. Acclimatization and adaptive capacity of marine species in a changing ocean. *Adv. Mar. Biol.* **2016**, *74*, 69–116. [PubMed]

49. Roller, R.A.; Stickle, W.B. Effects of salinity on larval tolerance and early developmental rates of 4 species of echinoderms. *Can. J. Zool.* **1985**, *63*, 1531–1538. [CrossRef]

50. Metaxas, A. The effect of salinity on larval survival and development in the sea urchin *Echinometra lucunter. Invertebr. Reprod. Dev.* **1998**, *34*, 323–330. [CrossRef]

51. Schiopu, D.; George, S.B. Diet and salinity effects on larval growth and development of the sand dollar *Mellita isometra. Invertebr. Reprod. Dev.* **2004**, *45*, 69–82. [CrossRef]

52. Cowart, D.A.; Ulrich, P.N.; Miller, D.C.; Marsh, A.G. Salinity sensitivity of early embryos of the Antarctic sea urchin *Sterechinus neumayeri. Pol. Biol.* **2009**, *32*, 435–441. [CrossRef]

53. Watts, S.A.; Scheibling, R.E.; Marsh, A.G.; McClintock, J.B. Effect of tempertature and salinity on larval development of sibling species of *Echinaster* (Echinodermata, Asteroidea) and their hybrids. *Biol. Bull.* **1982**, *163*, 348–354. [CrossRef]

54. Morgan, S.G. Life and death in the plankton: Larval mortality and adaptation. In *Ecology of Marine Invertebrate Larvae*; McEdward, L.R., Ed.; CRC Press: Boca Raton, FL, USA, 1995; pp. 279–321.

55. Lamare, M.; Barker, M.F. In situ estimates of larval development and mortality in the New Zealan sea urchin *Evechinus chloroticus* (Echinodermata: Echinoidea). *Mar. Ecol. Prog. Ser.* **1999**, *180*, 197–211. [CrossRef]

56. Vaughn, D.; Allen, J.D. The peril of the plankton. *Int. Comp. Biol.* **2010**, *50*, 552–570. [CrossRef] [PubMed]

57. Uthicke, S.; Pecorino, D.; Albright, R.; Negri, A.P.; Cantin, N.; Liddy, M.; Dworjanyn, S.; Kamya, P.; Byrne, M.; Lamare, M. Impacts of ocean acidification on early life-history stages and settlement of the coral-eating sea star *Acanthaster planci. PLoS ONE* **2013**, *8*, e8293. [CrossRef] [PubMed]

58. Byrne, M. Global change ecotoxicology: Identification of early life history bottlenecks in marine invertebrates, variable species responses and variable experimental approaches. *Mar. Environ. Res.* **2012**, *76*, 3–15. [CrossRef] [PubMed]

59. Kamya, P.Z.; Dworjanyn, S.A.; Hardy, N.; Mos, B.; Uthicke, S.; Byrne, M. Larvae of the coral eating crown-of-thorns starfish, *Acanthaster planci* in a warmer high CO_2 ocean. *Glob. Chang. Biol.* **2014**, *20*, 3365–3376. [CrossRef] [PubMed]

60. Kamya, P.Z.; Byrne, M.; Graba-Landry, A.; Dworjanyn, S.A. Near-future ocean acidification enhances the feeding rate and development of the herbivorous juveniles of the Crown of Thorns Starfish, *Acanthaster planci*. *Coral Reefs* **2016**. [CrossRef]

61. Benzie, J.; Black, K.; Moran, P.; Dixon, P. Small-Scale Dispersion of Eggs and Sperm of the Crown-of Thorns Starfish (*Acanthaster planci*) in a Shallow Coral Reef Habitat. *Biol. Bull.* **1994**, *186*, 153–167. [CrossRef]

62. Uthicke, S.; Logan, M.; Liddy, M.; Francis, D.; Hardy, N.; Lamare, M. Climate change as an unexpected co-factor promoting coral eating seastar (*Acanthaster planci*) outbreaks. *Sci. Rep.* **2015**, *5*, 8402. [CrossRef] [PubMed]

63. Wolanski, E.; Van Senden, D. Mixing of Burdekin river flood waters in the Great Barrier Reef. *Mar. Freshw. Res.* **1983**, *34*, 49–63. [CrossRef]

64. Schroeder, T.; Devlin, M.J.; Brando, V.E.; Dekker, A.G.; Brodie, J.E.; Clementson, L.A.; McKinna, L. Inter-annual variability of wet season freshwater plume extent into the Great Barrier Reef lagoon based on satellite coastal ocean colour observations. *Mar. Pollut. Bull.* **2012**, *65*, 210–223. [CrossRef] [PubMed]

65. Sarantchova, O.L. Research into tolerance for the environment salinity in sea starfish *Asterias rubens* L. from populations of the White Sea and Barentz Sea. *J. Exp. Mar. Biol. Ecol.* **2001**, *264*, 15–28. [CrossRef]

66. Casties, I.; Clemmesen, C.; Melzner, F.; Thomsen, J. Salinity dependence of recruitment success of the sea star *Asterias rubens* in the brackish western Baltic Sea. *Helgol. Mar. Res.* **2015**, *69*, 169–175. [CrossRef]

67. Pia, T.S.; Johnson, T.; George, S.B. Salinity-induced morphological changes in *Pisaster ochraceus* (Echinodermata: Asteroidea) larvae. *J. Plankton Res.* **2012**, *34*, 590–601. [CrossRef]

68. Strathmann, R.R. The feeding behavior of planktotrophic echinoderm larvae: mechanisms, regulation, and rates of suspension-feeding. *J. Exp. Mar. Biol. Ecol.* **1971**, *6*, 109–160. [CrossRef]

69. George, S.B. Phenotypic plasticity in the larvae of *Luidia foliolata* (Echinodermata: Asteroidea). In *Echinoderms through Time*; David, B., Guille, A., Roux, M., Eds.; A.A. Balkema: Rotterdam, The Netherlands, 1994; pp. 297–307.

70. George, S.B. Egg quality, larval growth and phenotypic plasticity in a forcipulate seastar. *J. Exp. Mar. Biol. Ecol.* **1999**, *237*, 203–224. [CrossRef]

71. Sameoto, J.A.; Metaxas, A. Can salinity-induced mortality explain larval vertical distribution with respect to a halocline? *Biol. Bull.* **2008**, *214*, 329–338. [CrossRef] [PubMed]

72. Sameoto, J.A.; Metaxas, A. Interactive effects of haloclines and food patches on the vertical distribution of 3 species of temperate invertebrate larvae. *J. Exp. Mar. Biol. Ecol.* **2008**, *367*, 131–141. [CrossRef]

73. Bashevkin, S.M.; Lee, D.; Driver, P.; Carrington, E.; George, S.B. Prior exposure to low salinity affects the vertical distribution of *Pisaster ochraceus* (Echinodermata: Asteroidea) larvae in haloclines. *Mar. Ecol. Prog. Ser.* **2016**, *542*, 123–140. [CrossRef]

diversity

MDPI

Article

Selective Feeding and Microalgal Consumption Rates by Crown-Of-Thorns Seastar (*Acanthaster* cf. *solaris*) Larvae

Camille Mellin [1,2,†,*], Claire Lugrin [1,3,†], Ken Okaji [4], David S. Francis [1,5] and Sven Uthicke [1]

[1] Australian Institute of Marine Science PMB No 3, Townsville, QLD 4810, Australia;
    claire.lugrin@epfl.ch (C.L.); d.francis@deakin.edu.au (D.S.F.); S.Uthicke@aims.gov.au (S.U.)
[2] The Environment Institute and School of Biological Sciences, The University of Adelaide, Adelaide,
    SA 5005, Australia
[3] AgroParisTech, Paris 75005, France
[4] Coralquest Inc., 1-34-10 Asahicho, Atsugi 2430014, Japan; cab67820@pop06.odn.ne.jp
[5] Deakin University, Geelong, Australia, School of Life and Environmental Sciences, Warrnambool Campus,
    Princes Hwy, Sherwood Park, PO Box 423, Warrnambool, VIC 3280, Australia
* Corresponding author: camille.mellin@adelaide.edu.au; Tel.: +61-08-8313-5432
† These authors equally contributed to this work

Academic Editors: Morgan Pratchett and Michael Wink
Received: 24 November 2016; Accepted: 30 January 2017; Published: 7 February 2017

Abstract: Outbreaks of the crown-of-thorns seastar (CoTS) represent a major cause of coral loss on the Great Barrier Reef. Outbreaks can be explained by enhanced larval survival supported by higher phytoplankton availability after flood events, yet little is known about CoTS larvae feeding behaviour, in particular their potential for selective feeding. Here, single- and mixed-species feeding experiment were conducted on CoTS larvae using five algae (*Phaeodactylum tricornutum*, *Pavlova lutheri*, *Tisochrysis lutea*, *Dunaliella* sp. and *Chaetoceros* sp.) and two algal concentrations (1000 and 2500 algae·mL^{-1}). Cell counts using flow-cytometry at the beginning and end of each incubation experiment allowed us to calculate the filtration and ingestion rates of each species by CoTS larvae. In line with previous studies, CoTS larvae ingested more algae when the initial algal concentration was higher. We found evidence for the selective ingestion of some species (*Chaetoceros* sp., *Dunaliella* sp.) over others (*P. lutheri*, *P. tricornutum*). The preferred algal species had the highest energy content, suggesting that CoTS selectively ingested the most energetic algae. Ultimately, combining these results with spatio-temporal patterns in phytoplankton communities will help elucidate the role of larval feeding behaviour in determining the frequency and magnitude of CoTS outbreaks.

Keywords: electivity; feeding behaviour; filtration rate; Great Barrier Reef; phytoplankton

1. Introduction

Outbreaks of the crown-of-thorns seastar (*Acanthaster* cf. *solaris*, CoTS) represent a major threat for coral reefs and, in particular, for the Australian Great Barrier Reef (GBR). CoTS outbreaks are among the main causes of coral loss on the GBR since 1985 [1] and, unlike other causes, such as bleaching or cyclones, may be amenable to local or regional management. One of the most widely-accepted hypotheses explaining the increased frequency of CoTS outbreaks relates to the availability of phytoplankton, which is intimately linked to elevated terrestrial nutrient runoff [2–6]. Like most planktotrophic echinoderm larvae, CoTS larvae feed on nano-phytoplankton, and the larval development rate increases with food availability [7]; although high food concentrations may be detrimental to development and survival [8,9]. However, it is likely that phytoplankton diversity (or the dominance of certain species) is equally or more important than overall phytoplankton abundance with respect to larval development.

While documented descriptions of the feeding mechanisms of CoTS in the early larval stages are scarce, it can be assumed that there are similarities to the feeding of other asteroid taxa (*Luidia* spp., *Patiria* spp., *Evasterias* spp., *Pisaster* spp., *Pycnopodia* spp.) [10]. These asteroid larvae facilitate food ingestion via the beating of epidermal cilia that generate a feeding current for the concentration and ingestion of particles [10]. Particles accumulate in the mouth and are subsequently swallowed via the action of muscles surrounding the oesophagus. The sphincter between the oesophagus and stomach then opens, allowing the particles to enter the stomach where they can be sorted, digested or rejected by dorsal flexion [10,11]. Rejection of particles that have not reached the stomach is accomplished by a reverse beating of the oesophagus and mouth cilia, and the reverse sequence of muscle contractions. This might be a mechanism for food selection [11].

Studies of larval feeding in CoTS conducted to date have focused on size selectivity and the influence of food concentration on feeding behaviour [7,12,13]. In particular, these investigations have demonstrated the ability of CoTS larvae to discriminate particle ingestion based on size (feeding in the range of 5–20 μm in diameter) and a modulatory effect of algal species on feeding and development. It has also been demonstrated that CoTS larvae can feed on other organic matter, such as mucus derived from corals [14]. As suggested by Ayukai [12], ingestion is likely controlled by other factors than algae size and concentration, with food quality potentially influencing CoTS feeding behaviour. While each of the aforementioned studies contributed knowledge to our understanding of the larval phase of this species, the corresponding experiments were short in duration (approximately 30 min) [12,13] or conducted using high food concentrations of little relevance to natural conditions [7]. Thus, there is currently limited knowledge on the feeding ecology (e.g., preferred food items, consumption) of the larvae of this important coral predator, in particular in terms of potential preference for some algal species and corresponding consumption rates.

In the present study, larvae were fed with five different algae species that are commonly used in CoTS larval experiments or naturally occurring on the GBR (in isolation or in 1:1 choice experiments) and tested over 18 h of consumption. Initial and final algae concentrations were measured by flow cytometry. This technique uses the fluorescence properties of each algal species for discrimination [15,16]. It simultaneously computes cell concentrations and, by comparing concentrations before and after consumption by CoTS, allows the calculation of filtration and ingestion rates. Flow cytometry has been used to study feeding behaviour of several zooplankton species, including copepods [17], oyster larvae [18], zebra mussel larvae [19] and decapod larvae [20], but never for studying CoTS feeding behaviour (nor, to our knowledge, that of any other echinoderm). Therefore, flow cytometry represents an alternative to the more traditional methods implemented previously in CoTS feeding studies and potentially an efficient way to increase accuracy in algal counts [12]. Specifically, we used this technique to determine the filtration and ingestion rates of the different algal species under two different ecologically relevant algal cell concentrations. In addition, we investigated the ability of CoTS larvae to selectively ingest or reject individual algal species in food mixes. Finally, we related these patterns of food selection to the energetic content of the algae species tested.

2. Material and Methods

2.1. Spawning and Maintenance of CoTS Larvae

Adult CoTS were collected on Rudder Reef in the northern section of the GBR (16.21 °S, 145.47 °E) in mid-September 2014, transported to the Australian Institute of Marine Science and maintained under natural flow-through seawater (FSW) conditions. Adult CoTS were spawned and oocytes fertilized as described previously [9]. Briefly, a small (≈1 cm) incision was made near the proximal end of one of the arms, and 3–4 gonadal lobes were removed from each specimen. After macroscopic sex determination, testes of six males and ovaries of six females were collected. Testes were placed in covered 6-well plates to prevent desiccation. Ovary lobes were rinsed with filtered seawater

(FSW) and subsequently submerged in a 10^{-5} M 1-methyladenine/seawater solution to induce maturation. After 60–70 min, mature oocytes were washed through a 50-μM mesh and oocytes from all females combined. Oocytes were diluted in seawater to achieve a stock solution of ~400 eggs·mL^{-1}. Two millilitres of dry sperm from each male were combined and added to 2500 mL of the egg solution, resulting in a concentration of 10^6–10^7 sperm·mL^{-1}. This yielded a fertilization success of >99%. After 20 min, eggs were washed repeatedly using a 50-mm mesh to remove excess sperm. Fertilized eggs were then divided evenly across three holding tanks (70 L each) and kept at 27 °C in FSW (0.2 μm) under a 12 h:12 h light dark cycle. CoTS larvae were maintained at a density of 0.6–1.5 larvae·mL^{-1} and fed with Dunaliella sp. (3000 cells·mL^{-1}) once per day starting two days post-hatching. These algal species and concentration were chosen because they are known to promote optimal development in CoTS larvae (e.g. [7]). Because the feeding behaviour of CoTS larvae changes over the course of larval development [7], all experiments were conducted using larvae between 3 and 7 days old. Microscopic examination showed that all larvae were in the bipinnaria/early brachiolaria stage.

2.2. Phytoplankton Cultures and Analysis

CoTS larvae were fed cultured phytoplankton over the course of the feeding experiments. Five pure algae species were sourced from the Australian Algal Culture Collection (Hobart, Australia), consisting of two diatom species: *Phaeodactylum tricornutum* (Species No. CS-29) and *Chaetoceros* sp. (CS-256), two haptophytes: *Pavlova lutheri* (CS-182) and *Tisochrysis lutea* (CS-177) and the green algae *Dunaliella* sp. (CS-353). All algae were grown at 26 °C under a 12 h:12 h light dark cycle, in F/2 medium. Experimental algal species were selected because they have been frequently used in CoTS larval culture (*Dunaliella* sp., *T. lutea*, *P. tricornutum*, *P. lutheri*) [7,9,13] or are naturally occurring on the GBR (*Chaetoceros sp.*) [21] and can dominate microalgal community assemblage following flood events [22]. The length and width of the microalgae were determined for 30 cells of each species under a 400× magnification.

To quantify algal nutritional content, algae were analysed for total lipid content following standardized procedures described previously [23]. Briefly, lipids were extracted with dichloromethane: methanol (2:1). Protein content was determined via the application of a nitrogen-to-protein conversion factor of 5.0, taking into account the significant sources of chlorophyll, nucleic acids, free amino acids and inorganic nitrogen [24], while the energetic content of each algal species was determined on a pg/cell basis from the enthalpies of combustion values published by Bureau et al. [25]: lipid (39.5 kJ·g^{-1}) and protein (23.9 kJ·g^{-1}).

Organic carbon (OC), nitrogen (N) and chlorophyll content of each algal species were measured in duplicate using standard water quality methods (e.g., [26]). In brief, 100 mL (for chlorophyll) or 50 mL (for OC, N) of stock solution were filtered on pre-combusted 25-mm filters (Whatman GF/F) and stored frozen (-20 °C) until analysis. Nitrogen was analysed on an ANTEK 9000 NS analyser and OC on a Shimadzu TOC-V carbon analyser, equipped with a solid sample module (SSM-5000A) after removing inorganic carbon using 2 M hydrochloric acid. Chlorophyll-a concentrations were measured fluorometrically (Turner Designs 10 AU fluorometer) after grinding filters in cold (4 °C) acetone (90%).

2.3. Larval Feeding Experiments

Each experiment corresponded to either a single algal species being tested or a mixture of two species, at either the high- or low-level treatment. Prior to incubations and for each experiment, larvae were starved for approximately 20 h, concentrated and placed in six glass jars containing 200 mL of FSW. Larvae were stocked at an approximate concentration of 1 larvae·mL^{-1} and counted both at the beginning and the end of each experiment. Initial counts in each experiment showed that the actual numbers varied slightly (mean ± standard deviation: 1.00 ± 0.17 larvae·mL^{-1}; range: 0.57–1.50 larvae·mL^{-1}). We accounted for this variation by adding the actual initial larval density as a factor in the statistical model (see below). To document possible intrinsic changes in algal concentrations without larvae, six jars were used as no larvae controls and filled with 200 mL of FSW. Jars were then placed on a shaker plate and larvae starved a further five hours. Thus, larvae had sufficient time to acclimate to their new environmental conditions. Prior to each experiment, algae stock concentrations were obtained using an Accuri C6 flow cytometer. Feeding concentrations were then calculated and the respective number of algae added to each jar (experimental and control). Each jar was adjusted to a final concentration of either 1000 cells·mL^{-1} for the low-level treatment or 2500 cells·mL^{-1} for the high-level treatment. These concentrations were in the range reported by Devlin et al. [22] in flood plumes on the Great Barrier Reef lagoon. In total, we therefore used 24 jars for each algal species (or mixture of species), including 12 jars for the low-level treatment (6 with larvae and 6 without larvae) and 12 jars for the high-level treatment (6 with larvae and 6 without larvae). Larval age (between 3 and 7 days old) was randomly distributed across all experiments and also accounted for in the statistical models as a fixed effect (see below).

In addition to the aforementioned experimental and control jars, three experimental (with larvae) and three control jars (without larvae) were run with *Dunaliella* sp. at a density of 2500 cells·mL^{-1} during each incubation These served as an inter-experiment reference to control for possible variation in CoTS algal consumption between experiments. In other words, we obtained a feeding rate for *Dunaliella* sp. at a standard concentration for each experiment.

Following the addition of algae, each jar was gently agitated with a customized plunger and a 2-mL subsample of water taken through a 100-μm mesh (to avoid removing larvae). The jars were placed back on the shaker plate and left for 18 h. All subsamples were immediately analysed with the flow cytometer and the initial algal concentration calculated. This step ensured that the actual initial algal concentrations (as measured by flow cytometry) that slightly varied around the expected value both in the single and mixed-species experiments could be taken into account in the statistical model. This step also ensured that, in each case, low- and high-level treatments spanned distinct ranges of initial algal concentrations (Table A1). After 18 h, each jar was agitated again, and a final sample was taken and cell numbers analysed. Larvae were then counted in a 25-mL subsample from each replicate to determine larval densities for each replicate.

We used the initial experiments to investigate the feeding rates on individual algal species. Subsequently, we conducted choice experiments with mixtures of two algal species. The four mixtures each consisted of 50% of *Dunaliella* sp. and 50% of either of the remaining species. *Dunaliella* sp. was chosen as a reference because it was commonly used in past CoTS feeding experiments [7,9,27]. We also limited experiments to two species at a time to ensure their accurate distinction by the flow cytometer. Two cell concentrations were used for single food and choice experiments: 1000 cells·mL^{-1} for the low treatment and 2500 cells·mL^{-1} for the high treatment.

We calculated two important ecological parameters for plankton feeders, namely the filtration rate (i.e., the volume of water cleared by a larvae in a given amount of time) and the ingestion rate (i.e., the number of algae cells ingested per larvae per unit of time). First, we calculated the filtration rate (*Fr*) for each algal species according to Coughlan [28]. For each treatment, a growth constant k was calculated from C'_0 and C'_t, i.e., the initial and final algae concentrations in the control jars, respectively.

$$C'_t = C'_0 e^{kt} \iff k = \frac{1}{t} \ln\left(\frac{C'_0}{C'_t}\right) \tag{1}$$

where t is the duration of the experiment (18 h). A grazing coefficient g was calculated in the same way from C_0 and C_t, i.e., initial and final algae concentrations in each replicate, respectively

$$g = \frac{1}{t} \ln\left(\frac{C_0}{C_t}\right) + k \tag{2}$$

The filtration (or clearance) rate (μL larva^{-1} h^{-1}) was then calculated as:

$$Fr = \frac{V}{N_{larv}} g \tag{3}$$

where N_{larv} is the number of larvae in the jar and V the volume of water in the jar (200 mL).

Mean ingestion rate (cells larva$^{-1}\cdot$h^{-1}) was calculated from:

$$I_r = F_r \, C_{mean} \tag{4}$$

where C_{mean}, the mean concentration (cells μL^{-1}), is calculated as:

$$C_{mean} = \frac{C_0 \left(e^{(k-g)t} - 1 \right)}{(k-g)t} \tag{5}$$

Selective feeding by CoTS larvae in the choice experiments was then evaluated as described by Baldwin [18] and Fileman et al. [20] and through an electivity index proposed by Vanderploeg and Scavia [29]:

$$E_i = \frac{Fr_i}{\sum_j Fr_j} \tag{6}$$

where Fr_i is the filtration rate for algae i in the mixture and the sum of filtrations rates of all species present in the mixture. For each species i, E_i ranges between 0 and 1, with 0.5 meaning no selectivity, 0–0.5 representing avoidance and 0.5–1 representing selection. The utilization of this index is particularly relevant here, because it is unaffected by the initial proportions of each algal species in the mixture [18].

2.4. Modelling

We predicted filtration and ingestion rates of CoTS larvae as a function of algal species and initial concentration using a Poisson error-distributed generalized linear mixed-effect model (GLMM) [30]. Initial larval concentration and age and initial algal concentrations were included in the models as fixed effects. This allowed us to (i) account for the fact that larvae of different ages (3–7 days old) might have slightly different dietary requirement and (ii) test for the effect of initial algal concentration on ingestion and filtration rates as suggested by Marin et al. [31]. All combinations of fixed effects including their interactions were considered in three model sets for each response variable (i.e., filtration and ingestion rates): (i) in single-species experiments, (ii) for *Dunaliella* sp. in mixed-species experiment and (iii) for the second species in the mixed-species experiments. We included the date of the experiment as a random effect to account for the non-independence of replicates tested on the same day and from the same batch of larvae.

We assessed GLMM performance using the marginal R^2 (R_m, variance explained by the fixed effects) and the conditional R^2 (R_c, variance explained by both the fixed and random effects) to provide an index of the model's goodness-of-fit [32]. We also used Akaike's information criterion corrected for small sample sizes (AIC$_c$) that provides an index of Kullback–Leibler information loss and corresponding weights (wAIC$_c$) to assign the relative strengths of evidence to the different competing models [33]. This information-theoretic approach offers a more robust method than standard regression techniques for testing alternative hypotheses, because it uses a multi-model inference framework without discarding any models or predictors based on arbitrary thresholds such as p-values [34]. We fitted the GLMM using the function lmer (library lme4) in R 3.0.1 [35].

We then used the GLMM to compare the effect of algal species and initial concentration (high vs. low treatment) on the filtration and ingestion rates both in single- and mixed-species experiments. To do this, we predicted the partial effects of the different algae species and their initial concentrations on the filtration and ingestion rates of CoTS larvae. We used a model-averaging approach whereby predictions from each model in the model set were weighted by their relative strength of evidence (wAIC$_c$), then averaged. This procedure allowed the estimation of mean filtration and ingestion rates in each treatment while correcting for the known variation in initial algae concentration and the possible measurement errors due to non-independence in the original dataset. We then compared the predicted filtration and ingestion rates of each algal species in single vs. mixed species and high vs. low treatments using *t*-test pairwise comparisons.

3. Results

3.1. Single Species Experiment

In the single-species experiments, the model predicting filtration rate from larval age, initial algal concentration and their interaction with algal species received the strongest support based on AIC_c and explained over 80% of the variation in filtration rates (with 66% attributed to fixed effects) (Table 1). The same model without the interaction between algal species and initial concentration received the strongest AIC_c support in explaining ingestion rates, accounting for up to 90% of their variation (with 74% attributed to fixed effects).

Table 1. Generalized linear mixed-effects model (GLMM) results for filtration and ingestion rates of larvae fed with a single algal species. Filtration and ingestion rates are predicted as a function of algal species, larvae concentration (LC), larvae age (LA), initial algae concentration (IC) and their interaction (as indicated by the asterisk). All models include the experimental date as the random effect. Shown are the number of model parameters (*k*), maximum log-likelihood (*LL*), the information-theoretic Akaike's information criterion corrected for small samples (AIC_c), AIC_c weight ($wAIC_c$) marginal and conditional R-squared (R_m and R_c) as measures of the model's goodness-of-fit. Models are ordered by decreasing $wAIC_c$. Only models with $wAIC_c > 0.001$ are shown, in addition to the null model.

Model	*k*	*LL*	AIC_c	$wAIC_c$	R_m	R_c
Filtration rate						
Species * IC + Species * LA + (1 | date)	17	−324.8	744.7	0.920	65.70	81.80
Species * IC + Species * LA + LA + (1 | date)	18	−322.4	749.5	0.080	64.50	82.10
1 + (1 | date)	3	−426.1	858.4	0.000	0.00	39.80
Ingestion rate						
Species * LA + IC + (1 | date)	13	−339	740.3	0.996	73.70	90.90
Species + IC + LA + (1 | date)	9	−357.4	751.2	0.004	73.70	91.40
1 + (1 | date)	3	−447.3	900.9	0.000	0.00	41.20

Filtration rates of individual algal species were either not significantly affected by initial food concentration (Figure 1a; *P. tricornutum*, *P. lutheri* and *Dunaliella* sp., *T. lutea*; *t*-test; *p* > 0.05) or decreased with increasing algae concentration (*Chaetoceros* sp.; *p* < 0.001). When initial algae concentrations increased from 1000–2500 algae mL^{-1}, the filtration rate for *Dunaliella* sp. remained around 60 µL larvae^{-1}·h^{-1} for both concentrations (Figure 1a), while the ingestion rates for the same species increased from 36.3 cells larva^{-1}·h^{-1} on average (95% confidence interval: 31.8–40.8) to 87.0 (84.2–89.8) cells larva^{-1}·h^{-1} (Figure 1b). Similarly, ingestion rates significantly increased at higher algal concentrations for all algal species tested (Figure 1; *p* < 0.001). The lowest mean ingestion rate of 10.3 cells larvae^{-1}·h^{-1} (5.9–14.8) was observed for *P. lutheri* in the 1000 cells·mL^{-1} concentration, while the highest one, 104.3 cells larva^{-1} h^{-1} on average (96.4–112.1), was found for *T. lutea* under the 2500 cells·mL^{-1} concentration.

CoTS larvae filtered and ingested *T. lutea*, *Dunaliella* sp. and *Chaetoceros* sp. at higher rates (filtration > 63.1 µL·larva^{-1}·h^{-1} and ingestion >77.2 cells larva^{-1}·h^{-1} for the high treatment), compared to rates of filtration and ingestion for *P. tricornutum* and *P. lutheri* (28.5 and 24.6 µL·larva^{-1}·h^{-1} for filtration and 63.0 and 46.4 cells larva^{-1}·h^{-1} for ingestion, respectively) (Figure 1).

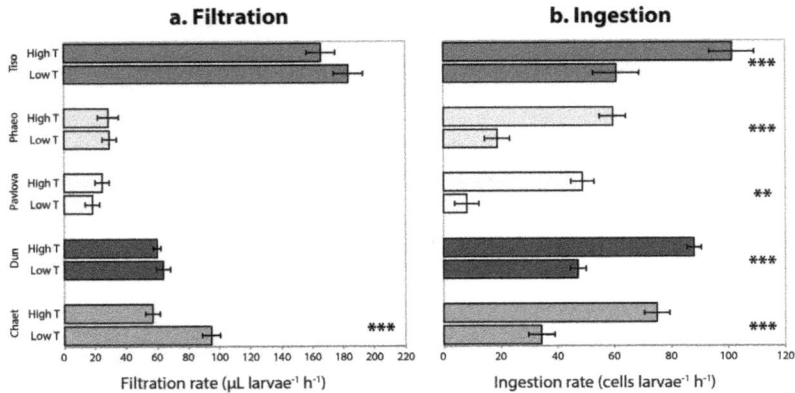

Figure 1. Filtration and ingestion rates for larvae fed with one algal species. Predicted filtration (**a**) and ingestion (**b**) rates of larvae fed with five different species. Predictions are made using GLMM models. High treatment represents an initial concentration of 2500 cells·mL^{-1} and low treatment an initial concentration of 1000 cells·mL^{-1}. The error bars represent the 95% confidence intervals. Asterisks represent significantly different rates between low- and high-concentration treatments (Student's *t*-test; *: $p < 0.05$, **: $p < 0.01$, ***: $p < 0.001$). Tiso: *Tisochrysis lutea*; Phaeo: *Phaeodactylum tricornutum*; Pavlova: *Pavlova lutheri*; Dun: *Dunaliella* sp.; Chaet: *Chaetoceros* sp.

3.2. Mixed Species Experiments

In the mixed-species experiments, the model including larvae age and the interaction between algal species and initial algal concentrations (both species) received the strongest AIC$_c$ support in explaining both the filtration and ingestion rates for *Dunaliella* sp. (Table 2). This model explained 63 and 87% of variation in *Dunaliella* sp. filtration and ingestion rates, respectively (with 27% and 71% attributed to fixed effects). The same model also received the strongest support in explaining filtration and ingestion rates for the second species (with 79% and 92% of variation explained in filtration and ingestion rates, respectively) (Table 3).

In a similar manner to the trends observed in single algal experiments, all larval ingestion rates increased with initial algal concentration ($p < 0.05$; Table 4) (Figure 2). Conversely, the filtration rates remained constant for *P. tricornutum* and *P. lutheri*, as well as for *Dunaliella* sp. when mixed with either *P. tricornutum* or *P. lutheri* ($p > 0.05$), but significantly decreased for *T. lutea* and *Chaetoceros* sp. ($p < 0.001$) (Table 4).

Table 2. Generalized linear mixed-effects model (GLMM) results for filtration and ingestion rates of *Dunaliella* sp. for larvae fed with 1:1 mixture. Filtration and ingestion rates of *Dunaliella* sp. are predicted as a function of algal species, larvae concentration (LC), larvae age (LA), initial algae concentration of *Dunaliella* sp. (IC dun), initial algal concentration for the second species (IC) and their interaction (as indicated by the asterisk). All models include the experimental date as the random effect. Shown are the number of model parameters (k), maximum log-likelihood (LL), the information-theoretic Akaike's information criterion corrected for small samples (AIC$_c$), AIC$_c$ weight (wAIC$_c$) marginal and conditional R-squared (R$_m$ and R$_c$) as measures of the model's goodness-of-fit. Models are ordered by decreasing wAIC$_c$. Only models with wAIC$_c$ > 0.001 are shown, in addition to the null model.

Model	k	LL	AIC$_c$	wAIC$_c$	R$_m$	R$_c$
Filtration rate						
Species * IC * IC dun + LA + (1 | date)	22	−317.1	779.9	0.980	27.0	63.5
Species * IC * IC dun + LA + LA + (1 | date)	23	−315	788.7	0.012	29.0	68.6
Species * IC + Species * IC dun + Species * LA + (1 | date)	21	−328.3	790.2	0.006	27.5	67.5

Table 2. Cont.

Model	k	LL	AIC_c	wAIC_c	R_m	R_c
Species * IC + Species * IC dun + LA + (1 | date)	17	−349.7	792.3	0.002	26.8	64.3
1 + (1 | date)	3	−422.4	851.1	0.000	0.0	38.0
Ingestion rate						
Species * IC * IC dun + LA + (1 | date)	22	−243.8	633.3	0.980	71.4	87.2
Species * IC + Species * IC dun + LA + (1 | date)	17	−274.6	642.3	0.011	69.2	86.7
Species * IC * IC dun + LA + LA + (1 | date)	23	−242.4	643.6	0.006	71.3	88.0
1 + (1 | date)	3	−399.4	805	0.000	0.0	27.8

Table 3. Generalized linear mixed-effects model (GLMM) results for filtration rates and ingestion rates of the second species for larvae fed with 1:1 mixture. Filtration and ingestion rates of the second species offered along *Dunaliella* sp. in a 1:1 mixture are predicted as a function of algal species, larvae concentration (LC), larvae age (LA), initial algae concentration of *Dunaliella* sp. (IC dun), initial algal concentration for the second species (IC) and their interaction (as indicated by the asterisk). All models include the experimental date as the random effect. Shown are the number of model parameters (k), maximum log-likelihood (LL), the information-theoretic Akaike's information criterion corrected for small samples (AIC_c), AIC_c weight (wAIC_c) and the marginal and conditional R-squared (R_m and R_c) as measures of the model's goodness-of-fit. Models are ordered by decreasing wAIC_c. Only models with wAIC_c > 0.001 are shown, in addition to the null model.

Model	k	LL	AIC_c	wAIC_c	R_m	R_c
Filtration rate						
Species * IC * IC dun + LA + (1 | date)	22	−317.3	780.4	0.985	75.0	79.4
Species * IC * IC dun + LA + LA + (1 | date)	23	−316	790.7	0.006	74.3	79.6
Species * IC + Species * IC dun + (1 | date)	16	−353.3	790.9	0.005	75.5	80.1
Species * IC + Species * IC dun + LA + (1 | date)	17	−349.7	792.4	0.002	75.1	80.0
Species * IC + Species * IC dun + Species * LA + (1 | date)	21	−329.6	793.0	0.002	74.7	79.6
1 + (1 | date)	3	−461.6	929.4	0.000	0.0	39.9
Ingestion rate						
Species * IC * IC dun + LA + (1 | date)	22	−243.2	632.2	0.980	83.4	91.9
Species * IC + Species * IC dun + LA + (1 | date)	17	−274	640.9	0.012	81.2	91.2
Species * IC * IC dun + LA + LA + (1 | date)	23	−242.4	643.1	0.004	83.1	92.1
Species * IC + Species * IC dun + (1 | date)	16	−279.6	643.5	0.003	78.9	91.4
1 + (1 | date)	3	−410.1	826.5	0.000	0.0	45.0

Table 4. Morphometry, chemical composition and energetic value of the five algal species on a per cell basis (mean and 95% confidence interval, in brackets). Values pertaining to lipid, carbon, nitrogen and chlorophyll concentration (pg/cell) were obtained via direct analytical measurement, whilst protein concentration and energetic content were calculated according to published conversion factors. For each species, a sample size of $N = 30$ was used to determine the length and width, and N = 2 for the other parameters.

Variable (unit)	*Dunaliella* sp.	*T. lutea*	*Chaetoceros* sp.	*P. lutheri*	*P. tricornutum*
Length (μm)	7.92	5.01	9.18	5.91	18.92
	(7.37–8.47)	*(4.77–5.25)*	*(8.65–9.72)*	*(5.51–6.3)*	*(17.94–19.9)*
Width (μm)	5.43	5.7	6.5	5.17	3.48
	(5.15–5.71)	*(5.46–5.94)*	*(6.1–6.9)*	*(4.94–5.4)*	*(3.31–3.64)*
Carbon (pg/cell)	33.29	15.39	49.87	22.97	24.2
	(33.01–33.58)	*(15.09–15.68)*	*(48.56–51.18)*	*(22.14–23.79)*	*(23.71–24.69)*
Nitrogen (pg/cell)	6.58	2.53	7.9	3.57	3.85
	(6.54–6.61)	*(2.44–2.62)*	*(7.44–8.36)*	*(3.52–3.62)*	*(3.73–3.97)*
C/N	5.06	6.07	6.31	6.42	6.27
	(4.99–5.13)	*(5.97–6.17)*	*(6.11–6.51)*	*(6.28–6.57)*	*(6.2–6.34)*
Chlorophyll (pg/cell)	1.05	0.35	0.75	0.58	0.31
	(0.99–1.12)	*(0.32–0.38)*	*(0.7–0.8)*	*(0.55–0.6)*	*(0.3–0.33)*

Table 4. *Cont.*

Variable (unit)	*Dunaliella* sp.	*T. lutea*	*Chaetoceros* sp.	*P. lutheri*	*P. tricornutum*
Lipid (pg/cell)	15.58	6.79	17.33	10.54	9.93
	(15.49–15.66)	*(5.48–8.09)*	*(16.53–18.13)*	*(8.96–12.12)*	*(9.5–10.36)*
Protein (pg/cell)	32.9	12.65	39.5	17.85	19.25
	(32.7–33.05)	*(12.2–13.1)*	*(37.2–41.8)*	*(17.6–18.1)*	*(18.65–19.85)*
Energy (10^{-9} kJ/cell)	1.39	0.57	1.62	0.84	0.85
	(1.38–1.40)	*(0.50–0.63)*	*(1.53–1.70)*	*(0.77–0.91)*	*(0.81–0.88)*

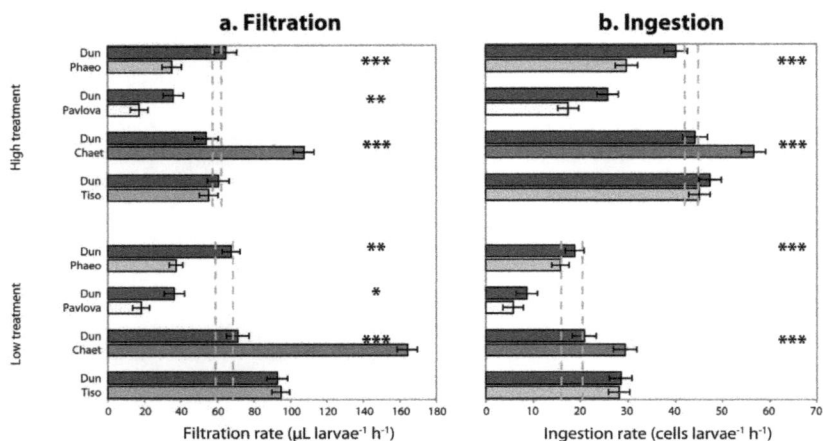

Figure 2. Filtration and ingestion rates for larvae fed with a 1:1 mixture of two algae species. Predicted filtration (**a**) and ingestion (**b**) rates for each of the two algae for the four food mixtures. Predictions are made using GLMM models. High treatment represents an initial concentration of 2500 cells·mL^{-1} (1250 cells mL^{-1} of each species) and low treatment an initial concentration of 1000 cells·mL^{-1} (500 of each species). The error bars represent the confidence intervals. The grey dotted lines represent the predicted 95% confidence interval of filtration and ingestion rates for larvae fed with *Dunaliella* sp. alone (filtration rate for 2500 and 1000 cells mL^{-1}, half of the ingestion rate for 2500 and 1000 cells·mL^{-1} of *Dunaliella* sp.). Asterisks represent significantly different rates between *Dunaliella* sp. and the second species (Student's *t*-test; *: $p < 0.05$, **: $p < 0.01$; ***: $p < 0.001$). Tiso: *Tisochrysis lutea*; Phaeo: *Phaeodactylum tricornutum*; Pavlova: *Pavlova lutheri*; Dun: *Dunaliella* sp.; Chaet: *Chaetoceros* sp.

Differences in filtration and ingestion rates between high and low treatments were significant for all species ($p < 0.05$), except for the filtration rate of *Dunaliella* sp. when mixed with *P. lutheri* ($p = 0.50$) or *P. tricornutum* ($p = 0.81$) or that of *P. lutheri* when mixed with *Dunaliella* sp. ($p = 0.90$). When offered in a 1:1 mixture, the ingestion rate for *Chaetoceros* sp. was higher than for *Dunaliella* sp. regardless of the initial concentration (56.6 and 44.2 cells larvae^{-1}·h^{-1}, respectively, in the high treatment) ($p < 0.001$; Figure 2b). In contrast, ingestion rates of *P. lutheri* (17.3 cells larvae^{-1}·h^{-1}) and *P. tricornutum* (29.7 cells larvae^{-1}·h^{-1}) in the high treatment were lower than those observed for *Dunaliella* sp. (25.8 and 40.2 cells larvae^{-1}·h^{-1} for each mixture, respectively), although the difference was only significant for *P. tricornutum*. *T. lutea* was ingested at a similar rate to *Dunaliella* sp. ($p > 0.05$) (Figure 2b). These results are further substantiated by the electivity indexes (Figure 3): *Dunaliella* sp. was preferably consumed over both *P. tricornutum* and *P. lutheri*, but not over *Chaetoceros* sp. There was no significant difference in electivity between *Dunaliella* sp. and *T. lutea*. These preferences for individual algae were independent of algal concentration at the two different concentrations measured.

Electivity indices

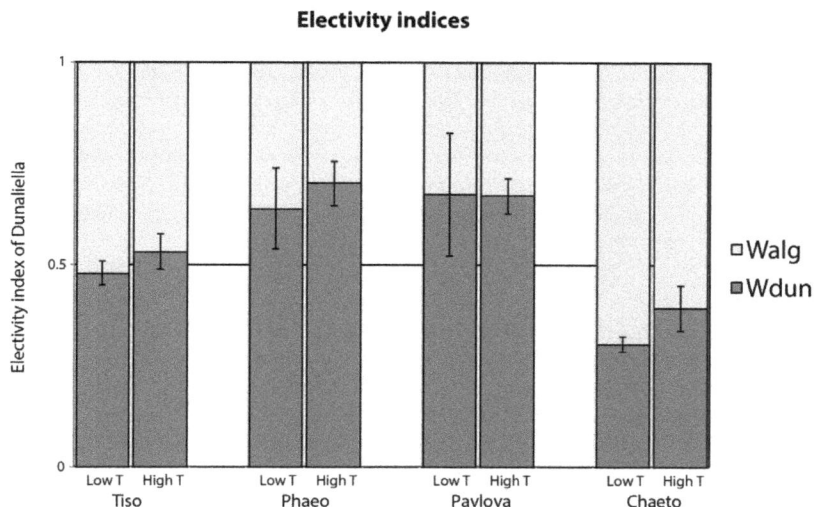

Figure 3. Electivity indices for larvae fed with a 1:1 mixture of two algal species. Mean electivity index for each of the for food mixtures. High treatment (High T) represents an initial concentration of 2500 cells·mL^{-1} (1250 cells mL^{-1} of each species). Low treatment (Low T) corresponds to an initial concentration of 1000 cells·mL^{-1} (500 of each species). The lower bars (Wdun, dark grey) represent the electivity for *Dunaliella* sp. and the upper bars (Walg, light grey) represent the electivity for the second algae present in the mixture. Electivity "Wdun" above 0.5 (grey dotted line) indicates that the larvae selectively ingested *Dunaliella* sp. over the other species. The error bars are 95% confidence interval.

The importance of the predictor "species" in all models of filtration and ingestion rates for *Dunaliella* sp. (Table 2) and for the second species (Table 3) showed that both rates were significantly influenced by the presence of the other species. This effect was mainly caused by *P. lutheri*: when *Dunaliella* sp. was offered combined with *P. lutheri*, its ingestion rate (25.8 algae larvae^{-1}·h^{-1}) was reduced compared to when *Dunaliella* sp. was offered in isolation (44.2 algae larvae^{-1}·h^{-1}) (dashed line on Figure 2b, top panel). The presence of other species did not have a significant effect on *Dunaliella* sp. consumption, with the exception of *T. lutea* at the low concentration. In that case, the consumption of *Dunaliella* sp. appeared slightly enhanced in the presence of *T. lutea*: indeed, we measured an ingestion rate of 28.5 algae larvae^{-1}·h^{-1} for 500 algae·mL^{-1} of *Dunaliella* sp. offered in a 1:1 mixture with *T. lutea* against 20.3 algae larvae^{-1}·h^{-1} for twice the same amount of *Dunaliella* sp. offered as a single species (Figure 2).

3.3. Algae Quality

The five algal species presented a variety of morphologies and crude nutritional compositions (Table 4). *P. tricornutum* was long and thin in shape, whereas the other species were roughly spherical. All species were of a different size, *Chaetoceros* sp. being the widest and, with a length of 9.2 µm, the second longest after *P. tricornutum* (18.9 µm in length). The content of each species varied with respect to the concentration (pg·cell^{-1}) of total lipid, total protein, carbon, nitrogen and chlorophyll. In accordance with its larger size, *Chaetoceros* sp. contained the highest cell-specific content of all constituents with the exception of chlorophyll, which was found in highest abundance in *Dunaliella* sp. (Table 4). Conversely, *T. lutea* (the smallest species) contained the lowest concentrations of these constituents followed by *P. lutheri* and *P. tricornutum*. Differences in crude nutritional composition subsequently manifested in clear trends relating to the energetic content of the algal cells. *Chaetoceros* sp. and *Dunaliella* sp. contained the highest energy levels with 1.6×10^{-9} and 1.4×10^{-9} kJ·cell^{-1}, respectively, while in comparison, *T. lutea* contained less than half the amount of total energy (0.6×10^{-9} kJ·cell^{-1}).

Based on algal energetic contents combined with effective ingestion rates, the greatest energy gain in the single species experiment resulted from the ingestion of *Dunaliella* sp. in the high concentration treatment ($12.2 \cdot 10^{-8}$·kJ·larvae^{-1}·h^{-1}). In the same experiment, the greater ingestion of *T. lutea* only resulted in $5.7 \cdot 10^{-8}$ kJ·larvae^{-1}·h^{-1} due to its lower energy content. In the mixed-species experiment, the greatest energy gain was obtained from the ingestion of *Dunaliella* sp. and *Chaetoceros* sp. in combination ($15.32 \cdot 10^{-8}$ kJ·larvae^{-1}·h^{-1}) followed by that of *Dunaliella* sp. mixed with *T. lutea* ($9.15 \cdot 10^{-8}$ kJ·larvae^{-1}·h^{-1}).

4. Discussion

Acanthaster cf. *solaris* (CoTS) larvae on the Great Barrier Reef are exposed to a diverse array of food choices to fuel their development through to settlement and metamorphosis [36]. The rate at which they select and consume these food items ultimately drives their rate of development and probability of survival, which is likely a major driver of CoTS population dynamics and probability of outbreak [37,38]. In the current experiments, algal consumption by CoTS larvae increased with algal concentration. As generally observed for other echinoderm larvae [10] and in CoTS subjected to short-term experiments with algal concentrations similar to those used here [7,12,13], this result is directly linked to an increased opportunity for larvae-algae interactions when algae are provided in abundance. However, clear differences appeared in larval consumption based on the algal species and quantity being offered (Figure 4).

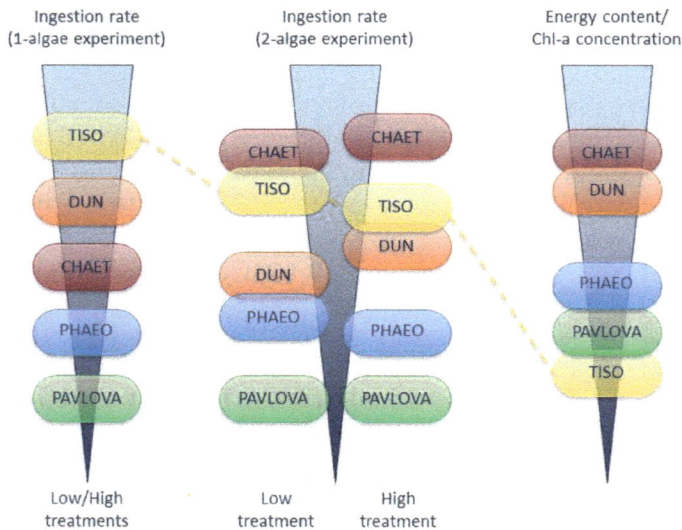

Figure 4. Conceptual synthesis of the results indicating relatively consistent ranking of the algal species based on ingestion rates (left and middle) and energy content (right), with the exception of *Tisochrysis lutea* (yellow). Overlapping bullets indicate non-significant differences (left and middle) or ambiguous ranking (right; e.g. *Chaetoceros* sp. had the highest energy content, but *Dunaliella* sp. had the greatest chlorophyll-a concentration; see Table 4 for details). Only *Tisochrysis lutea* showed an inverse ranking across gradients (as indicated by the dashed yellow line). TISO: *Tisochrysis lutea*; DUN: *Dunaliella* sp.; CHAET: *Chaetoceros* sp.; PHAEO: *Phaeodactylum tricornutum*; PAVLOVA: *Pavlova lutheri*.

The ability of CoTS larvae to discriminate and select the food they ingest was evident in the present study. Aside from the particular case of *T. lutea* (see below), *Dunaliella* sp. or *Chaetoceros* sp. were selectively ingested over *P. lutheri* and *P. tricornutum* in both the single- and mixed-species experiments (Figure 4). In particular, in the mixed-species experiment, a clear preference appeared for *Chaetoceros* sp., which was selectively ingested over *Dunaliella* sp. In contrast, *Dunaliella* sp. was preferred over both *P. lutheri* and *P. tricornutum*, as indicated by the lower ingestion rates and electivity indices observed for these two species. *Chaetoceros* sp. was the only species naturally present on the Great Barrier Reef, which could explain why it was selectively ingested over other species. Interestingly, these results also follow the trends observed in food quality and energy content, with *Chaetoceros* sp. having the highest content of energy, lipid, protein, carbon and nitrogen, whereas the lowest energy content was found for *P. lutheri* followed by *P. tricornutum*. This result indicates that, when offered a mixture of two algal species, CoTS larvae selectively ingested the species with higher energetic content when cells are present in abundant concentration. Consumption can be driven by the presence of primary metabolites (e.g., proteins, amino acids, sugars), with individual food items exhibiting unique primary metabolite ratios that distinguish one food source over another [39]. However, other factors relevant to cell morphometry, anti-nutritional factors and cell wall digestibility (among others) cannot be overlooked, and the exact mechanisms behind this discriminatory ability ultimately require further investigation.

Algal species that were selectively avoided by CoTS larvae (*P. lutheri, P. tricornutum*) were both characterized by lower energy content (Figure 4), which can lead to lower development and survival rates. Indeed, Lucas [7] found differences in development and survival for larvae fed with different algal species, with larvae fed with *P. lutheri* rarely making it through their development cycle. In addition to energy content, the size and/or shape of a particular algal species could also influence its selective ingestion by CoTS larvae, with the second least preferred species (*P. tricornutum*) being the only long and thin one. Okaji et al. [13] showed an effect of phytoplankton size on feeding rates, as well as a preference for eukaryotes over cyanobacteria (probably also resulting from their difference in size). Okaji et al. [13] also demonstrated that CoTS typically consume particles of 5–20 μm in equivalent spherical diameter. The length of *P. tricornutum* (19 μm), being close to the end of this range, is thus likely to hinder its ingestion by CoTS larvae. Furthermore, the thick cell walls of *P. tricornutum* could also reduce the rates of digestion and assimilation [7].

The case of *T. lutea* was somewhat particular in that no preference or avoidance was detected in the mixed-species experiment, yet *T. lutea* was ingested at a much higher rate when offered as a single species (Figure 4). Its low energy content meant that, despite more algal cells being ingested, the consumption of this species led to the lowest energy gain for CoTS larvae. Such an impact was demonstrated for other zooplankton species: for example, Baldwin [18] showed that oyster larvae ingestion rates were three times higher for algae with a low C/N ratio than for algae with a high ratio. Here, *T. lutea* was characterized by the lowest nitrogen content and second lowest C/N ratio, indicating that CoTS larvae would need to consume more of this species than the others to satisfy energy requirements.

The interaction between algal species affected their filtration and ingestion rates by CoTS larvae, with both rates for *Dunaliella* sp. being significantly influenced by the other species present in the mixture. Notably, the presence of *P. lutheri* reduced the ingestion of *Dunaliella* sp. by CoTS larvae, without increased uptake of the former species. A similar phenomenon was previously reported for larvae of the brittle star *Ophiopholis aculeata* and the sea urchin *Strongylocentrotus droebachiensis* [10]. In these species, the presence of the diatom *Ditylum brightwellii* reduced the feeding rate on smaller celled algae. Similarly, the inadvertent ingestion of *P. lutheri* by CoTS larvae may have stopped the larvae from feeding all together, but the exact mechanisms underpinning food selection (and how the ingestion of one species influences that of another) remain unclear. In fact, the filtration theory for suspension feeders is still controversial, as it does not explain the variable ingestion rates observed for different particles [40,41]. A study on sand dollar *Dendraster excentricus* larvae found that they selectively fed on specific particles and refused to capture or ingest others [42], while a more recent study showed that those larvae concentrated food particles without filtration [43]. Instead they use a different mechanism, actively capturing the particles by reversing the beat of a small portion of their cilia. In doing so, they reversed the movement of a small and targeted amount of water and were thus able to select which particle to consume. The ability of CoTS larvae to select their food could be explained by such a mechanism, but this hypothesis needs to be further investigated.

The ingestion and filtration rates found in this study were consistent with those reported by Lucas [7] based on incubation times of 24 h, but differed from the higher rates reported by Ayukai [12] and Okaji et al. [13] in short-term studies (5–30 min). In a copepod feeding experiment, Frost [44] found that copepods had abnormally high ingestion rates during the first few hours of feeding, which could be an artefact of the imposed period of starvation prior to the beginning of the experiment. As such, the high feeding rates reported by Ayukai [12] and Okaji et al. [13] likely correspond to the elevated cell intake that manifested only over a short experimental duration.

Preference for a particular algal species is likely to be taxon-specific. Together with *T. lutea*, *P. lutheri* represented the best diet for the larvae of *Saccostrea commercialis*, a commercially important oyster [45]. *Pecten maximus* larvae (scallop) have been shown to perform favourably when subjected to *P. lutheri* as a food source, but poorly when *Dunaliella tertiolecta* was offered [46]. In contrast, in the present study, CoTS larvae exhibited a significant preference for *Dunaliella sp.* over *P. lutheri*, a trend previously observed for *Crassostrea gigas* larvae (oyster) [47]. Indeed, oyster larvae fed with *P. lutheri* exhibited significantly lower growth and survival rates than the same larvae fed with *Chaetoceros calcitrans* [47]. Similar to CoTS, this trend might reflect the positive effect of a more energetic food source on ingestion and subsequently growth and survival. However, although facultative food intake in the early life stage of echinoids can influence individual condition, this does not necessarily translate into differences in growth and individual size [48]. Nutritional requirements seem to vary widely between species and life stages, meaning that the presence/absence of a single essential nutrient (specific to each species and development stage) will likely have a major impact on larval performance.

Phytoplankton concentration in the natural environment is often estimated using chlorophyll-*a* concentration as a proxy. On the Great Barrier Reef, long-term monitoring programs on the inner shelf region estimated a chlorophyll-*a* concentration of 0.5 μg·L^{-1} on average [4], ranging between 0.2 and 4.6 μg·L^{-1} in flood plumes [22]. In this experiment, chlorophyll-*a* concentrations ranged from 0.3–1.0 μg·L^{-1} in the low treatment (1000 cells·ml^{-1}) and from 0.8–2.6 μg·L^{-1} in the high treatment (2500 cells·mL^{-1}), which is consistent with the ranges reported on the Great Barrier Reef.

Phytoplankton composition is highly variable through different regions and flood conditions [22], yet little is known on the exact species being present, their seasonal succession and spatial distribution. Likewise, information pertaining to the energetic content and nutritional value of these species is minimal and requires in-depth investigation. This is especially apparent considering the discriminatory feeding nature of CoTS and the previous demonstration of nutritional augmentation of growth and survival in other larval echinoderm species in response to varying phytoplankton feed sources [49]. The highest phytoplankton concentrations in flood plumes of the Great Barrier Reef were dominated by *Skeletonema* sp. (a coastal diatom of 2–21 μm in diameter) that reached up to 4422 cells mL^{-1}, followed by *Chaetoceros* sp., which reached concentrations of 150 cells·mL^{-1} [22]. Although the other algal species considered in this study are commonly used in aquaculture and in the size range consumed by CoTS, they are not naturally present on the Great Barrier Reef, and future studies should focus on natural assemblages, pending sufficient knowledge of their composition. In particular, future studies should consider using *Skeletonema* sp., although its chain-forming nature indicates it is an unlikely food for echinoderm larvae and may hamper its quantification by flow cytometry. Consistent with previous studies (e.g., [7–9,38]), we used larval densities between 0.5 and 1.5 larvae·mL^{-1}, which is much higher than reported in the only published studies on CoTS natural larval densities [50]. Although studies at such low densities are near impossible to conduct, future studies should investigate how lower larval densities affect CoTS larval feeding ecology. High-throughput genetic techniques, such as next generation sequencing, now allow further insight into GBR phytoplankton dynamics and composition. Ultimately, improved knowledge of phytoplankton composition and its effect on food uptake by CoTS larvae will help elucidate the CoTS outbreak dynamics and triggers.

Acknowledgments: We are grateful to Steve Moon and the Association of Marine Park Tourism Operators for their ongoing assistance to obtain study animals. Research was carried out in the National Sea Simulator (SeaSim) and we are grateful for the support of the SeaSim staff. All collections have been conducted under Great Barrier Reef Marine Park Permit No. GBRMPA Permit G38062.1. CM was funded by an Australian Research Council grant (DE140100701).

Author Contributions: S.U., D.F. and K.O. conceived of and designed the experiments. C.L. performed the experiments. C.M. and C.L. analysed the data. S.U., D.S.F. and K.O. contributed materials and analysis tools. C.M. and C.L. wrote the paper, with contributions from all other authors.

Conflicts of Interest: The authors declare no conflict of interest.

Appendix A

Table A1. Initial algal concentrations determined by flow cytometry in comparison to expected values.

Experiment/Treatment	Expected Initial Concentration (cells·mL^{-1})	Actual Measured Concentration (cells·mL^{-1})	
		Mean ± Standard Deviation	Minimum-Maximum
Single-species			
Low concentration	1000	848 ± 180	520–1210
High concentration	3000	2563 ± 500	1530–3780
Mixed-species (total concentration)			
Low concentration	1000	947 ± 131	740–1310
High concentration	3000	848 ± 180	1860–3630
Mixed-species (individual concentrations)			
Low concentration	500	461 ± 110	340–760
High concentration	1500	1315 ± 252	850–2070

References

1. De'ath, G.; Fabricius, K.E.; Sweatman, H.; Puotinen, M. The 27-year decline of coral cover on the Great Barrier Reef and its causes. *Proc. Natl. Acad. Sci. USA* **2012**, *109*, 17995–17999. [CrossRef] [PubMed]
2. Birkeland, C. Terrestrial runoff as a cause of outbreaks of *Acanthaster planci* (Echinodermata: Asteroidea). *Mar. Biol.* **1982**, *69*, 175–185. [CrossRef]
3. Brodie, J. From the effects of terrestrial runoff: A review. *Mar. Freshw. Res.* **1992**, *43*, 539. [CrossRef]
4. Brodie, J.; Fabricius, K.; De'ath, G.; Okaji, K. Are increased nutrient inputs responsible for more outbreaks of crown-of-thorns starfish? An appraisal of the evidence. *Mar. Pollut. Bull.* **2005**, *51*, 266–278. [CrossRef] [PubMed]
5. Devlin, M.J.; Brodie, J. Terrestrial discharge into the Great Barrier Reef Lagoon: Nutrient behavior in coastal waters. *Mar. Pollut. Bull.* **2005**, *51*, 9–22. [CrossRef] [PubMed]
6. Fabricius, K.E.; Okaji, K.; De'ath, G. Three lines of evidence to link outbreaks of the crown-of-thorns seastar *Acanthaster planci* to the release of larval food limitation. *Coral Reefs* **2010**, *29*, 593–605. [CrossRef]
7. Lucas, J.S. Quantitative studies of feeding and nutrition during larval development of coral reef asteroid *Acanthaster planci* (L.). *J. Exp. Mar. Biol. Ecol.* **1982**, *65*, 173–193. [CrossRef]
8. Wolfe, K.; Graba-Landry, A.; Dworjanyn, S.A.; Byrne, M. Larval starvation to satiation: Influence of nutrient regime on the success of *Acanthaster planci*. *PLoS ONE* **2015**, *10*, e0122010. [CrossRef] [PubMed]
9. Uthicke, S.; Logan, M.; Liddy, M.; Francis, D.; Hardy, N.; Lamare, M. Climate change as an unexpected co-factor promoting coral eating seastar (*Acanthaster planci*) outbreaks. *Sci. Rep.* **2015**, *5*, 8402. [CrossRef] [PubMed]
10. Strathmann, R.R. The feeding behavior of planktotrophic echinoderm larvae: Mechanisms, regulation, and rates of suspensionfeeding. *J. Exp. Mar. Bio. Ecol.* **1971**, *6*, 109–160. [CrossRef]
11. Strathmann, R.R. Larval Feeding in Echinoderms. *Am. Zool.* **1975**, *15*, 717–730. [CrossRef]

12. Ayukai, T. Ingestion of ultraplankton by the planktonic larvae of the crown-of-thorns starfish, *Acanthaster planci*. *Biol. Bull.* **1994**, *186*, 90–100. [CrossRef]

13. Okaji, K.; Ayukai, T.; Lucas, J.S. Selective feeding by larvae of the crown-of-thorns starfish, *Acanthaster planci* (L.). *Coral Reefs* **1997**, *16*, 47–50. [CrossRef]

14. Nakajima, R.; Nakatomi, N.; Kurihara, H.; Fox, M.; Smith, J.; Okaji, K. Crown-of-thorns starfish larvae can feed on organic matter released from corals. *Diversity* **2016**, *8*, 18. [CrossRef]

15. Collier, J.L. Flow cytometry and the single cell in phycology. *J. Phycol.* **2000**, *36*, 628–644. [CrossRef]

16. Hofstraat, J.W.; van Zeijl, W.J.M.; de Vreeze, M.E.J.; Peeters, J.C.H.; Peperzak, L.; Colijn, F.; Rademaker, T.W.M. Phytoplankton monitoring by flow cytometry. *J. Plankton Res.* **1994**, *16*, 1197–1224. [CrossRef]

17. Cowles, T.J.; Olson, R.J.; Chisholm, S.W. Food selection by copepods: Discrimination on the basis of food quality. *Mar. Biol.* **1988**, *100*, 41–49. [CrossRef]

18. Baldwin, B.S. Selective particle ingestion by oyster larvae (*Crassostrea virginica*) feeding on natural seston and cultured algae. *Mar. Biol.* **1995**, *123*, 95–107. [CrossRef]

19. Dionisio Pires, L.M.; Jonker, R.R.; Van Donk, E.; Laanbroek, H.J. Selective grazing by adults and larvae of the zebra mussel (*Dreissena polymorpha*): Application of flow cytometry to natural seston. *Freshw. Biol.* **2004**, *49*, 116–126. [CrossRef]

20. Fileman, E.S.; Lindeque, P.K.; Harmer, R.A.; Halsband, C.; Atkinson, A. Feeding rates and prey selectivity of planktonic decapod larvae in the Western English Channel. *Mar. Biol.* **2014**, *161*, 2479–2494. [CrossRef]

21. Revelante, N.; Gilmartin, M. Dynamics of phytoplankton in the Great Barrier Reef lagoon. *J. Plankton Res.* **1982**, *4*, 47–76. [CrossRef]

22. Devlin, M.J.; Debose, J.; Ajani, P.; Petus, C.; da Silva, E.T.; Brodie, J.O.N. Phytoplankton in the Great Barrier Reef: Microscopy analysis of community structure in high flow events. Report to the National Environmental Research Program. Reef and Rainforest Research Centre Limited: Cairns, Australia, 2013; p. 68.

23. Conlan, J.A.; Jones, P.L.; Turchini, G.M.; Hall, M.R.; Francis, D. Changes in the nutritional composition of captive early-mid stage *Panulirus ornatus* phyllosoma over ecdysis and larval development. *Aquaculture* **2014**, *434*, 159–170. [CrossRef]

24. Angell, A.R.; Mata, L.; de Nys, R.; Paul, N.A. The protein content of seaweeds: A universal nitrogen-to-protein conversion. *J. Appl. Phycol.* **2016**, *28*, 511–524. [CrossRef]

25. Bureau, D.P.; Kaushik, S.J.; Cho, C.Y. Bioenergetics. In *Fish Nutrition*; Halver, J.E., Hardy, R.W., Eds.; Academic Press: San Diego, CA, USA, 2002.

26. Schaffelke, B.; Carleton, J.; Skuza, M.; Zagorskis, I.; Furnas, M.J. Water quality in the inshore Great Barrier Reef lagoon: Implications for long-term monitoring and management. *Mar. Pollut. Bull.* **2012**, *65*, 249–260. [CrossRef] [PubMed]

27. Okaji, K. Feeding ecology in the early life stages of the crown-of-thorns starfish, *Acanthaster planci* (L.). PhD dissertation, James Cook University, Townsville, Australia, 1996.

28. Coughlan, J. The estimation of filtering rate from the clearance of suspensions. *Mar. Biol.* **1969**, *2*, 356–358. [CrossRef]

29. Vanderploeg, H.a.; Scavia, D. Calculation and use of selectivity coefficients of feeding: Zooplankton grazing. *Ecol. Modell.* **1979**, *7*, 135–149. [CrossRef]

30. Bolker, B.M.; Brooks, M.E.; Clark, C.J.; Geange, S.W.; Poulsen, J.R.; Stevens, M.H.H.; White, J.S.S. Generalized linear mixed models: A practical guide for ecology and evolution. *Trends Ecol. Evol.* **2009**, *24*, 127–135. [CrossRef] [PubMed]

31. Marin, V.; Huntley, M.E.; Frost, B. Measuring feeding rates of pelagic herbivores: Analysis of experimental design and methods. *Mar. Biol.* **1986**, *58*, 49–58. [CrossRef]

32. Nakagawa, S.; Schielzeth, H. A general and simple method for obtaining R2 from generalized linear mixed-effects models. *Meth. Ecol. Evol.* **2013**, *4*, 133–142. [CrossRef]

33. Burnham, K.P.; Anderson, D.R. *Model selection and multimodel inference: A practical information theoretic approach*, 2nd ed.; Springer: New York, NY, USA, 2002.

34. Burnham, K.P.; Anderson, D.R.; Huyvaert, K.P. AIC model selection and multimodel inference in behavioral ecology: Some background, observations, and comparisons. *Behav. Ecol. Sociobiol.* **2011**, *65*, 23–35. [CrossRef]

35. R Development Core Team. *R: A language and environment for statistical computing*; R Foundation for Statistical Computing: Vienna, Austria; ISBN: 3–900051–07–0. 2014; Available online: http://www.R-project.org/. (accessed on 21 January 2015).

36. Pratchett, M.S.; Caballes, C.F.; Rivera-Posada, J.A.; Sweatman, H.P. Limits to understanding and managing outbreaks of crown-of-thorns starfish (*Acanthaster* spp.). *Oceanogr. Mar. Biol.* **2014**, *52*, 133–199.

37. Mellin, C.; Lurgi, M.; Matthews, S.; MacNeil, M.A.; Caley, M.J.; Bax, N.; Przeslawski, R.; Fordham, D.A. Forecasting marine invasions under climate change: Biotic interactions and demographic processes matter. *Biol. Conserv.* **2016**. [CrossRef]

38. Pratchett, M.S.; Dworjanyn, S.A.; Mos, B.; Caballes, C.F.; Thompson, C.A.; Blowes, S. Larval survivorship and settlement of crown-of-thorns starfish (*Acanthaster* cf. *solaris*) at varying algal cell densities. *Diversity* **2017**, *9*, 2. [CrossRef]

39. Hay, M.E. Marine chemical ecology: Chemical signals and cues structure marine populations, communities, and ecosystems. *Ann. Rev. Mar. Sci.* **2009**, *1*, 193–212. [CrossRef] [PubMed]

40. Labarbera, M. Feeding currents and particle capture mechanisms in suspension feeding animals. *Integr. Comp. Biol.* **1984**, *24*, 71–84. [CrossRef]

41. Rubenstein, D.I.; Koehl, M.A.R. The mechanisms of filter feeding: Some theoretical considerations. *Am. Nat.* **1977**, *111*, 981. [CrossRef]

42. Rassoulzadegan, F.; Fenaux, L.; Strathmann, R.R. Effect of flavor and size on selection of food by suspension-feeding plutei. *Limnol. Oceanog.* **1984**, *29*, 357–360. [CrossRef]

43. Strathmann, R.R. Time and extent of ciliary response to particles in a non-filtering feeding mechanism. *Biol. Bull.* **2007**, *212*, 93–103. [CrossRef] [PubMed]

44. Frost, B.W. Effects of size and concentration of food particles on the feeding behavior of the marine planktonic copepod *Calanus pacificus. Limnol. Oceanogr.* **1972**, *17*, 805–815. [CrossRef]

45. Nell, J.A.; Connor, W.A.O. The evaluation of fresh algae and stored algal concentrates as a food source for Sydney rock oyster, *Succostrea commercialis* (Iredale & Roughley), larvae. *Aquaculture* **1991**, *99*, 277–284.

46. Delaunay, F. The effect of monospecific algal diets on growth and fatty acid composition of *Pecten maximus* (L.) larvae. *J. Exp. Mar. Bio. Ecol.* **1993**, *173*, 163–179. [CrossRef]

47. Ponis, E.; Robert, R.; Parisi, G.; Tredici, M. Assessment of the performance of Pacific oyster (*Crassostrea gigas*) larvae fed with fresh and preserved *Pavlova lutheri* concentrates. *Aquaculture Inter.* **2003**, *11*, 69–79. [CrossRef]

48. Byrne, M.; Sewell, M.A.; Prowse, T.A.A. Nutritional ecology of sea urchin larvae: Influence of endogenous and exogenous nutrition on echinopluteal growth and phenotypic plasticity in *Tripneustes gratilla. Funct. Ecol.* **2008**, *22*, 643–648. [CrossRef]

49. Duy, N.D.Q.; Francis, D.S.; Pirozzi, I.; Southgate, P.C. Use of micro-algae concentrates for hatchery culture of sandfish, *Holothuria scabra. Aquaculture* **2016**, *464*, 145–152. [CrossRef]

50. Suzuki, G.; Yasuda, N.; Ikehara, K.; Fukuoka, K.; Kameda, T.; Kai, S.; Nagai, S.; Watanabe, A.; Nakamura, T.; Kitazawa, S.; et al. Detection of a high-density brachiolaria-stage larval population of Crown-of-Thorns sea star (*Acanthaster planci*) in Sekisei lagoon (Okinawa, Japan). *Diversity* **2016**, *8*, 9. [CrossRef]

diversity

MDPI

Communication

Larval Survivorship and Settlement of Crown-of-Thorns Starfish (*Acanthaster* cf. *solaris*) at Varying Algal Cell Densities

Morgan S. Pratchett [1,*], Symon Dworjanyn [2], Benjamin Mos [2], Ciemon F. Caballes [1], Cassandra A. Thompson [1] and Shane Blowes [1,3]

[1] ARC Centre of Excellence for Coral Reef Studies, James Cook University, Townsville, QLD 4811, Australia; ciemon.caballes@my.jcu.edu.au (C.F.C.); cassandra.thompson@jcu.edu.au (C.A.T.); shane.blowes@my.jcu.edu.au (S.B.)
[2] National Marine Science Centre, Southern Cross University, Coffs Harbour, NSW 2450, Australia; Symon.Dworjanyn@scu.edu.au (S.D.); benjamin.mos@scu.edu.au (B.M.)
[3] Department of Zoology, George S. Wise Faculty of Life Sciences, Tel Aviv University, Tel Aviv 69978, Israel
* Correspondence: morgan.pratchett@jcu.edu.au; Tel.: +61-747-81-5747

Academic Editors: Sven Uthicke and Michael Wink
Received: 14 November 2016; Accepted: 4 January 2017; Published: 10 January 2017

Abstract: The dispersal potential of crown-of-thorns starfish (CoTS) larvae is important in understanding both the initiation and spread of population outbreaks, and is fundamentally dependent upon how long larvae can persist while still retaining the capacity to settle. This study quantified variation in larval survivorship and settlement rates for CoTS maintained at three different densities of a single-celled flagellate phytoplankton, *Proteomonas sulcata* (1×10^3, 1×10^4, and 1×10^5 cells/mL). Based on the *larval starvation hypothesis*, we expected that low to moderate levels of phytoplankton prey would significantly constrain both survival and settlement. CoTS larvae were successfully maintained for up to 50 days post-fertilization, but larval survival differed significantly between treatments. Survival was greatest at intermediate food levels (1×10^4 cells/mL), and lowest at high (1×10^5 cells/mL) food levels. Rates of settlement were also highest at intermediate food levels and peaked at 22 days post-fertilization. Peak settlement was delayed at low food levels, probably reflective of delayed development, but there was no evidence of accelerated development at high chlorophyll concentrations. CoTS larvae were recorded to settle 17–43 days post-fertilization, but under optimum conditions with intermediate algal cell densities, peak settlement occurred at 22 days post-fertilization. Natural fluctuations in nutrient concentrations and food availability may affect the number of CoTS that effectively settle, but seem unlikely to influence dispersal dynamics.

Keywords: *Acanthaster*; coral reefs; food limitation; larval competency; planktonic larval duration (PLD)

1. Introduction

Sessile and benthic marine invertebrates are fundamentally dependent on the larval phase of their lifecycle for dispersal away from natal reefs, which is important for enabling colonization of new habitats, recolonization following population depletion, and genetic exchange among sub-populations [1,2]. Despite the short larval duration of most marine organisms (days to months), larvae may be dispersed over great distances [2]. Importantly, ecologically and evolutionarily significant levels of genetic exchange occur over very large (even oceanic) scales (e.g., [3]). There is however, evidence for some species that most of the larvae (up to 60%) settling on a given reef are of local origin [4], implying that most larvae may not even travel beyond the confines of a single reef. Ultimately, there may be two distinct modes (short versus long, retention versus dispersal, or self-recruitment versus departure) in the range of dispersal distances for individual species [5]. Variation in dispersal within and among cohorts of larvae will have a critical influence on population dynamics and persistence for widespread species [5].

For crown-of-thorns starfish (*Acanthaster* spp.), hereafter referred to as CoTS, the proportion of larvae that are retained versus dispersed is important in understanding both the initiation and spread of population outbreaks [6–8]. Strong larval retention is fundamental to the progressive accumulation of CoTS within a given location, which is increasingly viewed as the predominant mechanism that gives rise to primary outbreaks [6,7]. On the Great Barrier Reef (GBR). Woolridge and Brodie [7] suggested that rates of larval retention vary among years with changes in ENSO driven ocean current velocities, and that initiation of outbreaks is coincident with periods of "strong local clustering", where there are high levels of retention at individual reefs or within tightly packed reef clusters. Once established, initial outbreak populations then give rise to large numbers of larvae that often spread to other reefs. Moreover, dispersal to other downstream (generally southerly) reefs is facilitated by increases in ocean currents and relaxation of the strong local clustering [7]. This larval retention hypothesis [7] is particularly tenable because it explains why major flood events and nutrient enrichment in near reef waters do not necessarily initiate outbreaks of CoTS, but may facilitate rapid proliferation and spread of outbreaks across the GBR. Moreover, changes in rates of larval retention versus dispersal may be facilitated by shifts in larval development rates and settlement behaviour due to natural fluctuations in nutrients and food availability. Yet, little is known about how food availability may influence developmental rates for crown-of-thorns starfish larvae.

Aside from local hydrodynamics and larval behaviour, two fundamental parameters that influence the retention versus dispersal of marine larvae are (i) the minimum pre-competency period, which is the time taken (in hours or days) for larvae to complete necessary development prior to being physiologically and morphologically capable of settling; and (ii) the maximum competency period, which is the longest period that larvae can remain in the plankton (often dictated by their initial energy reserves and/or capacity to sustain themselves) and still retain the ability to successfully settle. The minimum pre-competency period has a strong bearing on the extent to which larvae are likely retained on their natal reef(s), whereas the maximum competency period is fundamental in establishing how far larvae could potentially spread. Marine fishes typically exhibit very narrow ranges in their planktonic larval duration (PLD) (i.e., only a few days separate the maximum competency period and minimum pre-competency period), especially for species with short pre-competency periods [9]. For fishes, therefore, individual variation in dispersal is largely attributed to traits that influence their physical capacity (e.g., swimming performance and navigation ability) to extend dispersal distances or return to their natal reefs despite limited differences in larval duration [5]. Marine invertebrates however, may be competent for highly protracted periods, such that individuals from the same cohort may settle anywhere from several hours to many months after fertilization [10,11] depending on their nutritional status and exposure to settlement cues.

Acanthaster spp. have been successfully cultured under laboratory conditions since the early 1970s [12], largely for the purposes of understanding their developmental biology. Initial rearing experiments, where larvae were maintained at 24–25 °C for the formative stages of development (until larvae reached late bipinnaria at day 21) resulted in PLD estimates of 30–47 days, with a mean of 38 days [12]. However, CoTS generally spawn only when sea surfaces temperatures are $\geq$27 °C [13], and larvae develop more quickly at higher temperatures [14,15]. Lamare et al. [14] showed that development rates of CoTS from Australia's GBR were maximized at or above 28.7 °C, at which temperature, larvae may complete development and be competent to settle within 12 to 14 days [16].

Aside from the effects of temperature, variation in larval survivorship and rates of larval development for CoTS are also influenced by nutrient concentrations and food availability (e.g., [17]). One of the foremost hypotheses asserting that outbreaks of CoTS are exacerbated (if not caused) by anthropogenic activities, the *larval starvation hypothesis*, links the initiation and/or spread of outbreaks to elevated nutrient concentrations caused by intensive agriculture in adjacent catchments [18,19]. This idea is largely predicated on marked differences in the proportion of CoTS larvae that complete development over relatively moderate ranges of chlorophyll concentrations [19]. Fabricius et al. [19] showed that developmental success is effectively zero at <0.25 $\mu g \cdot L^{-1}$ chlorophyll compared to 100% at 2–4 $\mu g \cdot L^{-1}$ chlorophyll. It is suggested therefore, that elevated chlorophyll concentrations caused by run-off from modified catchments during intensive storms will reduce constraints on larval survivorship and recruitment imposed by normally low-levels of nutrients and food [19]. More recent studies have, however, shown that very high levels of nutrients are deleterious to larval development for CoTS, while larval growth and development are maximized at close to normal background concentrations of chlorophyll [20]. These studies [17,20] compare the size and fate of CoTS larvae at specific time intervals (e.g., at 4 days old) across different nutrient concentrations, but it is unknown to what extent nutrient concentrations may influence the competency periods, and therefore dispersal potential, of crown-of-thorns starfish larvae.

In this study, we quantify variation in larval survivorship and settlement rates for CoTS maintained at three different food treatments. Based on the *larval starvation hypothesis* (e.g., [19]), we expected that starfish subject to very low food availability would develop more slowly and have lower survivorship compared to larval cultures maintained with higher cell densities of phytoplankton. Accordingly, we expected that increases food availability would increase rates of development, thereby reducing the minimum pre-competency period. High food availability meanwhile, are also expected to greatly extend the maximum competency period, both by increasing rates of survivorship and provisioning larvae with resources necessary for metamorphosis at settlement. It is known however, that excess food can inhibit filter feeding and digestion by CoTS larvae, ultimately leading to arrested development and high levels of mortality [17]. While previous studies (e.g., [17,20]) have compared the size, development and survival of larvae of CoTS at set intervals in their development (to explicitly test the effects of food availability among individuals of equivalent age), this study looks at overall longevity and competency of larvae to test whether increased food availability will expedite development, potentially leading to higher rates of larval retention. Alternatively, elevated nutrients and food availability may extend the maximum competency period for CoTS larvae and therefore, facilitate greater dispersal.

2. Materials and Methods

2.1. Larval Culture and Settlement Experiments

Adult *Acanthaster* cf. *solaris* [21] were collected from the GBR near Cairns, Australia (16°55′ S, 145°46′ E) by control divers employed by the Association of Marine Park Tourism Operators (AMPTO) and air freighted to the National Marine Science Centre, Southern Cross University at Coffs Harbour, NSW, Australia. Eggs from three female starfish were collected, pooled and fertilized using sperm from three males, following Kamya et al. [22]. Fertilized embryos were reared for two days in an aerated 300-L cylindro-conical tank at ~28 °C at a density of 2.5 larvae·mL^{-1}. After 2 days (coinciding with the onset of feeding), larvae were distributed among replicate 1-L cylindro-conical containers and subject to one of three different feeding/nutrient treatments. Larvae were reared in 1-L cylindro-conical containers in filtered (5 μm) seawater at a density of 1.2 larvae·mL^{-1} and fed *Proteomonas sulcata* at 1, 10 and 100 × 10^3 cells·mL^{-1}, respectively (Wolfe et al. 2015 [20]). *Proteomonas sulcata* is a single-celled flagellated phytoplankton that is 7–10 μm long, and within the size range of food particles readily consumed by CoTS [23,24]. There were 10 replicates for each distinct algal density and chlorophyll concentration. Containers were maintained in a climate-controlled room at 28.5 °C (±0.2 °C). Larvae were fed daily after a 100% water exchange. Larval cultures were moved to new containers every 2–3 days.

To follow survival, the number of larvae in each container was quantified every 2–3 days by extrapolating the number of larvae counted in 20 mL subsamples taken from each container.

However, when larval density was <0.1 larvae·mL^{-1}, all of the larvae were counted. Survival rates were calculated as the number of larvae remaining out of the total added to the container at the start of the experiment, while also adjusting for larvae removed from the containers for settlement assays and for larvae that spontaneously settled in the containers, from 17 days onwards. Importantly, estimates of survivorship in this laboratory-based study are likely to be much higher than expected in the field, where sustained predation on gametes and larvae [25] may result in very high rates of mortality [26].

Settlement assays were undertaken from 17 days post-fertilization, which was the first day at which late-stage brachiolaria with well-developed rudiments were observed across all treatments, thereby enabling comparisons of settlement rates across treatments. Settlement assays were conducted in 36 mm plastic petri dishes containing 4 mL of fresh seawater, and a 1 cm^2 crustose coralline algae (CCA) encrusted polycarbonate plate to induce settlement [27]. For each assay, there were three replicates from each larval container with 10 larvae per replicate, although this was reduced to one replicate when there were less than 100 larvae in a container. Settlement rates were quantified after 48 h. All newly settled, 5-arm juveniles were photographed in seawater, aboral side up, and flat to the plane of focus using a camera (Olympus DP26) mounted on a stereomicroscope. Area of the juveniles was measured from the photographs using Image-J image analysis software (National Institutes of Health, Bethesda, MD, USA). Settlement assays were conducted 17, 22, 29, 36 and 43 days post-fertilization. Thereafter, there were too few larvae in any of the treatments to effectively test for settlement.

2.2. Statistical Analyses

Larval survival was estimated from sub-sampling rather than a complete census, so traditional methods of survival analysis were not appropriate. Instead, we modeled counts using non-linear splines (GAMMs; [28]) assuming Poisson error and a log link function. More specifically, we examined counts of surviving larvae using models with a categorical treatment effect, and treatment specific cubic regression splines; splines were estimated with shrinkage, meaning that smoothness selection can zero a term completely [28], and the number of knots was set to 4 for all smooths to prevent over-fitting. Additionally, replicate was fitted as a random effect. All models were estimated using the gamm4 package [29]. Overall goodness-of-fit was checked using standard techniques (e.g., residuals plotted against all explanatory variables and fitted values from the models).

Settlement assay data were reduced to observations where the numbers of larvae available per settlement assay were known before analysis. This resulted in a data set of 292 observations for all age-treatment combinations, with replication at each combination ranging from 1 to 30. Due to this unbalanced design, we modeled the probability of successful settlement using a generalized linear mixed effects model (GLMM) with binomial error and a logit link function. Age (continuous) and treatment (categorical: three levels) were modeled as fixed effects; replicate and assay were modeled as random effects, with assay (up to three per replicate) nested within replicate (up to 10 replicates per age-treatment combination); age was centered by subtracting the mean age from all observed ages before model fitting. To examine potential nonlinearities in settlement probability with larval age and age-by-treatment interactions, we fit a model with a second-order term for age, and an age-treatment interaction term. Initial model fits suggested that data were over-dispersed, and the beta-binomial was supported over the binomial distribution (likelihood ratio test: $X^2 = 84.62$, $df = 1$, $p < 0.001$). Models were estimated using maximum likelihood in the glmmADMB package [30,31] in R 3.3.1 [32]. We assessed the significance of fixed effects using type II Wald's chi-square tests, and the fit of models and their conformity to statistical assumptions were visually inspected using standard techniques (i.e., plots of residuals versus fitted values, and all covariates).

3. Results

3.1. Larval Survival

Larval crown-of-thorns starfish were successfully maintained in 1-L cylindro-conical containers for up to 50 days post-fertilization, though most larvae had died or settled within 30 to 40 days. Mean larval survival differed significantly between treatments (i.e., the parametric treatment component was significant: $X^2 = 470$, $df = 2$, $p < 0.001$; see Table 1 for individual terms), and was greatest at intermediate (1.0 μg·L^{-1}) chlorophyll concentrations, followed by the low and the high chlorophyll concentrations (Table 1; Figure 1). There were initially high rates of larval mortality across all treatments, but mortality rate decreased (between approximately 15 and 22 days) at low to intermediate chlorophyll concentrations, before increased rates of mortality after 30 days (Figure 1). Larval survival initially decreased at similar rates for the high and low food treatments, but the initial loss of larvae was slower in the medium food treatment. At high algal concentrations, high rates of early post-fertilization mortality continued in an exponential-like decay throughout the experiment, such that the maximum longevity of larvae was only 36 days.

Table 1. Estimates of the (**a**) parametric coefficients and (**b**) the approximate significance of the smooth terms of the survival model (Poisson error distribution, log link functions). Parametric intercept term represents the low food treatment.

(a) Parametric Coefficients	Estimate	Std. Error	p
Intercept	4.59	0.05	$<2 \times 10^{-16}$
treatment (medium)	0.51	0.04	$<2 \times 10^{-16}$
treatment (high)	−3.50	0.23	$<2 \times 10^{-16}$
(b) Smooth Terms	edf	Ref.df	Chi.sq	p
s(time) treatment (low)	2.997	3	12,436	$<2 \times 10^{-16}$
s(time) treatment (medium)	2.998	3	9927	$<2 \times 10^{-16}$
s(time) treatment (high)	2.985	3	10,680	$<2 \times 10^{-16}$

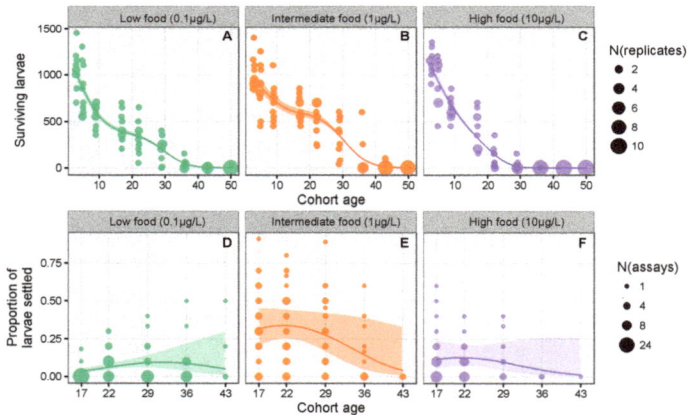

Figure 1. Non-linear effects of increasing food availability on larval survivorship and settlement for crown-of-thorns starfish (*Acanthaster* cf. *solaris*) from Australia's Great Barrier Reef: (**A–C**) Absolute larval survivorship between 1 and 50 days post-fertilization for each of ten replicate cultures maintained at low (0.1 μg·L^{-1}), medium (1.0 μg·L^{-1}) or high (10 μg·L^{-1}) chlorophyll concentrations; (**D–E**) Proportion of larvae (*n* = 10–30 per culture) that successfully settled 17, 22, 29, 36 and 43 days post-fertilization for each of the ten replicate cultures maintained at low (0.1 μg·L^{-1}), medium (1.0 μg·L^{-1}) or high (10 μg·L^{-1}) chlorophyll concentrations. Lines show the predictions from the GAMMs fit to the survival data (**A–C**), and from the GLMMs fit to the settlement data (**D–F**); shaded areas depict 95% confidence intervals (all panels).

3.2. Probability of Successful Settlement

Although larvae were maintained for up to 50 days post-fertilization, too few larvae persisted beyond 43 days post-fertilization to enable effective tests of competency. Even at 43 days post-fertilization, few of the larvae tested were capable of settling, and mainly from the low algal food treatment. The probability of a larva successfully settling depended on both age and food treatment (Figure 1). We found strong evidence for settlement probability being a non-linear function of age (age^2 term was significant: X^2 = 6.35, *df* = 1, *p* = 0.01), and there was strong support for an interaction between food treatment and age (X^2 = 15.45, *df* = 2, *p* < 0.001). Settlement probability was highest for the medium food treatment (1 µg/L) and peaked at around 22 days. Settlement probability was also greatest around the 22-day point for the high food (10 µg/L) treatment, but declined precipitously thereafter and none of the larvae from this treatment settled on days 36 and 43. For larvae from the low food treatment (0.1 µg/L chlorophyll A), settlement probability peaked at around 30 days post-fertilization, and some larvae successfully settled at day 43. However, peak settlement probabilities in the low food treatment were much lower than those observed in the intermediate food treatment (Figure 1).

The algal concentrations to which larvae were exposed appeared to affect the size of CoTS that settled at 17 days, whereby starfish from the low (0.1 µg·L^{-1}) chlorophyll concentration were one-third of the size (0.13 mm^2 ± 0.03 SE; *n* = 2) of the settled starfish from intermediate (0.43 mm^2 ± 0.60 SE; *n* = 10) and high (0.44 mm^2 ± 0.56 SE; *n* = 8) algal concentrations. However, too few larvae settled in the low food treatment to enable robust statistical comparisons of size at settlement, and any effect of algal treatment on size at settlement disappeared after 22 days post-fertilization.

4. Discussion

4.1. Non-Linear Effects of Increasing Food Availability on Larval Survivorship and Settlement

This study shows that variation in phytoplankton concentrations has a significant effect on both larval survivorship and settlement rates for CoTS, as would be expected for planktotrophic larvae [17,20]. It was also apparent that increasing concentrations of phytoplankton can expedite development such that peak settlement rates occurred 22 days post-fertilisation at intermediate and high food levels but were delayed (until 29 days) at low food levels. However, at least some larvae were recorded to settle at 17 days post fertilization across all treatments. Overall survivorship and settlement rates were maximized at intermediate food levels (1 × 10^4 cells·mL^{-1} of *Proteomonas sulcata*) algal concentrations, and lowest at high (1 × 10^5 cells·mL^{-1}) food levels. There is little doubt that very low chlorophyll concentrations (0–0.01 µg·L^{-1} chlorophyll) constrain food intake by CoTS [17,20], and thereby potentially constrain growth, development and survival. However, CoTS larvae may be very resilient to limited food availability [16] and/or derive nutrients from prior maternal provisioning [33] or from other exogenous food sources [34,35], especially in the wild. Unlike Fabricius et al. [19] which showed that CoTS larvae failed to develop at chlorophyll concentrations <0.25 µg·L^{-1}, we found limited effects on larval survivorship at chlorophyll concentrations of even 0.1 µg·L^{-1}, and at these low levels of food availability we still recorded moderate levels of settlement. These differences may be explained by variation in the chlorophyll concentrations of different microalgae, whereby cell densities used in the current study (1–100 × 10^3 cells·mL^{-1}) are very high relative to the cell densities used by Fabricius et al. [19], suggesting that *P. sulcata* has a very low cell specific chlorophyll concentration. Future studies should therefore focus on comparing specific cell densities, rather than using chlorophyll concentrations as a proxy for food availability. In the current study, high cell densities (1 × 10^5 cells·mL^{-1}) were more deleterious for larval survival and settlement than the lowest cell densities (1 × 10^3 cells·mL^{-1}) tested.

Deleterious effects of excessive cell densities on planktotrophic larvae of CoTS were first apparent based on sustained declines in filtration rates with increasing cell densities above 100 cells·mL^{-1} [17]. At very high cell densities (>5 × 10^3 cells·mL^{-1}), Lucas [17] noted that ingestion rates tended to plateau and only a fraction of algal cells were actually digested. Accordingly, CoTS larvae exposed to cell densities of >5 × 10^3 cells·mL^{-1} for *Dunaliella primolecta* and >1 × 10^4 cells·mL^{-1} for *Phaeodactylum tricornutum* had arrested development and poor survival [17]. Larvae may be overfed at extremely high cell densities, and the alimentary canal is so congested with cells that digestion is impeded and rates of assimilation are substantially reduced [17]. In bivalve cultures, overfeeding of larvae has resulted in increased incidence of bacterially related diseases [36]. However, it was considered very unlikely that CoTS larvae would ever be exposed to supra-optimal algal cell densities, and much more likely that larval development and survival would be constrained by low food availability [17]. Moreover, the highest algal concentrations used in this study (10 µg·L^{-1} chlorophyll at cell densities of 1 × 10^5 cells·mL^{-1} of *P. sulcata*) are well beyond what would normally be expected on mid-shelf reefs of the GBR [37].

Chlorophyll concentrations recorded on or adjacent to coral reefs generally range from 0.2 to 0.6 µg·L^{-1} [35,37], which are well below optimum levels for growth and survival of CoTS larvae (1.0–4.0 µg·L^{-1} chlorophyll; [19,20]). At mid-shelf reefs on the GBR, for example, chlorophyll concentrations generally range from 0.2 to 0.5 µg·L^{-1} (e.g., [37,38]) except on inshore (coastal) reefs or during extreme flood events. Moreover, chlorophyll concentrations in the wild do not accurately reflect the availability of phytoplankton (nanoplankton) prey for CoTS larvae, but are bolstered by significant quantities of picoplankton and microplankton (e.g., Furnas and Mitchell 1986). It is unclear however, whether these low levels of suitable prey would necessarily constrain larval survival and settlement. In our study, rates of survival for larvae up until 29 days post-fertilization (which is well beyond the period of peak settlement) at low phytoplankton concentrations (0.1 µg·L^{-1} chlorophyll) was still >50% of that for intermediate food levels (Figure 1). This suggests that even during "normal" conditions, a significant portion of CoTS larvae could effectively settle on mid-shelf reefs on the GBR. While moderate increases in algal concentrations may enhance survival and settlement of CoTS, thereby potentially contributing to the initiation [19] and/or spread of population outbreaks [13], it seems unlikely that normally low nutrient concentrations and algal densities would be sufficiently low to actually prevent any CoTS larvae from completing development and effectively settling.

During extreme flood events, which are often considered to be potential triggers for initiating outbreaks on the GBR [7,19], chlorophyll concentrations on inshore and mid-shelf reefs may exceed 10 µg·L^{-1} [39]. In the extreme, chlorophyll concentrations ≥18 µg·L^{-1} have been recorded along the leading edge of flood plumes [39], where conditions are optimal for phytoplankton production. Given increasing evidence of deleterious effects of high algal densities and chlorophyll concentrations, such events may actually constrain larval development and settlement for *Acanthaster* sp., rather than initiate outbreaks. Major flood events will also reduce salinity and increase levels of suspended sediment, which are also deleterious for development and survival of CoTS larvae (Caballes et al., Unpublished manuscript). It is clear however, that survival and settlement of CoTS larvae increases with moderate increases in chlorophyll concentrations (though limited data to resolve the optimum levels for performance of CoTS larvae) and given their exceptional reproductive potential [40], even small changes in the proportion of larvae that complete development and effectively settle will significantly affect the incidence, if not severity, of population outbreaks.

4.2. Minimum and Maximum Competency Periods

This study extends the maximum-recorded longevity (to 50 days) for CoTS larvae, and provides the first explicit evidence of extended competency for *Acanthaster* spp., up until 43 days post-fertilization. However, previous studies have tended to focus on the minimum pre-competency period [15] or more specifically, the minimum time taken for larvae to develop into late-stage brachiolaria which are assumed to be competent to settle [12], and generally do not consider the maximum competency period. In the current experiment, declines in the number of larvae after 30 days were caused by sustained settlement of competent larvae in experimental containers, as well as punctuated mortality events ("crashes") likely caused by instabilities in conditions. It is possible, therefore, under ideal conditions and with better constraints on spontaneous settlement, that larval longevity for CoTS might extend well beyond 50 days. Prolonged survival and competency will theoretically extend the maximum distances over which larvae can be dispersed [41], though maximum rates of settlement will nonetheless occur after relatively finite periods and probably close to natal reefs. At the medium (optimal) algal concentrations, after accounting for both larval survival and settlement rates, we found that peak settlement for *A.* cf. *solaris* tended to occur within 22 days post-settlement. It is possible, therefore, that few larvae would ever travel beyond the confines of individual reef systems, especially on the GBR, where there is high water retention and extensive reef habitat [42]. Rather, relatively rapid development of CoTS may promote high levels of self-recruitment, or at least significantly constrain dispersal among reefs.

This study had limited capacity to resolve differences in the minimum pre-competency period for CoTS larvae exposed to three different algal concentrations. Rather, competency was established only when late-stage brachiolaria larvae were observed across all treatments, thereby underestimating the minimum pre-competency period (17 days). Previous studies have shown that CoTS larvae can reach the late-stage brachiolaria stage within 9 to 11 days [43], though settlement has only ever been documented $\geq$14 days post-fertilization [16]. This minimum pre-competency period is short relative to other echinoderms with planktotrophic larvae [44], which will enhance larval retention and relatively short-distance dispersal. On the GBR, the step-wise southerly progression of outbreaks suggests that dispersal is limited to 1–2 degrees of latitude (~100–200 km) per generation [45,46], potentially representing the maximum geographic extent of dispersal for individual larvae.

Genetic differentiation apparent when comparing among populations of CoTS in different geographic regions (1000s of kilometers apart) suggest there are definitive constraints on larval dispersal at oceanic scales [47,48]. In the extreme, it is now clear that there are several distinct species of *Acanthaster* spp. [49,50], which must be maintained by limited larval and genetic exchange at these ocean scales. Within regions however, there is often very limited genetic structure [47,51–53], indicative of extensive genetic exchange via widespread dispersal of larvae. Timmers et al. [51] investigated genetic connectivity among CoTS populations in the northern Pacific, and revealed extensive gene flow along the 2500 km length of the Hawaiian archipelago, as well as between Hawai'i and Johnston Atoll separated by 865km. Similarly, Harrison et al. [53] found no genetic structure (using 27 optimized microsatellite loci) for outbreak populations of CoTS along the length of Australia's GBR. Distinct differences in levels of genetic structure within versus among regions, suggests that larvae are sufficiently long-lived so as to be effectively dispersed among reefs within regions, but largely incapable of dispersal among regions.

Existing connectivity models for CoTS [7,8] use idealized particle dispersal periods (generally equivalent to PLDs) ranging from 1 to 28 days. While these particle dispersal periods were largely determined by logistical constraints of the models, rather than the larval biology of the study species, our study suggests that most CoTS larvae will generally settle within 22 days post-fertilization. Low productivity of near-reef waters [37] may delay larval development and peak settlement periods, though extended PLDs would be expected to lead to even higher rates of mortality in the field, owing to sustained levels of predation. While it is possible that some larvae could settle >43 days post fertilization, high rates of larval mortality (even higher than recorded in the current study due to additional impacts of predation) will take an inevitable toll on the number of larvae that persist after this period. It will be interesting to test whether refining existing connectivity models to take account of constrained estimates of settlement windows will alter predictions of connectivity and patterns in the spread of outbreaks. However, it is also necessary to try and establish the typical PLD for CoTS larvae settling on the GBR, specifically comparing whether there are differences in settlement patterns during the initiation versus spread of outbreaks.

5. Conclusions

This study revealed significant differences in both larval survivorship and settlement rates for CoTS across two orders of magnitude variation in cell densities of the phytoplankton, *Proteomonas sulcata*. As expected, based on the *larval starvation hypothesis*, CoTS larvae exposed to very low food levels (1×10^3 cells·mL^{-1} of *Proteomonas sulcata*) had lower survivorship and delayed development compared to larvae maintained in intermediate food levels (1×10^4 cells·mL^{-1} of *Proteomonas sulcata*). Conversely, very high algal cell densities (1×10^5 cells·mL^{-1} of *Proteomonas sulcata*) did not accelerate development of CoTS larvae, nor extend their maximum competency period. Rather, very high algal concentrations had an overwhelmingly negative effect on CoTS larvae, resulting in much higher mortality rates from very early in larval development, compared to low and intermediate food levels. This study reaffirms that moderate increases in food availability (up to 1.0–4.0 µg·L^{-1} chlorophyll, after accounting for the moderate contribution of suitable prey to field-based chlorophyll concentrations; [19,20]) may lead to increases in rates of development, survival and settlement for CoTS larvae. However, low levels of food availability may still enable significant rates of population replenishment and settlement, while the greatest constraints on growth and survival of CoTS larvae occurred at very high chlorophyll concentrations and phytoplankton densities.

Acknowledgments: This research was supported by an Advance Queensland Accelerate Partnership Grant, administered by the Queensland Government's Department of Science, Information Technology and Innovation. The manuscript benefitted greatly from discussions with Jon Brodie regarding realistic environmental conditions to which CoTS larvae will be naturally exposed.

Author Contributions: M.S.P. and S.D. conceived and designed the experiments; B.M. performed the experiments; S.B. analyzed the data; M.P., C.F.C. and C.A.T. wrote the paper; All authors edited the paper.

Conflicts of Interest: The authors declare no conflict of interest.

References

1. Scheltema, R.S. Larval dispersal as a means of genetic exchange between geographically separated populations of shallow-water benthic marine gastropods. *Biol. Bull.* **1971**, *140*, 284–322. [CrossRef]
2. Scheltema, R.S. Long-distance dispersal by planktonic larvae of shoal-water benthic invertebrates among central Pacific islands. *Bull. Mar. Sci.* **1986**, *39*, 241–256.
3. Lawton, R.J.; Messmer, V.; Pratchett, M.S.; Bay, L.K. High gene flow across large geographic scales reduces extinction risk for a highly specialised coral feeding butterflyfish. *Mol. Ecol.* **2011**, *20*, 3584–3598. [CrossRef] [PubMed]
4. Almany, G.R.; Berumen, M.L.; Thorrold, S.R.; Planes, S.; Jones, G.P. Local replenishment of coral reef fish populations in a marine reserve. *Science* **2007**, *316*, 742–744. [CrossRef] [PubMed]
5. Nanninga, G.B.; Berumen, M.L. The role of individual variation in marine larval dispersal. *Front. Mar. Sci.* **2014**, *1*, 71. [CrossRef]
6. Pratchett, M.S. Dynamics of an outbreak population of *Acanthaster planci* at Lizard Island, northern Great Barrier Reef (1995–1999). *Coral Reefs* **2005**, *24*, 453–462. [CrossRef]
7. Wooldridge, S.A.; Brodie, J.E. Environmental triggers for primary outbreaks of crown-of-thorns starfish on the Great Barrier Reef, Australia. *Mar. Pollut. Bull.* **2015**, *101*, 805–815. [CrossRef] [PubMed]
8. Hock, K.; Wolff, N.H.; Condie, S.A.; Anthony, K.R.N.; Mumby, P.J. Connectivity networks reveal the risks of crown-of-thorns starfish outbreaks on the Great Barrier Reef. *J. Appl. Ecol.* **2014**, *51*, 1188–1196. [CrossRef]
9. Victor, B.C. Settlement strategies and biogeography of reef fishes. In *The Ecology of Fishes on Coral Reefs*; Sale, P.F., Ed.; Academic Press: San Diego, CA, USA, 1991; pp. 231–260.
10. Richmond, R.H. Energetics, competency, and long-distance dispersal of planula larvae of the coral *Pocillopora damicornis*. *Mar. Biol.* **1987**, *93*, 527–533. [CrossRef]
11. Graham, E.M.; Baird, A.H.; Connolly, S.R. Survival dynamics of scleractinian coral larvae and implications for dispersal. *Coral Reefs* **2008**, *27*, 529–539. [CrossRef]
12. Henderson, J.A.; Lucas, J.S. Larval development and metamorphosis of *Acanthaster planci* (Asteroidea). *Nature* **1971**, *232*, 655–657. [CrossRef] [PubMed]

13. Pratchett, M.S.; Caballes, C.F.; Rivera-Posada, J.A.; Sweatman, H.P.A. Limits to understanding and managing outbreaks of crown-of-thorns starfish (*Acanthaster* spp.). *Oceanogr. Mar. Biol. Annu. Rev.* **2014**, *52*, 133–200.

14. Lamare, M.D.; Pecorino, D.; Hardy, N.; Liddy, M.; Byrne, M.; Uthicke, S. The thermal tolerance of crown-of-thorns (*Acanthaster planci*) embryos and bipinnaria larvae: Implications for spatial and temporal variation in adult populations. *Coral Reefs* **2014**, *33*, 207–219. [CrossRef]

15. Uthicke, S.; Logan, M.; Liddy, M.; Francis, D.S.; Hardy, N.; Lamare, M.D. Climate change as an unexpected co-factor promoting coral eating seastar (*Acanthaster planci*) outbreaks. *Sci. Rep.* **2015**, *5*, 8402. [CrossRef] [PubMed]

16. Olson, R.R. In situ culturing as a test of the larval starvation hypothesis for the crown-of-thorns starfish, *Acanthaster planci*. *Limnol. Oceanogr.* **1987**, *32*, 895–904. [CrossRef]

17. Lucas, J.S. Quantitative studies of feeding and nutrition during larval development of the coral reef asteroid *Acanthaster planci* (L.). *J. Exp. Mar. Biol. Ecol.* **1982**, *65*, 173–193. [CrossRef]

18. Brodie, J.E.; Fabricius, K.E.; De'ath, G.; Okaji, K. Are increased nutrient inputs responsible for more outbreaks of crown-of-thorns starfish? An appraisal of the evidence. *Mar. Pollut. Bull.* **2005**, *51*, 266–278. [CrossRef] [PubMed]

19. Fabricius, K.E.; Okaji, K.; De'ath, G. Three lines of evidence to link outbreaks of the crown-of-thorns seastar *Acanthaster planci* to the release of larval food limitation. *Coral Reefs* **2010**, *29*, 593–605. [CrossRef]

20. Wolfe, K.; Graba-Landry, A.; Dworjanyn, S.A.; Byrne, M. Larval starvation to satiation: Influence of nutrient regime on the success of *Acanthaster planci*. *PLoS ONE* **2015**, *10*, 1–18. [CrossRef] [PubMed]

21. Haszprunar, G.; Spies, M. An integrative approach to the taxonomy of the crown-of-thorns starfish species group (Asteroidea: Acanthaster): A review of names and comparison to recent molecular data. *Zootaxa* **2014**, *3841*, 271–284. [CrossRef] [PubMed]

22. Kamya, P.Z.; Dworjanyn, S.A.; Hardy, N.; Mos, B.; Uthicke, S.; Byrne, M. Larvae of the coral eating crown-of-thorns starfish, *Acanthaster planci* in a warmer-high CO_2 ocean. *Glob. Chang. Biol.* **2014**, *20*, 3365–3376. [CrossRef] [PubMed]

23. Okaji, K.; Ayukai, T.; Lucas, J.S. Selective feeding by larvae of the crown-of-thorns starfish, *Acanthaster planci* (L.). *Coral Reefs* **1997**, *16*, 47–50. [CrossRef]

24. Mellin, C.; Lugrin, C.; Okaji, K.; Francis, D.S.; Uthicke, S. Selective feeding and microalgal consumption rates by crown-of-thorns seastar (*Acanthaster* cf. *solaris*) larvae. *Diversity* **2016**. under review.

25. Cowan, Z.L.; Dworjanyn, S.A.; Caballes, C.F.; Pratchett, M.S. Predation on crown-of-thorns starfish larvae by damselfishes. *Coral Reefs* **2016**, *35*, 1–10. [CrossRef]

26. Rumrill, S.S. Natural mortality of marine invertebrate larvae. *Ophelia* **1990**, *32*, 163–198. [CrossRef]

27. Mos, B.; Cowden, K.L.; Nielsen, S.J.; Dworjanyn, S.A. Do cues matter? Highly inductive settlement cues don't ensure high post-settlement survival in sea urchin aquaculture. *PLoS ONE* **2011**, *6*, e28054. [CrossRef] [PubMed]

28. Wood, S.N. *Generalized Additive Models: An Introduction with R*; CRC Press: Boca Raton, FL, USA, 2006.

29. Wood, S.N.; Scheipl, F. Gamm4: Generalized Additive Mixed Models Using mgcv and lme4. R Packag Version 0.2-3. Available online: http://CRAN.R-project.org/package=gamm4 (accessed on 5 May 2016).

30. Fournier, D.A.; Skaug, H.J.; Ancheta, J.; Ianelli, J.; Magnusson, A.; Maunder, M.N.; Nielsen, A.; Sibert, J. AD Model Builder: Using automatic differentiation for statistical inference of highly parameterized complex nonlinear models. *Optim. Methods Softw.* **2012**, *27*, 233–249. [CrossRef]

31. Skaug, H.J.; Fournier, D.A.; Nielsen, A.; Magnusson, A.; Bolker, B.M. Generalized Linear Mixed Models using AD Model Builder. *R Package* version 0.8.3.3. Available online: http://glmmadmb.r-forge.r-project.org (accessed on 19 January 2016).

32. R Core Team. *R: A Language and Environment for Statistical Computing*; R Foundation for Statistical Computing: Vienna, Austria. Available online: http://www.R-project.org/ (accessed on 21 June 2016).

33. Caballes, C.F.; Pratchett, M.S.; Kerr, A.M.; Rivera-Posada, J.A. The role of maternal nutrition on oocyte size and quality, with respect to early larval development in the coral-eating starfish, *Acanthaster planci*. *PLoS ONE* **2016**, *11*, e0158007. [CrossRef] [PubMed]

34. Hoegh-Guldberg, O. Uptake of dissolved organic matter by larval stage of the crown-of-thorns starfish *Acanthaster planci*. *Mar. Biol.* **1994**, *120*, 55–63.

35. Nakajima, R.; Nakatomi, N.; Kurihara, H.; Fox, M.; Smith, J.; Okaji, K. Crown-of-thorns starfish larvae can feed on organic matter released from corals. *Diversity* **2016**, *8*, 18. [CrossRef]

36. Helm, M.M.; Bourne, N.; Lovatelli, A. *Hatchery Culture of Bivalves: A Practical Manual*; Food and Agriculture Organization of the United Nations: Rome, Italy, 2004.

37. Brodie, J.E.; De'ath, G.; Devlin, M.J.; Furnas, M.J.; Wright, M. Spatial and temporal patterns of near-surface chlorophyll a in the Great Barrier Reef lagoon. *Mar. Freshw. Res.* **2007**, *58*, 342–353. [CrossRef]

38. Wolanski, E.; Jones, M.; Williams, W.T.; Wolanski, E.; Jones, M.; Williams, W. Physical properties of the Great Barrier reef Lagoon waters near Townsville. II. Seasonal variations. *Mar. Freshw. Res.* **1981**, *32*, 321–334. [CrossRef]

39. Brodie, J.E.; Schroeder, T.; Rohde, K.; Faithful, J.; Masters, B.; Dekker, A.G.; Brando, V.E. Dispersal of suspended sediments and nutrients in the Great Barrier Reef lagoon during river discharge events: Conclusions from satellite remote sensing and concurrent flood plume sampling. Brodie Jon. *Mar. Freshw. Res.* **2010**, *61*, 651–664. [CrossRef]

40. Babcock, R.C.; Milton, D.A.; Pratchett, M.S. Relationships between size and reproductive output in the crown-of-thorns starfish. *Mar. Biol.* **2016**, *163*, 234. [CrossRef]

41. Luiz, O.J.; Allen, A.P.; Robertson, D.R.; Floeter, S.R.; Kulbicki, M.; Vigliola, L.; Becheler, R.; Madin, J.S. Adult and larval traits as determinants of geographic range size among tropical reef fishes. *Proc. Natl. Acad. Sci. USA* **2013**, *110*, 16498–16502. [CrossRef] [PubMed]

42. Wolanski, E.; Spagnol, S. Sticky waters in the Great Barrier Reef. *Estuar. Coast. Shelf Sci.* **2000**, *50*, 27–32. [CrossRef]

43. Lucas, J.S. Environmental influences on the early development of *Acanthaster planci* (L.). In *Crown-of-Thorns Starfish Seminar Proceedings*; Austrlian Government Publishing Service: Canberra, Australia, 1974; pp. 109–121.

44. Strathmann, R.R. Length of pelagic period in echinoderms with feeding larvae from the Northeast Pacific. *J. Exp. Mar. Bio. Ecol.* **1978**, *34*, 23–27. [CrossRef]

45. Kenchington, R.A. Growth and recruitment of *Acanthaster planci* (L.) on the Great Barrier Reef. *Biol. Conserv.* **1977**, *11*, 103–118. [CrossRef]

46. Vanhatalo, J.; Hosack, G.R.; Sweatman, H.P.A. Spatio-temporal modelling of crown-of-thorns starfish outbreaks on the Great Barrier Reef to inform control strategies. *J. Appl. Ecol.* **2016**. [CrossRef]

47. Yasuda, N.; Nagai, S.; Hamaguchi, M.; Okaji, K.; Gérard, K.; Nadaoka, K. Gene flow of *Acanthaster planci* (L.) in relation to ocean currents revealed by microsatellite analysis. *Mol. Ecol.* **2009**, *18*, 1574–1590. [CrossRef] [PubMed]

48. Timmers, M.A.; Bird, C.E.; Skillings, D.J.; Smouse, P.E.; Toonen, R.J. There's no place like home: Crown-of-thorns outbreaks in the central pacific are regionally derived and independent events. *PLoS ONE* **2012**, *7*, e31159. [CrossRef] [PubMed]

49. Vogler, C.; Benzie, J.A.H.; Lessios, H.A.; Barber, P.H.; Wörheide, G. A threat to coral reefs multiplied? Four species of crown-of-thorns starfish. *Biol. Lett.* **2008**, *4*, 696–699. [CrossRef] [PubMed]

50. Haszprunar, G.; Vogler, C.; Wörheide, G. Taxonomy does matter—A plea for the use of DNA-barcoding and barcode identification numbers when studying crown-of-thorns seastars (*Acanthaster planci* species complex). *Diversity* **2016**, under review.

51. Timmers, M.A.; Andrews, K.R.; Bird, C.E.; DeMaintenton, M.J.; Brainard, R.E.; Toonen, R.J. Widespread dispersal of the crown-of-thorns sea Star, *Acanthaster planci*, across the Hawaiian Archipelago and Johnston Atoll. *J. Mar. Biol.* **2011**, *2011*, 1–10. [CrossRef]

52. Yasuda, N.; Taquet, C.; Nagai, S.; Yoshida, T.; Adjeroud, M. Genetic connectivity of the coral-eating sea star *Acanthaster planci* during the severe outbreak of 2006–2009 in the Society Islands, French Polynesia. *Mar. Ecol.* **2015**, *36*, 668–678. [CrossRef]

53. Harrison, H.B.; Pratchett, M.S.; Messmer, V.; Saenz-Agudelo, P.; Berumen, M.L. Microsatellites reveal genetic homogeneity among outbreak populations of crown-of-thorns starfish (*Acanthaster* cf. *solaris*) on Australia's Great Barrier Reef. *Diversity* **2016**, under review.

diversity

MDPI

Review

Potential Enhanced Survivorship of Crown of Thorns Starfish Larvae due to Near-Annual Nutrient Enrichment during Secondary Outbreaks on the Central Mid-Shelf of the Great Barrier Reef, Australia

Jon Brodie [1,2,*], Michelle Devlin [3] and Stephen Lewis [4]

[1]   ARC Centre of Excellence for Coral Reef Studies, James Cook University, Townsville, Queensland 4811, Australia
[2]   Coasts, Climate, Oceans Consulting (C2O), Townsville, Queensland 4811, Australia
[3]   The Centre for Environment, Fisheries and Aquaculture Science (Cefas), Lowestoft NR33 0HT, UK; michelle.devlin@cefas.co.uk
[4]   TropWATER, James Cook University, Townsville, Queensland 4811, Australia; stephen.lewis@jcu.edu.au
*   Correspondence: jon.brodie@jcu.edu.au; Tel.: +61-407-127-030

Academic Editors: Sven Uthicke and Michael Wink
Received: 7 December 2016; Accepted: 7 March 2017; Published: 12 March 2017

Abstract: The Great Barrier Reef (GBR) is currently experiencing widespread crown of thorns starfish (CoTS) outbreaks, as part of the fourth wave of outbreaks since 1962. It is believed that these outbreaks have become more frequent on the GBR and elsewhere in the Indo-Pacific and are associated with anthropogenic causes. The two widely accepted potential causes are (1) anthropogenic nutrient enrichment leading to the increased biomass of phytoplankton, the food of the planktonic stage of larval CoTS; and (2) the overfishing of predators in the juvenile to adult stages of CoTS, for example, commercially fished species such as coral trout. In this study, we show that the evidence for the nutrient enrichment causation hypothesis is strongly based on a large number of recent studies in the GBR. We also hypothesise that secondary outbreaks in the region between Cairns and Townsville can also be enhanced by nutrient enriched conditions associated with the annual nutrient discharge from Wet Tropics rivers.

Keywords: crown of thorns starfish; Great Barrier Reef; nutrient enrichment; larval survivorship

1. Introduction

The crown of thorns starfish (*Acanthaster* sp—CoTS) is a specialized coral-feeder and is found across the Indo-Pacific. Populations of CoTS display cyclic oscillations between extended periods of low-density with individuals sparsely distributed among large reef areas, and episodes of unsustainable high densities, commonly termed "outbreaks". The outbreaks result in mass mortalities of corals, with typically second-order and long-term consequences on various reef communities. CoTS outbreaks usually spread widely across reef ecosystems in the Indo-Pacific and are driven over large distances by larval transport, commonly leading to increases in benthic algae, a loss of coral-feeding assemblages, an overall collapse of reef structural complexity, and a decline in coral biodiversity and productivity [1,2].

CoTS is a naturally occurring animal on the Indo-Pacific coral reefs and seems to "normally" occur in low densities on most reefs. However, our knowledge of the density of CoTS on reefs before 1960 is very limited. CoTS have caused widespread damage to many coral reefs in the Indo-Pacific over the past six decades, as population"explosions" have occurred at regular intervals [1,3–5]. There have been several opposing views as to the origin of the outbreaks, including both a "natural occurrence" hypothesis and various anthropogenic causes' hypotheses [6]. However, given the huge number of animals which were found on reefs across the Indo-Pacific by the 1970s [1], it has been suggested that these "outbreaks" could not have occurred in the past, or alternatively, not at the frequency at which they were occurring by the 1990s. Therefore, a range of anthropogenic possible "causes" were postulated (reviewed in [1,5–7]).

The human causes which have been suggested for the CoTS population outbreaks include the removal of predators (especially fish, such as the commercially fished coral trout) [8,9] and/or increased larval survivorship due to increased phytoplankton food stocks associated with nutrient enrichment [6,10–12]. However, the cause (or causes) of the outbreaks remains a controversial issue and there are several opposing views in the literature [5]. One view postulates that population outbreaks are a natural phenomenon, due to the inherently unstable population sizes of highly fecund organisms such as CoTS [13]. Outbreaks have also been attributed to anthropogenic changes to the environment of the starfish, with a range of possible causes including: the removal of adult predators, particularly fish [8] and the gastropod *Charonia tritonis* [1]; changes to population structures of predators in larval and juvenile stages, caused by chemical (possibly pesticide) pollution [14,15]; the destruction of larval predators, particularly corals, by construction activities on reefs; and larval food supply (phytoplankton) enhancement from nutrient enriched terrestrial runoff [6].

Outbreaks of the pacific CoTS species, *A.* cf. *solaris*, on the Great Barrier Reef (GBR) have been traditionally divided into primary outbreaks and secondary outbreaks (Figure 1). Primary outbreaks originate in the "initiation area" between Cairns and Lizard Island. The conditions which are believed to be necessary for the initiation of an outbreak wave include a sufficient coral biomass for food [12], nutrient enrichment and a high biomass of phytoplankton to sustain larval survivorship [11], and suitable oceanographic conditions so as to retain high local connectivity and/or within-reef larval retention [16,17]. Following the initiation of a primary outbreak, massive larval production [18] leads to secondary outbreaks to the north and south of the initiation area, with a wave of secondary outbreaks occurring to the south of the initiation area from Cairns towards Townsville, over approximately eight to nine years after the primary outbreak and then offshore from Mackay (12 years after primary outbreak), after which the outbreaks appear to diminish [5]. This contribution examines the latest research on the nutrient enrichment hypothesis for CoTS outbreaks, with a specific focus on the GBR, north-eastern Australia. We highlight the importance of understanding the life history stages of CoTS development and the influence of nutrients on the larval stage, and for the first time, we examine the potential role of river-discharged nutrients on triggering and sustaining secondary CoTS outbreaks on the central GBR. We also reinterpret recent works on the potential causes of CoTS outbreaks in the southern GBR.

2. Mortality of COTS at Different Life History Stages

The nutrient enrichment hypothesis for CoTS outbreaks centers on the increased survivorship during the larval stage of development, although there is little robust knowledge on the rates of mortality at their different life stages. Keesing and Halford [19,20] studied mortality rates at the small settled juvenile stage (0.5–5.5 mm) and found high rates (6.49% d^{-1}) for one month old starfish (mean size = 1.1 mm). The mortality rates decreased considerably with increasing age, revealing a value of 1.24% d^{-1} for four month old starfish (2.7 mm) and 0.45% d^{-1} for seven month old starfish (5.5 mm) [19,20]. The results highlighted that mortality rates in these early life stages may be strongly influenced by both predators (epibenthic fauna) and food availability [21]. These experiments, involving the exclusion of predators, indicated that the major source of mortality was predation by epibenthic fauna for juvenile CoTS. Results also show that, in the presence of an adequate food supply, juvenile CoTS move very little, suggesting that juvenile survival is enhanced by settling in areas, whereas predation is minimal.

It has been suggested that the larval stage, especially at the brachiolaria stage where the larvae requires an external food source, is likely to display high mortality rates [6,10], but no field data exist to support this hypothesis [7]. However, laboratory experiments show that very low numbers of larvae are able to reach competency in food limited conditions [10,12,22–24]. In addition, predation would also be a factor contributing to the mortality rate, with fish, such as damselfish, preying on CoTS larvae [25] in experimental aquarium conditions; however, the extent of predation in reef waters is not known.

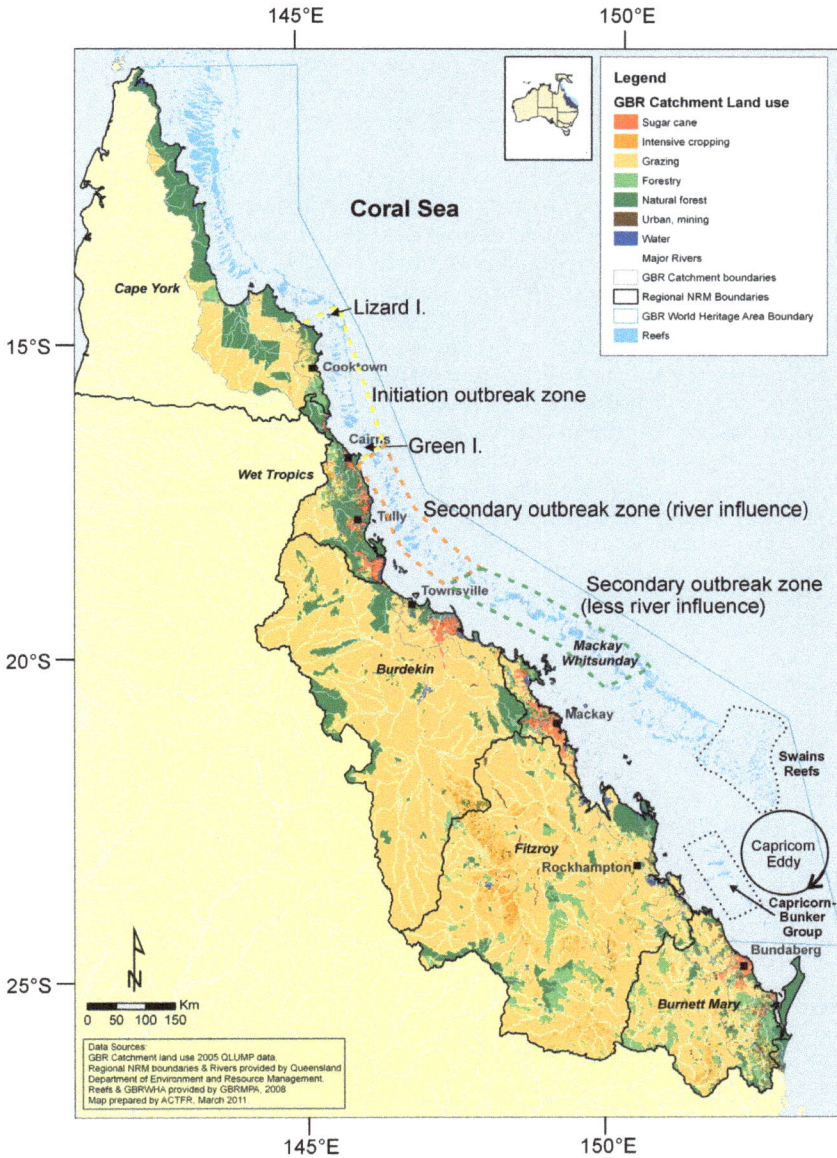

Figure 1. The Great Barrier Reef. The CoTS initiation zone is shown in yellow (dotted); the CoTS secondary outbreak zone strongly influenced by river discharge is shown in orange (dotted); the CoTS secondary outbreak zone weakly or not at all influenced by river discharge is shown in green (dotted).

3. Conceptual Understanding of the Nutrient Enrichment-CoTS Link

In the 1970s, Charles Birkeland formulated the hypothesis that CoTS population outbreaks were more common on high islands in the Pacific islands due to the terrestrial runoff of nutrients (nitrogen and phosphorus), than they were on atoll islands, where there is no natural runoff [26]. This hypothesis was founded on the fact that CoTS have a planktonic larval stage (of a few weeks) which requires feeding on a minimum amount of phytoplankton to reach a level of viability, such that they can settle on a reef. In "normal" nutrient conditions, the larvae have insufficient phytoplankton food to reach competence [10] and thus cannot provide the conditions required for the survival of high population numbers. However, in the areas where nutrient runoff in river discharge from the high islands occurs, the conditions required for phytoplankton to bloom can be achieved, thus providing the food source for this larval stage to develop [6,27].

The anthropogenic causation theory of nutrient enrichment is strongly linked to the increase in human activities such as sewage discharge, fertiliser runoff, and increased erosion of nutrient-rich soil. These activities lead to the enrichment of adjacent coastal waters [28], a higher phytoplankton biomass, and a shift to larger species of phytoplankton which are more suitable as CoTS larval food [29], resulting in better rates of survivorship of CoTS larvae, ultimately leading to more frequent outbreaks [6,11].

The analysis of larval growth, the effects of environmental factors, and the experimental testing of the nutrient enrichment hypothesis were first carried out in the 1970s [10,30–32], and showed that COTS larvae were food limited and did not develop well in low-nutrient and phytoplankton biomass conditions. In contrast, experimental testing using more sophisticated apparatus [33,34] tested the growth of larvae in a range of nutrient conditions, with results suggesting no nutrient effect. However, the experiments were repeated, this time with more care given to strict protocols regarding nutrient supply, with results supporting the nutrient enrichment hypothesis [12,20,21,35]. CoTS outbreaks associated with anthropogenic nutrient enrichment are now seen as one of the "signals" of partial eutrophication of the GBR [28,36]. Recently, refinements of the experiments have further supported the hypothesis [37–39], by establishing that larval survivorship increases with an increased food supply of phytoplankton biomass. Optimal survival is represented by chlorophyll a concentrations of about 1 $\mu g \cdot L^{-1}$ [38,39] or at algal food densities above 1000 cells·mL^{-1} [34]. Both low chlorophyll a concentrations (0.1 $\mu g \cdot L^{-1}$) and higher chlorophyll a concentrations (~10 $\mu g \cdot L^{-1}$) appear to retard the development of CoTS larvae [38–40], indicating that there is an optimum range of phytoplankton biomass or cell counts for larval survivorship.

Concentrations of chlorophyll a in the GBR lagoon are generally less than 0.5 $\mu g \cdot L^{-1}$ in non-river discharge conditions [41], but can be in the range of 0.5–3 $\mu g \cdot L^{-1}$ where influenced by major river discharge, particularly in the Burdekin—Wet Tropics regions, where flood plumes reach the mid- and outer shelf reef areas [42–46]. The phytoplankton biomass, measured as chlorophyll a during these peak and post flow conditions, is able to support the high larval survivorship of CoTS juveniles if the high nutrient conditions intersect with periods of high larval counts.

4. Crown of Thorns Starfish and the Great Barrier Reef

The cycles (waves) of outbreaks on the GBR have occurred from 1962 to 1976, 1979 to 1991, 1993 to 2005 [11,12], and 2009—current [5,16]. Each wave has severely reduced the coral cover on the GBR, especially in the mid-shelf reefs of the central section [47] (Figure 1). CoTS outbreaks have been less frequent on the inner and outer shelf reefs, for not-well understood reasons. There is evidence that CoTS outbreaks have occurred on the GBR over the past 7000 years [48], although historical records suggest that the waves of outbreaks post 1962 appear to be an unusual phenomenon and that outbreaks have become more frequent in recent times (see below) [11,12].

Oceanographic conditions associated with the ENSO cycle [49] also play a part in initiating CoTS outbreaks in the Cairns area of the GBR, with nutrient enrichment from river runoff and increased local connectivity and/or within-reef larval retention being due to ENSO conditions, both of which provide conditions that result in the initiation of waves of outbreaks [16]. Indeed, this new climate association suggests that the "natural" frequency of CoTS outbreaks may have also been highly variable on the GBR over the past 7000 years, given changes in the strength of ENSO forcing during this period. However, the modern conditions of strong ENSO coupled with increased nutrient sources, likely creates a scenario where CoTS outbreaks become an almost continuous feature on the GBR.

It is also important to note that CoTS outbreaks can occur at sites with no obvious nutrient enrichment [50], but where, in reality, upwelling may be present but not documented and at sites with possible nutrient enrichment associated with natural upwelling systems (e.g., in the Chagos archipelago—[51]) or at oceanic nutrient/productivity fronts [52,53].

There has also been a persistent outbreak area in the Swains reefs (Figure 1), but this is not considered to be associated with the outbreaks to the north and the cycle of anthropogenic drivers and higher larval survivorship. This long-standing situation of CoTS outbreaks in the Swains reefs in the southern GBR is likely associated with the known upwelling systems in that region [11]. This has been identified as an area of upwelling associated with the Capricorn Eddy [54–56] and the resultant phytoplankton blooms are associated with other enrichment phenomena, such as manta ray feeding aggregations [57] and plankton enrichment [58]. It is likely that these factors help to explain the CoTS outbreaks in the Swains and Capricorn-Bunker Group reefs of the southern GBR [50]. In addition, the recent outbreaks in the southern Capricorn Bunker Group [50] may be associated with record river discharge from the Burnett and Mary Rivers in early 2011 and 2013 [59].

Miller et al. [50] considered the role of one large southern river, the Fitzroy, but did not show that the discharge from this river had a significant effect on the CoTS outbreaks in the Capricorn Bunker group. However, work on the extent of all southern rivers, i.e., Mary, Burrum, Burnett, Kolan and Baffle in the Burnett Mary NRM Region(Figure 1), clearly showed the discharge plume extent of these rivers impinged on the southern Capricorn Bunkers [59]. While the Fitzroy River discharge only influences the northern section of the Capricorn-Bunker Group reefs during rare large flood events that are coupled with favorable wind conditions [42,60], the five rivers to the south of the Fitzroy are known to influence phytoplankton dynamics in the southern Capricorn-Bunker group [57] and may influence the survivorship of CoTS larvae in this region.

5. Nutrient Enrichment and Secondary Outbreaks

While it often assumed that nutrient enrichment is primarily involved in producing "primary" CoTS outbreaks in the initiation area (Cairns—Lizard Island), there is now evidence that nutrient enrichment can enhance "secondary" outbreaks, especially in the offshore Wet Tropics region. Nutrient loading from these rivers has greatly increased with the proliferation of agriculture and associated fertiliser use in the region [61,62]. The period in which CoTS larvae are in the plankton stages, during November to February, coincides with the point at which the Wet Tropics rivers exhibit high discharge levels each year (Figure 2), with high levels of intra-annual variability, and regularly produce phytoplankton blooms on the GBR shelf in this region [42,43,63–67]. Measurements of high chlorophyll a concentrations during this period represent an extended period of time, in which conditions of high phytoplankton biomass are optimal for CoTS larval development (see also [38,67]).

Images of river discharge are easily identifiable using remote sensed imagery and are available on a daily basis, where the brown turbid, coastal waters are clearly contrasted with the greener offshore waters associated with increased phytoplankton production, as well as the blue oceanic waters (Figures 3 and 4); although it is worth noting that these images are subject to cloud cover. In Figure 4, we show a series of MODIS images in true colour from 2007, demonstrating the formation and dispersal of a phytoplankton bloom driven by high discharge from Wet Tropics rivers over a five day period. By the 13 February, the plume/bloom has extended across the outer-shelf reefs of this area and into the Coral Sea.

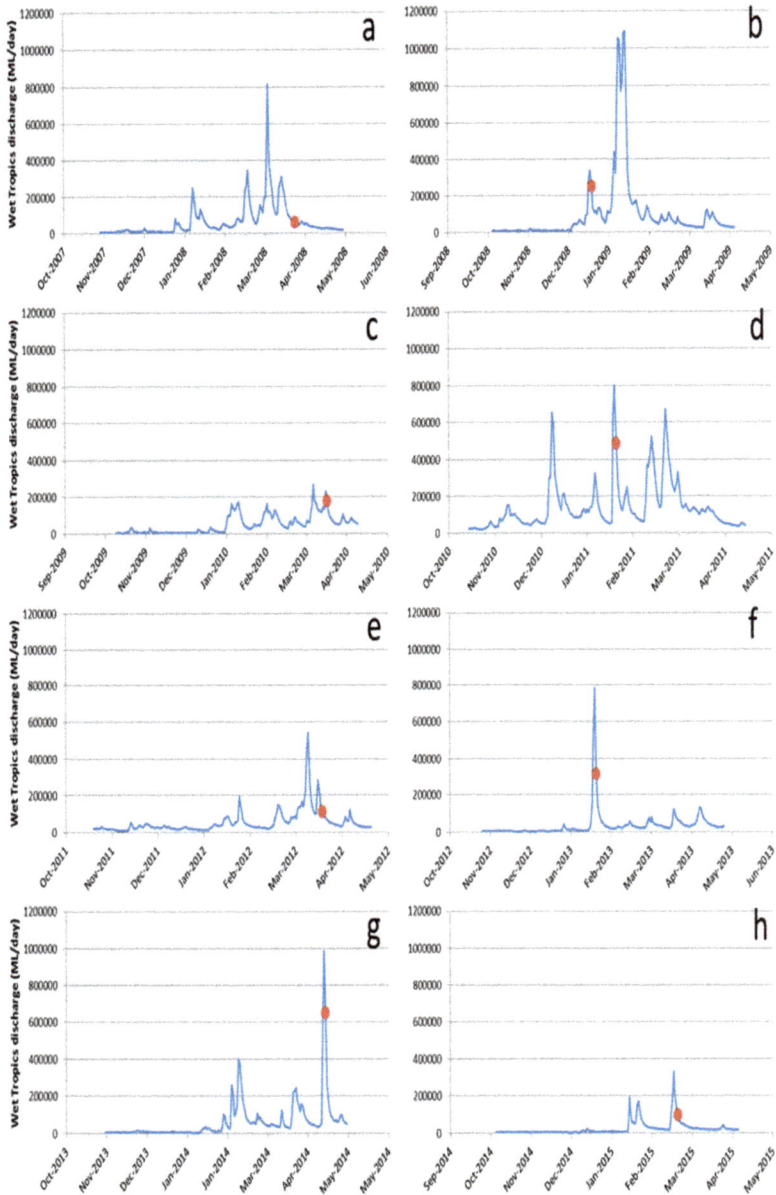

Figure 2. Total daily wet season discharge (November to April, inclusive, ML/day) for the Wet Tropics rivers for the wet seasons of 2007–2008 (**a**), 2008–2009 (**b**), 2009–2010 (**c**), 2010–2011 (**d**), 2011–2012 (**e**), 2012–2013 (**f**), 2013–2014 (**g**), 2014–2015 (**h**). Red dots in parts (**d**), (**f**), and (**h**) stand for the day on which the satellite images shown in Figure 3 were taken and are within the CoTS spawning period.

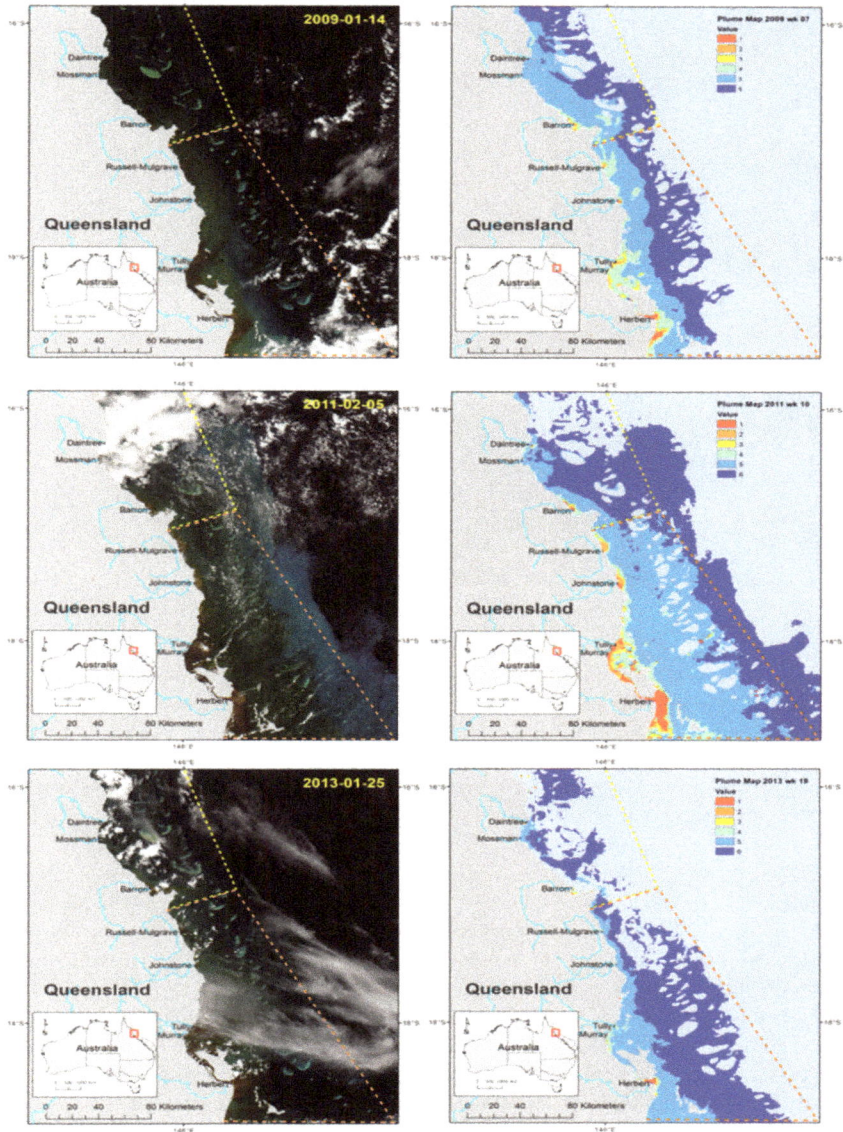

Figure 3. Selected satellite images showing plume water intrusion into the CoTS initiation zone (south) and the region south to the Palm Island Group over three years (left column) and the corresponding weekly plume colour class map (right column). Dashed yellow lines on maps stand for the south delimitation of the CoTS initiation zone and the secondary outbreak zone is shown by the red dashed lines. Plume colour classes vary from one to six, as described in the text. The stage of river discharge on each occasion is shown in Figure 2.

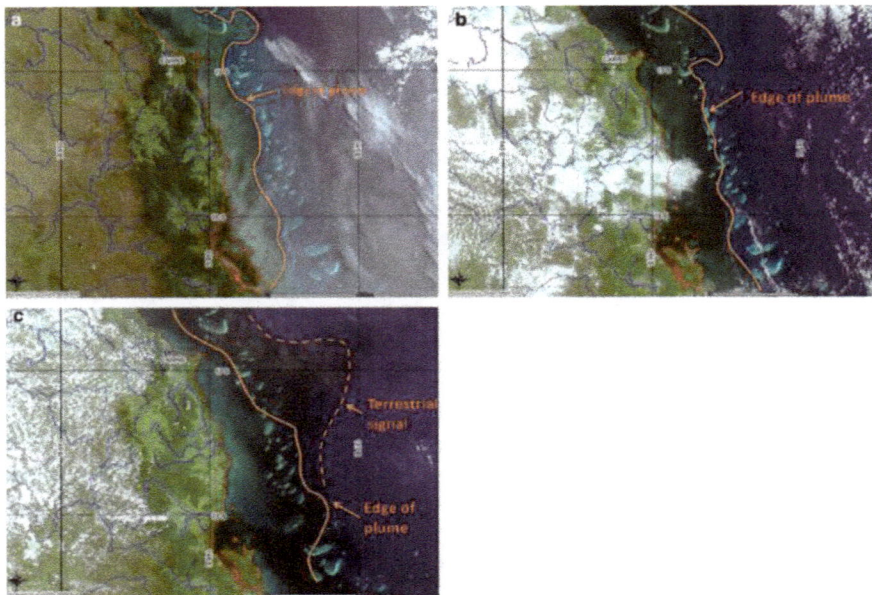

Figure 4. Progression (**a–c**) of multiple river plumes in the Wet Tropics (9, 11, 13 February 2007, respectively) extending from the coast to beyond the outer reef. The lines show the outer edge of the plume made visible due to Colored Dissolved Organic Matter and phytoplankton. Images (**a–c**) show the transformation from a plume dominated by terrestrial particulate matter (brown colour near river mouths) into a plume dominated by a dissolved nutrient, driven phytoplankton bloom (green colour). A proportion of the contained nutrients in the plume may be seen "escaping" to the Coral Sea in part c. The images show a period of enrichment on the mid- and outer- shelf of the Wet Tropics GBR lasting at least five days. Images courtesy of CSIRO.

The mapping and modelling of the extent and composition of flood plumes has been employed as a monitoring tool since 1991. More recently, the classification of plume water types has allowed a more detailed modelling of the extent and composition of plume waters at a greater frequency. The frequency and extent of the plumes, correlated with water quality values, provides mechanisms to predict the mean chlorophyll values within the plume waters across the wet season [44]. Work focused on the Tully region [68] showed that plume waters were mapped on a weekly basis and secondary waters extended out into the mid shelf areas for 14 out of the 22 weeks of the wet season in 2010/2011, highlighting the extended period of time that optimal conditions existed for supporting phytoplankton growth.

A daily image of flood plumes was selected from remote sensed imagery (MODIS, Aqua, 250 m) (Figure 3), at a period of elevated discharge (or in the days following) (Figure 2) and at periods of low clouds, to ensure that the plume was visible. The three selectedimages from plumes associated with the Wet Tropics rivers during the period of CoTS spawning, show an elevated phytoplankton biomass on the mid-shelf reef area of the GBR between Cairns and Townsville, with examples presented for the period between November to March (Figure 3) for each year.

These conditions of elevated phytoplankton biomass are seen both visually, and by water type analysis [44,64,65,67] and ascribed water types (right column panels of Figure 3), following the methodology of Devlin et al. [44] and Petus et al. [69,70].

Each image represents a period of time of optimal conditions for high phytoplankton biomass, and potentially a source of increased CoTS larval survivorship. Chlorophyll a concentrations measured on the Cairns—Townsville mid-shelf area during major river discharge periods (December to March), are 0.5–3 µg·L^{-1} at salinities in the range 30–35 [42,64,68]. Water type descriptions [44,65,70] are derived from true colour MODIS imagery, calibrated against the direct measurement of salinity, temperature, particulate and dissolved nutrients, phytoplankton, total suspended solids (TSS), the diffuse attenuation coefficient of photosynthetically active radiation (K$_d$(PAR)), colored dissolved organic matter (CDOM), chlorophyll a, and herbicides [70]. The ranges and mean values of chlorophyll a for the water types are summarized in Table 1, drawn from Devlin et al. [44].

Table 1. Chlorophyll a concentrations in the different water types and colour classes.

Water Type and Class	Mean Chlorophyll a (µg·L^{-1})	Range Chlorophyll a (µg·L^{-1})
Primary, Colour Classes one–four	0.9	0.6–3.7
Secondary, Colour Class five	1.3	0.7–2.7
Tertiary, Colour Class six	0.7	0.2–1.6

Water types and classes are a necessary step in estimating likely chlorophyll a concentrations in high river discharge conditions, as the algorithms for estimated chlorophyll a from satellites such as MODIS do not work well in the Case 2 waters of the GBR [71], with errors, even in the deeper, clearer waters of the outer GBR lagoon, of typically 100%. While other researchers [38,72] have used estimates of chlorophyll a concentrations from MODIS, it should be noted these estimates can have very large errors. It should also be noted that regular direct measurements of chlorophyll in the GBR lagoon ceased in 2006. The results of the 1991–2006 program can be found in Brodie et al. [41]. Since 2006, only irregular direct measurements of chlorophyll a have occurred during flood plume studies [44] and estimates from remote sensing algorithms, which can be problematic in the Case 2 coastal waters of the GBR [71].

The concentrations of chlorophyll a in colour class six (and to some extent five) waters are in the range at which CoTS larval development is most favorable, estimated to be at 1 µg·L^{-1} (compared to 0.1 and 10 µg·L^{-1}) in Pratchett et al. [40]; at 1 µg·L^{-1} compared to much lower and higher concentrations in Wolfe et al. [38]; and at 1 µg·L^{-1} in Fabricius et al. [12]. While Wolfe et al. [38] claim *"it seems that mid-shelf regions on the GBR have a high background level of phytoplankton (0.4–0.6 µg chl a·L^{-1}; Table 1, S1 Table), which may support successful development of A. planci larvae irrespective of flood or storm events"*, these results for chlorophyll a are estimated from the MODIS sensor using algorithms which are known to have serious errors and which generally overestimate chlorophyll a concentrations (see above). From the measured results, the only periods when chlorophyll a concentrations are in this range on the mid-shelf of the GBR between Townsville and Lizard Island is during high river discharge conditions.

At higher chlorophyll a concentrations, larvae develop less strongly [38,40]. CoTS larvae generally also grow slowly and do not easily reach a competent stage at lower concentrations of chlorophyll a (<0.2 µg·L^{-1}) or low phytoplankton biomass [12,37,38,40]. An exception to this situation seems to be in the case where female CoTS have been fed on Acropora corals. In these circumstances, the weight of the females increased, as well as the gonads, and also produced larger oocytes compared to Porites-fed and starved females [73]. The CoTS larvae under these conditions may be able to reach competency faster, without significant phytoplankton feeding [73], but this has not actually been demonstrated.

The remote sensed imagery only examines a single day during the high discharge period and a further analysis to extend the period over the primary four months associated with CoTS spawning i.e., November to March inclusive, would be needed to further advance this speculative hypothesis. However, it is worth noting that the river flow conditions for all of the images were variable, and represent periods of flow from 80th to 99th percentile conditions. Optimal growth conditions can still occur at the lower flow rates, indicating that high phytoplankton biomass can be stimulated in first flush, medium to high flow conditions, and may represent a period of time from days to weeks. Extended periods of high biomass conditions can increase the probability of time that larval bloom and enhanced plume conditions can intersect and increase survivorship, supporting the growth associated with secondary outbreaks.

6. Conclusions

The range of phytoplankton biomass, as measured via the chlorophyll a concentration, for the favorable development of CoTS, is 0.6–1 µg·L^{-1}. It appears that well-nourished (from their mother) [73] larvae may be able to reach competency at a lower phytoplankton biomass, although this has not been demonstrated. This may correspond to the low chlorophyll results of Wolfe et al. [72], where larvae were able to reach competency at chlorophyll a concentrations in the range of 0.1 to 0.4 µg·L^{-1}; conversely, in this study, the CoTS larvae were fed the microalga *Proteomonas*, an algae with hardly any chlorophyll, and so the results are hard to interpret compared to other phytoplankton food. However, most larvae need chlorophyll a concentrations in the range 0.6 to 1.0 µg·L^{-1} for adequate development [12,38,40]. Larvae develop poorly when chlorophyll a concentrations are above about 1.5 µg·L^{-1} [38,40].

This chlorophyll a range (0.6–1.0 µg·L^{-1}) has been shown to occur commonly in the wet season on the mid-shelf waters of the GBR between Cairns and Townsville, where secondary CoTS outbreaks occur. As such, it is possible that the enhanced survivorship of CoTS larvae is occurring in this region on a regular basis, as a result of increased nutrient delivery during periods of high discharge from the Wet Tropics rivers [42], and also during the much more irregular high discharge periods from the Burdekin River (e.g., [46,69]). Thus, these secondary outbreaks of CoTS, while mainly driven by massive larval supply, may also be accelerated by suitable phytoplankton conditions provided by increased nutrient discharges from the relevant rivers [61]. This finding also explains the query in Babcock et al. [74], expressed as follows: *"Moreover, if the productivity of mid-shelf waters on the GBR are consistently below levels (0.25 µg·L^{-1}) at which there is almost zero survival of COTS larvae, it is hard to explain how the southward propagating waves of outbreaks (that subsequently cause widespread devastation) are maintained."* In fact, chlorophyll a concentrations in the mid-shelf waters of the GBR (at least in the area between Townsville and Lizard Island) are well above 0.25 µg·L^{-1} for extended periods during the wet season.

Further studies are required to assess the strength of the nutrient enrichment element in secondary outbreaks of CoTS, especially in the mid-shelf area between Townsville and Cairns, where river plumes reach this area on a regular basis. Further south from Townsville, river plumes (from rivers south of the Burdekin) rarely reach the mid-shelf of the GBR and it may be speculated that this effect does not play a part in enhancing CoTS outbreaks. An exception may be associated with the very large discharge events from the Mary and Burnett Rivers and CoTS outbreaks in the southern Capricorn Bunker Group reefs, but further research is required to confirm this connection.

Unfortunately, the direct measurement of chlorophyll a in the GBR lagoon is still limited by sample numbers and locations of sampling. Estimates of chlorophyll a concentrations can be made from water type analysis [44,70], and by using the eReefs model in conjunction with direct measurements [75]. However, a more intensive direct measurement program is still required to be able to answer questions regarding the influence of nutrient enrichment on populations of CoTS.

Acknowledgments: We would like to acknowledge the help of Dieter Tracey and Eduardo da Silva in preparing Figures 2 and 3. The paper also benefitted greatly from the suggested changes and additions recommended by two reviewers and the comments of the Academic Editor.

Author Contributions: Jon Brodie conceived the intent and structure of the paper, while all authors contributed to writing the text.

Conflicts of Interest: The authors declare no conflict of interest.

References

1. Birkeland, C.; Lucas, J.S. *Acanthaster planci: Major Management Problem of Coral Reefs*; CRC Press, Inc.: Boca Raton, FL, USA, 1990.
2. Moran, P.J. The Acanthaster phenomenon. *Oceanogr. Mar. Biol. Ann. Rev.* **1986**, *24*, 379–480.

3. Zann, L.; Brodie, J.; Berryman, C.; Naqasima, M. Recruitment, ecology, growth and behavior of juvenile *Acanthaster planci* (L.)(Echinodermata: Asteroidea). *Bull. Mar. Sci.* **1987**, *41*, 561–575.
4. Zann, L.; Brodie, J.; Vuki, V. History and dynamics of the crown-of-thorns starfish *Acanthaster planci* (L.) in the Suva area, Fiji. *Coral Reef.* **1990**, *9*, 135–144. [CrossRef]
5. Pratchett, M.S.; Caballes, C.F.; Rivera-Posada, J.A.; Sweatman, H.P.A. Limits to understanding and managing outbreaks of crown-of-thorns starfish (Acanthaster spp.). *Oceanogr. Mar. Biol. Ann. Rev.* **2014**, *52*, 133–200.
6. Brodie, J.E. Enhancement of larval and juvenile survival and recruitment in *Acanthaster planci* from the effects of terrestrial runoff: A review. *Aust. J. Mar. Freshw. Res.* **1992**, *43*, 539–554. [CrossRef]
7. Caballes, C.F.; Pratchett, M.S. Reproductive biology and early life history of the crown-of-thorns starfish. In *Echinoderms: Ecology, Habitats and Reproductive Biology*; Eric, W., Ed.; Nova Science Publishers: New York, NY, USA, 2014; pp. 101–146.
8. Sweatman, H.P.A. A field study of fish predation on juvenile crown of- thorns starfish. *Coral Reefs* **1995**, *14*, 47–53. [CrossRef]
9. Mendonça, V.M.; Al Jabri, M.M.; Al Ajmi, I.; Al Muharrami, M.; Al Areimi, M.; Al Aghbari, H.A. Starfish *Acanthaster planci* in the Northwestern Indian Ocean: Are they really a consequence of unsustainable starfish predator removal through overfishing in coral reefs, or a response to a changing environment? *Zool. Stud.* **2010**, *49*, 108–123.
10. Lucas, J.S. Quantitative studies of feeding and nutrition during larval development of the coral reef asteroid *Acanthaster planci* (L.). *J. Exp. Mar. Biol. Ecol.* **1982**, *65*, 173–193. [CrossRef]
11. Brodie, J.; Fabricius, K.; De'ath, G.; Okaji, K. Are increased nutrient inputs responsible for more outbreaks of crown-of-thorns starfish? An appraisal of the evidence. *Mar. Pollut. Bull.* **2005**, *51*, 266–278. [CrossRef] [PubMed]
12. Fabricius, K.E.; Okaji, K.; De'ath, G. Three lines of evidence to link outbreaks of the crown-of-thorns seastar *Acanthaster planci* to the release of larval food limitation. *Coral Reefs* **2010**, *29*, 593–605. [CrossRef]
13. Potts, D.C. Crown-of-thorns starfish—Man-induced pest or natural phenomenon. In *the Ecology of Pests*; Commonwealth Scientific and Industrial Research Organization: Melbourne, Australia, 1981; pp. 55–86.
14. Chesher, R.H. Destruction of the Pacific Corals by the Sea Star *Acanthaster planci*. *Science* **1969**, *165*, 280–283. [CrossRef] [PubMed]
15. Randall, J.E. Chemical pollution in the sea and the crown-of-thorns starfish (*Acanthaster planci*). *Biotropica* **1972**, *4*, 132–144. [CrossRef]
16. Wooldridge, S.; Brodie, J. Environmental triggers for primary outbreaks of crown-of-thorns starfish on the Great Barrier Reef, Australia. *Mar. Pollut. Bull.* **2015**, *101*, 805–815. [CrossRef] [PubMed]
17. Hock, K.; Wolff, N.H.; Condie, S.A.; Anthony, K.; Mumby, P.J. Connectivity networks reveal the risks of crown-of-thorns starfish outbreaks on the Great Barrier Reef. *J. Appl. Ecol.* **2014**, *51*, 1188–1196. [CrossRef]
18. Uthicke, S.; Doyle, J.; Duggan, S.; Yasuda, N.; McKinnon, A.D. Outbreak of coral-eating Crown-of-Thorns creates continuous cloud of larvae over 320 km of the Great Barrier Reef. *Sci. Rep.* **2015**, *5*, 16885. [CrossRef] [PubMed]
19. Keesing, J.K.; Halford, A.R. Importance of post settlement processes for the population dynamics of *Acanthaster planci* (L.). *Mar. Freshw. Res.* **1992**, *43*, 635–651. [CrossRef]
20. Keesing, J.K.; Halford, A.R. Field measurement of survival rates of juvenile *Acanthaster planci*: Techniques and preliminary results. *Mar. Ecol. Prog. Ser.* **1992**, *85*, 107–114. [CrossRef]
21. Keesing, J.K.; Wiedermeyer, W.L.; Okaji, K.; Halford, A.R.; Hall, K.C.; Cartwright, C.M. Mortality rates of juvenile starfish *Acanthaster planci* and *Nardoa* spp. measured on the Great Barrier Reef, Australia and in Okinawa, Japan. *Oceanol. Acta* **1996**, *19*, 441–448.
22. Ayukai, T. Ingestion of ultraplankton by the planktonic larvae of the crown-of-thorns starfish *Acanthaster planci*. *Biol. Bull.* **1994**, *186*, 90–100. [CrossRef]
23. Okaji, K.; Ayukai, T.; Lucas, J. Selective feeding by larvae of the crown-of-thorns starfish, *Acanthaster planci* (L.). *Coral Reefs* **1997**, *16*, 47–50. [CrossRef]
24. Okaji, K.; Ayukai, T.; Lucas, J. Are *Acanthaster planci* larvae food limited in the Great Barrier Reef waters. In Proceedings of the 8th International Coral Reef Symposium, Panama, Panama, 24–29 June 1996.
25. Cowan, Z.L.; Dworjanyn, S.A.; Caballes, C.F.; Pratchett, M.S. Predation on crown-of-thorns starfish larvae by damselfishes. *Coral Reefs* **2016**, *35*, 1253–1262. [CrossRef]
26. Birkeland, C. Acanthaster in the cultures of high islands. *Atoll Res. Bull.* **1981**, *255*, 55–58.

27. Birkeland, C. Terrestrial runoff as a cause of outbreaks of *Acanthaster planci* (Echinodermata: Asteroidea). *Mar. Biol.* **1982**, *69*, 175–185. [CrossRef]

28. Brodie, J.E.; Devlin, M.; Haynes, D.; Waterhouse, J. Assessment of the eutrophication status of the Great Barrier Reef lagoon (Australia). *Biogeochemistry* **2011**, *106*, 281–302. [CrossRef]

29. Mellin, C.; Lugrin, C.; Okaji, K.; Francis, D.S.; Uthicke, S. Selective feeding and microalgal consumption rates by Crown-of-Thorns Seastar (*Acanthaster cf. solaris*) larvae. *Diversity* **2017**, *9*, 8. [CrossRef]

30. Henderson, J.A.; Lucas, J.S. Larval development and metamorphosis of *Acanthaster planci* (Asteroidea). *Nature* **1971**, *232*, 655–657. [CrossRef] [PubMed]

31. Lucas, J.S. Reproductive and larval biology of *Acanthaster planci* (L.) in Great Barrier Reef waters. *Micronesica* **1973**, *9*, 197–203.

32. Lucas, J.S. Growth, maturation and effects of diet *in Acanthaster planci* (L.)(Asteroidea) and hybrids reared in the laboratory. *J. Exp. Mar. Biol. Ecol.* **1984**, *79*, 129–147. [CrossRef]

33. Olson, R.R. In situ culturing as a test of the larval starvation hypothesis for the crown-of-thorns starfish, *Acanthaster planci*. *Limnol. Oceanogr.* **1987**, *32*, 895–904. [CrossRef]

34. Olson, R.R.; Olson, M.H. Food limitation of planktotrophic marine invertebrate larvae: Does it control recruitment success? *Ann. Rev. Ecol. Syst.* **1989**, *20*, 225–247. [CrossRef]

35. Ayukai, T.; Okaji, K.; Lucas, J.S. Food limitation in the growth and development of crown-of-thorns starfish in the Great Barrier Reef. In Proceedings of the 8th International Coral Reef Symposium, Panama, Panama, 24–29 June 1996; pp. 621–626.

36. Fabricius, K.E. Factors determining the resilience of coral reefs to eutrophication: A review and conceptual model. In *Coral Reefs: An Ecosystem in Transition*; Dubinski, Z., Stambler, N., Eds.; Springer: New York, NY, USA, 2011; pp. 493–509.

37. Uthicke, S.; Logan, M.; Liddy, M.; Francis, D.; Hardy, N.; Lamare, M. Climate change as an unexpected co-factor promoting coral eating seastar (*Acanthaster planci*) outbreaks. *Sci. Rep.* **2015**, *5*. [CrossRef] [PubMed]

38. Wolfe, K.; Graba-Landry, A.; Dworjanyn, S.A.; Byrne, M. Larval starvation to satiation: Influence of nutrient regime on the success of *Acanthaster planci*. *PLoS ONE* **2015**, *10*, e0122010. [CrossRef] [PubMed]

39. Wolfe, K.; Graba-Landry, A.; Dworjanyn, S.A.; Byrne, M. Larval phenotypic plasticity in the boom-and-bust crown-of-thorns seastar, *Acanthaster planci*. *Mar. Ecol. Prog. Ser.* **2015**, *539*, 179–189. [CrossRef]

40. Pratchett, M.S.; Dworjanyn, S.; Mos, B.; Caballes, C.F.; Thompson, C.A.; Blowes, S. Larval Survivorship and Settlement of Crown-of-Thorns Starfish (*Acanthaster cf. solaris*) at Varying Algal Cell Densities. *Diversity* **2017**, *9*, 2. [CrossRef]

41. Brodie, J.; De'ath, G.; Devlin, M.; Furnas, M.; Wright, M. Spatial and temporal patterns of near-surface chlorophyll a in the Great Barrier Reef lagoon. *Mar. Freshw. Res.* **2007**, *58*, 342–353. [CrossRef]

42. Devlin, M.J.; Brodie, J. Terrestrial discharge into the Great Barrier Reef Lagoon: Nutrient behaviour in coastal waters. *Mar. Pollut. Bull.* **2005**, *51*, 9–22. [CrossRef] [PubMed]

43. Devlin, M.; Schaffelke, B. Spatial extent of riverine flood plumes and exposure of marine ecosystems in the Tully coastal region, Great Barrier Reef. *Mar. Freshw. Res.* **2009**, *60*, 1109–1122. [CrossRef]

44. Devlin, M.J.; Da Silva, E.T.; Petus, C.; Wenger, A.; Zeh, D.; Tracey, D.; Álvarez-Romero, J.G.; Brodie, J. Combining in-situ water quality and remotely sensed data across spatial and temporal scales to measure variability in wet season chlorophyll-a: Great Barrier Reef lagoon (Queensland, Australia). *Ecol. Process.* **2013**, *2*, 1–22. [CrossRef]

45. Schroeder, T.; Devlin, M.J.; Brando, V.E.; Dekker, A.G.; Brodie, J.E.; Clementson, L.A.; McKinna, L. Inter-annual variability of wet season freshwater plume extent into the Great Barrier Reef lagoon based on satellite coastal ocean colour observations. *Mar. Pollut. Bull.* **2012**, *65*, 210–223. [CrossRef] [PubMed]

46. Bainbridge, Z.T.; Wolanski, E.; Álvarez-Romero, J.G.; Lewis, S.E.; Brodie, J.E. Fine sediment and nutrient dynamics related to particle size and floc formation in a Burdekin River flood plume, Australia. *Mar. Pollut. Bull.* **2012**, *65*, 236–248. [CrossRef] [PubMed]

47. De'ath, G.; Fabricius, K.E.; Sweatman, H.; Puotinen, M. The 27-year decline of coral cover on the Great Barrier Reef and its causes. *Proc. Natl. Acad. Sci. USA* **2012**, *109*, 17995–17999. [CrossRef] [PubMed]

48. Walbran, P.D.; Henderson, R.A.; Faithful, J.W.; Polach, H.A.; Sparks, R.J.; Wallace, G.; Lowe, D.C. Crown of thorns starfish outbreaks on the Great Barrier Reef: A geological perspective based upon the geological record. *Coral Reefs* **1989**, *8*, 67–78. [CrossRef]

49. Hock, K.; Wolff, N.H.; Beeden, R.; Hoey, J.; Condie, S.A.; Anthony, K.; Possingham, H.P.; Mumby, P.I. Controlling range expansion in habitat networks by adaptively targeting source populations. *Conserv. Biol.* **2016**, *30*, 856–866. [CrossRef] [PubMed]

50. Miller, I.; Sweatman, H.; Cheal, A.; Emslie, M.; Johns, K.; Jonker, M.; Osborne, K. Origins and implications of a primary crown-of-thorns starfish outbreak in the southern Great Barrier Reef. *J. Mar. Biol.* **2015**, *2015*, 809624. [CrossRef]

51. Roche, R.C.; Pratchett, M.S.; Carr, P.; Turner, J.R.; Wagner, D.; Head, C.; Sheppard, C.R.C. Localized outbreaks of *Acanthaster planci* at an isolated and unpopulated reef atoll in the Chagos Archipelago. *Mar. Biol.* **2015**, *162*, 1695–1704. [CrossRef]

52. Houk, P.; Raubani, J. *Acanthaster planci* outbreaks in Vanuatu coincide with ocean productivity, furthering trends throughout the Pacific Ocean. *J. Oceanogr.* **2010**, *66*, 435–438. [CrossRef]

53. Houk, P.; Bograd, S.; van Woesik, R. The transition zone chlorophyll front can trigger *Acanthaster planci* outbreaks in the Pacific Ocean: Historical confirmation. *J. Oceanogr.* **2007**, *63*, 149–154. [CrossRef]

54. Weeks, S.; Bakun, A.; Steinberg, C.; Brinkman, R.; Hoegh-Guldberg, O. The Capricorn Eddy: A prominent driver of the ecology and future of the southern Great Barrier Reef. *Coral Reefs* **2010**, *29*, 975–985. [CrossRef]

55. Mao, Y.; Luick, J.L. Circulation in the southern Great Barrier Reef studied through an integration of multiple remote sensing and in situ measurements. *J. Geophys. Res.* **2014**, *119*, 1621–1643. [CrossRef]

56. Condie, S.; Condie, R. Retention of plankton within ocean eddies. *Glob. Ecol. Biogeogr.* **2016**, *25*, 1264–1277. [CrossRef]

57. Weeks, S.J.; Magno-Canto, M.M.; Jaine, F.R.A.; Brodie, J.; Richardson, A.J. Unique sequence of events triggers manta ray feeding frenzy in the southern Great Barrier Reef, Australia. *Remote Sens.* **2015**, *7*, 3138–3152. [CrossRef]

58. Alongi, D.M.; Patten, N.L.; McKinnon, D.; Köstner, N.; Bourne, D.G.; Brinkman, R. Phytoplankton, bacterioplankton and virioplankton structure and function across the southern Great Barrier Reef shelf. *J. Mar. Syst.* **2015**, *142*, 25–39. [CrossRef]

59. Da Silva, E.T.; Devlin, M.; Wenger, A.; Petus, C. *Burnett-Mary Wet Season 2012–2013: Water Quality Data Sampling, Analysis and Comparison against Wet Season 2010–2011 Data.*; Centre for Tropical Water & Aquatic Ecosystem Research (TropWATER) Publication: James Cook University, Townsville, Australia, 2013; p. 31.

60. Brodie, J.E.; Mitchell, A.W. Nutrient composition of the January (1991) fitzroy river flood plume. In *Workshop on the Impacts of Flooding*; GBRMPA Workshop Series No. 17; Byron, G.T., Ed.; Great Barrier Reef Marine Park Authority: Townsville, Australia, 1992; pp. 56–74.

61. Kroon, F.J.; Kuhnert, P.M.; Henderson, B.L.; Wilkinson, S.N.; Kinsey-Henderson, A.; Abbott, B.; Brodie, J.E.; Turner, R.D.R. River loads of suspended solids, nitrogen, phosphorus and herbicides delivered to the Great Barrier Reef lagoon. *Mar. Pollut. Bull.* **2012**, *65*, 167–181. [CrossRef] [PubMed]

62. Waterhouse, J.; Brodie, J.; Lewis, S.; Mitchell, A. Quantifying the sources of pollutants in the Great Barrier Reef catchments and the relative risk to reef ecosystems. *Mar. Pollut. Bull.* **2012**, *65*, 394–406. [CrossRef] [PubMed]

63. Furnas, M.; Mitchell, A.; Skuza, M.; Brodie, J. In the other 90%: Phytoplankton responses to enhanced nutrient availability in the Great Barrier Reef Lagoon. *Mar. Pollut. Bull.* **2005**, *51*, 253–265. [CrossRef] [PubMed]

64. Devlin, M.J.; McKinna, L.W.; Álvarez-Romero, J.G.; Petus, C.; Abott, B.; Harkness, P.; Brodie, J. Mapping the pollutants in surface riverine flood plume waters in the Great Barrier Reef, Australia. *Mar. Pollut. Bull.* **2012**, *65*, 224–235. [CrossRef] [PubMed]

65. Álvarez-Romero, J.G.; Devlin, M.; Teixeira da Silva, E.; Petus, C.; Ban, N.C.; Pressey, R.L.; Kool, J.; Roberts, J.J.; Cerdeira-Estrada, S.; Wenger, A.S.; et al. A novel approach to model exposure of coastal-marine ecosystems to riverine flood plumes based on remote sensing techniques. *J. Environ. Manag.* **2013**, *119*, 194–207. [CrossRef] [PubMed]

66. Blondeau-Patissier, D.; Schroeder, T.; Brando, V.E.; Maier, S.W.; Dekker, A.G.; Phinn, S. ESA-MERIS 10-year mission reveals contrasting phytoplankton bloom dynamics in two tropical regions of Northern Australia. *Remote Sens.* **2014**, *6*, 2963–2988. [CrossRef]

67. Devlin, M.; Petus, C.; da Silva, E.; Tracey, D.; Wolff, N.; Waterhouse, J.; Brodie, J. Water quality and river plume monitoring in the Great Barrier Reef: An overview of methods based on ocean colour satellite data. *Remote Sens.* **2015**, *7*, 12909–12941. [CrossRef]

68. Devlin, M.; Waterhouse, J.; Taylor, J.; Brodie, J. *Flood Plumes in the Great Barrier Reef: Spatial and Temporal Patterns in Composition and Distribution*; GBRMPA Research Publication No. 68; Great Barrier Reef Marine Park Authority: Townsville, Australia, 2001.

69. Petus, C.; Collier, C.; Devlin, M.; Rasheed, M.; McKenna, S. Using MODIS data for understanding changes in seagrass meadow health: A case study in the Great Barrier Reef (Australia). *Mar. Environ. Res.* **2014**, *98*, 68–85. [CrossRef] [PubMed]

70. Petus, C.; Devlin, M.; Thompson, A.; McKenzie, L.; Teixeira da Silva, E.; Collier, C.; Tracey, D.; Martin, K. Estimating the Exposure of Coral Reefs and Seagrass Meadows to Land-Sourced Contaminants in River Flood Plumes of the Great Barrier Reef: Validating a Simple Satellite Risk Framework with Environmental Data. *Remote Sens.* **2016**, *8*, 210. [CrossRef]

71. Waterhouse, J.; Brodie, J.; Petus, C.; Devlin, M.; da Silva, E.; Maynard, J.; Heron, S.; Tracey, D. *Recent Findings of an Assessment of Remote Sensing Data for Water Quality Measurement in the Great Barrier Reef: Supporting Information for the GBR Water Quality Improvement Plans*; Centre for Tropical Water & Aquatic Ecosystem Research (TropWATER) Publication: James Cook University, Townsville, Australia, 2015; p. 71.

72. Wolfe, K.; Graba-Landry, A.; Dworjanyn, S.A.; Byrne, M. Superstars: Assessing nutrient thresholds for enhanced larval success of *Acanthaster planci*, a review of the evidence. *Mar. Pollut. Bull.* **2017**, in press. [CrossRef] [PubMed]

73. Caballes, C.F.; Pratchett, M.S.; Kerr, A.M.; Rivera-Posada, J.A. The role of maternal nutrition on oocyte size and quality, with respect to early larval development in the coral-eating starfish, *Acanthaster planci*. *PLoS ONE* **2016**, *11*, e0158007. [CrossRef] [PubMed]

74. Babcock, R.C.; Dambacher, J.M.; Morello, E.B.; Plaganyi, É.E.; Hayes, K.R.; Sweatman, H.P.A.; Pratchett, M.S. Assessing Different Causes of Crown-of-Thorns Starfish Outbreaks and Appropriate Responses for Management on the Great Barrier Reef. *PLoS ONE* **2016**, *11*, e0169048. [CrossRef] [PubMed]

75. Baird, M.E.; Cherukuru, N.; Jones, E.; Margvelashvili, N.; Mongin, M.; Oubelkheir, K.; Ralph, P.J.; Rizwi, F.; Robson, B.J.; Schroeder, T.; et al. Remote-sensing reflectance and true colour produced by a coupled hydrodynamic, optical, sediment, biogeochemical model of the Great Barrier Reef, Australia: Comparison with satellite data. *Environ. Model. Softw.* **2016**, *78*, 79–96. [CrossRef]

diversity

MDPI

Article

Benthic Predators Influence Microhabitat Preferences and Settlement Success of Crown-of-Thorns Starfish (*Acanthaster* cf. *solaris*)

Zara-Louise Cowan [1,*], Symon A. Dworjanyn [2], Ciemon F. Caballes [1] and Morgan Pratchett [1]

[1] ARC Centre of Excellence for Coral Reef Studies, James Cook University, Townsville, QLD 4811, Australia; ciemon.caballes@my.jcu.edu.au (C.F.C.); morgan.pratchett@my.jcu.edu.au (M.P.)
[2] National Marine Science Centre, Southern Cross University, Coffs Harbour, NSW 2450, Australia; Symon.Dworjanyn@scu.edu.au
* Correspondence: zaralouise.cowan@my.jcu.edu.au; Tel.: +61-7-4781-5747

Academic Editors: Sven Uthicke and Michael Wink
Received: 28 October 2016; Accepted: 6 December 2016; Published: 9 December 2016

Abstract: Like most coral reef organisms, crown-of-thorns starfish (*Acanthaster* spp.) are expected to be highly vulnerable to predation as they transition from a planktonic larval phase to settling among reef habitats. Accordingly, crown-of-thorns starfish might be expected to exhibit behavioural adaptations which moderate exposure to predation at this critical stage in their life history. Using pairwise choice experiments and settlement assays, we explored the ability of competent larvae of *Acanthaster* cf. *solaris* to first detect and then actively avoid benthic predators during settlement. Pairwise choice experiments revealed that late stage brachiolaria larvae are able to detect predators in the substrate and where possible, will preferentially settle in microhabitats without predators. Settlement assays (without choices) revealed that larvae do not necessarily delay settlement in the presence of predators, but high levels of predation on settling larvae by benthic predators significantly reduce the number of larvae that settle successfully. Taken together, these results show that crown-of-thorns starfish are highly vulnerable to benthic predators during settlement, and that variation in the abundance of benthic predators may exert a significant influence on patterns of settlement for crown-of-thorns starfish.

Keywords: behaviour; coral reefs; predation; resilience

1. Introduction

As for many benthic reef organisms, settlement is expected to represent one of the major bottlenecks in the life history of crown-of-thorns starfish (*Acanthaster* spp.), whereby relatively naïve planktonic larvae will be exposed to an entirely new suite of potential predators as they transition to living in benthic reef habitats [1]. Reef-based predators include both planktivorous fishes and sessile invertebrates (e.g., corals) that intercept larvae as they swim towards benthic habitats [2,3], as well as infaunal invertebrate predators that will feed on starfish that settle to specific microhabitats [4]. Both pre- and post-settlement mortality play important roles in structuring populations of marine organisms (e.g., [5]), but predation rates are generally highest ($\geq$30% day^{-1}) immediately after settlement (reviewed by Gosselin and Qian [6]). Importantly, high rates of early post-settlement mortality can significantly augment patterns of larval supply, having a major bearing on the distribution and abundance of benthic marine organisms (e.g., [7,8]). Moreover, there will be strong selection for settling larvae to choose microhabitats that minimise predation risk [9], either by avoiding habitats with high abundance of potential predators or preferentially settling in complex microhabitats that provide greater refuge from predators.

Predation on crown-of-thorns starfish may be moderated by high concentrations of saponins and other toxins that are presumed to deter potential predators (e.g., [10]). Notably, three-day-old larval crown-of-thorns starfish have more than two times higher concentrations of saponins than adult starfish [11], which may reflect their increased vulnerability to predation due to limited physical predator defences (e.g., spines). Even so, larval crown-of-thorns starfish are readily consumed by a range of planktivorous reef fishes (e.g., [2]), as well as corals, such as *Pocillopora damicornis* [3]. Accordingly, in laboratory-based experiments, >50% of larvae are lost during settlement, and this may be a result of predation by benthic animals which could not be removed from the settlement substrates [3]. Even after settlement, juvenile crown-of-thorns starfish experience significant rates of mortality (up to 6.49% day^{-1} for one-month-old starfish), which decreases with size and age, and is largely attributed to predation (e.g., [4]).

The purpose of this study was to test whether crown-of-thorns starfish can detect the presence of benthic predators within potential settlement substrates (largely based on chemoreception, *sensu* [12]), and thereafter, explore the extent to which larval crown-of-thorns starfish preferentially settle in microhabitats with and without predators present. There has been much work on the role of chemoreception in the selection of settlement substrates by marine larvae (reviewed by Pawlik [13]). For example, larval fishes and corals use chemical cues to discriminate between settlement substrates [14] and degraded and healthy reefs [15], as well as respond to the presence of conspecifics [16]. Similarly, brachiolaria larvae of crown-of-thorns starfish respond to chemical cues associated with specific bacterial films, causing them to discriminate between different substrates during settlement (e.g., [12,17,18]). Moreover, adult crown-of-thorns starfish have been shown to use chemoreception both to locate and orientate towards potential prey [19] and feeding conspecifics [20]. Given the inherent ability of crown-of-thorns starfish to respond to chemical stimuli, combined with potentially intense predation pressure at settlement, we expect that larval crown-of-thorns starfish are able to both detect and actively avoid predators during settlement.

2. Materials and Methods

2.1. Collection and Maintenance of Study Species

Adult *Acanthaster* cf. *solaris* [21] were collected from reefs around Lizard Island (14°40' S; 145°27' E) in the northern Great Barrier Reef between October and November 2015. All experiments were conducted at the Lizard Island Research Station. Spawning was induced following Cowan et al. [2]. One mL 10^{-4} M 1-methyladenine (1-MA) was injected into the gonads in each arm, through the aboral side of each starfish, which immediately induced spawning in males. Females, meanwhile, spawned within 20–30 min of administering 1-MA. To ensure that a mix of genotypes was used, gametes were collected using glass pipettes from at least 3 individuals of each sex. Sperm was rinsed in 0.2-μm filtered seawater (FSW) and refrigerated at 4 °C prior to use. Eggs were collected from around the arms of females following their release from gonopores and rinsed with 0.2-μm FSW. Fertilization was achieved by adding sperm to reach a final sperm-egg ratio of approximately 100:1. Fertilised eggs were transferred to 16 L larval rearing chambers at a density of approximately 1–2 larvae mL^{-1}. Chambers were maintained at 28.4 ± 1.1 °C (mean ± SD). Larvae were fed twice daily on a mixture of cultured algal species (*Dunaliella tertiolecta* and *Chaetoceros muelleri* at a concentration of 5000 cells mL^{-1} of each species). Water in rearing chambers was changed daily. Larval development was monitored daily, and late brachiolaria stage larvae were placed in separate rearing containers prior to use in experiments.

2.2. Preparation of Cues

Settlement experiments were conducted using coral rubble encrusted with crustose coralline algae (CCA), which was collected from shallow reef environments (<3 m depth) on the sheltered (north-west) side of Lizard Island. Rubble was broken into 1–2 cm pieces. Prior to experiments, all motile fauna were removed from coral rubble by manually removing fauna and repeatedly rinsing small fragments in saltwater. To ensure that rubble fragments were free of any potential infaunal predators, we also immersed fragments in freshwater for 30 s prior to using them in experiments. Although potential epibenthic predators were physically removed from the rubble, CCA and other encrustations were left intact. The predominant motile invertebrates removed from freshly collected rubble were polychaetes (mainly Nereididae and Amphinomidae). Amphinomidae polychaetes, *Pherecardia striata*, are known to prey upon newly settled *Acanthaster* cf. *planci* [22,23], and thus extracted polychaetes were retained for predation experiments. Trapeziid crabs (*Trapezia flavopunctata*, *Trapezia bidentata*, and *Trapezia cymodoce*) were collected by manually removing crabs with plastic forceps from *Pocillopora* corals collected from lagoonal reefs at Lizard Island. Corals were then maintained in a separate aquarium with flow-through seawater, for a minimum period of one week, prior to use in experiments. We did not distinguish between specific crab species or polychaete species in any of the experiments.

2.3. Predation Rates by Benthic Predators

To quantify predation rates by polychaetes and trapeziid crabs on brachiolaria larvae of crown-of-thorns starfish, individual predatory organisms ($n = 16$ for polychaetes; $n = 8$ for crabs) were placed in 70 mL specimen containers with 0.2 μm FSW and 10 brachiolaria larvae. No habitat was added in order to minimise the possibility of larval mortality occurring due to factors other than predation by the study organism. Predators were allowed to feed for 12 h through the night (18:30–06:30), and the number of starfish larvae that remained after this period was recorded. Controls were also conducted in which 10 brachiolaria larvae were added to 70 mL specimen containers with no predators.

2.4. Static Choice Chambers

To test the ability of settling *Acanthaster* sp. larvae to detect and respond to olfactory cues associated with potential settlement substrates and/or potential predators, we used static choice chambers consisting of two 10 L aquaria (chambers) connected by 150-mm diameter clear acrylic pipe. Substrates, with and without potential predators, were added to 0.2 μm FSW 24 h prior to the onset of experiments. Cues offered were: (i) the coral, *Pocillopora damicornis* without any infaunal organisms; (ii) *P. damicornis* with commensal trapeziid crabs; (iii) cleaned rubble; (iv) rubble with predators (polychaetes); and (v) adult *A.* cf. *solaris*. These cues were added to one or both chambers, in the combinations: (a) Cleaned rubble vs. FSW; (b) Rubble with polychaete predators vs. FSW; (c) Cleaned rubble vs. rubble with polychaete predators; (d) Cleaned rubble vs. coral; (e) Coral vs. FSW; (f) Coral with commensal crabs vs. FSW; (g) Coral vs. coral with commensal crabs; (h) Adult *A.* cf. *solaris* vs. FSW, offering larvae a pairwise choice of water sources. Larvae ($n = 10$) were individually placed in the centre of the pipe connecting both chambers, allowing horizontal movement towards one or the other aquaria. During trials, both aquaria were covered to minimise wind-driven water movement. Trials were conducted on a single larva, and larvae were not re-used. A choice was scored when the larvae moved well outside of the connecting tube and into one or the other of the two aquaria; there were no instances of larvae swimming back into the tube after entering an aquarium. If no choice was made after one hour, the larva was removed and "no choice" was recorded. After five replicates the chamber was cleaned and the water sources were switched to the opposite side to ensure that preferences were not biased for one side of the chamber. The response of larvae to each pairwise choice of cues was analysed using a Chi-square goodness of fit test against equal expected proportions, using Yates's correction, as expected counts were ≤ 5.

2.5. Settlement Assays

Settlement assays were conducted to determine whether the presence of benthic predators causes differences in rates of larval settlement. Ten larvae were introduced into 250 mL beakers filled with 0.2 μm FSW and containing one of two treatments: cleaned rubble ($n = 7$); or rubble with predators (polychaetes) ($n = 7$). Polychaetes were used as the predators in these experiments (cf. trapeziid crabs), because they naturally associate with rubble (cf. trapeziid crabs that generally associate with live coral). Controls were conducted in which larvae were introduced to 250 mL beakers (*n* = 7) containing only 0.2 μm FSW. Beakers were visually examined for indication of settlement/predation at 1, 2 and 6 h: when larvae were no longer swimming in the water column, they were assumed to have settled, or been consumed by predators. At 12, 24, 36 and 48 h, the number of larvae still swimming and the number that had settled out of the water column were counted. Larvae detected on the bottom of the beaker were counted as settled and checked under a microscope at the end of the experiment for evidence of metamorphosis. At 48 h, all substrates were also examined under a microscope for evidence of settled and metamorphosed individuals.

A repeated measures permutational analysis of variance (PERMANOVA) with "Treatment" (3 levels, fixed) and "Time" (4 levels, random) was run to test whether the biological habitat and presence or absence of predators had an effect on settlement success of larvae. PERMANOVA is a non-parametric technique that can be used in analysing univariate data [24]. Analyses were conducted using the PERMANOVA+ add-on for PRIMER v.6 (Primer-E Ltd., Plymouth, UK), using the Euclidian distance measure and 9999 permutations of the residuals under a reduced model, to calculate the significance of the pseudo-F statistic. Post-hoc pair-wise comparisons used Monte-Carlo asymptotic *p*-values (p_{MC}), as the number of unique permutations was low.

3. Results

3.1. Potential Predators

All three categories of potential benthic predators tested in this study (polychaetes, trapeziid crabs and scleractinian corals) caused elevated rates of mortality among late stage (competent) brachiolaria larvae of crown-of-thorns starfish (Figure 1). Mortality rates for starfish larvae exposed to individual polychaete worms averaged 1.9 larvae (out of 10) $\pm$ 0.6 (SE). For starfish larvae exposed to individual trapeziid crabs (ca. 10 mm carapace diameter), mortality rates were 4.5 larvae (out of 10) $\pm$ 0.9 (SE). By comparison, 100% of larvae survived across all controls. Mortality rates for starfish larvae exposed to the scleractinian corals, *Pocillopora damicornis*, were not explicitly measured, but all larvae that came into contact with the polyps were immediately consumed.

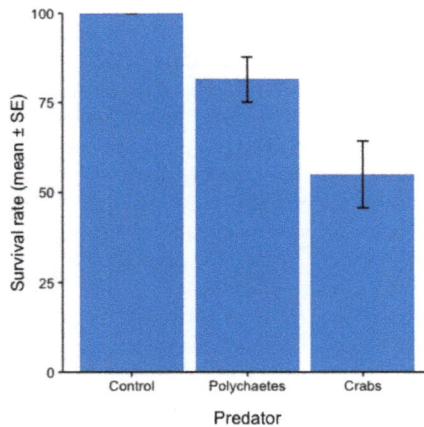

Figure 1. Mean survival ($\pm$ SE) of brachiolaria larvae during a nocturnal, 12-h period, in the absence of predators versus when exposed to polychaetes and trapeziid crabs.

3.2. Static Choice Chambers

A total of 80 brachiolaria larvae were used in static choice experiments, across 8 different treatments (Figure 3). Of these 80 larvae, a total of 55 larvae moved outside of the connecting tube, actively swimming towards one or other of the adjoining aquaria containing alternative settlement substrates or cues. Starfish larvae did exhibit significant avoidance of predators (both trapeziid crabs associated with *Pocillopora* coral ($\chi^2 = 6.2$, df = 2, $p = 0.05$), and polychaete predators naturally associated with freshly collected pieces of rubble ($\chi^2 = 7.4$, df = 2, $p = 0.02$) relative to FSW (Figure 2). However, starfish larvae did not discriminate when comparing settlement substrates (rubble with conspicuous CCA and the scleractinian coral, *P. damicornis*) with and without predators (polychaetes: $\chi^2 = 0.2$, df = 2, $p = 0.90$, crabs: $\chi^2 = 3.8$, df = 2, $p = 0.15$; Figure 2). When comparing corals with and without trapeziid crabs, 6 (out of 10) of the starfish larvae remained within the connecting tube (and did not venture into either of the adjoining aquaria) for the entire period of observation (60-min).

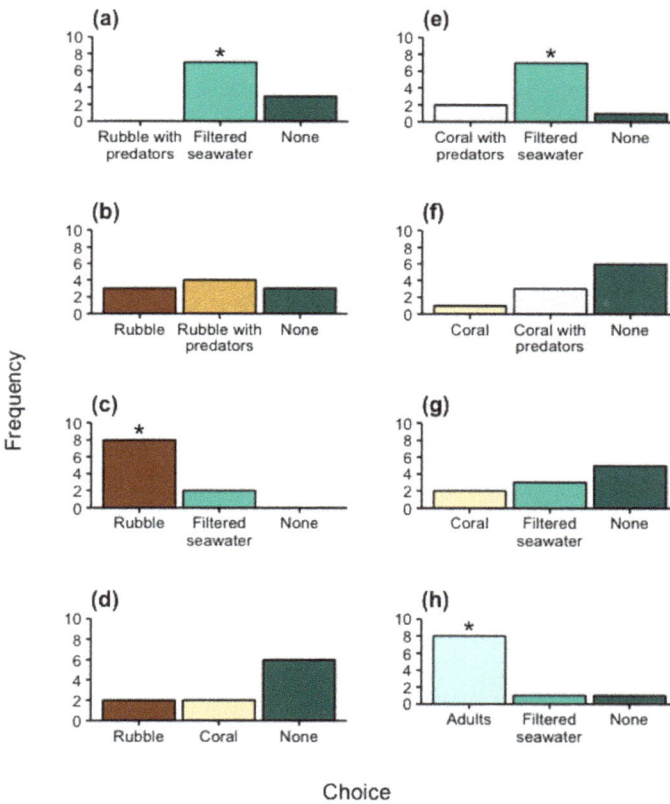

Figure 2. Behavioural response of brachiolaria larvae when offered a pairwise choice of cues: (**a**) Cleaned rubble vs. rubble with polychaete predators; (**b**) Rubble with polychaete predators vs. filtered seawater; (**c**) Cleaned rubble vs. filtered seawater; (**d**) Cleaned rubble vs. coral; (**e**) Coral vs. coral with commensal crabs; (**f**) Coral with commensal crabs vs. filtered seawater; (**g**) Coral vs. filtered seawater; (**h**) Adult crown-of-thorns starfish vs. filtered seawater. * Indicates a significant difference in larval choice frequency indicated by Chi-square goodness of fit test.

In the absence of predators (i.e., using rubble and live coral from which all predators had been removed), brachiolaria larvae of crown-of-thorns starfish exhibited strong and significant preference for cleaned rubble with conspicuous CCA over FSW ($\chi^2 = 10.4$, df = 2, $p < 0.01$). However, there was no significant difference in response when larvae were offered coral versus FSW ($\chi^2 = 1.4$, df = 2, $p = 0.50$; Figure 2e). Starfish larvae also did not significantly discriminate when offered a choice of cleaned rubble (with CCA) versus coral ($\chi^2 = 3.2$, df = 2, $p = 0.20$; Figure 2c). Larvae exhibited the strongest preference when comparing adult *Acanthaster* cf. *solaris* with FSW ($\chi^2 = 9.8$, df = 2, $p < 0.01$), with 8 (out of 10) larvae swimming towards adult conspecifics (Figure 2h).

3.3. Settlement Assays

A total of 210 larvae were used in the settlement assay, across three treatments (Figure 3). Of these 210 larvae, a total of 126 larvae were recovered after 48 h, either still swimming, or in contact with the substrate. There was a significant effect of treatment on the number of larvae still swimming in the settlement assay (Pseudo-$F_{(2,54)} = 13.27$, $p_{perm} < 0.01$). However, there was no significant effect of time (Pseudo-$F_{(3,54)} = 2.17$, $p_{perm} > 0.05$) and no significant interaction between treatment and time (Pseudo-$F_{(6,54)} = 1.26$, $p_{perm} > 0.05$). In the presence of a settlement substrate, significantly fewer larvae were recorded as swimming, compared to the control ($t = 4.88$, $p_{MC} < 0.01$), and this was regardless of the presence of polychaete predators ($t = 5.28$, $p_{MC} < 0.01$). However, there was no significant difference in the number of larvae still swimming in the presence of a substrate with, versus without, polychaete predators ($t = 0.69$, $p_{MC} > 0.05$; Figure 3a).

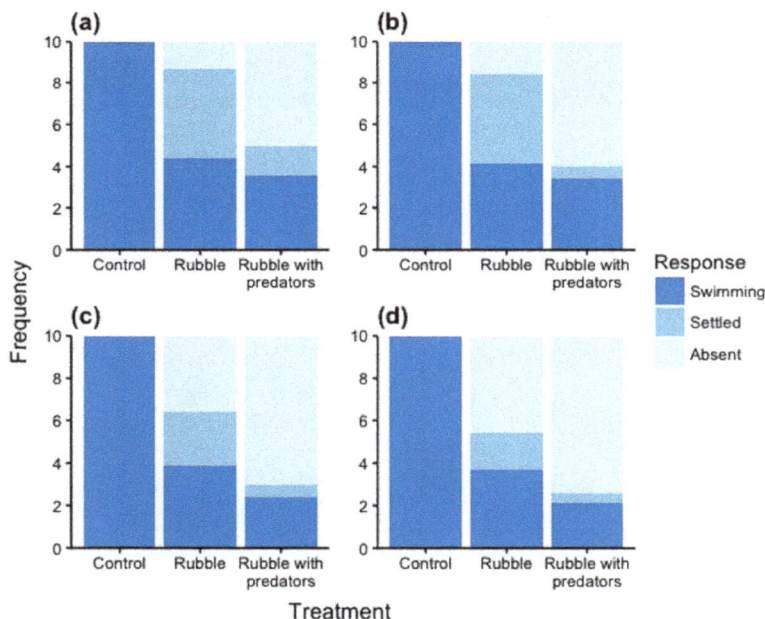

Figure 3. Mean number of larvae still swimming, settled and absent in each treatment after (**a**) 12, (**b**) 24, (**c**) 36, and (**d**) 48-h.

A total of 12 larvae were recorded as settled after 48 h, across the three treatments. In the absence of a substrate, 0% of larvae were induced to settle. There was a significant effect of treatment (Pseudo-$F_{(2,54)} = 13.60$, $p_{perm} < 0.01$) and time (Pseudo-$F_{(3,54)} = 3.80$, $p_{perm} = 0.01$), and a significant interaction between treatment and time (Pseudo-$F_{(6,54)} = 2.29$, $p_{perm} = 0.05$) on the number of larvae that were recorded as settled in the settlement assay (Figure 3b). The number of settled larvae that were recovered from the cleaned rubble treatment was significantly greater than for the control at 12 (t = 3.67, p_{MC} < 0.01), 24 (t = 3.43, p_{MC} < 0.01), 36 (t = 3.58, p_{MC} < 0.01), and 48 (t = 2.46, p_{MC} = 0.03) hours. There was no significant difference in the number of settled larvae in the cleaned rubble treatment, compared to the rubble with predators treatment after 12 h (t = 2.01, p_{MC} > 0.05); however, a significant decrease in the number of settled larvae recorded in the predator treatment from 12 to 24 h (t = 2.52, p < 0.05) meant that the number of settled larvae in the cleaned rubble treatment was significantly greater than for the rubble with predators treatment at 24 h (t = 2.71, p_{MC} = 0.02) and 36 h (t = 2.57, p_{MC} = 0.03) (Figure 3b). Although not significant, the number of settled larvae in the cleaned rubble treatment decreased from 12 to 48 h (t = 2.26, p_{MC} > 0.05), so the number of settled larvae in the cleaned rubble treatment, versus rubble with predators, was not significantly different after 48 h (t = 1.53, p_{MC} > 0.05). Notably however, 55% of larvae that settled in the absence of predators had undergone metamorphosis at 48 h, compared to 0% when predators were present.

4. Discussion

Given limited capacity for inter-reef movement of adult crown-of-thorns starfish [22], their abundance on any given reef is fundamentally dependent on rates of successful settlement and recruitment. To maximise settlement success, it would be expected that crown-of-thorns starfish preferentially settle to locations and microhabitats that minimise exposure to potential predators (e.g., [25]), either settling in locations that are naturally depauperate of potential predators or selecting microhabitats that provide refugia from predators. This study shows that settling larvae of *Acanthaster* cf. *solaris* are highly vulnerable to a range of different benthic predators, including scleractinian corals and their commensals (e.g., trapeziid crabs), as well as polychaete worms that were commonly found on freshly collected pieces of coral rubble, adding to the wide range of predators known to feed on early life-stages of *Acanthaster* spp. (e.g., [2–4,26]). Given the vulnerability of crown-of-thorns starfish to predators (reviewed in Cowan et al. this issue), it is likely that there are many more predators within the coral reef benthos capable of feeding on settling or post-metamorphic starfish. Once a more complete range of potential predators is established, or those that have the most significant impact on settlement and recruitment success of crown-of-thorns starfish, it will be important to test for spatial variation in abundance and diversity of these predators. For highly fecund species, such as *Acanthaster* spp., small changes to the rates of mortality at these early life stages are likely to have significant knock-on effects, and may account for spatial and temporal variation in the incidence of outbreaks [27,28].

Larvae of coral reef organisms, including fishes and corals, use chemical cues to distinguish between healthy and degraded reefs [15]. The attraction of brachiolaria larvae of *Acanthaster* cf. *solaris* to cues from cleaned rubble and deterrence from these substrates when polychaete predators were present suggests that larvae are able to detect these predators and may have the capacity to avoid settling to environments with high densities of benthic predators. This indicates a mechanism by which crown-of-thorns larvae can similarly distinguish between healthy and degraded reefs, as degraded reefs can have reduced densities of benthic predators. Polychaetes are amongst the most numerous and abundant component of the macrofauna found within the reef matrix (e.g., [29]) and are indicator species for marine degradation [30]. Both abundance and species richness of polychaetes are reduced in fished sites, compared to marine protected areas, as a result of trophic cascades [31]. Reefs that have been damaged by cyclones also have reduced polychaete abundance [32]. This may be an important factor in shaping recruitment patterns of *Acanthaster* spp. to reefs, with degraded reefs being more attractive to settling larvae.

Settling *Acanthaster* spp. larvae were not attracted to live coral substrates, and given that corals will prey upon settling larvae, this may explain the rejection of these substrates regardless of the presence, or absence, of commensal predators. Rather, settling crown-of-thorns starfish may actually avoid areas with high coral cover [33]. It has been suggested that larvae of crown-of-thorns starfish preferentially settle in areas with high abundance of adult conspecifics because the feeding activities of these adult starfish provide areas of recently dead coral, representing suitable settlement habitats [34] as well as minimizing the risk of predation by live corals [33]. Supporting this, our analyses of settlement preferences based on static choice chambers confirmed that larvae are significantly attracted to adult conspecifics. While other invertebrates including echinoderms settle on, or near to, conspecifics [13,35], this is the first time that it has been demonstrated for *Acanthaster* spp. and work is required to establish the mechanistic basis of this behaviour.

Whilst *Acanthaster* spp. larvae may preferentially settle in habitats with fine-scale topographic complexity to minimise mortality in the early life stages [9], numerous presumed predators are abundant in the reef matrix (e.g., [23,36,37]) and results of the settlement assay suggest that settlement is induced in the presence of a cue from substrates (cleaned rubble with naturally attached crustose coralline algae), with the presence of benthic predators unlikely to cause larvae to delay settlement (Figure 3a). Lack of difference in the number of settled larvae recovered from the rubble with predators treatment compared to the control, despite significant difference in the numbers of larvae still swimming in the settlement assays, indicates high levels of predation upon settling larvae, and is supported by our measurements of benthic predation rates. Additionally, no fully metamorphosed juveniles were recovered from treatments containing predators, compared to cleaned rubble treatments, in which 55% of settled starfish had metamorphosed. These data suggest that on a local scale, healthy benthic fauna is highly important in reducing successful recruitment of *Acanthaster* spp. through predation on settling larvae. Settlement substrate and predation by benthic predators may explain high variability in recruitment rates of *Acanthaster* sp. observed by Nakamura et al. [38]. Acroporids and *Acanthaster* sp. likely prefer to settle on similar substrate types, but variation in local abundance of benthic predators may have a more significant effect on recruitment of *Acanthaster* sp. compared to corals, for which recruitment was observed approximately 1 month after spawning [38].

In addition to the methods outlined in this study, we tested the ability of *Acanthaster* cf. *solaris* larvae to respond to olfactory cues using two-channel Atema flumes [39]. The flow rate in the flume chamber was set to the lowest possible speed (0.1 L·min^{-1}), which allowed larvae to maintain their position and move across the current, whilst also ensuring laminar flow. Food dye was used to test for laminar flow and confirm separation of water sources. Individual larvae were placed in the centre of the test chamber, allowing choice between the two streams and movement towards a preferred water source. Whilst larvae were able to maintain position and move across the flow of water in the Atema flume chamber, we could not determine with confidence whether larvae were actively choosing one cue over the other: larvae were observed to move from the central starting position; however, they settled against either the outer edges or inner partition of the flume chamber and appeared to become trapped. Adaptation of the flume chamber to exclude edges in which larvae could become trapped would likely overcome this; thus, this may be an efficient method to test the response of *Acanthaster* spp. larvae to a range of environmental cues. Further, this demonstrates an ability of, at least the late-brachiolaria stage, larvae to move against a light current. This could be important in enabling larvae to manoeuvre across the reef matrix when searching for a suitable settlement habitat.

5. Conclusions

Variation in the abundance of benthic predators is likely to have a significant influence on settlement patterns of the crown-of-thorns starfish, as indicated by the avoidance of late-stage brachiolaria larvae of substrates that contain predators and low rates of settlement and metamorphosis on these substrates. Healthy benthic fauna is therefore likely to be important in regulating abundance and moderating settlement success of *Acanthaster* spp. Any extrinsic threat to benthic communities, including disruptive effects that may lead to trophic cascades, are likely to reduce the buffering capacity of coral reefs, increasing susceptibility to devastating outbreaks of crown-of-thorns starfish. Demographic studies of marine invertebrates often reveal particularly intense mortality in the period immediately following settlement (e.g., [6,40]). Studies described herein may also be extended to newly settled, post-metamorphic juveniles, with the aim of determining predation rates and further examining their behavioural responses to environmental cues.

Acknowledgments: This project was funded by an Ian Potter Foundation 50th Anniversary Commemorative Grant awarded to Z.L.C. by the Australian Museum's Lizard Island Research Station. We would like to thank Alexander Buck, Alexia Graba-Landry, Sam Matthews and Ben Mos for their assistance. Invertebrate collection was permitted under Great Barrier Reef Marine Park Authority permit number G15/38002.1. *Acanthaster* sp. were collected under Great Barrier Reef Marine Park Authority permit number G13/35909.1.

Author Contributions: All authors contributed equally to the concept and layout of this paper. Z.L.C. and S.D. compiled and analysed the data; Z.L.C. and M.P. wrote the paper; S.D. and C.C. edited the paper.

Conflicts of Interest: The authors declare no conflict of interest.

References

1. Almany, G.R.; Webster, M.S. The predation gauntlet: Early post-settlement mortality in reef fishes. *Coral Reefs* **2006**, *25*, 19–22. [CrossRef]
2. Cowan, Z.-L.; Dworjanyn, S.A.; Caballes, C.F.; Pratchett, M.S. Predation on crown-of-thorns starfish larvae by damselfishes. *Coral Reefs* **2016**, *35*, 1253–1262. [CrossRef]
3. Yamaguchi, M. Early life histories of coral reef asteroids, with special reference to *Acanthaster planci* (L.). In *Biology and Geology of Coral Reefs*; Jones, O., Endean, R., Eds.; Academic Press: New York, NY, USA, 1973; pp. 369–387.
4. Keesing, J.; Halford, A. Field measurement of survival rates of juvenile *Acanthaster planci*, techniques and preliminary results. *Mar. Ecol. Prog. Ser.* **1992**, *85*, 107–114. [CrossRef]
5. Morgan, S.G. The larval ecology of marine communities. In *Marine Community Ecology*; Bertness, M.D., Gaines, S.D., Hay, M.E., Eds.; Sinauer Associates: Sunderland, MA, USA, 2001; pp. 159–181.
6. Gosselin, L.A.; Qian, P.-Y. Juvenile mortality in benthic marine invertebrates. *Mar. Ecol. Prog. Ser.* **1997**, *146*, 265–282. [CrossRef]
7. Connell, J.H. The consequences of variation in initial settlement vs. post-settlement mortality in rocky intertidal communities. *J. Exp. Mar. Bio. Ecol.* **1985**, *93*, 11–45. [CrossRef]
8. Rowley, R.J. Settlement and recruitment of sea urchins (*Strongylocentrotus* spp.) in a sea-urchin barren ground and a kelp bed: Are populations regulated by settlement or post-settlement processes? *Mar. Biol.* **1989**, *100*, 485–494. [CrossRef]
9. Lucas, J.S. Environmental influences on the early development of *Acanthaster planci* (L.). In *Crown-Thorns Starfish Seminar Proceedings*; Australian Government Publishing Service: Canberra, Australia, 1975; pp. 109–121.
10. Lucas, J.; Hart, R.; Howden, M.; Salathe, R. Saponins in eggs and larvae of *Acanthaster planci* (L.) (Asteroidea) as chemical defences against planktivorous fish. *J. Exp. Mar. Bio. Ecol.* **1979**, *40*, 155–165. [CrossRef]
11. Barnett, D.; Dean, P.W.; Hart, R.J.; Lucas, J.S.; Salathe, R.; Howden, M.E.H. Determination of contents of steroidal saponins in starfish tissues and study of their biosynthesis. *Comp. Biochem. Physiol. Part B Comp. Biochem.* **1988**, *90*, 141–145. [CrossRef]
12. Johnson, C.R.; Sutton, D.C.; Olson, R.R.; Giddins, R. Settlement of crown-of-thorns starfish: Role of bacteria on surfaces of coralline algae and a hypothesis for deepwater recruitment. *Mar. Ecol. Prog. Ser.* **1991**, *71*, 143–162. [CrossRef]
13. Pawlik, J.R. Chemical ecology of the settlement of benthic marine invertebrates. *Oceanogr. Mar. Biol.* **1992**, *30*, 273–335.
14. Dixson, D.L.; Munday, P.L.; Pratchett, M.; Jones, G.P. Ontogenetic changes in responses to settlement cues by anemonefish. *Coral Reefs* **2011**, *30*, 903–910. [CrossRef]
15. Dixson, D.L.; Abrego, D.; Hay, M.E. Chemically mediated behavior of recruiting corals and fishes: A tipping point that may limit reef recovery. *Science* **2014**, *345*, 892–897. [CrossRef] [PubMed]
16. Sweatman, H. Field evidence that settling coral reef fish larvae detect resident fishes using dissolved chemical cues. *J. Exp. Mar. Bio. Ecol.* **1988**, *124*, 163–174. [CrossRef]
17. Johnson, C.R.; Muir, D.G.; Reysenbach, A.L. Characteristic bacteria associated with surfaces of coralline algae: A hypothesis for bacterial induction of marine invertebrate larvae. *Mar. Ecol. Prog. Ser.* **1991**, *74*, 281–294. [CrossRef]

18. Johnson, C.R.; Sutton, D.C. Bacteria on the surface of crustose coralline algae induce metamorphosis of the crown-of-thorns starfish *Acanthaster planci*. *Mar. Biol.* **1994**, *120*, 305–310. [CrossRef]
19. Brauer, R.W.; Jordan, M.J. Triggering of the stomach eversion reflex of *Acanthaster planci* by coral extracts. *Nature* **1970**, *228*, 344–346. [CrossRef] [PubMed]
20. Ormond, R.; Campbell, A.; Head, S.; Moore, R.; Rainbow, P.; Saunders, A. Formation and breakdown of aggregations of the crown-of-thorns starfish, *Acanthaster planci* (L.). *Nature* **1973**, *246*, 167–169. [CrossRef]
21. Haszprunar, G.; Spies, M. An integrative approach to the taxonomy of the crown-of-thorns starfish species group (Asteroidea: *Acanthaster*): A review of names and comparison to recent molecular data. *Zootaxa* **2014**, *3841*, 271–284. [CrossRef] [PubMed]
22. Glynn, P.W. *Acanthaster* population regulation by a shrimp and a worm. In Proceedings of the 4th International Coral Reef Symposium, Manila, Philippines, 18–22 May 1981; Gomez, E.D., Birkeland, C.E., Buddemeier, R.W., Johannes, R.E., Marsh, J.A., Jr., Tsuda, R.T., Eds.; 1982; Volume 2, pp. 607–612.
23. Glynn, P.W. An amphinoid worm predator of the crown-of-thorns sea star and general predation on asteroids in eastern and western pacific coral reefs. *Bull. Mar. Sci.* **1984**, *35*, 54–71.
24. Anderson, M.J.; Gorley, R.N.; Clarke, K.R. *PERMANOVA+ for PRIMER: Guide to Software and Statistical Methods*; PRIMER-E Ltd.: Plymouth, UK, 2008.
25. Mileikovsky, S.A. On predation of pelagic larvae and early juveniles of marine bottom invertebrates by adult benthic invertebrates and their passing alive through their predators. *Mar. Biol.* **1974**, *26*, 303–311. [CrossRef]
26. Zann, L.; Brodie, J.; Berryman, C.; Naqasima, M. Recruitment, ecology, growth and behavior of juvenile *Acanthaster planci* (L.) (Echinodermata: Asteroidea). *Bull. Mar. Sci.* **1987**, *41*, 561–575.
27. McCallum, H.I. Effects of predation on organisms with pelagic larval stages: Models of metapopulations. In Proceedings of the 6th International Coral Reef Symposium, Townsville, Australia, 8–12 August 1988; Choat, J.H., Barnes, D., Borowitzka, M.A., Coll, J.C., Davies, P.J., Flood, P., Hatcher, B.G., Hopley, D., Hutchings, P.A., Kinsey, D., et al., Eds.; 1988; Volume 2, pp. 101–106.
28. McCallum, H.I. Effects of predation on *Acanthaster*: Age-structured metapopulation models. In *Acanthaster and the Coral Reef: A Theoretical Perspective*; Bradbury, R., Ed.; Springer-Verlag: Berlin/Heidelberg, Germany, 1990; pp. 208–219.
29. Grassle, J.F. Variety in coral reef communities. *Biol. Geol. Coral Reefs* **1973**, *2*, 247–270.
30. Dean, H. The use of polychaetes (Annelida) as indicator species of marine pollution: A review. *Rev. Biol. Trop.* **2008**, *56*, 11–38.
31. Pinnegar, J.K.; Polunin, N.V.C.; Francour, P.; Badalamenti, F.; Chemello, R.; Harmelin-Vivien, M.-L.; Hereu, B.; Milazzo, M.; Zabala, M.; D'Anna, G.; et al. Trophic cascades in benthic marine ecosystems: Lessons for fisheries and protected-area management. *Environ. Conserv.* **2000**, *27*, 179–200. [CrossRef]
32. Sukumaran, S.; Vijapure, T.; Kubal, P.; Mulik, J.; Rokade, M.A.; Salvi, S.; Thomas, J.; Naidu, V.S. Polychaete community of a marine protected area along the west coast of India-prior and post the tropical cyclone, *Phyan*. *PLoS ONE* **2016**, *11*, e0159368. [CrossRef] [PubMed]
33. Chesher, R.H. Destruction of Pacific corals by the sea star *Acanthaster planci*. *Science* **1969**, *165*, 280–283. [CrossRef] [PubMed]
34. Pratchett, M.S.; Caballes, C.F.; Rivera-Posada, J.A.; Sweatman, H.P.A. Limits to understanding and managing outbreaks of crown-of-thorns starfish (*Acanthaster* spp.). *Oceanogr. Mar. Biol. Annu. Rev.* **2014**, *52*, 133–200.
35. Dworjanyn, S.A.; Pirozzi, I.; Liu, W. The effect of the addition of algae feeding stimulants to artificial diets for the sea urchin *Tripneustes gratilla*. *Aquaculture* **2007**, *273*, 624–633. [CrossRef]
36. Keesing, J.K.; Wiedemeyer, W.L.; Okaji, K.; Halford, A.R.; Hall, K.C.; Cartwright, C.M. Mortality rates of juvenile starfish *Acanthaster planci* and *Nardoa* spp. measured on the Great Barrier Reef, Australia and in Okinawa, Japan. *Oceanol. Acta* **1996**, *19*, 441–448.
37. Rivera-Posada, J.; Caballes, C.F.; Pratchett, M.S. Size-related variation in arm damage frequency in the crown-of-thorns sea star, *Acanthaster planci*. *J. Coast. Life Med.* **2014**, *2*, 187–195.
38. Nakamura, M.; Kumagai, N.H.; Sakai, K.; Okaji, K.; Ogasawara, K.; Mitarai, S. Spatial variability in recruitment of acroporid corals and predatory starfish along the Onna coast, Okinawa, Japan. *Mar. Ecol. Prog. Ser.* **2015**, *540*, 1–12. [CrossRef]

39. Atema, J.; Kingsford, M.J.; Gerlach, G. Larval reef fish could use odour for detection, retention and orientation to reefs. *Mar. Ecol. Prog. Ser.* **2002**, *241*, 151–160. [CrossRef]

40. Pineda, J.; Riebensahm, D.; Medeiros-Bergen, D. *Semibalanus balanoides* in winter and spring: Larval concentration, settlement, and substrate occupancy. *Mar. Biol.* **2002**, *140*, 789–800.

diversity

MDPI

Article

Modelling Growth of Juvenile Crown-of-Thorns Starfish on the Northern Great Barrier Reef

Jennifer Wilmes [1,2,*], Samuel Matthews [1], Daniel Schultz [2], Vanessa Messmer [1], Andrew Hoey [1] and Morgan Pratchett [1]

[1] ARC Centre of Excellence for Coral Reef Studies, James Cook University, Townsville QLD 4811, Australia; sammatthews990@gmail.com (S.M.); vanessa.messmer@gmail.com (V.M.); andrew.hoey1@jcu.edu.au (A.H.); morgan.pratchett@jcu.edu.au (M.P.)
[2] Red Fish Blue Fish Marine, Cairns QLD 4870, Australia; schultz.dj@gmail.com
*   Correspondence: wilmes.jennifer@gmail.com; Tel.: +61-476-421-811

Academic Editors: Sven Uthicke and Michael Wink
Received: 18 November 2016; Accepted: 26 December 2016; Published: 29 December 2016

Abstract: The corallivorous crown-of-thorns starfish (*Acanthaster* spp.) is a major cause of coral mortality on Indo-Pacific reefs. Despite considerable research into the biology of crown-of-thorns starfish, our understanding of the early post-settlement life stage has been hindered by the small size and cryptic nature of recently settled individuals. Most growth rates are derived from either laboratory studies or field studies conducted in Fiji and Japan. The Great Barrier Reef (GBR) is currently experiencing its fourth recorded outbreak and population models to inform the progression of outbreaks lack critical growth rates of early life history stages. High numbers of 0+ year juveniles (n = 3532) were measured during extensive surveys of 64 reefs on the northern GBR between May and December 2015. An exponential growth model was fitted to the size measurement data to estimate monthly ranges of growth rates for 0+ year juveniles. Estimated growth rates varied considerably and increased with age (e.g., 0.028–0.041 mm·day^{-1} for one-month-old juveniles versus 0.108–0.216 mm·day^{-1} for twelve-month-old juveniles). This pioneering study of 0+ year juveniles on the GBR will inform population models and form the basis for more rigorous ongoing research to understand the fate of newly settled *Acanthaster* spp.

Keywords: juvenile crown-of-thorns starfish; growth rates; growth curve analysis

1. Introduction

Crown-of-thorns starfish (*Acanthaster* spp.) are among the most intensively studied of all coral reef organisms [1]. This starfish species is renowned for its extreme temporal and spatial variation in abundance, which can result in extensive destruction of coral reef habitats if starfish occur at high densities (during outbreaks). Outbreaks of crown-of-thorns starfish are considered to be one of the foremost causes of significant and sustained declines in live hard coral cover on Indo-Pacific reefs [2–5]. On Australia's Great Barrier Reef (GBR), there have been four distinct episodes of outbreaks since the early 1960's, with the latest outbreak first apparent in 2010 [6,7]. Over a period of 27 years (1985 to 2012), the GBR has lost approximately half of its initial coral cover, with 42% of this loss attributable to recurrent outbreaks of crown-of-thorns starfish [2].

Outbreaks of crown-of-thorns starfish are fundamentally caused by changes in key demographic rates and population dynamics [8], resulting in either progressive accumulation of starfish over several successive cohorts [9] or the rapid onset of outbreaks following a single mass-recruitment event [10]. However, the extent to which these demographic changes are caused by anthropogenic degradation of reef ecosystems (e.g., overfishing and/or eutrophication) versus inherent environmental changes and stochasticity (e.g., cycles of food availability for larvae, juveniles, and/or adult starfish) is largely unknown and widely debated [6,11]. Our understanding of the proximal and ultimate causes of outbreaks has been hampered at least in part by difficulties in studying the early life stages of crown-of-thorns starfish in natural environments [6,11]. Current models [7,12,13] that aim to understand and predict outbreak dynamics lack critical demographic rates, such as estimates of growth rates and survival for early life history stages in the field.

Field-based studies of early life stages of crown-of-thorns starfish have largely been constrained by the small size, cryptic nature, and largely nocturnal habits of recently settled individuals [10], making them difficult to detect and sample [14–16]. Attempts to locate 0+ year old juveniles on the GBR have remained largely unsuccessful [14–16]. However, studies conducted in Fiji and Japan that followed individual cohorts of crown-of-thorns starfish through time on single island reefs effectively sampled newly settled individuals (0+ year old) [10,17]. Resulting estimates of growth rates for coralline algae feeding juveniles (2–3 mm/month) [10,17] were consistent with early post-settlement growth estimates for laboratory reared juveniles [18,19].

As early stages of newly settled juveniles (i.e., 1–3 months-old starfish after settlement, size = 0.3–5 mm) have rarely been detected in the field [17], demographic rates for these early stages are largely derived from aquarium-based studies [18,19]. Crown-of-thorns starfish have been reared in captivity since 1973, providing important insights into their early development and life history [18–21]. Once fully developed competent larvae (i.e., at the late brachiolaria stage) find a suitable settlement substrate (i.e., coralline algae), they metamorphose within two days into five-armed juvenile starfish that measure between 0.3 and 0.8 mm in diameter [18,19,22,23]. Yamaguchi (1973) [18] found that laboratory-reared juveniles grew 0.076 mm·day^{-1} in the 20 weeks following settlement, while Lucas (1974) [19] estimated that 3-month-old juveniles grew 0.048 mm·day^{-1} in the laboratory. Field-based estimates of juvenile growth rates averaged 0.10–0.15 mm·day^{-1} in the coralline algae feeding phase and 0.40–0.84 mm·day^{-1} in the coral feeding phase [10,17]. So far, existing growth data has been fitted with logistic or Gompertz growth equations to describe the sigmoidal growth pattern of crown-of-thorns starfish [10,17,20,21,24]. However, these equations have been acknowledged to be limited in accurately describing growth during distinct life stages [21], and so alternative equations have been suggested for distinct stages, such as for coralline algae feeding juveniles [25] and coral-eating 1+ year old juveniles [21,26].

The purpose of this study was to reconstruct growth curves for juvenile (0+ year old) crown-of-thorns starfish on the northern GBR, based on intensive (near monthly) field sampling of newly settled individuals. The exact ages of juvenile starfish cannot be verified, but may be inferred for 0+ year old starfish by assuming that settlement occurs within a relatively narrow period [10,17]. Here we provide, for the first time, monthly ranges of growth rates for 0+ year juveniles to inform crown-of-thorns starfish population models. In addition, we compare these results to previous field studies and present ranges of predicted mean sizes for different age classes that can be used to inform the planning of future juvenile monitoring studies on the GBR.

2. Materials and Methods

2.1. Field Collection

All field sampling was conducted in conjunction with the Association of Marine Park Tourism Operators' (AMPTO) crown-of-thorns starfish control vessels during the fourth recorded outbreak of *Acanthaster* cf. *solaris* [27] on Australia's GBR. One hundred and eleven sites on 64 reefs within six geographic locations (or reef complexes) located between 14.72° S and 17.67° S were sampled between 7 May and 15 December 2015 (Figure 1). At each site, 1–2 SCUBA divers searched coralline algae encrusted pieces of dead coral and live coral colonies for juvenile crown-of-thorns starfish and their feeding marks. Individual pieces of live and dead coral were thoroughly inspected for juvenile starfish if feeding marks were sighted on exposed surfaces. Each diver was able to cover up to 250 m^2 during a typical 40 min dive. However, if juvenile abundances were high (e.g., 40 individuals collected

during a 40 min dive), search effort was restricted to a much smaller area (≈50 m²). All starfish
(target size ≤ 50 mm) were collected and placed in sampling jars underwater. This size threshold
was selected, as previous growth models [17,21,24] predict that the mean size of starfish from the
previous year's cohort would be >50 mm at the time of our first sampling (May 2015). The size of
individuals was not measured in situ, and as such, two of the 3532 juveniles were slightly larger than
50 mm (52.5 mm and 64 mm). These larger individuals were collected in November/December 2015,
and were retained in the analysis, as they were likely to have come from the same cohort. After each
sampling dive, juvenile starfish were kept alive in containers filled with seawater, and their maximum
diameter was measured to the nearest half millimetre with a stainless steel ruler. Once starfish were
measured, all individuals were preserved in 95% ethanol for future analyses.

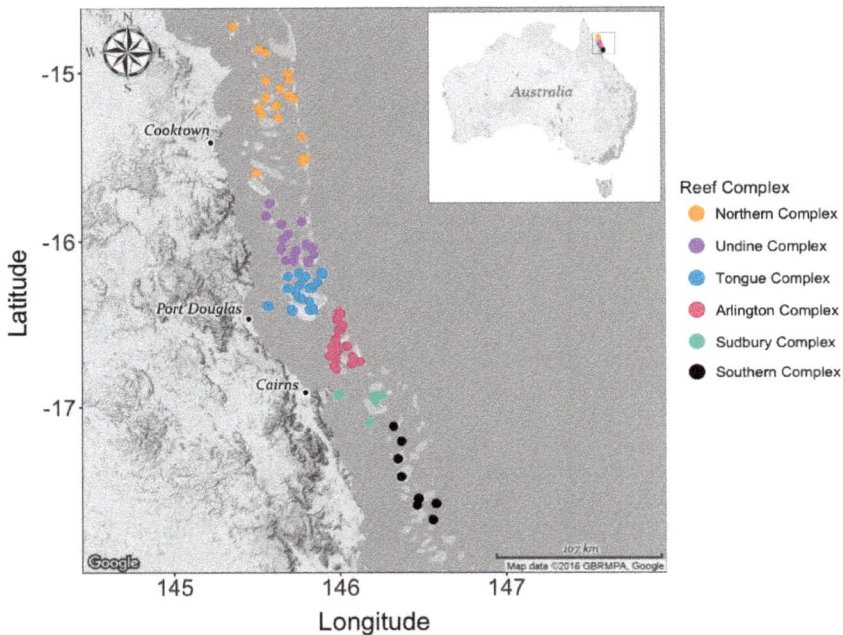

Figure 1. Map of the north Queensland coast showing the sampling locations (*n* = 111) and designated
reef complexes (*n* = 6) for the collection of 0+ year crown-of-thorns starfish.

2.2. Growth Curve Analysis

On the GBR, spawning has been reported to occur throughout summer months, but is concentrated
in December and January [6]. In the year of sampling, Uthicke et al. (2015) [28] reported that spawning
occurred between 10 and 21 of December 2014 in the area of the northern GBR relevant to our study
(i.e., 1 to 7 days prior to sampling), and that there was no subsequent spawning throughout December
or January. Assuming a planktonic larval duration of 10 to 40 days [22,23,29–32], settlement is
likely to have occurred predominantly—if not exclusively—in early January 2015. As age of sampled
individuals cannot be verified, age of sampled juveniles for the model was based on the assumption that
all juveniles settled on 1 January 2015 (i.e., 10–21 days after the reported spawning, which incorporates
the 17–22 peak settlement window determined by Pratchett et al., in review [33]). Ultimately, all growth
estimates are presented as monthly growth rates, and so even if crown-of-thorns starfish settled over
several days to weeks, it would have limited influence on our results.

All exploratory and growth curve analyses were conducted using R [34]. A series of preliminary models (e.g., logistic growth model and exponential growth model) were fitted to the 2015-juvenile size data to determine which type of model was most suitable to describe the relationship between size and age. Visual inspection and exploratory analysis of the fitted growth curves (based on least residual sum of squares) identified an exponential growth model (as suggested by Okaji (1996) [25]) in the following form as the most accurate in describing the shape of the size measurement data:

$$Size_{Age} = \beta_0 \times e^{k \times Age} + \beta_1 \qquad (m1)$$

where *Size* is the diameter of starfish in mm, *Age* is number of days since 1 January, and β_0, k, and β_1 are parameters to be estimated. As size measurements were missing for the initial four months of development (i.e., sampling period: January to April 2015), the y-intercept (i.e., $Size_{Age=0}$) was fixed to a biologically meaningful size (see below), representative of the range of predicted mean sizes for newly metamorphosed juvenile starfish in the laboratory ([23]; see Appendix A, Table A1). The growth model may therefore be represented as:

$$Size_{Age} = \beta_0 \times e^{k \times Age} + (Size_{Age=0} - \beta_0) \qquad (m2)$$

Three estimates of size after metamorphosis (i.e., $Size_{Age=0}$ = 0.30 mm, 0.56 mm, and 0.82 mm) from Fabricius et al. (2010) [23] were used to generate three different *m2* models ($m2_{Size\ at\ Age0=0.30}$, $m2_{Size\ at\ Age0=0.56}$, $m2_{Size\ at\ Age0=0.82}$). Best-fit parameter estimates for *m1* (β_0, k, and β_1) and *m2* (β_0, k) models were obtained with associated 95% confidence intervals using the *nlstools* package [35]. Residuals from the preliminary models were found to display a distinct wedge pattern, indicating that variance was related to age of juveniles, and thus multiple variance functions (Age^2, Age^3, Age^4) were trialled to optimize best-fit parameter estimates for each model (*m1*, $m2_{Size\ at\ Age0=0.30}$, $m2_{Size\ at\ Age0=0.56}$, $m2_{Size\ at\ Age0=0.82}$). Best-fit parameter estimates for all 12 models were then used to perform linear mixed effect analysis of the relationship between *Size* and *Age* using the *nlme* package [36]. Reef complex and survey sites were included in the linear mixed effect models as nested random factors to account for variation at the site and reef complex scale, and variance structures (Age^2, Age^3, Age^4) were included to account for the variation in size with age. Best-fit models were selected based on lowest Akaike Information Criterion (AIC). No evidence of spatial autocorrelation was found after both visual (variograms and bubble plots) and statistical inspection (incorporating spatial correlation structures into mixed effects models).

A selection of best-fit models and corresponding 95% confidence intervals was then plotted using the *ggplot2* package [37]. The best-fit model of $m2_{Size\ at\ Age0=0.56}$ was plotted with the upper bound of the 95% confidence interval of $m2_{Size\ at\ Age0=0.82}$ and the lower bound of the 95% confidence interval of $m2_{Size\ at\ Age0=0.30}$, forming a so-called "combined 95% confidence interval". Growth rates (i.e., growth increment per unit time) were calculated for different time spans (i.e., 30-day intervals) for the upper bound of the 95% confidence interval of $m2_{Size\ at\ Age0=0.82}$, and the lower bound of the 95% confidence interval of $m2_{Size\ at\ Age0=0.30}$ to provide ranges of modelled growth rates.

3. Results

Size ranges of juvenile starfish increased as sampling progressed through the year (Figure 2a). Juveniles sampled in May measured between 3 and 15.5 mm in size (size range = 12.5 mm), while those sampled in December ranged from 8.5 to 52.5 mm (size range = 44 mm) (Figure 2a), representing a 3.5-fold increase in size range within this time period.

For *m1* models (variable intercept), growth was indeed exponentially related to age (t_{3420} = 59.79, $p < 0.0001$), while including *site* ($\sigma \approx 0.0007$) and *reef complex* ($\sigma \approx 0.0003$) as random factors. Best model performance was achieved with a variance structure of Age^3 (AIC = 18,057), indicating that variance increased cubically with age. However, the best-fit *m1* model predicted mean size after metamorphosis ($Size_{Age=0}$) to be 5.42 mm. This appears to be erroneous, as it is in stark contrast with the range of expected mean sizes for newly settled juveniles (i.e., 0.30–0.82 mm). Consequently, the *m2* models seemed to describe growth more accurately because their intercept was fixed to a biologically relevant size after metamorphosis. Again, for *m2* models (fixed intercept), growth was indeed exponentially related to age ($m2_{Size\ at\ Age0=0.82}$, $t_{3420} = 59.55$, $p < 0.0001$), and a variance structure of Age^3 gave the best model performance ($m2_{Size\ at\ Age0=0.82}$, AIC = 18,071). Although overall model performance was slightly reduced using the fixed intercept models (based on a higher AIC), fixing the intercept to a biologically meaningful size provides a better characterisation of the growth curves for juvenile crown-of-thorns starfish.

The growth curve analysis highlights increasing variation in size among older individuals. As shown by the distribution of size–frequency data for different sampling periods (Figure 2a) and the gradual widening of the combined 95% confidence interval (Figure 2b), variance increased considerably as juveniles grew older. The increased variation in size with age was further reflected by the increase of monthly ranges of modelled growth rates (Table 1). Ranges of modelled growth rates increased from 0.028–0.041 mm·day^{-1} for one-month-old juveniles to 0.108–0.216 mm·day^{-1} for twelve-month-old juveniles.

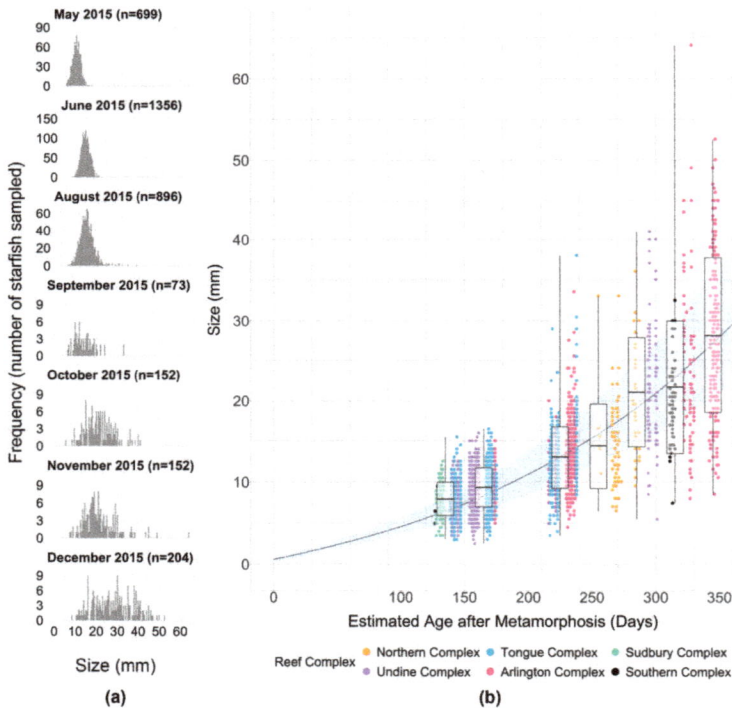

Figure 2. (**a**) Monthly size–frequency distributions of juvenile crown-of-thorns starfish sampled during May–December 2015; (**b**) Fitted growth curve ($m2_{Size\ at\ Age0=0.56}$, black line) and combined 95% confidence interval (light blue) for the 0+ year crown-of-thorns starfish cohort sampled in the northern Great Barrier Reef (GBR). The combined 95% confidence interval is formed by the lower and upper bounds of the 95% confidence interval of *m2* growth models with size at age 0 set to 0.30 mm and 0.82 mm, respectively. Individual starfish of the 2015 cohort are depicted as points, and the colours indicate the reef complex (see Figure 1). Each boxplot (by sampling month) is characterised by a mean size (horizontal middle line), ±one standard deviation (rectangle), and minimum and maximum size (vertical line). The vertical lines of the grid indicate 30-day intervals.

Table 1. Range of modelled mean sizes and growth rates for estimated age (in months after metamorphosis) and different time spans (30-day intervals).

Estimated Age after Metamorphosis (Months)	Time Span (30-Day Interval)	Range of Modelled Mean Sizes (mm)	Range of Modelled Growth Rates (mm·Day^{-1})
1	0–30	0.30–2.04	0.028–0.041
2	30–60	1.15–3.46	0.032–0.047
3	60–90	2.11–5.12	0.036–0.055
4	90–120	3.19–7.04	0.041–0.064
5	120–150	4.41–9.28	0.046–0.075
6	150–180	5.79–11.89	0.052–0.087
7	180–210	7.35–14.93	0.059–0.101
8	210–240	9.11–18.47	0.066–0.118
9	240–270	11.09–22.58	0.075–0.137
10	270–300	13.33–27.37	0.084–0.160
11	300–330	15.87–32.94	0.095–0.186
12	330–360	18.73–39.43	0.108–0.216

4. Discussion

This study provides the first estimates of monthly growth rates for 0+ year crown-of-thorns starfish on the GBR based on intensive field sampling of newly settled juvenile starfish. Juvenile crown-of-thorns starfish exhibited exponential growth over the first year on the reef and reached a size of up to 52.5 mm by mid-December. Size ranges of juveniles increased by a factor of 3.5 from May to December. The marked increase in size range was also reflected in the results of the growth curve analysis, which showed that the variation in size increased cubically with age in the best-fit models. Similarly, ranges of modelled growth rates increase with age; while one month old juveniles grow 0.028–0.041 mm·day^{-1}, twelve-month-old juveniles grow 0.108–0.216 mm·day^{-1}.

The increasing variation in size among older individuals may be attributed to variation in the availability of suitable coral prey within specific study sites and microhabitats. Although laboratory studies have shown that same-aged newly settled crown-of-thorns starfish vary in size from the beginning of their post-metamorphic life [23], marked variations in growth rates of juvenile starfish are generally attributed to the availability of suitable coral prey [20,38]. At settlement, crown-of-thorns starfish feed on coralline algae, and it is thought that they preferentially settle in locations (and microhabitats) where coralline algae are ubiquitous [18,39]. After an obligatory period of feeding on coralline algae (4.5—Yamaguchi 1973 [18]), juvenile starfish may or may not transition to feeding on Scleractinian corals, depending upon the local availability of suitable coral prey [10]. The fastest growth rates recorded in both field and laboratory settings are for individuals that make rapid transitions to coral feeding, resulting in accelerated growth and are reported to be significantly larger than siblings that continue to feed on algae [21]. Individual crown-of-thorns starfish that continue to feed on coralline algae after 4–5 months are thought to be more vulnerable to predation by epibenthic fauna [18], whereas fast-growing juveniles that make a rapid transition to feeding on coral—and thereby escape this predation pressure—are more likely to survive [40]. As sampling was conducted on 64 reefs across 111 sites, the availability of suitable coralline algae and coral prey would have differed considerably among microhabitats within and between sampling sites. Individual sampled starfish are likely to have been exposed to different environmental pressures (e.g., predation and food availability) in each of these microhabitats, which would have shaped their growth in the first year of development differently. Consequently, the gradual widening of the 95% confidence interval likely reflects differential growth rates between individuals that have transitioned to feeding on live coral versus those continuing to feed on coralline algae. This appears to be driven by a varying availability of suitable coral prey within microhabitats.

Ranges of modelled growth rates were broadly comparable to both laboratory-reared individuals released into the field [25] and the growth of juvenile cohorts in Japan and Fiji [10,17]. Observed mean sizes, standard deviations, and size ranges of juveniles sampled in these studies [10,17,25] were plotted to the fitted growth curve for comparison (Figure 3). Given the uncertainties related to the approach taken to estimate age in Zann et al. (1987) [10] and Habe et al. (1989) [17], and the discrepancies related to ages being estimated in months instead of days, estimated ages were not standardised across different studies. Care should therefore be taken when interpreting the results of this comparison. While Zann et al. (1987) [10] and Habe et al. (1989) [17] followed distinct cohorts of juvenile starfish in the field, Okaji (1996) [25] deployed four groups of laboratory-reared juveniles of different ages (i.e., 2 × 0-, 2-, and 3-month old juveniles) for varying periods of time (i.e., 49, 37, 57, and 92 days, respectively). For each of these groups, he calculated an initial and final mean size before and after deployment in the field (Figure 3).

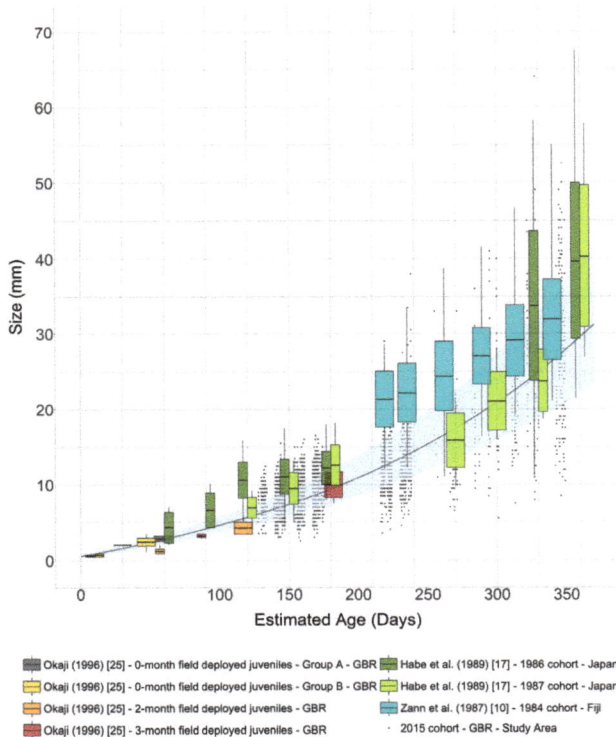

Figure 3. Fitted growth curve (black line) and combined 95% confidence interval for the 2015 crown-of-thorns starfish cohort (GBR). Individual observations of the 2015 cohort are depicted as points, while boxplots represent the results of previous field studies (Zann et al., 1987 [10], n = 651; Habe et al., 1989 [17], n_{1986} = 683 and n_{1987} = 125; Okaji, 1996 [25], $n_{initial}$ = 1137 and n_{final} = 138). Each boxplot is characterised by a mean size (horizontal middle line), ±one standard deviation (rectangle), and minimum and maximum size (vertical line). Note that the estimated age in Zann et al. (1987) [10] and Habe et al. (1989) [17] likely represents the age after fertilisation without taking into account potentially long planktonic larval durations (PLDs); ages are therefore likely to be underestimated. In contrast, the estimated age in Okaji (1996) [25] and the present study represents the age after metamorphosis (i.e., after settlement).

Ten-day old juveniles set out in the field on the GBR for 37–49 days grew on average 0.045–0.046 mm·day^{-1}, mostly within the range of the modelled growth rates for the same time span (i.e., 0.030–0.045 mm·day^{-1}) [25]. Similarly, growth rates for 2-month-old starfish (deployment time = 57 days) averaged 0.053 mm·day^{-1} compared to 0.038–0.059 mm·day^{-1}; while 3-month-old starfish (deployment time = 92 days) grew at a rate of 0.074 mm·day^{-1} compared to 0.046–0.076 mm·day^{-1} [25]. Growth rates derived from the findings of field studies conducted in Fiji [10] and Japan [17] are also broadly comparable to the range of modelled growth rates of this study. In Fiji, growth rates averaged 0.087 mm·day^{-1} for juveniles between 7 and 12 months, largely falling within the modelled range of 0.073–0.135 mm·day^{-1}. Similarly, field studies conducted in Japan showed that juveniles (between 4 and 6 months old) from the 1986-cohort and 1987-cohort grew at 0.121 mm·day^{-1} and 0.139 mm·day^{-1} respectively, also within the modelled range of growth rates (i.e., 0.083–0.158 mm·day^{-1}) for the same time span. These results show that growth rates of juvenile *Acanthaster* spp. in the field appear to be remarkably consistent over large geographic scales. However, mean sizes of juveniles of the 1984-cohort (Fiji) and the 1986-cohort (Japan) deviate considerably from the modelled mean sizes (Figure 3). In both studies, age was estimated based on the date of spawning without taking into account of potentially long planktonic larval durations (i.e., up to 43 days [33]). For instance, juveniles that were sampled in Fiji in July were estimated to be seven months old, based on the assumption of a January spawning [10]. Similarly in Japan, newly settled starfish sampled in July were assumed to be one month old, based on the assumption of a mid-June spawning [17]. Assuming a mid-June spawning and a pelagic larval duration of two weeks, settlement would have occurred at the beginning of July at the earliest. However, juveniles that were collected in July and August were already 2 and 4.30 mm, respectively [17]. These sizes seem considerably large, as we know from laboratory studies that juveniles measure between 0.3 and 0.82 mm after settlement and that it takes approximately 40–45 days to grow to 2 mm, and 80–90 days to grow to 4.3 mm [18]. As a result, ages in Zann et al. (1987) [10] and Habe et al. (1989) [17] are likely to be underestimated by 0.5–1.5 months, resulting in overestimates of growth rates.

An implicit assumption of this study was that spawning occurred within a relatively limited period (in December 2014) across the entire study area. This assumption appears valid, given that the estimated growth rates for field-deployed juveniles of known age are broadly comparable to the range of modelled growth rates. The assumption of a single spawning event or narrow spawning window is also supported by the fact that smaller juveniles became rare as sampling progressed through the year (see Figure 2), indicating that settlement and so spawning did not occur throughout the entire year. Furthermore, spawning occurred in the study area between the 10 and 21 of December 2014 (temperature recorded at Lizard Island at 0.6 m between 12 and 30 December 2014: 28.58 °C–29.29 °C [41]) according to Uthicke et al. (2015) [28]. Assuming a planktonic larval phase of 10 to 40 days [22,23,29–31,33], settlement would have occurred between the 20 of December 2014 and 30 of January 2015. However, new research is suggesting that peak settlement likely occurs within 22 days after spawning and fertilization [33], and few larvae persist beyond 30 days post-fertilization. Accordingly, Uthicke et al. (2015) [28] failed to detect larvae in plankton samples from 13 of January onwards. If so, estimated ages (in days) of juveniles (measured from settlement) would have an accuracy of ±12 days.

Low ocean current velocities linked to El Niño-Southern Oscillation (ENSO) hydrodynamics that cause larval retention around reefs or within reef groups are thought to increase survival of crown-of-thorns starfish larvae when they co-occur with enhanced phytoplankton concentrations [42]. According to Wooldridge and Brodie (2015) [42], the 2009 spawning event—which led to the onset of the current outbreak in 2010/2011—coincided with nutrient-enriched stagnant shelf currents. Larvae of the studied cohort would have been present in the water column between 10 December 2014 and 13 January 2015 (temperature range recorded at Lizard Island weather station between 12 December 2014 and 15 January 2015: 27.9 °C–29.5 °C [41]). Mean phytoplankton concentrations (i.e., chlorophyll *a*) during this period averaged in the Wet Tropics 0.55 μg·L^{-1} (range: 0–7.35 μg·L^{-1})

for coastal areas, 0.31 μg·L^{-1} (range: 0.01–12.42 μg·L^{-1}) for midshelf areas, and 0.16 μg·L^{-1} (range: 0–8.09 μg·L^{-1}) for offshore areas [43]. In comparison, mean chlorophyll *a* concentrations in December 2009 averaged in the Wet Tropics 0.67 μg·L^{-1} (range: 0–11.61 μg·L^{-1}) for coastal areas, 0.31 μg·L^{-1} (range: 0.01–13.82 μg·L^{-1}) for midshelf areas, and 0.19 μg·L^{-1} (range: 0–2.90 μg·L^{-1}) for offshore areas [43]. While the Southern Oscillation Index (SOI)—which provides an indication of ocean current velocity (neutral SOI = surrogate for low shelf currents)—was −5.5 in December 2014 (compared to −7 in December 2009) and −7.8 in January 2015 (compared to −10.1 in January 2010) [43]. Subsequent SOI values continued to decrease in 2015 to a −20.2 low in October, indicating an El Niña event. Moderate increases in chlorophyll concentrations (up to 1.0–4.0 μg·L^{-1}) that significantly increase rates of larval development, survival, and settlement [23,32,33] would have occurred on multiple occasions in the study area during the pelagic larval phase of the studied cohort. Consequently, larval development, survival, and settlement may have been enhanced in the study area during this time period, leading to high settlement rates and the development of this seemingly conspicuous cohort in 2015.

5. Conclusions

Demographic rates are fundamental to understanding population dynamics and creating meaningful population models. However, attempts to estimate these rates for 0+ year juvenile crown-of-thorns starfish have been hindered by the ability to detect them in the field [14–16]. The present study not only demonstrates that 0+ year juveniles can be sampled in high numbers (*n* = 3532), it also provides ranges of monthly growth rates to inform population models. Here, we also provide evidence that supports the assumption that spawning and subsequent settlement occurred in 2014–2015 during a relatively narrow period of time across a vast geographic area (i.e., 350 km) on the GBR. In addition, the predicted ranges of mean sizes for different sampling periods deliver valuable information to research and management bodies for the planning of juvenile monitoring studies. Rigorous ongoing monitoring should be conducted in the future on a number of selected sites to collect field-based data on demographic rates (e.g., growth, recruitment, mortality, and reproductive output rates) that can be related to variables such as food availability and adult population densities in order to inform population models and improve our understanding of population and outbreak dynamics.

Acknowledgments: We would like to thank the Association for Marine Tourism Operators (AMPTO) for supporting this research. We would also like to thank Ken Okaji for his assistance with the literature research.

Author Contributions: J.W. conceived and designed the experiments; J.W. and D.S. performed the experiments; J.W. and S.M. analysed the data; J.W. produced the figures and tables; J.W., S.M., V.M., A.H., M.P., wrote the paper.

Conflicts of Interest: J.W. and D.S. provide marine consultancy services for crown-of-thorns starfish control and monitoring. They declare no conflict of interest.

Appendix A

Table A1. Mean sizes for newly-metamorphosed starfish reared in the laboratory under naturally occurring chlorophyll *a* concentrations [23].

Chlorophyll *a* (μg·L^{-1})	Mean Size (mm)	Standard Error (SE)	95% Confidence Interval (mm)
0.28	0.44	0.07	0.30–0.58
2.90	0.66	0.05	0.56–0.76
5.20	0.64	0.09	0.46–0.82

References

1. Moran, P.J. The *Acanthaster* phenomenon. In *Oceanography and Marine Biology: An Annual Review*; Aberdeen University Press: Townsville, Australia, 1986.
2. De'ath, G.; Fabricius, K.E.; Sweatman, H.; Puotinen, M. The 27-year decline of coral cover on the Great Barrier Reef and its causes. *Proc. Natl. Acad. Sci. USA* **2012**, *109*, 17995–17999. [CrossRef] [PubMed]
3. Osborne, K.; Dolman, A.M.; Burgess, S.C.; Johns, K.A. Disturbance and the dynamics of coral cover on the Great Barrier Reef (1995–2009). *PLoS ONE* **2011**, *6*, e17516. [CrossRef] [PubMed]
4. Baird, A.H.; Pratchett, M.S.; Hoey, A.S.; Herdiana, Y.; Campbell, S.J. *Acanthaster planci* is a major cause of coral mortality in Indonesia. *Coral Reefs* **2013**, *32*, 803–812. [CrossRef]
5. Chesher, R.H. Destruction of Pacific Corals by the Sea Star *Acanthaster planci. Science* **1969**, *165*, 280–283. [CrossRef] [PubMed]
6. Pratchett, M.S.; Caballes, C.F.; Posada, J.A.R.; Sweatman, H.P.A. Limits to understanding and managing outbreaks of crown-of-thorns starfish (*Acanthaster* spp.). *Oceanogr. Mar. Biol. Annu. Rev.* **2014**, *52*, 133–200.
7. Vanhatalo, J.; Hosack, G.R.; Sweatman, H. Spatiotemporal modelling of crown-of-thorns starfish outbreaks on the Great Barrier Reef to inform control strategies. *J. Appl. Ecol.* **2016**. [CrossRef]
8. Moore, R.J. Persistent and transient populations of the crown-of-thorns starfish, *Acanthaster planci*. In *Acanthaster and the Coral Reef: A Theoretical Perspective, Proceedings of a Workshop Held at the Australian Institute of Marine Science, Townsville, Australia, 6–7 August 1988*; Bradbury, R., Ed.; Springer: Berlin/Heidelberg, Germany, 1990; pp. 236–277.
9. Pratchett, M.S. Dynamics of an outbreak population of *Acanthaster planci* at Lizard Island, northern Great Barrier Reef (1995–1999). *Coral Reefs* **2005**, *24*, 453–462. [CrossRef]
10. Zann, L.; Brodie, J.; Berryman, C.; Naqasima, M. Recruitment, ecology, growth and behavior of juvenile *Acanthaster Planci* (L.) (Echinodermata: Asteroidea). *Bull. Mar. Sci.* **1987**, *41*, 561–575.
11. Caballes, C.F.; Pratchett, M.S. Reproductive biology and early life history of the crown-of-thorns starfish. In *Echinoderms—Ecology, Habitats and Reproductive Biology*; Whitmore, E., Ed.; Nova Science Publishers, Inc.: New York, NY, USA, 2014; pp. 101–146.
12. Morello, E.B.; Plagányi, É.E.; Babcock, R.C.; Sweatman, H.; Hillary, R.; Punt, A.E. Model to manage and reduce crown-of-thorns starfish outbreaks. *Mar. Ecol. Prog. Ser.* **2014**, *512*, 167–183. [CrossRef]
13. MacNeil, M.A.; Mellin, C.; Pratchett, M.S.; Hoey, J.; Anthony, K.R.N.; Cheal, A.J.; Miller, I.; Sweatman, H.; Cowan, Z.L.; Taylor, S.; et al. Joint estimation of crown of thorns (*Acanthaster planci*) densities on the Great Barrier Reef. *PeerJ* **2016**, *4*, e2310. [CrossRef] [PubMed]
14. Doherty, P.J.; Davidson, J. Monitoring the distribution and abundance of juvenile *Acanthaster planci*. In Proceedings of the 6th International Coral Reef Symposium, Townsville, Australia, 8–12 August 1988; Volume 2, pp. 131–136.
15. Pearson, R.G.; Endean, R. *A Preliminary Study of the Coral Predator Acanthaster Planci (L) (Asteroidea) on the Great Barrier Reef*; Fisheries Notes, Department of Harbours and Marine Queensland: Brisbane, Australia, 1969; pp. 27–68.
16. Johnson, D.B.; Moran, P.J.; Baker, V.J.; Christie, C.A.; Miller, I.R.; Miller-Smith, B.A.; Thompson, A.A. *Report on Field Surveys to Locate High Density Populations of Juvenile Crown-of-Thorns Starfish (Acanthaster Planci) within the Central Great Barrier Reef*; Australian Institute of Marine Science: Townsville, Australia, 1991.
17. Habe, T.; Yamamoto, G.; Nagai, A.; Kosaka, M.; Ogura, M.; Sawamoto, S.; Ueno, S.; Yokochi, H. *Studies on the Conservation and Management of Coral Reefs and the Control of Acanthaster Planci Juveniles*; Report of Grant-in-Aid for Scientific Research, Ministry of Education, Science and Culture; Japan, 1989; pp. 158–186.
18. Yamaguchi, M. Early life histories of coral reef asteroids, with special reference to Acanthaster planci (L.). In *Biology and Geology of Coral Reefs*; Jones, O.A., Endean, R., Eds.; Academic Press, Inc.: New York, NY, USA, 1973; pp. 369–387.
19. Lucas, J.S. Environmental influences on the early development of *Acanthaster planci* (L.). In Proceedings of the Crown-of-Thorns Starfish Seminar Proceedings, Brisbane, Australia, 6 September 1974; Australian Government Publishing Service: Canberra, Australia, 1974; pp. 109–121.
20. Yamaguchi, M. Growth of juvenile *Acanthaster planci* (L.) in the laboratory. *Pac. Sci* **1974**, *28*, 123–138.
21. Lucas, J.S. Growth, maturation and effects of diet in *Acanthaster planci* (L.) (Asteroidea) and hybrids reared in the laboratory. *J. Exp. Mar. Biol. Ecol.* **1984**, *79*, 129–147. [CrossRef]

22. Henderson, J.A.; Lucas, J.S. Larval development and metamorphosis of *Acanthaster planci* (Asteroidea). *Nature* **1971**, *232*, 655–657. [CrossRef] [PubMed]

23. Fabricius, K.E.; Okaji, K.; De'ath, G. Three lines of evidence to link outbreaks of the crown-of-thorns seastar *Acanthaster planci* to the release of larval food limitation. *Coral Reefs* **2010**, *29*, 593–605. [CrossRef]

24. Zann, L.; Brodie, J.; Vuki, V. History and dynamics of the crown-of-thorns starfish *Acanthaster planci* (L.) in the Suva area, Fiji. *Coral Reefs* **1990**, *9*, 135–144. [CrossRef]

25. Okaji, K. Feeding Ecology in the Early Life Stages of the Crown-of-Thorns Starfish, Acanthaster planci (L.). Ph.D. Thesis, James Cook University, Townsville, Australia, February 1996; p. 140.

26. Kenchington, R.A. Growth and recruitement of *Acanthaster planci* (L.) on the Great Barrier Reef. *Biol. Conserv.* **1977**, *11*, 103–118. [CrossRef]

27. Haszprunar, G.; Spies, M. An integrative approach to the taxonomy of the crown-of-thorns starfish species group (Asteroidea: *Acanthaster*): A review of names and comparison to recent molecular data. *Zootaxa* **2014**, *3841*, 271–284. [CrossRef] [PubMed]

28. Uthicke, S.; Doyle, J.; Duggan, S.; Yasuda, N.; McKinnon, A.D. Outbreak of coral-eating Crown-of-Thorns creates continuous cloud of larvae over 320 km of the Great Barrier Reef. *Sci. Rep.* **2015**, *5*. [CrossRef] [PubMed]

29. Lucas, J.S. Quantitative studies of feeding and nutrition during larval development of the coral reef asteroid *Acanthaster planci* (L.). *J. Exp. Mar. Biol. Ecol.* **1982**, *65*, 173–193. [CrossRef]

30. Uthicke, S.; Pecorino, D.; Albright, R.; Negri, A.P.; Cantin, N.; Liddy, M.; Dworjanyn, S.; Kamya, P.; Byrne, M.; Lamare, M. Impacts of ocean acidification on early life-history stages and settlement of the coral-eating sea star *Acanthaster planci*. *PLoS ONE* **2013**, *8*, e82938. [CrossRef] [PubMed]

31. Uthicke, S.; Logan, M.; Liddy, M.; Francis, D.; Hardy, N.; Lamare, M. Climate change as an unexpected co-factor promoting coral eating seastar (*Acanthaster planci*) outbreaks. *Sci. Rep.* **2015**, *5*. [CrossRef] [PubMed]

32. Wolfe, K.; Graba-Landry, A.; Dworjanyn, S.A.; Byrne, M. Larval starvation to satiation: Influence of nutrient regime on the success of *Acanthaster planci*. *PLoS ONE* **2015**, *10*, e0122010. [CrossRef] [PubMed]

33. Pratchett, M.S.; Dworjanyn, S.A.; Mos, B.; Caballes, C.F.; Thompson, C.; Blowes, S. Larval survivorship and settlement of crown-of-thorns starfish (*Acanthaster* cf. *solaris*) at varying chlorophyll concentrations. *Diversity* **2016**. [CrossRef]

34. R Core Team. *A Language and Environment for Statistical Computing*; R Foundation for Statistical Computing: Vienna, Austria, 2016.

35. Baty, F.; Ritz, C.; Charles, S.; Brutsche, M.; Flandrois, J.-P.; Delignette-Muller, M.-L. A toolbox for nonlinear regression in R: The package nlstools. *J. Stat. Softw.* **2015**, *66*, 21. [CrossRef]

36. Pinheiro, J.; Bates, D.; DebRoy, S.; Sarkar, D.; Team, R.C. *nlme: Linear and Nonlinear Mixed Effects Models*; R Package Version 3.1-128; R Foundation for Statistical Computing: Vienna, Austria, 2016.

37. Wickham, H. *ggplot2: Elegant Graphics for Data Analysis*; Springer-Verlag: New York, NY, USA, 2009.

38. Birkeland, C.; Lucas, J.S. *Acanthaster planci: Major Management Problem of Coral Reefs*; CRC Press, Inc.: Boca Raton, FL, USA, 1990; p. 257.

39. Yokochi, H.; Ogura, M. Spawning period and discovery of juvenile *Acanthaster planci* (L.) (Echinodermata: Asteroidea) at northwestern Iriomote-Jima, Ryukyu Islands. *Bull. Mar. Sci.* **1987**, *41*, 611–616.

40. Keesing, J.; Halford, A. Field measurement of survival rates of juvenile *Acanthaster planci*, techniques and preliminary results. *Mar. Ecol. Prog. Ser.* **1992**, *85*, 107–114. [CrossRef]

41. Australian Institute of Marine Science. The Future of the Reef. Data Centre, AIMS. Available online: http://data.aims.gov.au/aimsrtds/yearlytrends.xhtml (accessed on 22 December 2016).

42. Wooldridge, S.A.; Brodie, J.E. Environmental triggers for primary outbreaks of crown-of-thorns starfish on the Great Barrier Reef, Australia. *Mar. Pollut. Bull.* **2015**, *101*, 805–815. [CrossRef] [PubMed]

43. Australian Bureau of Meteorology (BOM). eReefs. Available online: http://www.bom.gov.au/marinewaterquality/ (accessed on 22 December 2016).

diversity

MDPI

Article

Microsatellites Reveal Genetic Homogeneity among Outbreak Populations of Crown-of-Thorns Starfish (*Acanthaster* cf. *solaris*) on Australia's Great Barrier Reef

Hugo B. Harrison [1,*], Morgan S. Pratchett [1], Vanessa Messmer [1], Pablo Saenz-Agudelo [2] and Michael L. Berumen [3]

[1]   Australian Research Council Centre of Excellence for Coral Reef Studies, James Cook University, Townsville, QLD 4811, Australia; Morgan.Pratchett@jcu.edu.au (M.S.P.); vanessa.messmer@gmail.com (V.M.)
[2]   Instituto de Ciencias Ambientales y Evolutivas, Universidad Austral de Chile, Valdivia 5090000, Chile; pablo.saenzagudelo@gmail.com
[3]   Red Sea Research Center, King Abdullah University of Science and Technology, Thuwal 23955-6900, Saudi Arabia; Michael.Berumen@kaust.edu.sa
*   Correspondence: hugo.harrison@my.jcu.edu.au; Tel.: +61-747-81-6358

Academic Editors: Sven Uthicke and Michael Wink
Received: 30 November 2016; Accepted: 7 March 2017; Published: 10 March 2017

Abstract: Specific patterns in the initiation and spread of reef-wide outbreaks of crown-of-thorns starfish are important, both to understand potential causes (or triggers) of outbreaks and to develop more effective and highly targeted management and containment responses. Using analyses of genetic diversity and structure (based on 17 microsatellite loci), this study attempted to resolve the specific origin for recent outbreaks of crown-of-thorns on Australia's Great Barrier Reef (GBR). We assessed the genetic structure amongst 2705 starfish collected from 13 coral reefs in four regions that spanned ~1000 km of the GBR. Our results indicate that populations sampled across the full length of the GBR are genetically homogeneous ($G'_{ST} = -0.001$; $p = 0.948$) with no apparent genetic structure between regions. Approximate Bayesian computational analyses suggest that all sampled populations had a common origin and that current outbreaking populations of crown-of-thorns starfish (CoTS) in the Swains are not independent of outbreak populations in the northern GBR. Despite hierarchical sampling and large numbers of CoTS genotyped from individual reefs and regions, limited genetic structure meant we were unable to determine a putative source population for the current outbreak of CoTS on the GBR. The very high genetic homogeneity of sampled populations and limited evidence of inbreeding indicate rapid expansion in population size from multiple, undifferentiated latent populations.

Keywords: coral reefs; Great Barrier Reef Marine Park; population genetics; approximate Bayesian computation

1. Introduction

Crown-of-thorns starfish (CoTS; *Acanthaster spp.*) naturally occur on coral reefs throughout the Indo-Pacific [1,2] While normally found at low densities [3], sporadic population outbreaks of CoTS cause significant localised coral loss and are a major contributor to the ongoing degradation of coral reefs throughout the Indo West-Pacific [4–7]. Numerous hypotheses have been put forward to explain the occurrence of CoTS outbreaks (reviewed by [1,2,8]), most of which incite an anthropogenic basis for the purportedly recent and increasing occurrence of outbreaks. While their inherent life-history characteristics (most notably their high fecundity [9]) predisposes CoTS to major fluctuations in abundance [10], there are two prominent theories proposed to trigger outbreaks; the larval survival hypothesis suggests that anthropogenic eutrophication of nearshore waters dramatically increases the survival of planktonic larvae [11–13], whereas the predator removal hypothesis postulates that overfishing of natural predators has allowed more CoTS to reach sexual maturity [1]. Tests of either hypothesis require improved knowledge of when and where outbreaks start and corresponding research on environmental conditions and population demographics of CoTS at these locations.

Outbreaks of CoTS on Australia's Great Barrier Reef (GBR) were first documented in 1962 [14], though there are earlier reports of high densities of starfish (which may or may not have constituted an "outbreak") on the GBR e.g., [15,16]. Since 1962, there have been three additional outbreak episodes on the GBR, starting in 1979, 1993 and 2009. However, each outbreak has followed a reasonably consistent pattern where primary outbreaks were first recorded on mid-shelf reefs between Lizard Island and Cairns (the 'initiation box'; Fabricius et al. [13]), followed by a wave of 'secondary outbreaks' that tend to propagate southwards [17,18]. Biophysical modelling of larval dispersal patterns suggests that reefs within the initiation box are highly connected [19], thereby explaining why outbreaks that initiate in this region inevitably lead to reef-wide outbreaks. However, limited temporal and spatial resolution of monitoring e.g., [20], as well as inevitable delays in responding to new outbreaks mean that it is still unclear where exactly outbreaks arise. It is also unknown whether outbreaks start from a small cluster of reefs within this area or arise simultaneously on widely separated reefs [2]. Resolving the exact timing and location where outbreaks start is important to establish environmental triggers [13] or changes in population demographics [21] that cause outbreaks. This could also lead to improved management and containment strategies to stop outbreaks before they spread.

Although native to the GBR, the spatial distribution of CoTS following an outbreak closely resembles the spread of invasive species and infectious diseases [22]. Historical and observational data have previously identified invasion routes or disease vectors e.g., [23], but direct observations have proven ineffective for CoTS e.g., [20], where it is still unclear whether outbreaks start from a single reef or arise simultaneously from separate locations [24]. The rapid expansion of populations following biological invasions can, however, lead to distinct patterns of genetic structure and diversity. Genetic data have increasingly been used as an indirect method to describe the spread of invasive species [25,26] and infer the relationship between discrete populations or possible migration routes [27]. Elaborating on these patterns, model-based statistics can provide probabilistic estimations of the demographic and genetic history that are necessary to generate observed patterns of genetic structure e.g., [28,29]. Approximate Bayesian computation (ABC) approaches that incorporate the divergence and admixture of populations, as well as changes in population size and structure [30] can provide important information on the likely initiation and spread of species e.g., [31,32].

This study examines genetic diversity and structure, based on sampling of crown-of-thorns starfish during the current outbreak on the GBR. Over four thousand starfish were sampled at 13 reefs spanning ~1000 km. The spatial genetic structure of a CoTS outbreak will depend on the history of the source population(s), the size of the initial population(s), the dispersal of individuals that led to a primary outbreak and successive secondary outbreaks. While the demographic factors contributing to each stage of an outbreak are unclear, a recent review clarifies many aspects of their population dynamics and life-history characteristics [2]. A model-based approach would therefore allow us to take into account the stochasticity of these demographic processes and test multiple scenarios that would have generated the observed spatial population structure of CoTS on the GBR. Specifically, we test whether the outbreak was generated from a single source population in the 'initiation box' or multiple populations. If a small number of individuals from a single source population caused a localised primary outbreak, we would expect successive secondary outbreaks to be affected by a single bottleneck, be composed of highly related individuals and have low genetic diversity. If the primary outbreak originated from multiple populations, we would expect multiple bottlenecks, founding a widespread admixed population, and high genetic diversity. Using a model-based Bayesian approach, we explore the likelihood of competing outbreak scenarios against the observed spatial genetic structure of CoTS to determine the most parsimonious origin of primary outbreaks on the GBR.

2. Materials and Methods

2.1. Sample Collection

Crown-of-thorns starfish (*Acanthaster* cf. *solaris*) were collected between April 2013 and May 2015 from 13 reefs between Lizard Island (S 14.7; E 145.4) and the Swains reefs (S 22.3; E 1527) (Figure 1). Starfish were collected whilst snorkelling or SCUBA diving (depending on working depths). All starfish were kept alive in 500 L tanks connected to high flow-through seawater systems on live-aboard boats or at the Lizard Island research station for a maximum of 20 h before being processed. Starfish were placed on their aboral surface to remove tube feet for genetic analyses; 5–10 tube feet (depending on size) were removed using fine scissors to cut the tube feet close to their base. Multiple tube feet from each individual were placed together in 2-mL vials of 100% ethanol. Later, a single foot was taken, frozen and transported in dry ice for processing at King Abdullah University of Science and Technology (KAUST). All equipment used during sampling and processing was sterilised using a three-step rinse procedure involving bleach, water and ethanol.

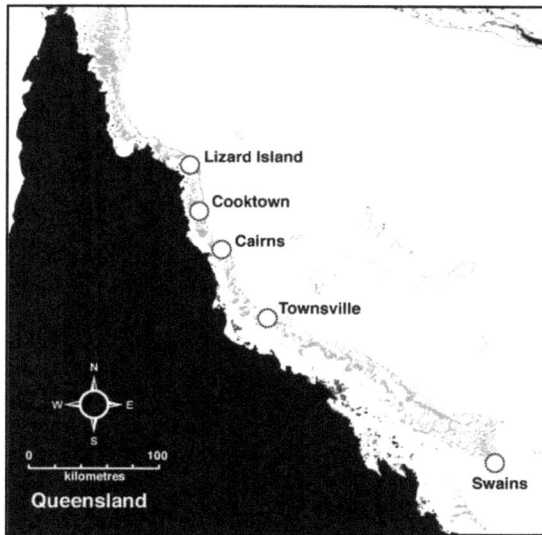

Figure 1. Sampling locations of crown-of-thorns starfish (CoTS) on the Great Barrier Reef. Over four thousand CoTS were collected between April 2013 and May 2015 from reefs between Lizard Island (S 14.7; E 145.4) and the Swains reefs (S 22.3; E 1527).

2.2. Microsatellite Genotyping and Locus Characteristics

The genetic diversity amongst sample populations was assessed using 26 previously-described microsatellite loci [33–36]. Full details of microsatellite development and screening are available in Harrison et al. [36]. DNA extractions were performed from single tube feet following procedures described in the Nucleospin-96 Tissue kit (Macherey-Nagel, Germany), and microsatellites were amplified in four multiplex reactions of 6 or 7 loci. All PCR were performed using the QIAGEN Microsatellite Type-it kit (QIAGEN, Germany) and screened on an ABI 3370xl DNA Analyzer (Applied Biosystems). PCRs were repeated on 96 individuals to estimate locus-specific genotyping error. Individual genotypes were scored in Genemapper v4.0 (Applied Biosystems), and unique alleles were distinguished using marker specific bin sets in the R package 'msatallele' [37].

The numbers of genotyped individuals (*N*), number alleles (*Na*), observed heterozygosity (*Ho*) and expected heterozygosity (*He*) were estimated for each locus using GenAlEx v6.5 [38,39]. The exact test of Hardy–Weinberg and the score test for heterozygote deficiency were performed in Genepop on the web [40,41] alongside Weir and Cockerham's estimate of F_{IS} [42]. The probability test for linkage disequilibrium was performed for each pair of loci in Genepop on the web based on 10,000 dememorisations, 5000 batches and 10,000 iterations. Significance levels of 0.05 were adjusted for a given false discovery rate of 10% to account for multiple testing [43]. The presence of null alleles was determined from 700 randomly selected individuals in Microchecker [44] based on the estimator of Brookfield [45].

2.3. Diversity Analysis of Sampled Populations

Allelic richness, observed and expected heterozygosities within populations were calculated in GenoDive 2.0 [46]. To assess population structure, we estimate global and pairwise genetic differences among sampled reefs within regions. Given the high level of heterozygosity in microsatellite markers, the degree of global genetic differentiation among populations was estimated from Hedrick's standardised fixation index G'_{ST} [47]. Population structure among reefs was estimated from an analysis of molecular variance (AMOVA) with reefs nested in regions. Significance was assessed from 9999 permutations, and the standard deviation of genetic variance was obtained through jack-knifing over loci. The significance of population differentiation between all pairs of populations was assessed from 9999 permutations over populations. We tested for isolation by distance (IBD) by comparing the pairwise matrix of linearised genetic distance (Hedrick's G'_{ST}) and geographic distance (km) between all sampled reefs. Statistical significance was assessed using a Mantel test with 9999 permutations, and significance values are shown.

2.4. Spatial Genetic Clustering

We applied a model-based Bayesian clustering method implemented in Structure v2.3.2 [48,49] to evaluate the most parsimonious allocation of samples to distinct genetic clusters following the method described by Evanno et al. [50]. If population outbreaks of CoTS stem from the successful fertilisation of multiple populations, we would expect the likelihood probability of the data to depart from its expected distribution of a single homogenous population. We performed three short runs for each number of populations K, from K = 1 to 10, and calculated the mean posterior probability for each value of K. Each run assumed population admixture for correlated allele frequencies and no a priori population assignment. The burn-in length was 30,000 followed by a Markov chain Monte Carlo (MCMC) length of 50,000 repetitions. The initial Dirichlet parameter for the degree of admixture 'alpha' was fixed to 1.0 in all simulations.

The statistical power to detect genetic structure amongst sampled populations was evaluated in the software Powsim v4.1 [51]. Three tests were performed to determine whether the number of sampled individuals, the number and diversity of loci could detect F_{ST} values of 0.0010 and 0.0005 and 0.0001. Each test assumed an effective population size of 5000 individuals and divergence times of 10, 5 and 1 generations, respectively. Due to software limitations in the number of alleles per locus, *Apl19* was excluded from these analyses. The numbers of dememorisations, batches and iterations were set at 1000, 100 and 1000, respectively. A total of 1000 replicates were run for each test.

2.5. Approximate Bayesian Computation of Source Populations

We developed four scenarios to represent the possible origins and subsequent spread of CoTS outbreaks on the GBR and used an approximate Bayesian computation approach to determine the probability of each scenario to generate observed patterns of genetic diversity. All analyses were developed using DIYABC 2.1.0 [52,53]. Each scenario was defined to represent possible historical events and demographic changes, including possible population bottlenecks, the effective population sizes of ancestral populations, as well the number of individuals that would have contributed to primary outbreaks, the effective population size of sampled populations and possible divergence times. The four scenarios represent: (i) independent primary outbreaks in the northern and southern GBR, which led to separate secondary outbreaks in the northern and southern GBR (Figure 2a); (ii) a single primary outbreak, which led to secondary outbreaks with a common origin (Figure 2b); (iii) divergence of the northern and southern GBR populations followed by independent primary outbreaks in each of the five focal regions and subsequent secondary outbreaks (Figure 2c); and (iv) sequential secondary outbreaks starting from a common population at Lizard island (Figure 2d).

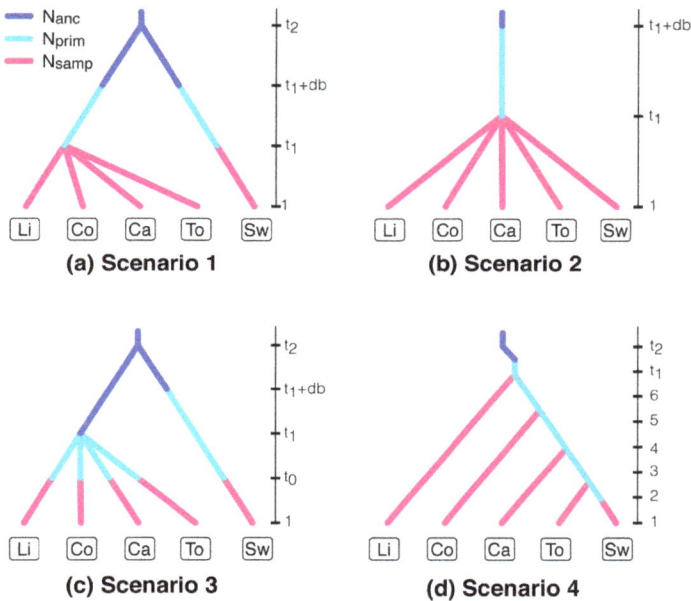

Figure 2. Graphical representation of competing scenarios used in an approximate Bayesian computation (ABC) framework. Each scenario represents a possible outbreak history of crown-of-thorns starfish (CoTS) on the Great Barrier Reef. (**a**) Independent primary outbreaks in the northern and southern GBR, which led to separate secondary outbreaks in the northern and southern GBR; (**b**) a single primary outbreak, which led to secondary outbreaks with a common origin; (**c**) divergence of the northern and southern GBR populations followed by independent primary outbreaks in each of the five focal regions and subsequent secondary outbreaks; (**d**) sequential secondary outbreaks starting from a common population at Lizard island. The Y-axis indicates the number of generations between events (not to scale). Populations have been abbreviated as follows: Lizard Island (Li), Cooktown (Co), Cairns (Ca), Townsville (To) and Swains (Sw).

In order to make the ABC approach computationally feasible, we performed all tests on a subset of 68 randomly selected individuals from Lizard Islands, Cooktown, Cairns and Townsville and all 68 individuals from the Swains. All 17 loci were kept in the analysis. Parameters of mutation models of microsatellite loci were drawn from a set of 10,000 simulations generated for each putative scenario. The posterior probability value of each scenario was determined following 100,000 simulations of each scenario. The details of the ABC parameters and analyses are described in Figure S3.

2.6. Inbreeding

The mating of closely-related individuals, or closely-related groups of individuals, can result in an increase in the frequency of homozygote loci and, thus, heterozygote deficiency. The coefficient of inbreeding f is the probability that two alleles at any one locus are identical by descent (IBD) [54] Wright 1943) so that the progeny of close relatives will have many autozygous genotypes. We measured the multilocus heterozygosity (MLH) [55,56] to investigate the occurrence of inbred individuals in our sample. MLH is the proportion of loci at which an individual was heterozygous divided by the mean heterozygosity across all samples. If individuals are in fact inbred, all loci should exhibit signs of excess heterozygosity deficiency, and estimates of MHL from two random sets of loci should be similar. Furthermore, our sample should exhibit lower estimates of MLH than a random population generated from the observed allelic frequencies. We estimated MLH of CoTS sampled and tested the correlation between randomly selected sets of loci in the R package 'Rhh' [57]. Ninety-five percent confidence intervals around the mean were estimated from 999 iterations. Simulated genotypes were generated from observed allelic frequencies using R packages 'gstudio' [58] and 'adegenet' [59], and MLH between groups were compared used a paired *t*-test. Inbreeding should result in a significant over-representation of individuals with a low MLH when compared to randomly-assorted genotypes.

3. Results

A total of 4082 individual CoTS were collected between April 2013 and May 2015 from 13 reefs between Lizard Island and the Swains reefs (Figure 1). All individuals were genotyped at 26 microsatellite loci, and the final dataset was curated to focus on five focal regions (Table 1). Individuals removed from the data included 642 individuals sampled prior to September 2013 or less than 80 mm in length and 121 individuals that were sampled from reefs where less than 50 individuals were sampled.

Table 1. Sampling locations of 2705 crown-of-thorns starfish collected on the Great Barrier Reef that were then genotyped with 17 microsatellite loci. Table includes collection period, the number of genotyped individuals (N), allelic richness (AR) and the observed (Ho) and expected heterozygosity of each sampled population (Hs).

Region	Sampling Site	Latitude	Longitude	Date of Collection	N	AR	Ho	Hs
Lizard Is	Lizard Island reefs	−14.6916	145.4479	Oct 2013–Feb 2015	385	5.6	0.673	0.685
Lizard Is	MacGillivray Reef	−14.6524	145.4892	Oct 2013	192	5.4	0.675	0.685
Lizard Is	Nth Direction	−14.7445	145.5399	Oct 2013	247	5.6	0.674	0.690
Lizard Is	Sth Direction	−14.8562	145.4825	Oct 2013	182	5.3	0.672	0.684
Cooktown	Emily Reef	−15.6325	145.6511	Feb 2014	278	5.5	0.687	0.688
Cooktown	Endeavour Reef	−15.7823	145.5847	Feb 2014	208	5.6	0.683	0.692
Cooktown	Pickersgill Reef	−15.8838	145.5640	Feb 2014	151	5.6	0.670	0.693
Cooktown	Spitfire Reef	−16.1148	145.6424	Oct 2013	155	5.4	0.673	0.688
Cairns	Arlington Reef	−16.7749	145.9767	Sept 2014	262	5.4	0.670	0.685
Cairns	Hedley Reef	−17.2474	146.4637	Sept 2014	275	5.4	0.666	0.684
Cairns	McCulloch Reef	−17.2996	146.4257	Sept 2014	265	5.4	0.673	0.686
Townsville	Townsville reefs	−18.4303	146.8209	Nov 2014	137	5.5	0.651	0.682
Swains	Swains reefs	−22.3112	152.6720	May 2015	68	5.5	0.685	0.694

Poor DNA quality or the possible presence of PCR inhibitors in DNA extractions led to a large amount of missing data and genotyping error. A training dataset [36] identified four loci (*Apl21*, *Apl36*, *Apl39* and *Maki12*) with over 3% genotyping error and five loci (*Apl01*, *Apl25*, *Apl27*, *Apl29* and *Apl37*) with over 30% missing data, which were all discarded prior to analysis. In addition, 600 individuals with four or more missing loci were discarded from further analyses along with 14 duplicate genotypes associated with the presence of missing data.

Of an initial 4082 samples, our final data included 2705 unique individuals from 13 reefs from Lizard Island (*n* = 1002), Cooktown (*n* = 692), Cairns (*n* = 802), Townsville (*n* = 137) and the Swains (n = 68). Individuals ranged from 90 mm to 510 mm (mean = 282 mm $\pm$ 66 mm SD) with no statistical difference in the size of individuals between regions. Amongst all genotyped individuals, 1925 were identified as either mature male or mature female with 1.34 males for every female. This ratio was used to parameterise the ABC models.

3.1. Microsatellite Data

Data presented here contained 17 polymorphic loci with 1.7% missing data (Table 2). The mean number of alleles per locus was 17.4 and ranged from four to 62 alleles. Similarly, the average observed heterozygosity was 0.67 and ranged from 0.21 to 0.95. Each locus was tested for departure from Hardy-Weinberg equilibrium, and nine loci showed non-random association of alleles after correction for multiple testing. A specific test for heterozygous deficiency highlighted 12 loci with a higher than expected frequency of heterozygotes after correction for multiple testing. Estimates of Weir and Cockerham's F_{IS} were positively skewed with an average F_{IS} of 0.017 $\pm$ 0.004 SE across all loci, which suggests some degree of inbreeding in these populations. Amongst 136 pairwise comparisons, two locus pairs measured linkage disequilibrium after correction for multiple testing, and no locus showed evidence of null alleles.

Table 2. Characteristics of 17 microsatellite used in the analysis of genetic diversity in crown-of-thorns starfish, *Acanthaster* cf. *solaris*. The number of genotyped individuals (N), number of alleles (Na), observed (H_O) and expected (H_E) heterozygosity were measured from 2705 unique individual from the Great Barrier Reef. The exact test of Hardy–Weinberg (HWE-p) and a specific test for heterozygosity deficiency (HWE-h) were measured alongside Weir and Cockerham's measure of F_{IS} (F). Significant values following correction for multiple comparisons are shown in bold.

Locus name	N	Na	Ho	He	Missing Data (%)	HWE-p	HWE-h	F
AP1	2705	4	0.648	0.657	0.0	0.399	0.159	0.0091
AP12QS	2703	6	0.490	0.514	0.1	0.002	0.001	0.0321
AP654	2689	11	0.685	0.702	0.6	0.389	0.047	0.0089
AP9	2702	23	0.845	0.843	0.1	0.471	0.360	0.0009
Hisayo01	2688	32	0.804	0.829	0.6	0.010	0.003	0.0179
Maki03	2701	28	0.775	0.792	0.1	0.001	0.005	0.0123
Sayo03	2659	9	0.492	0.519	1.7	0.010	0.006	0.0273
AP11QS	2703	11	0.751	0.762	0.1	0.171	0.119	0.0056
AP30QS	2705	9	0.617	0.625	0.0	0.010	0.001	0.0635
AyU03	2704	29	0.885	0.876	0.0	0.276	0.917	−0.0041
Yukina06	2681	32	0.878	0.917	0.9	0.000	0.000	0.0287
AP5QS	2705	7	0.590	0.600	0.0	0.000	0.014	0.0211
Apl07	2440	7	0.390	0.409	9.8	0.040	0.009	0.0269
Apl19	2474	62	0.949	0.965	8.5	0.644	0.010	0.0066
Etsuko01	2657	10	0.772	0.777	1.8	0.001	0.183	0.0045
Sayo01	2704	7	0.214	0.220	0.0	0.095	0.035	0.0148
Apl02	2577	10	0.653	0.669	4.7	0.335	0.032	0.0108

3.2. Spatial Patterns of Genetic Diversity

The mean allelic richness (AR) and genetic diversity (H_S) within sampled reefs were high (AR = 5.5; H_S = 0.687) and consistent among reefs (Table 1). There was no evidence for genetic differentiation amongst sampled reefs with global estimates of G'_{ST} = -0.001 (p = 0.948) and no genetic variance amongst regions (F_{CT} = 0.000 $\pm$ 0.000 SD) or amongst populations nested in regions (F_{SC} = 0.000 $\pm$ 0.000 SD; Table 3). Pairwise genetic differences amongst populations were not significantly different from zero (Table 4), and showed no evidence of isolation by distance (Mantel test: p = 0.9).

Table 3. Analysis of molecular variance amongst crown-of-thorns starfish collected from 12 reefs in 4 regions of the Great Barrier Reef. The significance of F-statistics was measured from 9999 permutations.

Source of Variation	Nested in	% var [1]	F-stat	F-value	SD [2]	CI 2.5%	CI 97.5%	p-Value [3]
Within Individual		0.98	Fit	0.020	0.004	0.012	0.028	-
Among Individual	Population	0.02	Fis	0.020	0.004	0.012	0.028	0.000
Among Reefs	Region	0	Fsc	0.000	0.000	0.000	0.000	0.903
Among Region		0	Fct	0.000	0.000	0.000	0.000	0.830

[1] Per cent of total genetic variance; [2] Standard deviations of F-statistics were obtained through jack-knifing over loci; [3] 95% confidence intervals of F-statistics were obtained through bootstrapping over loci.

Table 4. Pairwise differentiation between crown-of-thorns starfish collected from 13 reefs of the Great Barrier Reef (upper diagonal). The significance of p-values (lower diagonal) was measured from 9999 permutations.

	Lizard Island	MacGillivray	North Direction	South Direction	Emily	Endeavour	Pickersgill	Spitfire	Arlington	Hedley	McCulloch	Townsville	Swains
Lizard Island		0.210	0.958	0.391	0.721	0.990	0.707	0.302	0.752	0.333	0.826	0.730	0.952
MacGillivray	0.000		0.380	0.749	0.380	0.657	0.988	0.407	0.358	0.210	0.764	0.585	0.634
North Direction	0.000	0.000		0.316	0.662	0.992	0.784	0.158	0.331	0.480	0.700	0.954	0.949
South Direction	0.000	0.000	0.000		0.827	0.764	0.248	0.380	0.622	0.960	0.821	0.980	0.857
Emily	0.000	0.000	0.000	0.000		0.912	0.678	0.570	0.672	0.425	0.845	0.948	0.854
Endeavour	-0.001	-0.001	-0.001	0.000	-0.001		0.876	0.551	0.637	0.850	0.889	0.991	1.000
Pickersgill	0.000	0.000	0.000	0.000	0.000	-0.001		0.349	0.553	0.096	0.484	0.601	0.745
Spitfire	0.000	0.000	0.001	0.000	0.000	0.000	0.000		0.060	0.202	0.712	0.703	0.874
Arlington	0.000	0.000	0.000	0.000	0.000	0.000	0.000	0.001		0.717	0.865	0.854	0.739
Hedley	0.000	0.000	0.000	-0.001	0.000	0.000	0.000	0.000	0.000		0.892	0.948	0.952
McCulloch	0.000	0.000	0.000	0.000	0.000	0.000	0.000	0.000	0.000	0.000		0.822	0.895
Townsville	0.000	0.000	-0.001	0.000	-0.001	-0.001	0.000	0.000	-0.001	-0.001	0.000		0.986
Swains	-0.001	0.000	-0.001	-0.001	-0.001	-0.002	-0.001	-0.001	-0.001	-0.001	-0.001	-0.002	

There was, however, evidence of low, but significant heterozygote deficiency within the sample. Individual heterozygosity was lower than would be expected in a large, randomly-mating population with a global estimate of G_{IS} = 0.022 (p = 0.001) and amongst individuals within individuals reefs (F_{IS} = 0.021 ± 0.004 SD). Such patterns are commonly associated with evidence of the mixing of different source populations (Wahlund effect), inbreeding due to the mating of close relatives or the non-random sampling of a limited number of familial pools. The level of heterozygote deficiency also varied amongst reefs, ranging from −0.004 to 0.048 and an average G_{IS} of 0.021 ± 0.012 SD amongst reefs.

Using the whole sample of individuals without prior information of sampling location, the population could not be partitioned into independent populations, indicating that the sample has a common origin. The largest mean log-likelihood values of the data were for K = 1 population and decreased with increasing values of K (Figures S1 and S2). The result provided by the analysis of spatial genetic clustering could not unambiguously detect separate groups of individuals in the sample, indicating a homogenous population, and does not support a Wahlund effect as a source of heterozygote deficiency. The power of the tests suggest that the number of sampled individuals, the number of loci and the allelic diversity at these loci were sufficient to detect genetic structure amongst sampled populations. At a level of differentiation of F_{ST} = 0.0005, the power to detect genetic heterogeneity with 95% confidence was 100%. At the lowest level of differentiation (F_{ST} = 0.0001), the power was reduced to 53.7%. Analyses were repeated after excluding nine loci that did not meet equilibrium assumptions (*AP12QS*, *Hisayo01*, *Maki03*, *Sayo03*, *AP30QS*, *Yukina06*, *AP5QS*, *Apl07*, *Etsuko01*). Results from these runs were not different from runs with the full dataset.

3.3. Inbreeding

We measured low, but significant levels of heterozygote deficiency in most screened loci, which could indicate the presence of inbred individuals in our sample. However, the mean multilocus heterozygosity correlation was −0.003 (CI 95% −0.025 to 0.019) indicating that the heterozygote deficiency was not significantly different across loci. To confirm that inbreeding was not the cause of heterozygote deficiency in sampled CoTS, we compared the MLH between sampled and simulated genotypes. Amongst all CoTS sampled in the GBR, the average MLH was 0.29 ± 0.11 SD, which was not significantly different from simulated genotypes (0.26 ± 0.10 SD, t = −9.41, df = 2704, *p*-value = 1).

3.4. ABC Framework

The most likely scenario supported by the ABC analyses was Scenario 4 (Figure 2d), whereby all sampled populations originated sequentially from a single common primary outbreak (Figure S4). However, the outcomes of this scenario could not be distinguished from Scenario 2, whereby sampled populations originated from a single common primary outbreak at the same time point (Figure 2b). Both scenarios were more representative of the observed data than alternative scenarios that incorporated a divergence between the Swains and the northern populations (Figure 2a,c). In all simulated scenarios, ABC analyses were particularly sensitive to the effective population size for the ancestral population (N_{anc}). The posterior distribution of N_{anc} was however consistent amongst scenarios, and estimates suggest that the effective population size that would have led to a primary outbreak would be ~5000 individuals (Table 5; Figure S3).

Table 5. Prior distribution of historic and genetic data associated with evolutionary scenarios in the ABC analysis.

Parameter	Definition	Distribution (Interval, Mean, SD)
Nanc	Effective population size of an ancestral population	Normal (10^3 to 10^4, 5×10^3, 10^3)
Nprim	Effective population size of a primary outbreak	Normal (10^3 to 10^4, 4×10^3, 8×10^2)
Nsamp	Effective population size of a sampled population	Normal (5×10^4 to 2×10^5, 8×10^4, 10^4)
t1	Divergence time of independent outbreaks	Uniform (1 to 10)
db	Foundation time of a primary outbreak	Uniform (1 to 10)
t2	Divergence time of northern and southern populations	Uniform (10^2 to 1.5×10^3)
Mean μ	Mean mutation rate	Gamma (10^{-4} to 10^{-3}, 10^{-4}, 2)
Mean P	Mean of the geometric distribution of the number of repeats	Gamma (10^{-2} to 10^{-0}, 5.5×10^{-1}, 3)
Mean μSNI	Mean single nucleotide insertion/deletion mutation rate	Gamma (10^{-8} to 5×10^{-5}, 1.5×10^{-5}, 3)

4. Discussion

Assessing a species' dispersal ability and capacity to colonise new habitats is critical for our understanding of their population biology and ecology [60,61], particularly where species have ecological or economic importance. CoTS outbreaks represent the single most important biological disturbance on coral reefs throughout the Indo West-Pacific [62] and often account for up to 50% of coral loss recorded on coral reefs over the last few decades [5–7]. For these starfish, knowledge of population structure and the movement of individuals among reefs can greatly influence management decisions e.g., [19], leading to improved detection and understanding of the patterns of outbreaks, as well as prioritisation of reefs for direct intervention (culling) in an attempt to contain outbreaks. In the current study, we investigated the genetic diversity and structure of CoTS on the GBR with the specific intention of identifying the origin and the direction of subsequent spread for current outbreaks apparent at reefs between Cooktown and Townsville, as well as at Swains reefs, in the southernmost portion of the GBR. This would have further important management ramifications, whereby containment of future outbreaks would be most effective by concentrating monitoring and control on the reef(s) where outbreaks initiate. Our results indicate that populations sampled across the full length of the GBR are genetically homogeneous, highly diverse and have no apparent genetic structure. Furthermore, model-based Bayesian analyses showed that the current outbreaking population of CoTS in the Swains is not independent of the outbreak populations in the northern GBR, but share a common origin.

We found no evidence of genetic structure amongst CoTS genotyped from 13 reefs and five regions spanning over 1000 km along the GBR. While the genetic diversity in our sample was high, there was no variance in diversity among reefs. Power analysis confirms that our data were sufficient to detect even very low degrees of genetic differentiation ($F_{ST} > 0.0005$) had there been any structure in the sampled population. The results indicate that CoTS from Lizard Island to the Swains are genetically homogenous. There was a small but significant deficiency in heterozygosity amongst individuals that could have arisen from: (i) the successful reproduction among differentiated populations (Wahlund effect); (ii) the mating of close relatives (inbreeding) or (iii) the sampling of close relatives. We could not distinguish independent clusters in our sample, and therefore, a Wahlund effect is unlikely (sensu [63]). Moreover, multilocus genotypes did not provide consistent evidence of inbreeding. We did find evidence that some individuals were highly related, which would result in a small but significant deficit in heterozygosity (data not shown). The strength and number of relationships between individuals from the same or different reefs could not, however, distinguish between alternative explanations for deficiencies in heterozygosity. Furthermore, a spatial autocorrelation analysis did not identify any relationship between the relatedness of individuals and their spatial distribution. Previous sampling during non-outbreak periods reported small but significant genetic structure among latent CoTS populations on the GBR ($F_{ST} = 0.003$, [64]), even though they also found limited structure when sampling each outbreak population.

Our ability to determine a putative source population for the current outbreak of CoTS on the GBR is significantly constrained by the limited genetic structure (sensu [65,66]), despite hierarchical sampling and large numbers of CoTS genotyped from individual reefs and regions. Very high levels of homogeneity across the 2705 individual starfish points to the recent and very rapid increase in population size of CoTS across the GBR, which has arisen from either a single source population or multiple undifferentiated populations. A single origin would be expected to generate some inbreeding and highly related individuals, which the data do not support. A more parsimonious explanation, therefore, is that the current outbreak arose almost simultaneously across a number of reefs, but with largely undifferentiated latent populations. This is consistent with reports during the last major outbreak in 1994, whereby increasing CoTS densities occurred almost simultaneously at Lizard Island and several nearby reefs (including Linnet Reef, North Direction Island and Rocky Islet) before being reported on reefs to the south [20]. High levels of gene flow amongst these closely-positioned populations would have likely resulted in admixture and high levels of genetic homogeneity, as recorded in this study. However, given the lack of genetic structure between regions, we are unable to unequivocally state whether the current outbreak did or did not originate in the northernmost sector of the initiation box, nor can we establish the directionality in the spread of individuals between regions. It is also likely that the latent population of CoTS is sufficiently large and sufficiently diverse to prevent genetic drift from occurring within regions and therefore maintaining the genetic homogeneity in the latent population. We cannot, therefore, dismiss the possibility that outbreaks arise almost simultaneously and/or independently across the entire area of the initiation box, as suggested by Fabricius et al. [13].

While our genetic analyses did not resolve competing hypotheses about the initiation and spread of CoTS outbreaks on the GBR, these data could be used in conjunction with demographic information and/or fine-scale monitoring data to better resolve the patterns of initiation and spread. Extensive and intensive monitoring of CoTS populations was undertaken across the northern GBR throughout the period of this study, recording the extent and severity of outbreaks at every reef between Lizard Island and Cairns (P. Doherty, unpublished data), as well as documenting the size-structure of CoTS populations at select reefs throughout this range [67]. However, these surveys were undertaken (in 2014 to 2015) only after outbreaks had become well established throughout the entire area, such that sequential (cf. simultaneous) occurrence of outbreaks will only be apparent based on spatial variation in the size and abundance of CoTS. In the future, systematic and intensive monitoring should be undertaken across a range of reefs within the initiation box to unequivocally establish the sequence and inter-dependence of outbreaks within this area [2]. It is also possible (but not certain) that next generation sequencing might reveal greater genetic structure among existing samples and thereby provide meaningful differences among sub-populations to explicitly test for directionality in spread. The recent compilation and publication of an entire mitochondrial genome for *Acanthaster* cf. *solaris* collected from Japan [68] certainly paves the way for much more detailed studies of population genetics for CoTS. Ongoing genetic sampling and re-analyses of existing genetic samples from the GBR are underway.

An unexpected outcome of this study was that outbreak populations of CoTS in the southern GBR (Swains reefs) were not significantly differentiated and have a similar origin to outbreak populations sampled in the northern and central GBR during 2014/2015. For the most part, CoTS outbreaks in the Swains have been thought to occur independently of outbreaks in the northern GBR and have an altogether different origin (e.g., [19]), though the appearance of high CoTS densities at Swains reefs in 2014 is consistent with continual and progressive southerly spread of the outbreak that started at and near Lizard Island in 1993/1994 [69]. Model-based Bayesian analyses resulted in a higher posterior probability of selected scenarios that considered a common origin for outbreaks in both the Swains reefs and northern GBR, as opposed to two independent primary outbreaks (sensu [19]). Both scenarios that represent a single common primary outbreak indicate a strong goodness-of-fit to our genetic data, though we could not distinguish whether secondary outbreaks originated from a single time point or sequentially. It is possible that the limited time (in generations) elapsed between primary and secondary outbreaks, and even between successive waves of outbreaks, would result in minimal observable genetic differences.

5. Conclusions

Resolving the specific location where reef-wide outbreaks of CoTS actually originate on the GBR remains a high priority, both to understand potential causes or triggers of outbreaks and to develop more effective and highly targeted management responses [2]. However, our capacity to establish the origin of the current outbreak was significantly constrained by the limited genetic structure apparent based on 17 microsatellite loci. Very high homogeneity observed within the current outbreak population, with limited evidence of inbreeding, suggests that rapid expansion in population size most likely arose from multiple and undifferentiated latent populations. Indeed, our data suggest that CoTS outbreaks may have occurred almost simultaneously and independently across the entire area of the 'initiation box', from Cairns to Lizard Island. The priority, therefore, is to undertake intensive sampling on reefs throughout the 'initiation box' in the lead up to the next outbreak of CoTS on the GBR. However, ongoing sampling should be combined with testing of alternative molecular markers (e.g., SNPs) and the application of new techniques to sample larvae [70] and newly settled starfish [71] to further resolve the origin of reef wide outbreaks.

Supplementary Materials: The following are available online at www.mdpi.com/1424-2818/9/1/16/s1, Figure S1: Mean likelihood probability of describing the population structure of CoTS in the Great Barrier Reef into K clusters with standard deviation around the mean. Three runs were performed for each value of K and compiled in Structure Harvester (Earl et al. [1]); Figure S2: Change in the mean likelihood probability of K clusters describing the population structure of CoTS in the Great Barrier Reef. Three runs were performed for each value of K and compiled in Structure Harvester (Earl et al. [1]; Figure S3. Parameter posterior density estimates from Scenario 4. Nanc: the effective population size of an ancestral population. NP and NS correspond to Nprim and Samp in table 5, the the effective population size of the primary and secondary outbreaks, respectively. t1: the divergence time of outbreaks. db: the foundation time of primary outbreaks. All time priors are represented in number of generations. μmic: the mean mutation rate. pmic the mean distribution of the number of repeats of microsatellite markers. snimic: mean rate of single nucleotide insertions and deletions. Figure S4. Comparing the posterior probabilities of modelled scenarios using direct estimates—the number of times a given scenario is chosen to best represents the observed data.

Acknowledgments: We are grateful to those who assisted us in the collection of crown-of-thorns starfish: Kristen Anderson, Zara-Louise Cowan, Karen Chong-Seng, Jacob Johansen, Elmar Messmer, Kirsty Nash, Laura Richardson, Robert Streit, Stuart Watson, Simon Wever; Lizard Island Research Staff; Tim Godfrey; and the crew of Reef Connection and Capricorn Star. We also wish to thank Manalle Al-Salamah for assistance in processing samples. This research was supported by the Commonwealth Government of Australia, through the Caring for Country and Reef Rescue Program. P.S.-A. and H.B.H. were partially funded by CONICYT, programa FONDECYT Grant 11140121.

Author Contributions: H.B.H. and M.S.P. conceived of and designed the experiments. M.S.P. and V.M. collected samples. H.B.H. and P.S.-A. analysed the data. M.L.B. contributed reagents/materials/analysis tools. All authors contributed to writing the paper.

Conflicts of Interest: The authors declare no conflict of interest.

References

1. Birkeland, C.E.; Lucas, J.S. *Acanthaster. planci: Major Management Problem of Coral Reefs*; CRC Press: Boca Raton, FL, USA, 1990.
2. Pratchett, M.; Caballes, C.; Posada, J.R.; Sweatman, H. Limits to Understanding and Managing Outbreaks of Crown-of-Thorns Starfish (*Acanthaster.* spp.). *Oceanogr. Mar. Biol.* **2014**, *52*, 133–200.
3. Moran, P.J.; De'ath, G. Estimates of the abundance of the crown-of-thorns starfish *Acanthaster planci* in outbreaking and non-outbreaking populations on reefs within the Great Barrier Reef. *Mar. Biol.* **1992**, *113*, 509–515. [CrossRef]
4. Bruno, J.F.; Selig, E.R. Regional Decline of Coral Cover in the Indo-Pacific: Timing, Extent, and Subregional Comparisons. *PLoS ONE* **2007**, *2*, e711. [CrossRef] [PubMed]

5. Trapon, M.L.; Pratchett, M.S.; Penin, L. Comparative Effects of Different Disturbances in Coral Reef Habitats in Moorea, French Polynesia. *J. Mar. Biol.* **2011**, 1–11. [CrossRef]

6. De'ath, G.; Fabricius, K.E.; Sweatman, H.; Puotinen, M. The 27-year decline of coral cover on the Great Barrier Reef and its causes. *PNAS* **2012**, *104*, 17995–17999. [CrossRef] [PubMed]

7. Pisapia, C.; Burn, D.; Yoosuf, R.; Najeeb, A.; Anderson, K.D.; Pratchett, M.S. Coral recovery in the central Maldives archipelago since the last major mass-bleaching, in 1998. *Sci. Rep.* **2016**, *6*, 34720. [CrossRef] [PubMed]

8. Moran, P.J. The Acanthaster phenomenon. *Oceanogr. Mar. Biol.* **1986**, *24*, 379–480.

9. Babcock, R.C.; Milton, D.A.; Pratchett, M.S. Relationships between size and reproductive output in the crown-of-thorns starfish. *Mar. Biol.* **2016**, *163*, 234. [CrossRef]

10. Uthicke, S.; Schaffelke, B.; Byrne, M. A boom–bust phylum? Ecological and evolutionary consequences of density variations in echinoderms. *Ecol. Monogr.* **2009**, *79*, 3–24. [CrossRef]

11. Lucas, J.S. Reproductive and larval biology of *Acanthaster planci* (L.) in Great Barrier Reef waters. *Micronesica* **1973**, *9*, 197–203.

12. Birkeland, C. Terrestrial runoff as a cause of outbreaks of *Acanthaster. planci* (Echinodermata: Asteroidea). *Mar. Biol.* **1982**, *69*, 175–185. [CrossRef]

13. Fabricius, K.E.; Okaji, K.; De'ath, G. Three lines of evidence to link outbreaks of the crown-of-thorns seastar Acanthaster planci to the release of larval food limitation. *Coral Reefs* **2010**, *29*, 593–605. [CrossRef]

14. Pearson, R.G.; Endean, R. *A Preliminary Study of the Coral Predator Acanthaster. planci (L.)(Asteroidea) on the Great Barrier Reef*; Fisheries Notes; Queensland Department of Harbours and Marine: Brisbane, Australia, 1969; Volume 3, pp. 27–55.

15. Vine, P.J. Crown of thorns (*Acanthaster planci*) plagues: The natural causes theory. *Atoll Res. Bull.* **1973**, *166*, 1–10. [CrossRef]

16. Ganter, R. *Oral History of Human Use and Experience of Crown of Thorns Starfish on the Great Barrier Reef*; Report to the Great Barrier Reef Marine Park Authority; Great Barrier Reef Marine Park Authority: Townsville, Australia, 1987.

17. Kenchington, R.A. Acanthaster planci on the Great Barrier Reef: Detailed surveys of four transects between 19° and 20°S. *Biol. Conserv.* **1976**, *9*, 165–179. [CrossRef]

18. Kenchington, R.A. Growth and recruitment of *Acanthaster planci* (L.) on the Great Barrier Reef. *Biol. Conserv.* **1977**, *11*, 103–118. [CrossRef]

19. Hock, K.; Wolff, N.H.; Condie, S.A.; Anthony, K.; Mumby, P.J. Connectivity networks reveal the risks of crown-of-thorns starfish outbreaks on the Great Barrier Reef. *J. Appl. Ecol.* **2014**, *51*, 1188–1196. [CrossRef]

20. Sweatman, H.; Bass, D.K.; Cheal, A.J.; Coleman, G.; Miller, I.R.; Ninio, R.; Osborne, K.; Oxley, W.G.; Ryan, D.A.J.; Thompson, A.A.; et al. *Long-Term Monitoring of the Great Barrier Reef. Status Report Number 3*; Australian Institute of Marine Science: Townsville, Australia, 1998.

21. Moore, R.J. Persistent and Transient Populations of the Crown-of-Thorns Starfish, Acanthaster Planci. In *Acanthaster and the Coral Reef: A Theoretical Perspective*; Springer: Berlin/Heidelberg, Germany, 1990.

22. Crowl, T.A.; Crist, T.O.; Parmenter, R.R.; Belovsky, G.; Lugo, A.E. The spread of invasive species and infectious disease as drivers of ecosystem change. *Front. Ecol. Environ.* **2008**, *6*, 238–246. [CrossRef]

23. Suarez, A.V.; Holway, D.A.; Case, T.J. Patterns of spread in biological invasions dominated by long-distance jump dispersal: Insights from Argentine ants. *PNAS* **2001**, *98*, 1095–1100. [CrossRef] [PubMed]

24. Pratchett, M.S. Dynamics of an outbreak population of *Acanthaster planci* at Lizard Island, northern Great Barrier Reef (1995-1999). *Coral Reefs* **2005**, *24*, 453–462. [CrossRef]

25. Guillemaud, T.; Ciosi, M.; Lombaert, E.; Estoup, A. Biological invasions in agricultural settings: Insights from evolutionary biology and population genetics. *Comptes Rendus Biol.* **2011**, *334*, 237–246. [CrossRef] [PubMed]

26. Handley, L.J.L.; Estoup, A.; Evans, D.M.; Thomas, C.E.; Lombaert, E.; Facon, B.; Aebi, A.; Roy, H.E. Ecological genetics of invasive alien species. *BioControl* **2011**, *56*, 409–428. [CrossRef]

27. Lowe, W.H.; Allendorf, F.W. What can genetics tell us about population connectivity? *Mol. Ecol.* **2010**, *19*, 3038–3051. [CrossRef] [PubMed]

28. Lombaert, E.; Guillemaud, T.; Thomas, C.E.; Lawson Handley, L.J.; Li, J.; Wang, S.; Pang, H.; Goryacheva, I.; Zakharov, I.A.; Jousselin, E.; et al. Inferring the origin of populations introduced from a genetically structured native range by approximate Bayesian computation: Case study of the invasive ladybird Harmonia axyridis. *Mol. Ecol.* **2011**, *20*, 4654–4670. [CrossRef] [PubMed]

29. Estoup, A.; Lombaert, E.; Marin, J.-M.; Guillemaud, T.; Pudlo, P.; Robert, C.P.; Cornuet, J.-M. Estimation of demo-genetic model probabilities with Approximate Bayesian Computation using linear discriminant analysis on summary statistics. *Mol. Ecol. Res.* **2012**, *12*, 846–855. [CrossRef] [PubMed]

30. Beaumont, M.A.; Zhang, W.; Balding, D.J. Approximate Bayesian Computation in Population Genetics. *Genetics* **2002**, *162*, 2025–2035. [PubMed]

31. Yalcindag, E.; Elguero, E.; Arnathau, C.; Durand, P.; Akiana, J.; Anderson, T.J.; Aubouy, A.; Balloux, F.; Besnard, P.; Bogreau, H.; et al. Multiple independent introductions of *Plasmodium falciparum* in South America. *PNAS* **2012**, *109*, 511–516. [CrossRef] [PubMed]

32. Fontaine, M.C.; Austerlitz, F.; Giraud, T.; Labbé, F.; Papura, D.; Richard Cervera, S.; Delmotte, F. Genetic signature of a range expansion and leap-frog event after the recent invasion of Europe by the grapevine downy mildew pathogen *Plasmopara. viticola*. *Mol. Ecol.* **2013**, *22*, 2771–2786. [CrossRef] [PubMed]

33. Yasuda, N.; Nagai, S.; Hamaguchi, M.; Lian, C.L.; Nadaoka, K. Development of microsatellite markers for the crown-of-thorns starfish Acanthaster planci. *Mol. Ecol. Notes* **2006**, *6*, 141–143. [CrossRef]

34. Yasuda, N.; Nagai, S.; Hamaguchi, M.; Nadaoka, K. Seven new microsatellite markers for crown-of-thorns starfish Acanthaster planci. *Plank* **2007**, *2*, 103–106. [CrossRef]

35. Wainwright, B.J.; Arlyza, I.S.; Karl, S.A. Eighteen microsatellite loci for the crown-of-thorns starfish, *Acanthaster planci. Conserv. Genet. Res.* **2012**, *4*, 861–863. [CrossRef]

36. Harrison, H.B.; Saenz-Agudelo, P.; Al-Salamah, M. Microsatellite multiplex assay for the coral-eating crown-of-thorns starfish, *Acanthaster. cf. planci. Conserv. Genet.* **2015**, *7*, 627–630. [CrossRef]

37. Peakall, R.; Smouse, P.E. GENALEX 6: Genetic analysis in Excel. Population genetic software for teaching and research. *Mol. Ecol. Notes* **2006**, *6*, 288–295. [CrossRef]

38. Alberto, F. MsatAllele_1.0: An R package to visualize the binning of microsatellite alleles. *J. Heredity* **2009**, *100*, 394–397. [CrossRef] [PubMed]

39. Peakall, R.; Smouse, P.E. GenAlEx 6.5: Genetic analysis in Excel. Population genetic software for teaching and research-an update. *Bioinformatics* **2012**, *28*, 2537–2539. [CrossRef] [PubMed]

40. Raymond, M.; Rousset, F. An exact test for population differentiation. *Evolution* **1995**, *49*, 1280–1283. [CrossRef]

41. Rousset, F. genepop'007: A complete re-implementation of the genepop software for Windows and Linux. *Mol. Ecol. Res.* **2008**, *8*, 103–106. [CrossRef] [PubMed]

42. Weir, B.; Cockerham, C. Estimating F-statistics for the analysis of population structure. *Evolution* **1984**, *38*, 1358–1370. [CrossRef]

43. Benjamini, Y.; Hochberg, Y. Controlling the False Discovery Rate: A Practical and Powerful Approach to Multiple Testing. *J. R. Stat. Soc. B* **1995**, *57*, 289–300.

44. Van Oosterhout, C.; Hutchinson, W.; Wills, D.; Shipley, P. MICRO-CHECKER: Software for identifying and correcting genotyping errors in microsatellite data. *Mol. Ecol. Notes* **2004**, *4*, 535–538. [CrossRef]

45. Brookfield, J.F. A simple new method for estimating null allele frequency from heterozygote deficiency. *Mol. Ecol.* **1996**, *5*, 453–455. [CrossRef] [PubMed]

46. Meirmans, P.G.; van Tienderen, P.H. genotype and genodive: Two programs for the analysis of genetic diversity of asexual organisms. *Mol. Ecol. Notes* **2004**, *4*, 792–794. [CrossRef]

47. Hedrick, P. Large variance in reproductive success and the Ne/N ratio. *Evolution* **2005**, *59*, 1596–1599. [CrossRef] [PubMed]

48. Pritchard, J.K.; Stephens, M.; Donnelly, P. Inference of population structure using multilocus genotype data. *Genetics* **2000**, *155*, 945–959. [PubMed]

49. Falush, D.; Stephens, M.; Pritchard, J.K. Inference of population structure using multilocus genotype data: Dominant markers and null alleles. *Mol. Ecol. Notes* **2007**, *7*, 574–578. [CrossRef] [PubMed]

50. Evanno, G.; Regnaut, S.; Goudet, J. Detecting the number of clusters of individuals using the software structure: A simulation study. *Mol. Ecol.* **2005**, *14*, 2611–2620. [CrossRef] [PubMed]

51. Ryman, N.; Palm, S. POWSIM: A computer program for assessing statistical power when testing for genetic differentiation. *Mol. Ecol. Notes* **2006**, *6*, 600–602. [CrossRef]

52. Cornuet, J.-M.; Pudlo, P.; Veyssier, J.; Dehne-Garcia, A.; Gautier, M.; Leblois, R.; Marin, J.-M.; Estoup, A. DIYABC v2.0: A software to make approximate Bayesian computation inferences about population history using single nucleotide polymorphism, DNA sequence and microsatellite data. *Bioinformatics* **2014**, *30*, 1187–1189. [CrossRef] [PubMed]

53. Cornuet, J.-M.; Ravigné, V.; Estoup, A. Inference on population history and model checking using DNA sequence and microsatellite data with the software DIYABC (v1.0). *BMC Bioinform.* **2010**, *11*, 401. [CrossRef] [PubMed]
54. Wright, S. Isolation by Distance. *Genetics* **1943**, *28*, 114–138. [PubMed]
55. Coltman, D.W.; Pilkington, J.G.; Smith, J.A.; Pemberton, J.M. Parasite-Mediated Selection against Inbred Soay Sheep in a Free-Living, Island Population. *Evolution* **1999**, *53*, 1259. [CrossRef]
56. Amos, W.; Worthington Wilmer, J.; Fullard, K.; Burg, T.M.; Croxall, J.P.; Bloch, D.; Coulson, T. The influence of parental relatedness on reproductive success. *Proc. R. Soc. Lond. Biol.* **2001**, *268*, 2021–2027. [CrossRef] [PubMed]
57. Alho, J.S.; Välimäki, K.; Merilä, J. Rhh: An R extension for estimating multilocus heterozygosity and heterozygosity-heterozygosity correlation. *Mol. Ecol. Res.* **2010**, *10*, 720–722. [CrossRef] [PubMed]
58. Dyer, R.J.; Nason, J.D. Population Graphs: The graph theoretic shape of genetic structure. *Mol. Ecol.* **2004**, *13*, 1713–1727. [CrossRef] [PubMed]
59. Jombart, T. adegenet: A R package for the multivariate analysis of genetic markers. *Bioinformatics* **2008**, *24*, 1403–1405. [CrossRef] [PubMed]
60. Clobert, J.; Danchin, E.; Dhondt, A.A.; Nichols, J.D. *Dispersal*; Oxford University Press: Oxford, UK, 2001.
61. Sax, D.F.; Stachowicz, J.J.; Gaines, S.D. *Species Invasions: Insight into Ecology, Evolution and Biogeography*; Sinauer Associates Incorporated: Sunderland, Massachusetts, 2005.
62. Pearson, R.G. Recovery and Recolonization of Coral Reefs. *Mar. Ecol. Prog. Ser.* **1981**, *4*, 105–122. [CrossRef]
63. Francois, O.; Durand, E. Spatially explicit Bayesian clustering models in population genetics. *Mol. Ecol. Res.* **2010**, *10*, 773–784. [CrossRef] [PubMed]
64. Benzie, J.A.H.; Wakeford, M. Genetic structure of crown-of-thorns starfish (*Acanthaster. planci*) on the Great Barrier Reef, Australia: Comparison of two sets of outbreak populations occurring ten years apart. *Mar. Biol.* **1997**, *129*, 149–157. [CrossRef]
65. Muirhead, J.R.; Gray, D.K.; Kelly, D.W.; Ellis, S.M.; Heath, D.D.; Macisaac, H.J. Identifying the source of species invasions: Sampling intensity vs. genetic diversity. *Mol. Ecol.* **2008**, *17*, 1020–1035. [CrossRef] [PubMed]
66. Geller, J.B.; Darling, J.A.; Carlton, J.T. Genetic perspectives on marine biological invasions. *Ann. Rev. Mar. Sci.* **2010**, *2*, 367–393. [CrossRef] [PubMed]
67. MacNeil, M.A.; Chong-Seng, K.M.; Pratchett, D.J.; Thompson, C.A.; Messmer, V.; Pratchett, M.S. Age and Growth of An Outbreaking *Acanthaster* cf. *solaris* Population within the Great Barrier Reef. *Diversity* **2017**. accepted.
68. Yasuda, N.; Hamaguchi, M.; Sasaki, M.; Nagai, S.; Saba, M.; Nadaoka, K. Complete mitochondrial genome sequences for Crown-of-thorns starfish *Acanthaster planci* and *Acanthaster brevispinus*. *BMC Genom.* **2006**, *7*, 17. [CrossRef] [PubMed]
69. Vanhatalo, J.; Hosack, G.R.; Sweatman, H. Spatiotemporal modelling of crown-of-thorns starfish outbreaks on the Great Barrier Reef to inform control strategies. *J. Appl. Ecol.* **2016**, 1–10. [CrossRef]
70. Uthicke, S.; Doyle, J.; Duggan, S.; Yasuda, N.; McKinnon, A.D. Outbreak of coral-eating Crown-of-Thorns creates continuous cloud of larvae over 320 km of the Great Barrier Reef. *Sci. Rep.* **2015**, *5*, 16885. [CrossRef] [PubMed]
71. Wilmes, J.; Matthews, S.; Schultz, D.; Messmer, V.; Hoey, A.; Pratchett, M. Modelling Growth of Juvenile Crown-of-Thorns Starfish on the Northern Great Barrier Reef. *Diversity* **2016**, *9*, 1. [CrossRef]

diversity

MDPI

Article

Variation in Incidence and Severity of Injuries among Crown-of-Thorns Starfish (*Acanthaster* cf. *solaris*) on Australia's Great Barrier Reef

Vanessa Messmer *, Morgan Pratchett and Karen Chong-Seng

Australian Research Council Centre of Excellence for Coral Reef Studies, James Cook University, Townsville, QLD 4811, Australia; morgan.pratchett@jcu.edu.au (M.P.); karen.mcs@gmail.com (K.C.-S.)
* Correspondence: vanessa.messmer@gmail.com; Tel.: +61-7-4781-5531

Academic Editors: Sven Uthicke and Michael Wink
Received: 8 December 2016; Accepted: 10 February 2017; Published: 21 February 2017

Abstract: Despite the presence of numerous sharp poisonous spines, adult crown-of-thorns starfish (CoTS) are vulnerable to predation, though the importance and rates of predation are generally unknown. This study explores variation in the incidence and severity of injuries for *Acanthaster* cf. *solaris* from Australia's Great Barrier Reef. The major cause of such injuries is presumed to be sub-lethal predation such that the incidence of injuries may provide a proxy for overall predation and mortality rates. A total of 3846 *Acanthaster* cf. *solaris* were sampled across 19 reefs, of which 1955 (50.83%) were injured. Both the incidence and severity of injuries decreased with increasing body size. For small CoTS (<125 mm total diameter) >60% of individuals had injuries, and a mean 20.7% of arms ($\pm$2.9 SE) were affected. By comparison, <30% of large (>450 mm total diameter) CoTS had injuries, and, among those, only 8.3% of arms ($\pm$1.7 SE) were injured. The incidence of injuries varied greatly among reefs but was unaffected by the regulations of local fisheries.

Keywords: predator removal hypothesis; sub-lethal predation; arm damage; body size

1. Introduction

Predation has long been considered a key process in population regulation [1], contributing to long-term stability in the abundance of prey species. By inference, organisms that exhibit rapid and pronounced increases in abundance (often termed outbreaks; [2]) are considered to be released (or otherwise free) from predatory regulation (e.g., [3,4]). Predation may nonetheless be an important and often major cause of mortality but simply has no discernable influence on prey abundance [5]. Population outbreaks of prey species may occur when predators fail to react to increases in prey densities [6]. However, population outbreaks may also arise completely independently of predation, due to intrinsic and extrinsic processes. Most notably, outbreaks may result from steep changes in rates of population replenishment, especially where organisms have exceptional reproductive potential but generally low fertilisation rates and reproductive success [7]. Moreover, the abundance and reproductive success of outbreaking species is often influenced by marked changes in environmental conditions and resources (e.g., food availability) [8].

Crown-of-thorns starfish (CoTS; *Acanthaster* spp.) have gained considerable notoriety due to their propensity for population outbreaks, as well as their corresponding impacts on local assemblages of prey corals [9]. Very few marine organisms show changes in abundance of the magnitude or rate shown by crown-of-thorns starfish. In the extreme, 10-fold increases in localised (within reef) densities of CoTS have been documented within one year (e.g., [10]). In Moorea (French Polynesia), CoTS densities ranged from 11,500 individuals per km^2 to up to 151,650 individuals per km^2 around the circumference of the reef and were both spatially and temporally variable [10]. During the course of this outbreak, coral cover declined by up to 93% in approximate accordance with the cumulative number of CoTS recorded at each location [10].

One of the foremost hypotheses put forward to account for outbreaks of *Acanthaster* spp. is the predator removal hypothesis, initially proposed by Endean [11]. Endean [11] noted that shell collectors had removed ~10,000 giant tritons (*Charonia tritonis*), leading to significant declines in their abundance in the lead up to the first major outbreak of *Acanthaster* sp. recorded on Australia's Great Barrier Reef. At that time, *C. tritonis* was also regarded as one of the only effective predators of CoTS [11]. There are now a large number of coral reef organisms known to prey upon CoTS during different stages of their life cycle [12], including fish and other invertebrates. These may be important in regulating the abundance of CoTS, if not in actually preventing outbreaks. Coral reef fishes that prey on CoTS are receiving particular attention [13–15], given that localised levels of fishing seem to correspond with inter-reef variation in the severity [15] or incidence [14] of CoTS outbreaks. The most intuitive explanation for these patterns is that the overexploitation of particular fishes relaxes the top-down control necessary to regulate populations of *Acanthaster* spp. [11], thereby leading to population outbreaks. However, these studies have not identified the specific target species that predate on CoTS nor have they explicitly compared densities of CoTS predators or quantified rates of predation of CoTS along gradients of fishing pressure. Trophic cascades induced by fisheries, resulting in fewer invertebrates preying on juvenile starfish, may be another mechanism releasing predation pressure on CoTS [14].

One of the main limitations to testing the predator removal hypothesis is the inherent difficulty in quantifying predation rates on *Acanthaster* spp. in the field. This is particularly difficult for small and juvenile *Acanthaster* spp. due to their cryptic nature [16]. One possible proxy for measuring variation (spatial, temporal, taxonomic, and ontogenetic) in the susceptibility to predation among *Acanthaster* spp. is the incidence of recent injuries. These are most apparent as missing or regenerating arms (Figure 1), which are often attributed to sub-lethal or partial predation [16,17]. Although sub-lethal predation is also generally considered a good proxy for mortality due to predation, or overall predation pressure [16,17], this has not been explicitly tested, and predators causing injuries may not cause outright mortality of CoTS. Relatively few predators are known to consume adult CoTS in their entirety (but see [18]), while CoTS survive and can escape from predators (e.g., fishes) that only remove a portion of the body mass [19]. In previous studies, up to 67% of CoTS in some locations exhibit recent or sustained injuries [16], and high incidences of injuries appear to be generally reflective of a higher intensity of predation [16,17]. In the Philippines, for example, Rivera-Posada et al. [16] showed that the incidence of injuries was higher inside rather than outside of marine protected areas (MPAs) where fishing is prohibited, which would be consistent with a higher abundance of potential predators. The incidence of injuries also tends to decrease with the increasing body size of *Acanthaster* sp. [20] as well as for several other species of starfishes [21,22], which probably reflects their increased susceptibility to predation when small [22]. Even if the predators that cause a high incidence of injuries among *Acanthaster* spp. do not kill these starfish outright, they may nonetheless have important effects on the behaviour and fitness of starfish, thereby contributing to population regulation [23].

The purpose of this study was to test for variation in the incidence and severity of injuries among crown-of-thorns starfish (*Acanthaster* cf. *solaris*) from Australia's Great Barrier Reef (GBR). More specifically, we wanted to test whether the incidence of injuries is higher inside versus outside MPAs, where fishing is prohibited, as would be expected if fisheries' target species impose significant predatory regulation on CoTS on the GBR and the abundance of these key predators vary significantly in accordance with spatial management zones [14]. We also tested for size-based variation in the incidence of injuries among CoTS, ranging in size from 60 mm total diameter (TD) to 510 mm TD. These injuries are presumably caused mostly by sub-lethal predation [17] (but see [22]).

Figure 1. Small and regenerating arm of *Acanthaster* cf. *solaris* (as indicated by arrow), which is indicative of past injury, presumably caused by sub-lethal predation (see also [16]).

2. Materials and Methods

A total of 3846 crown-of-thorns starfish (*Acanthaster* cf. *solaris*) were collected between October 2012 and May 2015 along the Great Barrier Reef (GBR). Sampling was conducted at 19 reefs, spanning 1150 km of the GBR (Figure 2). All reefs, except Centipede Reef, Davies Reef, Michaelmas Cay, and Sweetlip Reef, were considered to have an active CoTS outbreak at the time of sampling. All starfish were collected while snorkelling or SCUBA diving using large purpose-built tongs to carefully extract starfish from among the reef matrix. Starfish were kept alive in 500 L tanks connected to high flow-through sea-water systems on live-aboard boats or at the research station on Lizard Island for a maximum of 20 h before they were processed and disposed. During processing, the starfish were removed from the water and placed on a flat surface for 30–90 s before measuring the total diameter across opposite arms that were ostensibly undamaged. The severity of injury was then assessed by counting the number of missing or damaged arms, which was then expressed as a percentage of the total number of arms (also referred to as "severity"). Missing arms were apparent where the ambulacral groove terminated at the edge of the oral disk. All arms that were less than 75% the length of the adjacent arms were considered to have been damaged. Recent injuries, apparent due to fresh tears in the surface integument, were ignored, as they likely occurred during collection. We also determined the sex of each individual starfish (where possible) based on visual inspection of the gonads that were exposed following the removal of a few arms using a paint scraper [9]. Starfish that were either immature or spent (virtually no gonad tissue left after spawning) could not be sexed, resulting in 1078 and 1475 individuals identified as females and males respectively.

Data Analyses

The probability of an individual CoTS being injured (injury incidence) was analysed using binary logistic mixed models (logit link), with "zone" (no fishing, restricted fishing, open to fishing) and "size" (total diameter: mm) as possible influential factors. The influence of zone and size on the severity of injuries received during predation events was also investigated using a subset of the data ($n = 1797$), which only included individuals that had experienced arm damage (injury = 1) and was modeled as the proportion of injured arms (in relation to the total no. arms) using binomial mixed models (logit link). The potential influence of adult gender (male or female) on CoTS injuries was also investigated using the subset of starfish for which sex could be established (*n* = 2553 for influence on injury incidence and *n* = 1263 for influence on injury severity) using generalized mixed models (logit link: binary logistic and binomial, respectively).

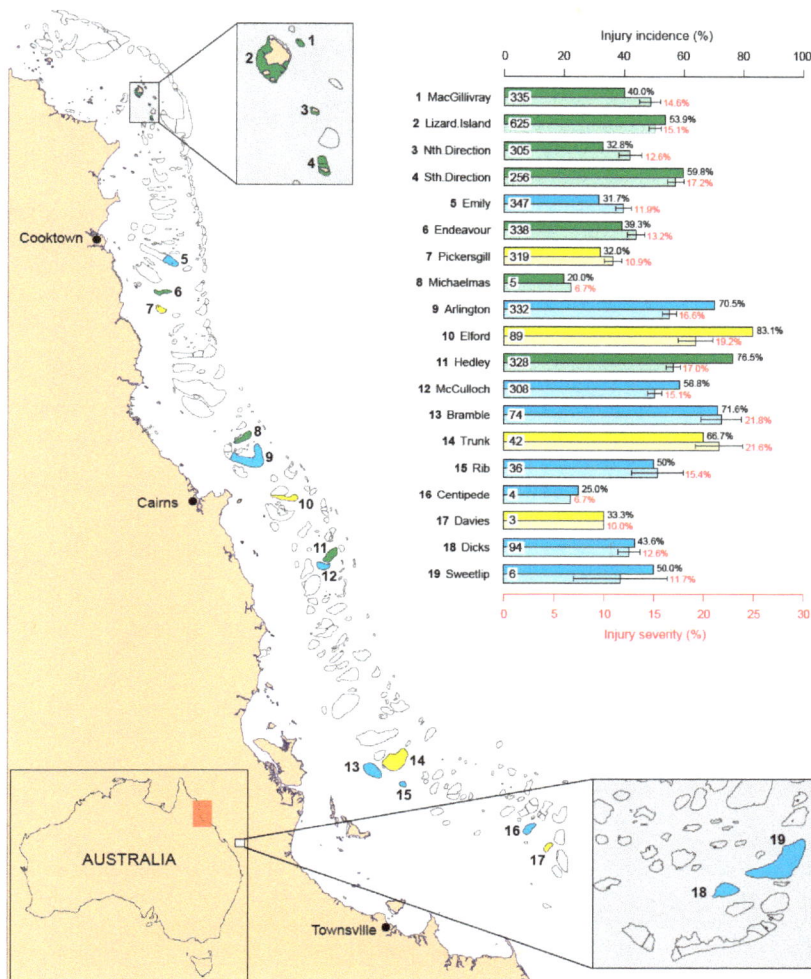

Figure 2. Map indicating sampled reefs along the Great Barrier Reef (GBR). The red box on the map of Australia shows the extent of the main map, whilst the reefs sampled in the Swains Region on the south-eastern end of the GBR are magnified in the bottom right rectangle. The colour of sampled reefs designates fishing regulations: green = no fishing (marine reserve), yellow = restricted fishing allowed, blue = open to fishing. The bar graph shows injury incidence per reef (top paired bar with solid fill and black values) and mean ± SE injury severity per reef (lower paired bar with semi-transparent fill and red values). Numbers in the white boxes within the bars represent the total number of individuals collected per reef. Injury severity values were calculated based on injured individuals only.

For all models, "reef" was included as a random effect (19 levels), while "observer" was included as a potential fixed effect (3 levels) to account for possible artifacts of the sampling design and different data collectors. Full models, with all the potential relevant interactions, were fit first, and then model selection procedures were applied, comparing models using likelihood ratio tests [24,25]. The model assumptions were checked graphically, investigating residuals and random effects before interpretation of the final model. The confidence intervals around the coefficient estimates of the final models were generated by parametric bootstrapping. All statistical analyses were conducted using the R statistical software program (R Core Team 2016) and the lme4 package [26].

3. Results

3.1. Incidence of Injuries

In all, 1955 out of 3846 (50.83%) CoTS collected from the GBR exhibited evidence of recent or sustained injuries, based on the number of arms that were missing or evidently shorter and thinner and generally covered in shorter and finer spines (Figure 1). The incidence of injury varied considerably between reefs, ranging from 20% (1 starfish injured out of 5) at Michelmas Cay near Cairns up to 83% (74 starfish injured out of 89) at Elford Reef, also located very close to Cairns (Figure 2). The mean severity of injuries (calculated based on the percentage of arms affected) also varied among reefs, ranging from 6.7% at Centipede Reef, where only one starfish that was injured was collected, up to 21.8% ($\pm$2.0% SE) at Bramble Reef. Interestingly, there was a positive linear correlation ($R^2 = 0.783$, $p < 0.001$) between the incidence and the severity of injuries at the scale of individual reefs (Figure 3).

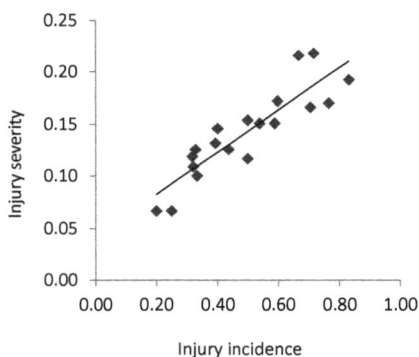

Figure 3. Linear regression ($R^2 = 0.783$; slope = 0.2023 ($F_{(1,17)} = 61.29$, $p < 0.001$); intercept is not significantly different from 0) between incidence (proportion of starfish with any evidence of injury) and severity (mean proportion of arms affected among injured starfish) at the scale of reefs.

Despite marked inter-reef differences in the incidence and severity of injuries, there was no obvious effect of the regulations of local fisheries (Tables 1 and 2). When averaged across all reefs in each of the three distinct management zones, the mean incidence of injury was non-significantly higher for yellow zones (53.78% $\pm$ 12.66% SE), followed by blue zones (50.15% $\pm$ 5.92% SE) and lowest in green zones (46.05% $\pm$ 7.10% SE). However, these differences are negligible compared to the variation observed among 'reefs' within each group, as evident based on large standard errors (Figure 4a). There was also no influence of fishing regulation (zone) in any of the models. Likelihood ratio tests with and without zone as a fixed effect can be seen in Table 1. The frequency distributions of the number of arms missing were also very similar across the three management zones, with a single missing arm affecting the majority of injured individuals (65%–70%) (Figure 4b–d).

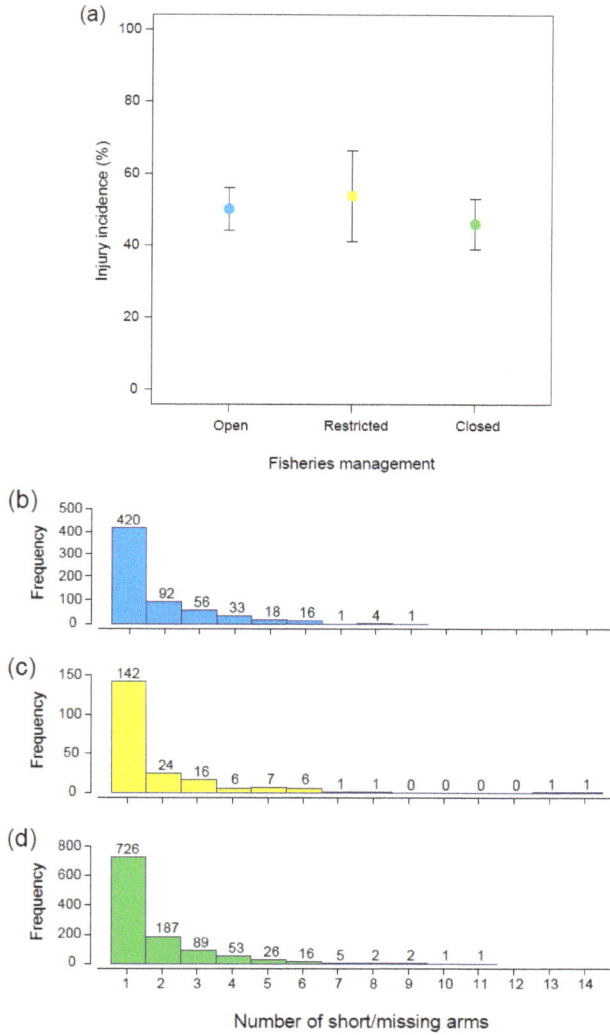

Figure 4. (**a**) Incidence of injury (percentage of individuals with damaged arms) averaged across the reefs in each of three different management zones (the colour of the circle designates fishing regulations: see (**b**–**d**)); and (**b**) frequency distributions of short or missing arms for each zone ((**b**) blue = open to fishing; (**c**) yellow = restricted fishing; (**d**) green = no fishing).

Table 1. Overview of the model selection process, starting with the full model including all relevant interactions, compared with Akaike Information Criterion (AIC) and Likelihood-ratio tests.

Data	Candidate Model Effects	AIC	Likelihood Test *
Incidence (*n* = 3846)	Zone * Size + Observer + (1 l Reef)	4937.8	
	Zone * Size + Observer + (Size l Reef)	4935.4	*Chi* = 1.0, *df* = 2, *p* = 0.617
	Zone + Size + Observer + (Size l Reef)	4933.2	
	Size + Observer + (Size l Reef)	4930.1	
Severity ** (*n* = 1797)	Zone * Size + Observer	5847.6	
	Zone + Size + Observer	5846.0	*Chi* = 3.1, *df* = 4, *p* = 0.541
	Size	5841.1	
Gender (Incidence) ** (*n* = 2553)	Zone + Size * Sex + Observer	3287.6	
	Zone + Size + Sex + Observer	3285.8	*Chi* = 3.2, *df* = 2, *p* = 0.204
	Size + Sex	3285.0	
Gender (Severity) ** (*n* = 1263)	Zone + Size * Sex + Observer	4210.1	
	Zone + Size + Sex + Observer	4204.1	*Chi* = 1.2, *df* = 4, *p* = 0.814
	Size + Sex	4177.2	

* Influence of factors checked with Likelihood-ratio tests with and without the variable as a fixed effect; ** All models had "reef" as a random effect (1 l Reef), and the inclusion of a random slope (e.g., Size l Reef) did not improve model fit.

Table 2. Model coefficients for the best fitting models describing the relationship between injury incidence and severity with size, whilst accounting for the variability between the reefs in all models and observer bias for models of incidence (Figures 5 and 6).

Data	Variable	Estimate ± SE	*z*-Value	*p*-Value
Incidence	(Intercept)	1.0708 ± 0.3239	3.3055	0.0009
	Size	−0.0043 ± 0.0009	−4.8358	0.0000
	Observer2	1.1351 ± 0.3793	2.9929	0.0028
	Observer3	−0.0872 ± 0.3248	−0.2685	0.7883
Severity	(Intercept)	−1.3512 ± 0.0925	−14.6038	0.0000
	Size	−0.0020 ± 0.0003	−7.0685	0.0000
Gender (incidence)	(Intercept)	1.0882 ± 0.3639	2.9907	0.0028
	Size	−0.0042 ± 0.0008	−5.4448	0.0000
	Male	−0.0343 ± 0.0874	−0.3922	0.6949
	Observer2	0.9585 ± 0.3945	2.4300	0.0151
	Observer3	0.0255 ± 0.3693	0.0690	0.9450
Gender (severity)	(Intercept)	−1.1265 ± 0.1258	−8.9549	0.0000
	Size	−0.0025 ± 0.0004	−6.5331	0.0000
	Male	−0.0623 ± 0.0394	−1.5818	0.1137

3.2. Size-Based Variation

The incidence of injuries showed a negative relationship with the increasing size of CoTS (z = −4.836, p < 0.001) (Figure 5a). The probability of a small starfish (60 mm) exhibiting any level of injury was 0.70 (95% CI = [0.57, 0.79] by parameter bootstrap [PB]), decreasing to 0.25 (95% PB CI = [0.15, 0.38]) in the largest individual (510 mm) when observed by the median observer. Although injury incidence was affected by the observer (Tables 1 and 2), Observer identity did not influence the shape of the relationship (Likelihood-ratio tests with and without the observer influencing slope: *Chi* = 1.0, *df* = 5, *p* = 0.959) (Figure 5a). The severity of injuries showed a similar decline with increasing size (z = −7.069, p < 0.001) (Figure 5b). Small starfish (60 mm) generally exhibited injuries to 19% of their arms [16%, 21% CI], decreasing to 9% [8%, 11% CI] in large starfish (~470 mm). Neither the observer or the reef had any effect on the relationship between the severity of the injuries and the size of the CoTS (Figure 5b).

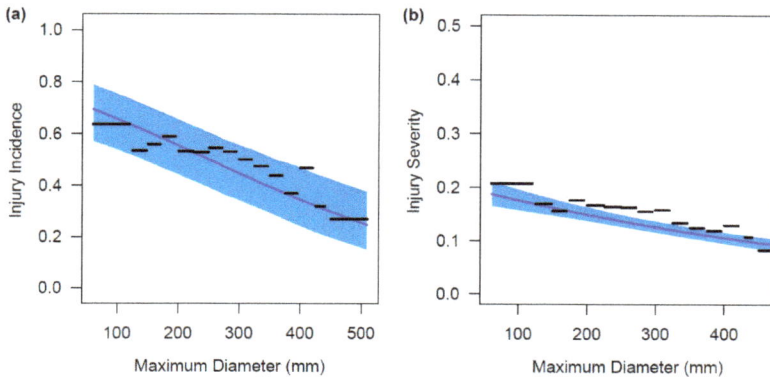

Figure 5. Relationship (with 95% confidence intervals (CI) generated by parametric bootstrap) between (**a**) CoTS maximum diameter and the probability of the individual to have experienced arm damage (injury incidence; binary data), as measured by the median observer on an average reef; (**b**) CoTS maximum diameter and the severity of an individual's injury on an average Reef (note y-axis has been truncated to aid visualization of the relationship). Black horizontal lines represent the mean probability of individuals in 25mm size classes (<125, 125–150, 150-175, ... , >450 mm) superimposed over the relationship.

3.3. Gender-Based Differences in Incidence and Severity of Injuries

Gender did not influence the incidence or severity of starfish arm damage statistically (Tables 1 and 2). Female and Male starfish incidence of arm damage was 0.64 [0.49, 0.77] and 0.63 [0.49, 0.76], respectively, when small (~125 mm), decreasing to 0.26 [0.16, 0.40] and 0.26 [0.15, 0.40] when large (~510 mm), respectively (Figure 6a). However, the severity of arm damage was slightly reduced in males in comparison to females; 0.18 [0.16, 0.21] when small (~125 mm) and 0.09 [0.07, 0.10 CI] when large (~470 mm) vs. 0.19 [0.17, 0.22 CI] when small and 0.09 [0.08, 0.11] when large, respectively (Figure 6b).

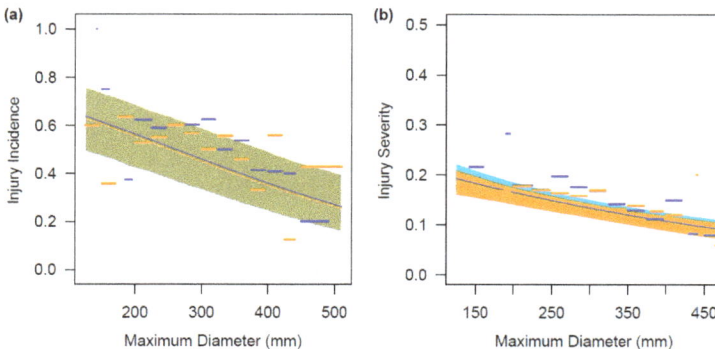

Figure 6. Relationships (with 95% confidence intervals (CI) generated by parametric bootstrap) between female (blue) and male (orange) CoTS maximum diameter and injury (**a**) incidence (arm damage as measured by the median observer on an average reef; binary data) and (**b**) severity on an average reef. Horizontal lines represent the mean probability of female (blue) and male (orange) individuals in 25 mm size classes (<125, 125–150, 150-175, ... , >450 mm) superimposed over the relationship. Note that the y-axis has been truncated to aid visualization of the relationship for severity.

4. Discussion

Outbreaks of *Acanthaster* spp. are a major contributor to coral loss and reef degradation throughout the Indo-Pacific region (e.g., [10,27]), and a major concern for coral reef management. Understanding the possible role of predation in regulating populations of *Acanthaster* spp. is fundamental in establishing whether increased regulation of fisheries will mitigate or prevent ongoing outbreaks [12]. Although the overall incidence of injury was high (50%) amongst the CoTS sampled along much of the length of the GBR, there was no difference in the incidence or severity of injuries for CoTS from reefs where fishing is prohibited (green zones) versus reefs where fishing activities are permitted (yellow and blue zones). However, there was strong variation in the incidence of injuries among reefs, ranging from 20% to 83%. There was also a strong negative correlation between body size (CoTS diameter) and arm damage, with individuals ≤125 mm in diameter being twice as likely to be injured than individuals ≥400 mm. The severity of injuries (the proportion of the number of injured arms over the total number of arms) showed a very similar reduction with increasing body size. There was only a very weak gender trend, with females being slightly more likely to suffer increased arm damage (severity) than males.

Whether predators are able to regulate CoTS abundances remains the topic of ongoing debate. Across the GBR, the occurrence of CoTS outbreaks is lower on reefs where fishing is prohibited, compared to reefs where fishing is permitted [14,28], which suggests that the higher abundance of large target species results in greater predation of CoTS. In the current study, fishing restrictions had no effect on the incidence of injuries, which are presumably caused mostly by sub-lethal predation [17] (but see [22]). However, even if injuries are caused mainly by predation, overall mortality and predation rates may still vary with the management of fisheries. The sub-lethal predation rate is generally considered to be a proxy for overall predation pressure [16,17], but this has not been explicitly tested. It may be that the fishes that cause injuries (loss of individual arms) are altogether different from those that kill CoTS outright, such that direct tests of the predator-removal hypothesis still require measurements of actual mortality rates alongside explicit consideration of the abundance and composition of potential predators, irrespective of zoning. Except for some lethrinid species, most other known CoTS predators (e.g., the stellar pufferfish *Arothron stellatus*) [12] are not generally targeted by recreational and commercial finfish fisheries on the GBR, which mainly target large piscivorous fishes such as coral trout [29].

Irrespective of whether injuries are a valid proxy for overall predation and predation rates, higher incidences and severity of sub-lethal predation will have a significant impact on the individual fitness and population dynamics of CoTS [23,30,31]. CoTS, like many other echinoderms, have the capacity to regenerate parts of their central disc and missing arms [19,22]. However, regeneration comes at an energetic cost and affects the fitness of individuals [31,32]. Injuries and regeneration can reduce feeding and growth, delay maturation, or compromise the reproductive output. Given that each arm contains gonads in CoTS, fewer arms directly affect the reproductive capacity of a female. In the sea star *Heliaster helianthus*, the energetic content in their pyloric caeca and gonads showed a 5 to 7 fold decrease following autotomy [31]. Similarly, regenerating tails in lizards were found to affect clutch and egg size [33], and maternal effects were observed in CoTS, with reduced egg size resulting in lower survivorship in developing larvae [34]. As a result, even relatively minor injuries may have considerable negative impacts on the reproductive success of individual females.

The overall incidence of injury recorded in this study (50%) is at the higher end of rates reported previously from the GBR (33% [35]; 40% [17]; 50% [36]). However, our study clearly showed that the incidence of injury varies among reefs on the GBR (20%–83%), and the range of variation recorded in this study is consistent with differences in the estimates of the incidence of injuries from previous studies, all of which considered only a single or few sites. The highest incidences of injury recorded (81.3%) on some reefs exceeded the highest reported incidence of damaged arms (67%) recorded in the Philippines [16]. Although the observer had an effect on the incidence of arm damage, substantial variation between reefs was still clearly observed irrespective of who counted the number of short or missing arms. The high rates on some reefs may represent a regional effect (this could not be tested due to the covariation with the observer in some cases). The Cairns sector (Reefs 8–12, Figure 2) had above average rates (59%–83%), whereas the Cooktown sector (Reefs 5–7, Figure 2) had below average incidences of arm damage (32%–39%). The high injury rates (50%–72%) on reefs with outbreaks in the Townsville sector (Reefs 13–17, Figure 2) are likely to reflect the on average smaller individuals collected on these reefs. The variation in injury rates between reefs may also be due to differences in the local abundance of predators. Future studies should therefore explicitly test if there is a relationship between the incidence of arm damage and the abundance of predators.

Vulnerability to predation is known to vary with ontogeny in many organisms [37,38]. Younger and smaller individuals are often subject to higher predation rates than older and larger conspecifics [39–41]. Large CoTS exhibit an impressive defense against predators through the presence of large, very sharp, and poisonous spines, which may limit the number of species able to predate on large adult CoTS. It is therefore not surprising to see a 50% reduction in the incidence and severity (measured as the number of injured arms over the total number of arms) of arm damage with increasing body size (the relationship was not affected by the observers). Size has previously been identified as an important factor in predation rates in other echinoderms and CoTS, but patterns are not always consistent. Sub-lethal injuries generally decline with increasing body size in echinoderms [21], asteroids [22], sea urchins [42,43], and CoTS [16,17]. McCallum et al. [17] found a similar but weak linear relationship in CoTS, whereas a hump-shaped relationship was observed in the Philippines, which covered a similar size range as the present study [16]. Although the discrepancy between the studies could be due to sample size (both studies had relatively low sample sizes of individuals ≤10 cm), the discrepancy may also be due to differences in local predation pressure. A reduction in arm damage in individuals ≤10 cm is possible and could be explained by behavioural changes or changes in the ratio between lethal and sub-lethal predation rates [16]. Increased sampling in these smaller size classes, including very small juvenile CoTS [44], should clarify the effect of size on sub-lethal predation rates in these early life stages. Nevertheless, the high incidence of predation rates in smaller CoTS (≤20 cm) across both studies suggests that substantial predation levels are likely to occur at night due to the nocturnal feeding habits and cryptic nature of small CoTS during the day.

There were indications that gender may have a weak effect on predation severity. Although not statistically significant, female CoTS exhibited slightly higher levels of injuries compared to males, which may be explained by the relatively higher energy content in oocytes compared to spermaries. Similarly, egg-bearers in other marine organisms were found to be more susceptible to predation, with the higher nutritional value of oocytes proposed as a mechanism [38,45]. Interestingly, there was no difference in the injury incidence between males and females, suggesting that predators do not discriminate between males and females but may feed more intensively on females.

5. Conclusions

Predation has the potential to play a significant role in regulating populations of CoTS, though the effect is most likely to dampen fluctuations in the local abundance of CoTS rather than to prevent outbreaks per se. Ormond et al. [13] suggested that predation levels by fish on CoTS could maintain CoTS in sufficiently low densities to avoid outbreaks. Our study showed that at least sub-lethal predation on CoTS is common, with 50% of the 3846 individuals studied showing evidence of predation events, and that size plays a major role in the frequency and severity of predation events. Although the predator removal hypothesis remains controversial, with studies showing variable results, predation of CoTS is a common event that warrants further investigation. It is critical to determine predation rates in small juveniles, identify all possible predators, and assess the effects of sub-lethal predation on growth, fitness, and reproductive output to better inform population models. There is no stopping the current outbreak on the GBR, but any attempts to prevent future outbreaks in light of the increasing threats to coral reefs is possibly our easiest chance to improve the outlook for coral reefs.

Acknowledgments: We are grateful to Hugo Harrison, Simon Wever, Kirsty Nash, Zara-Louise Cowan, Laura Richardson, Robert Streit, Kristen Anderson, Stuart Watson, Jacob Johansen, and Elmar Messmer for their assistance in the collection of CoTS and to the Lizard Island Research Station staff, Tim Godfrey, and the crew of Reef Connections and the Capricorn Star for logistical support. This project was supported by the Commonwealth Government of Australia through the Caring for Country and Reef Rescue Program to M.P. and V.M. and the Australian Museum's Lizard Island Research Station with specific awards to M.P. and V.M from the Ian Potter Foundation 50th Anniversary Commemorative Grant Scheme.

Author Contributions: M.P. conceived and designed the experiments; V.M., M.P., and K.C-S. performed the experiments; V.M. and K.C-S. analysed the data; V.M., M.P., and K.C-S. wrote the paper.

Conflicts of Interest: The authors declare no conflict of interest.

References

1. Murdoch, W.W.; Oaten, A. Predation and population stability. *Adv. Ecol. Res.* **1975**, *9*, 1–131.
2. Carpenter, J.R. Insect outbreaks in Europe. *J. Anim. Ecol.* **1940**, *9*, 108–147. [CrossRef]
3. Holling, C.S. The functional response of predators to prey density and its role in mimicry and population regulation. *Mem. Entomol. Soc. Can.* **1965**, *45*, 5–60. [CrossRef]
4. Rosenzweig, M.L. Paradox of enrichment: Destabilization of exploitation ecosystems in ecological time. *Science* **1971**, *171*, 385–387. [CrossRef] [PubMed]
5. Stiling, P. Density-dependent processes and key factors in insect populations. *J. Anim. Ecol.* **1988**, *57*, 581–594. [CrossRef]
6. Clark, L.R. Predation by birds in relation to the population density of *Cardiaspina albitextura* (Psyllidae). *Aust. J. Zool.* **1964**, *12*, 349–361. [CrossRef]
7. Uthicke, S.; Schaffelke, B.; Byrne, M. A boom-bust phylum? Ecological and evolutionary consequences of density variations in echinoderms. *Ecol. Monogr.* **2009**, *79*, 3–24.
8. Singleton, G.R. Population dynamics of an outbreak of house mice (*Mus domesticus*) in the Mallee Wheatlands of Australia—Hypothesis of plague formation. *J. Zool.* **1989**, *219*, 495–515. [CrossRef]
9. Pratchett, M.S.; Caballes, C.F.; Rivera-Posada, J.A.; Sweatman, H.P.A. Limits to understanding and managing outbreaks of crown-of-thorns stafish (*Acanthaster* spp.). *Oceanogr. Mar. Biol. Annu. Rev.* **2014**, *52*, 133–199.
10. Kayal, M.; Vercelloni, J.; de Loma, T.L.; Bosserelle, P.; Chancerelle, Y.; Geoffroy, S.; Stievenart, C.; Michonneau, F.; Penin, L.; Planes, S.; et al. Predator Crown-of-Thorns Starfish (*Acanthaster planci*) Outbreak, Mass Mortality of Corals, and Cascading Effects on Reef Fish and Benthic Communities. *PLoS ONE* **2012**, *7*, e47363. [CrossRef] [PubMed]
11. Endean, R. *Report on Investigations Made into Aspects of the Current Acanthaster planci (Crown-of-Thorns) Infestations of Certain Reefs of the Great Barrier Reef*; Queensland Department of Primary Industries: Brisbane, Australia, 1969.
12. Cowan, Z.L.; Pratchett, M.S.; Messmer, V.; Ling, S. Known predators of crown-of-thorns starfish (*Acanthaster* spp.) and their role in mitigating, if not preventing, population outbreaks. *Diversity* **2017**, *9*, 7.
13. Ormond, R.F.G.; Bradbury, R.H.; Brainbridge, S.; Fabricius, K.; Kessing, J.K.; de Vantier, L.M.; Medlay, P.; Steven, A.D.L. Test of a model for regulation of crown-of-thorns starfish by fish predators. In *Acanthaster and the Coral Reef: A Theoretical Perspective*; Bradbury, R.H., Ed.; Springer: Heidelberg, Germany, 1990; pp. 189–207.
14. Sweatman, H. No-take reserves protect coral reefs from predatory starfish. *Curr. Biol.* **2008**, *18*, R598–R599. [CrossRef] [PubMed]
15. Dulvy, N.K.; Freckleton, R.P.; Polunin, N.V.C. Coral reef cascades and the indirect effects of predator removal by exploitation. *Ecol. Lett.* **2004**, *7*, 410–416. [CrossRef]
16. Rivera-Posada, J.; Caballes, C.F.; Pratchett, M.S. Size-related variation in arm damage frequency in the crown-of-thorns sea star, *Acanthaster planci*. *J. Coast. Life Med.* **2014**, *2*, 187–195.
17. McCallum, H.I.; Endean, R.; Cameron, A.M. Sublethal damage to *Acanthaster planci* as an index of predation pressure. *Mar. Ecol. Prog. Ser.* **1989**, *56*, 29–36. [CrossRef]
18. Bos, A.R.; Gumanao, G.S.; Salac, F.N. A newly discovered predator of the crown-of-thorns starfish. *Coral Reefs* **2008**, *27*, 581. [CrossRef]
19. Messmer, V.; Pratchett, M.S.; Clark, T.D. Capacity for regeneration in crown of thorns starfish, *Acanthaster planci*. *Coral Reefs* **2013**, *32*, 461. [CrossRef]

20. Simpson, C.J.; Grey, K.A. *Crown-of-Thorns Starfish (Acanthaster planci) in the Dampier Archipelago, Western Australia*; Technical Series; Perth Environmental Protection Authority: Perth, Australia, 1988.
21. Lawrence, J.M.; Vasquez, J. The effect of sublethal predation on the biology of echinoderms. *Oceanol. Acta* **1996**, *19*, 431.
22. Lawrence, J.M. Arm loss and regeneration in Asteroidea (Echinodermata). In *Echinoderm Research*; Scalera-Liaci, L., Canicatti, C., Eds.; Balkema: Rotterdam, The Netherlands, 1991; pp. 39–52.
23. Harris, R.N. Nonlethal injury to organisms as a mechanism of population regulation. *Am. Nat.* **1989**, *134*, 835–847. [CrossRef]
24. Bolker, B.M.; Brooks, M.E.; Clark, C.J.; Geange, S.W.; Poulson, J.R.; Stevens, M.H.H.; White, J.-S.S. Generalized linear mixed models: A practical guide for ecology and evolution. *Trends Ecol. Evol.* **2009**, *24*, 127–135. [CrossRef] [PubMed]
25. Zuur, A.F.; Ieno, E.N.; Walker, N.; Saveliev, A.A.; Smith, G.M. *Mixed Effects Models and Extensions in Ecology with R*; Springer: New York, NY, USA, 2009.
26. Bates, D.; Maechler, M.; Bolker, B.; Walker, S. Fitting linear mixed-effects models using LME4. *J. Stat. Softw.* **2015**, *67*, 1–48. [CrossRef]
27. Pisapia, C.; Burn, D.; Yoosuf, R.; Najeeb, A.; Anderson, K.D.; Pratchett, M.S. Coral recovery in the central Maldives archipelago since the last major mass-bleaching, in 1998. *Sci. Rep.* **2016**, *6*. [CrossRef] [PubMed]
28. McCook, L.J.; Ayling, T.; Cappo, M.; Choat, J.H.; Evans, R.D.; De Freitas, D.M.; Heupel, M.; Hughes, T.P.; Jones, G.P.; Mapstone, B.; et al. Adaptive management of the Great Barrier Reef: A globally significant demonstration of the benefits of networks of marine reserves. *Proc. Natl. Acad. Sci. USA* **2010**, *107*, 18278–18285. [CrossRef] [PubMed]
29. Messmer, V.; Pratchett, M.S.; Hoey, A.S.; Tobin, A.J.; Coker, D.J.; Cooke, S.J.; Clark, T.D. Global warming may disproportionately affect larger adults in a predatory coral reef fish. *Glob. Chang. Biol.* **2016**. [CrossRef] [PubMed]
30. Zajac, R.N. Sublethal predation on *Polydora cornuta* (Polychaeta, Spionidae)—Patterns of tissue loss in a field population, predator functional response and potential demographic impacts. *Mar. Biol.* **1995**, *123*, 531–541. [CrossRef]
31. Barrios, J.V.; Gaymer, C.F.; Vasquez, J.A.; Brokordt, K.B. Effect of the degree of autotomy on feeding, growth, and reproductive capacity in the multi-armed sea star *Heliaster helianthus*. *J. Exp. Mar. Biol. Ecol.* **2008**, *361*, 21–27. [CrossRef]
32. Maginnis, T.L. The costs of autotomy and regeneration in animals: A review and framework for future research. *Behav. Ecol.* **2006**, *17*, 857–872. [CrossRef]
33. Bernardo, J.; Agosta, S.J. Evolutionary implications of hierarchical impacts of nonlethal injury on reproduction, including maternal effects. *Biol. J. Linn. Soc.* **2005**, *86*, 309–331. [CrossRef]
34. Caballes, C.F.; Pratchett, M.S.; Kerr, A.M.; Rivera-Posada, J.A. The role of maternal nutrition on oocyte size and quality, with respect to early larval development in the coral-eating starfish, *Acanthaster planci*. *PLoS ONE* **2016**, *11*, e0158007. [CrossRef]
35. Pearson, R.G.; Endean, R. A preliminary study of the coral predator *Acanthaster planci* (L.) (Asteroidea) on the Great Barrier Reef. *Fish Notes* **1969**, *3*, 27–68.
36. Stump, R.J.W. *An Investigation to Describe the Population Dynamics of Acanthaster planci (L.) around Lizard Island, Cairns Section, Great Barrier Reef Marine Park*; CRC Reef Research Centre: Townsville, Australia, 1996.
37. Scharf, F.S.; Juanes, F.; Rountree, R.A. Predator size-prey size relationships of marine fish predators: Interspecific variation and effects of ontogeny and body size on trophic-niche breadth. *Mar. Ecol. Prog. Ser.* **2000**, *208*, 229–248. [CrossRef]
38. Berglund, A.; Rosenqvist, G. Reproductive costs in the prawn Palaemon adspersus: Effects on growth and predator vulnerability. *Oikos* **1986**, *46*, 349–354. [CrossRef]
39. Nordström, M.C.; Aarnio, K.; Törnroos, A.; Bonsdorff, E. Nestedness of trophic links and biological traits in a marine food web. *Ecosphere* **2015**, *6*, 161. [CrossRef]
40. Fuiman, L.A. The interplay of ontogeny and scaling in the interactions of fish larvae adn their predators. *J. Fish Biol.* **1994**, *45* (Suppl. SA), 55–79. [CrossRef]
41. Lundvall, D.; Svanbäck, R.; Persson, L.; Byström, P. Size-dependent predation in piscivores: Interactions between predator foraging adn prey avoidance abilities. *Can. J. Fish. Aquat. Sci.* **1999**, *56*, 1285–1292. [CrossRef]

42. McClanahan, T.R.; Muthinga, N.A. Patterns of predation on a sea urchin, *Echinometra mathei* (de Blainville), on Kenyan coral reefs. *J. Exp. Mar. Biol. Ecol.* **1989**, *126*, 77–94. [CrossRef]

43. Ling, S.D.; Johnson, C.R.; Frusher, S.D.; Ridgway, K.R. Overfishing reduces resilience of kelp beds to climate-driven catastrophic phase shift. *Proc. Natl. Acad. Sci. USA* **2009**, *106*, 22341–22345. [CrossRef] [PubMed]

44. Wilmes, J.; Matthews, S.; Schultz, D.; Messmer, V.; Hoey, A.S.; Pratchett, M.S. Modelling growth of juvenile crown-of-thorns starfish (0+ year cohort) on the northern Great Barrier Reef. *Diversity* **2017**, *9*, 1. [CrossRef]

45. Svensson, I. Reproductive costs in two sex-role reversed pipefish species (Sygnathidae). *J. Anim. Ecol.* **1988**, *57*, 929–942. [CrossRef]

diversity

MDPI

Review

Known Predators of Crown-of-Thorns Starfish (*Acanthaster* spp.) and Their Role in Mitigating, If Not Preventing, Population Outbreaks

Zara-Louise Cowan [1,*], Morgan Pratchett [1], Vanessa Messmer [1] and Scott Ling [2]

[1] ARC Centre of Excellence for Coral Reef Studies, James Cook University,
     Townsville, QLD 4811, Australia; morgan.pratchett@jcu.edu.au (M.P.); vanessa.messmer@gmail.com (V.M.)
[2] Institute for Marine and Antarctic Studies, University of Tasmania, Hobart, TAS 7001, Australia;
     scott.ling@utas.edu.au
*   Correspondence: zaralouise.cowan@my.jcu.edu.au; Tel.: +61-7-4781-5747

Academic Editors: Sven Uthicke and Michael Wink
Received: 20 November 2016; Accepted: 17 January 2017; Published: 22 January 2017

Abstract: Predatory release has long been considered a potential contributor to population outbreaks of crown-of-thorns starfish (CoTS; *Acanthaster* spp.). This has initiated extensive searches for potentially important predators that can consume large numbers of CoTS at high rates, which are also vulnerable to over-fishing or reef degradation. Herein, we review reported predators of CoTS and assess the potential for these organisms to exert significant mortality, and thereby prevent and/or moderate CoTS outbreaks. In all, 80 species of coral reef organisms (including fishes, and motile and sessile invertebrates) are reported to predate on CoTS gametes (three species), larvae (17 species), juveniles (15 species), adults (18 species) and/or opportunistically feed on injured (10 species) or moribund (42 species) individuals within reef habitats. It is clear however, that predation on early life-history stages has been understudied, and there are likely to be many more species of reef fishes and/or sessile invertebrates that readily consume CoTS gametes and/or larvae. Given the number and diversity of coral reef species that consume *Acanthaster* spp., most of which (e.g., *Arothron* pufferfishes) are not explicitly targeted by reef-based fisheries, links between overfishing and CoTS outbreaks remain equivocal. There is also no single species that appears to have a disproportionate role in regulating CoTS populations. Rather, the collective consumption of CoTS by multiple different species and at different life-history stages is likely to suppress the local abundance of CoTS, and thereby mediate the severity of outbreaks. It is possible therefore, that general degradation of reef ecosystems and corresponding declines in biodiversity and productivity, may contribute to increasing incidence or severity of outbreaks of *Acanthaster* spp. However, it seems unlikely that predatory release in and of itself could account for initial onset of CoTS outbreaks. In conclusion, reducing anthropogenic stressors that reduce the abundance and/or diversity of potential predatory species represents a "no regrets" management strategy, but will need to be used in conjunction with other management strategies to prevent, or reduce the occurrence, of CoTS outbreaks.

Keywords: *Acanthaster* (Acanthasteridae); fisheries closures; marine parks; predation; predator removal hypothesis; chemical defences; saponins; population regulation; top-down control; trophic cascades

1. Introduction

Adult crown-of-thorns starfish (CoTS; *Acanthaster* sp.) have numerous long, very sharp and toxic spines (Figure 1). In addition, the dermal tissues of CoTS (and all of their organs) contain high concentrations of chemicals, including saponins [1,2] and plancitoxins [3], which are both unpalatable [4] and highly toxic [5–7]. Intuitively therefore, one might expect that these starfish are effectively protected and largely immune from predation (e.g., [8]). In reality, there are few organisms that are completely immune to predation at any or all stages of their life cycle. Rather, well-developed anti-predatory defences reduce the range of predators to which prey species are vulnerable [9], but may or may not affect overall rates of predation and the extent to which prey populations are controlled by predators. Accordingly, there is an increasing number of coral reef organisms (fishes and invertebrates) reported to predate on CoTS [10,11], including some predators (e.g., *Arothron* pufferfishes) that feed almost exclusively on adult CoTS when they are in abundant supply (e.g., during outbreaks). Such predators may be important in supressing the abundance of prey species [12] as well as influencing the behaviour, habitat-associations, and population dynamics of even well-armoured and/or chemically defended prey species (e.g., [13]).

Despite their physical and chemical defences, post-settlement stages (juvenile and adults) of CoTS often exhibit injuries, largely manifested as missing arms [11,14,15]. These injuries are believed to occur when predators are only able to remove one or a few arms before the starfish escapes or avoids further damage by hiding within the reef matrix [15]. If however, there are high rates of partial predation at specific reef locations [11] then it is expected at least some CoTS will also be killed outright and/or consumed in entirety. The cryptic nature and nocturnal behaviour of CoTS, especially when small (<12 cm diameter) or at low densities [11,16] further suggests that they must be highly vulnerable to predators. In controlled experiments, survivorship of laboratory reared *Acanthaster* spp. settled to natural substrates is effectively zero, owing to very high rates of predation by naturally occurring predators [17–19]. Recent research also demonstrates that CoTS larvae are highly vulnerable to predation [20], despite having the highest concentrations of anti-predator chemicals (discussed later). Cowan et al. [20] showed that CoTS larvae are readily consumed by many common planktivorous damselfishes, and often in preference to other asteroid larvae.

While there is now general acceptance that CoTS are vulnerable to predation (e.g., [11,21]), on-going controversies relate to whether known predators would ever be capable of regulating CoTS populations, and mitigating, if not preventing outbreaks. More specifically, attention is focussed on whether anthropogenic impacts (via fishing or habitat degradation) have supressed the abundance of key predators, thereby accounting for the seemingly recent and/or increasing occurrence of CoTS outbreaks [10].

Figure 1. Adult crown-of-thorns starfish are defended against predators by numerous long, very sharp and toxic spines. Photographic credit: Scott Ling, Dick's Reef, Swains Region, southern Great Barrier Reef (22°18′ S, 152°39′ E).

1.1. The Predator Removal Hypothesis

The *predator removal hypothesis* was one of the first hypotheses proposed to account for CoTS outbreaks [22]. Following trophic-cascade concepts as a result of ecological extinction of functional echinoderm predators such as sea otters (e.g., [23–25]), lobsters [26] and large benthic predatory fishes [27], this hypothesis (like many other hypotheses put forward in the 1960s and 1970s, such as the *nutrient enrichment hypothesis*, e.g., [28–30]) is predicated on the idea that CoTS outbreaks are an unnatural phenomenon, caused by anthropogenic modification and degradation of coral reef environments [31]. The initial formulation of the *predator removal hypothesis* related to apparent overfishing of the giant triton (*Charonia tritonis*) on the GBR in the decades immediately preceding the first documented outbreak of CoTS in 1962 [22]. Notably, ~10,000 giant tritons were removed from the GBR each year from 1947 to 1960 by trochus fishermen and commercial shell collectors [22]. Densities of triton must have been significant to sustain this level of removal, or at the very least, much higher than they are now. While there is no empirical data on their abundance, *C. tritonis* are exceedingly rare on the GBR, and have been since the 1960s, perhaps reflecting the legacy of excessive removals in the 1950s [32].

Endean [22] argued that the effective loss of giant triton from reefs in the northern GBR relaxed normally strong regulatory pressure on abundance of juvenile and sub-adult CoTS, leading to increased abundance of large adult starfish that were capable of initiating outbreaks by virtue of their massive combined reproductive output. Adding weight to this hypothesis, outbreaks of CoTS were reported from other locations (e.g., Fiji and Western Samoa) where *C. tritonis* had also been extensively harvested, whilst outbreaks had not been reported in areas (e.g., Malaysia, Philippines and Taiwan) where *C. tritonis* were abundant [22]. The ability of *C. tritonis* to provide the sufficient top-down control necessary to regulate CoTS populations has since been questioned (e.g., [33]) largely based on their generally low rates of feeding and the apparent reluctance to eat CoTS when provided with alternative prey.

Though the role of giant triton in regulating abundances of CoTS (past, present, or future) is still not resolved, the *predator removal hypothesis* has evolved through time to place increasing emphasis on fish predators. Attention has focused on large predatory fishes capable of consuming adult starfish (e.g., [33–35]), which are targeted by fisheries and/or have declined in abundance due to localized fishing activities. Explicit and direct evidence that any of the major fisheries target species (e.g., coral trout, *Plectropomus* sp.) are significant predators of crown-of-thorns starfish is meagre [34]. However, some studies [33,36,37] have reported increased incidence and/or severity of outbreaks of CoTS along gradients of increasing fishing effort. On the GBR, Sweatman [37] showed that reefs open to fishing were seven times more likely to experience an outbreak of crown-of-thorns starfish (57% of reefs affected) compared to reefs effectively closed to fishing within no-take marine reserves (8% of reefs affected). While the mechanistic basis of these patterns has not been critically tested, increasing evidence of links between fishing and starfish outbreaks [36,37] has fuelled significant interest in predation, both to understand the cause(s) and ultimately manage CoTS outbreaks.

1.2. Objectives of This Review

The purpose of this review is to synthesise existing knowledge of potentially important CoTS predators, considering their individual and collective capacity to influence population dynamics of CoTS. There is an ever-increasing list of putative predators (e.g., [11,38,39], Table 1), largely based on anecdotal observations of different coral reef organisms (mainly fishes) feeding on dead or dying CoTS within reef environments. Our intention in this review is to differentiate between organisms that opportunistically feed on dead or injured CoTS (scavengers), versus those predators that feed on live and healthy starfish and either kill them outright or reduce their individual fitness and/or reduce population level fitness by altering patterns in abundance and distribution. It is possible, for example, that the mere presence of benthic predators could disperse adult CoTS that might otherwise aggregate to spawn, and thereby reduce fertilization success. Moreover, this review will explicitly consider potential predators at different stages in the life cycle of CoTS, especially pre-settlement (e.g., gametes, larvae) and early post-settlement life stages, which is quite possibly the most significant bottleneck in their life-history [40–42]. Where possible, we report or derive estimates of the rates of mortality due to predation across different life-history stages of CoTS.

Having established the range of putative CoTS predators, this review will consider empirical and theoretical evidence that supports (or refutes) the potential role of predators in moderating (if not preventing) CoTS outbreaks. If predation underlies observed differences in the incidence or severity of outbreaks across gradients of fishing pressure [36,37], we would expect to find that the specific predators would be significantly more abundant in areas with little or no fishing, with corresponding increases in effective rates of predation on juvenile and/or adult CoTS within these areas. Persistent controversy around the role of predation in regulating abundance of CoTS (e.g., [11,36,37]) highlights many deficiencies in previous research approaches and points to the definitive need for experimental studies that explicitly test the mechanistic underpinnings of the *predator removal hypothesis*.

2. Known Predators of Crown-of-Thorns Starfish

A total of 80 species of coral reef organisms are reported to feed on CoTS, including 24 motile and sessile invertebrates versus 56 species of coral reef fishes (Table 1). However, most species have been observed feeding on moribund and dead individuals in the field, while observations of predation on healthy, uninjured starfish are comparatively rare. Similarly, field observations of species feeding on the gametes of CoTS are also extremely limited, and field observations of predation upon larvae are simply not feasible. Gut content analysis has also been largely unsuccessful in identifying putative predatory species (e.g., [43]). However, there have been significant advances in the field of environmental DNA (eDNA) analysis in recent years [44] and it is likely that this technique could be utilised to both identify previously unknown predators, and establish the frequency with which known predators actually consume CoTS. One suggested method to do this would be to collect faeces of presumed CoTS predators and test this for presence of CoTS DNA. However, there are limitations to this technique. Most notably, it is not possible to distinguish between particular life stages of prey species, nor whether specific prey species were alive or dead when consumed [44], which is important in understanding the role of predators in structuring populations of CoTS. Such experiments should be supplemented with benthic surveys to confirm presence of juvenile or adult starfish, and plankton tows to confirm presence/absence of CoTS larvae (see [45]). This could provide an indication of the CoTS life stage from which DNA found in predator faeces has originated, and would be particularly beneficial for predators such as damselfish which may prey upon both pre- and post-settlement life stages (e.g., [20,28]).

Table 1. Species that feed on different life stages and states of health of *Acanthaster* spp. "F" denotes that the particular predator has been directly observed feeding on a particular life stage in the field, which also includes where starfish were made unnaturally available; "L" denotes where feeding is inferred based on studies in a laboratory/aquarium; "G" denotes that *Acanthaster* remains have been recovered from the stomach of the predator; I = not directly witnessed.

Predator	Sperm	Eggs	Larvae	Juvenile	Adult–Healthy	Adult–Injured	Adult–Moribund/Dead	Reference
Fishes								
Angelfish								
Holacanthus passer						L		[39]
Pomacanthus semicirculatus							F	[38]
Pomacanthus sexstriatus							F	[38]
Bream								
Scolopsis bilineatus	F						F	[46]
Butterflyfish								
Chaetodon aureofasciatus							F	[46]
Chaetodon auriga						L	FL	[11,38,46]
Chaetodon auripes						L		[47]
Chaetodon citrinellus								[39]
Chaetodon plebeius							F	[46]
Chaetodon rafflesi							F	[46]
Chaetodon rainfordi							F	[46]
Chaetodon vagabundus							FL	[46,48]
Damselfish								
Abudefduf sexfasciatus		F	L					[20,22]
Acanthochromis polyacanthus			L				L	[20,48]
Amblyglyphidodon curacao		F	L					[20,28]
Chromis atripectoralis			L					[20]
Chromis caerulea						L	F	[11,38]
Chromis dimidiata			L					[49]
Chromis viridis			L					[20]
Chrysiptera rollandi			L					[20]
Dascyllus aruanus			L					[20]
Dascyllus reticulatus			L					[20]
Neoglyphidodon melas							F	[46]
Neoglyphidodon oxyodon							F	[46]

Table 1. *Cont.*

Predator	Sperm	Eggs	Larvae	Juvenile	Adult–Healthy	Adult–Injured	Adult–Moribund/Dead	Reference
Neopomacentrus azysron			L					[20]
Pomacentrus amboinensis			L					[20]
Pomacentrus chrysurus						L	F	[46]
Pomacentrus moluccensis			L				FL	[11,20,38,46]
Pomacentrus wardi							F	[46]
Stegastes acapulcoensis						F		[39]
Stegastes nigricans							F	[46]
Emperors								
Lethrinus atkinsoni				F			FL	[46,50]
Lethrinus miniatus				F				[50,51]
Lethrinus nebulosus					G		F	[46,51]
Goatfish								
Parupeneus multifasciatus							F	[46]
Gobies								
Cryptocentrus sp.							F	[38]
Groupers								
Epinephelus lanceolatus				FG				[52]
Pufferfish								
Arothron hispidus				F	FL	L	FL	[11,39,43,46, 48,53]
Arothron manilensis							L	[46]
Arothron meleagris							F	[39]
Arothron nigropunctatus							F	[38]
Arothron stellatus					F			[47]
Triggerfish								
Balistapus undulatus							F	[46]
Balistoides viridescens				L	FL		L	[11,43,46,48, 53]
Pseudobalistes flavimarginatus				L	FGL			[11,43,54]
Rhinecanthus aculeatus							L	[48]
Sufflamen verres							F	[39]

Table 1. Cont.

Predator	Sperm	Eggs	Larvae	Juvenile	Adult–Healthy	Adult–Injured	Adult–Moribund/Dead	Reference
Wrasses								
Cheilinus diagrammus							F	[38]
Cheilinus fasciatus							F	[38]
Cheilinus undulatus					FG			[43,51,55,56]
Coris caudomacula							F	[46]
Halichoeres melanurus							L	[48]
Thalassoma hardwicke							F	[38]
Thalassoma lucasanum						FL		[11,39]
Thalassoma lunare							FL	[46,48]
Thalassoma nigrofasciatum							F	[46]
Motile invertebrates								
Acanthaster planci							F	[38]
Alpheus sp.							F	[38]
Bursa rubeta				I	I			[57] in [21]
Cassis cornuta					L			[55]
Charonia tritonis				F	FL		F	[22,28,32,38]
Cymatorium lotorium					I			[43]
Dardanus sp.					I			[43]
Diadema mexicanum							F	[39]
Dromidiopsis dormia					I			[57] in [21]
Hymenocera elegans/picta				L	F	F		[39,58,59]
Murex sp.					L			[55]
Neaxius glyptocercus				I	I			[60] in [21]
Panulirus penicillatus				L				[18]
Pherecardia striata				F		F	F	[39,59]
Trapezia flavopunctata			L					[19]
Trapezia bidentata			L					[19]
Trapezia cymodoce			L					[19]
Trizopagurus magnificus							F	[39]
Xanthidae				L				[49]
Sessile invertebrates								
Paracorynactis hoplites				F	F			[61]
Platygyra sp.			L					Cowan Pers. obs.
Pocillopora damicornis			L					[8]
Pseudocorynactis sp.					F			[62]
Stoichactis sp.					L			[55]

2.1. Pre-Settlement Predation

Most of the putative CoTS predators feed or scavenge on post-settlement life stages (juveniles and adults) compared to pre-settlement stages (Figures 2 and 3). However, this probably reflects limited research directed at identifying potential predators on CoTS eggs and larvae and/or difficulties in documenting predation on these early life-history stages. Coral reefs typically support very high abundance and diversity of planktivorous species, including many different reef fishes (e.g., [20,28,47,63]) as well as sessile invertebrates, such as corals [8], which may consume CoTS propagules during spawning, as well as feeding on CoTS larvae when they settle. CoTS are one of the most fecund invertebrates [64], with very high fertilisation rates [65,66], but intuitively, most eggs and larvae must fail to survive. As for other marine species with planktonic larvae, significant rates of pre-settlement mortality are also likely to arise due to predation [67,68]. Yet, given their exceptional reproductive potential [64], even small changes in the proportion of larvae that survive and settle will lead to vast differences in the absolute number of juvenile and adult starfish.

Early studies suggested that CoTS eggs and larvae were largely immune from predation due to unpalatable chemical defences (saponins) contained within [4]. However, more recent examination of predation upon both eggs [69] and larvae [20] reveals that these are indeed readily consumed by a range of highly abundant, planktivorous damselfishes. Given that this group of predators can be extremely abundant on coral reefs, it is likely that they play an important role in reducing the proportion of CoTS that survive through to settlement, and high densities of damselfishes should be considered important for the buffering capacity of coral reefs. Given that all the planktivorous damselfishes tested in the recent studies [20,69] consumed CoTS material, albeit to varying extents, it is highly likely that there are more predators of the pre-settlement stages that are yet to be identified. Furthermore, the actual suite of predators that prey upon the early life stages of CoTS is likely to span a far greater taxonomic range, from benthic species such as those already identified (e.g., [8,19], Table 1) to large pelagic fishes, such as manta rays and whale sharks.

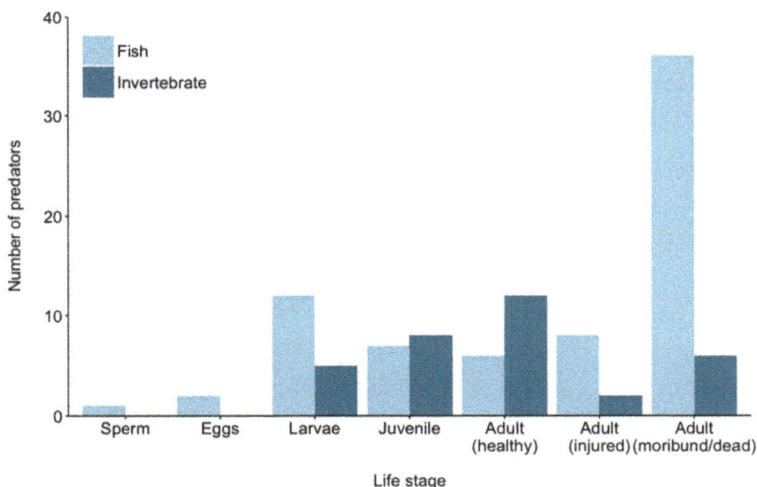

Figure 2. Putative predators of crown-of-thorns starfish across each major life stage. Predators at each life stage are not mutually exclusive. This figure is based on references from Table 1.

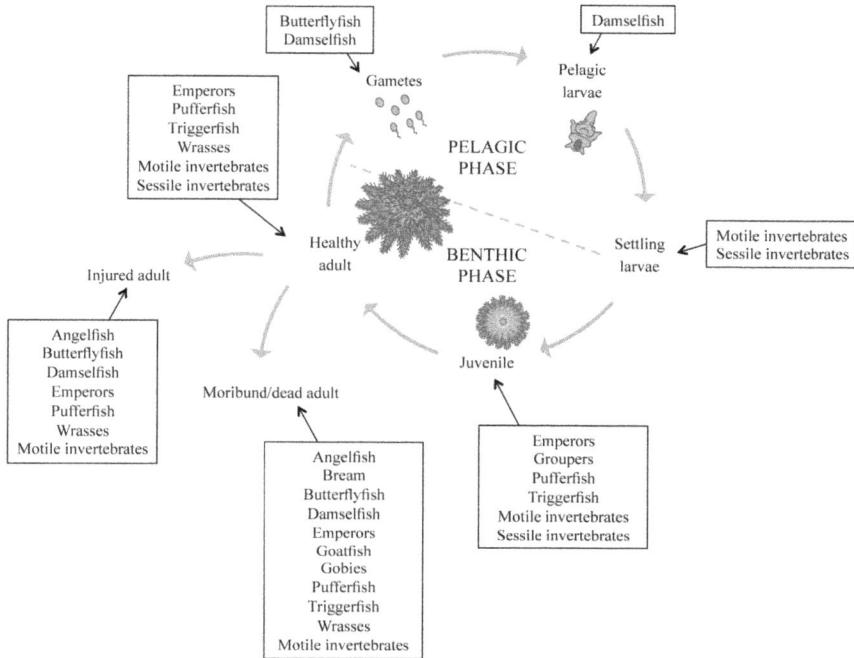

Figure 3. Main predatory groups acting at different life stages of crown-of-thorns starfish. This figure is based on references from Table 1.

2.2. Post-Settlement Predation

A key component of the vulnerability of larvae is their susceptibility to predation during settlement. Larvae preferentially settle in habitats with fine-scale topographic complexity [49], which is likely to be an adaptation to minimise early post-settlement mortality. However, a wide range of potential predators are abundant within the reef matrix (e.g., [11,17,39,70]). Benthic predators and filter feeders, including corals such as *Pocillopora damicornis*, may exact a heavy toll on the settling larvae of coral reef asteroids, including *Acanthaster* spp. [8]. Keesing and Halford [47] estimate mortality at settlement and during metamorphosis to be in excess of 85%. Furthermore, Cowan et al. [19] observed that 55% of brachiolaria larvae that settled to rubble with naturally attached crustose coralline had undergone metamorphosis after 48-h when substrates had been cleaned of potential predators, compared to 0% when polychaete predators were present. If predation represents a significant threat to settling larvae, this should have provided sufficient selective pressure for the evolution of behavioural mechanisms to evade predators, as seen in juveniles and adults in low-density populations. Cowan et al. [19] used static choice chambers to determine whether such mechanisms are present to assist larvae in avoiding settlement on or near predators, revealing that larvae are able to both detect, and respond to, the presence of predators within settlement substrates; larvae were attracted to rubble with naturally attached crustose coralline algae that had been cleaned of predators, but were deterred from these substrates when polychaete predators were present. Whether similar mechanisms are present to enable juveniles to detect and actively avoid predators in the reef matrix remains to be tested.

Following settlement, many marine organisms experience very high early post-settlement mortality (e.g., asteroids [8]; bivalves [71]; fish [72]; corals [73]), to the extent that this is a recognised demographic bottleneck for some taxa [73]. Although Sweatman's [50] measurements of predation by fishes on laboratory-reared juvenile CoTS (25–79 mm diameter) revealed very low predation rates (0.13% per day), predation on juvenile CoTS is due mainly to epifaunal invertebrate predators, which are highly abundant on coral reefs (e.g., [17,70]). For example, survivorship of laboratory reared CoTS that are settled to freshly collected rubble is effectively zero, owing to very high rates of predation on newly settled larvae by naturally occurring predators (e.g., polychaetes) within the rubble [17].

In the eastern Pacific, Glynn [39] demonstrated that the harlequin shrimp *Hymenocera picta* and a polychaete worm *Pherecardia striata* were significant predators of CoTS [39,59,74]. Moreover, there have not been any outbreaks, or persistently high numbers of CoTS on reefs in Panama, where these two predators are found in high abundance, and where populations of alternative prey (ophidiasterids) are scarce. Although both field [74] and laboratory studies [43,74,75] demonstrate that the shrimp has difficulties attacking larger and more active CoTS, it is expected that they are highly effective predators of smaller, more cryptic CoTS within the reef matrix [39]. The observed preference of *H. picta* for other asteroid prey species, particularly smaller ophidiasterids [43], and the strong preference of *P. striata* for crustacean tissue over tissue from CoTS [39] emphasises how the relative scarcity of alternative prey may be an important factor influencing the capacity of a potential predator to manipulate the population dynamics of CoTS [74]. If they are important in regulating populations of CoTS, this may occur within a relatively restricted geographical area.

The polyp *Pseudocorynactis* sp. may also play an important role in the population control of CoTS [62] in areas where it occurs in high abundances, such as Sogod Bay, Philippines. *Pseudocorynactis* sp. prey on a range of echinoderms and has been observed ingesting adult CoTS up to 250 mm in diameter [62]. Furthermore, *Pseudocorynactis* sp. preferentially settles under coral ledges and in reef crevices, where it likely predates on cryptic juvenile and sub-adult CoTS [62].

3. Rates of Predation on Crown-of-Thorns Starfish

Understanding of the importance of predation in the population dynamics of CoTS is significantly constrained by a lack of empirical data on background mortality rates and natural predation rates. Quantifying mortality rates of CoTS in the field is tractable, but requires significant investment to follow the fate of a large number of uniquely tagged or recognizable individuals (*sensu* [76]) across a broad size range of individuals and in different habitats. The biggest limitation to such experiments is the limited capacity to tag CoTS (e.g., [77]). Previous methods used to distinguish individual starfish (staining, branding, tagging and dyeing) were only effective for days to weeks, limiting any capacity to get meaningful long-term data on rates of mortality [77]. Identifying more permanent tagging solutions is essential, but even short-term tagging and tethering experiments could yield important information about the vulnerability of CoTS to predation.

Short-term tagging and tethering experiments have been used to effectively estimate or compare predation vulnerability for a range of echinoderms, especially echinoids or urchins (e.g., [76,78,79]). These studies typically use very short observations (≤3 days) to estimate predation rates, though some recent studies have been conducted over several months; Ling and Johnson [76] successfully tagged and tethered the urchin *Centrostephanous rodgersii*, and measured subsequent survivorship >100 days. At present, there are few direct, quantitative estimates of predation or mortality rates for CoTS in the field, which are critical for establishing the importance of predation in limiting the densities of starfish at individual reefs [21].

Previous research on CoTS predators has focused on estimating maximum rates of starfish consumption by individual predators and extrapolating this to account for natural densities of these predators (e.g., [28,80,81]). This is based on the belief that effective control of outbreak populations is fundamentally reliant on high rates of predation to compensate for rapid population growth of CoTS when outbreaks begin. However, it is not the rate of feeding per se that is important in determining whether a predator can effectively regulate prey densities [12], but changes in the rates of predation in response to spatial and temporal gradients in prey abundance. Predators that are capable of consuming large numbers of CoTS, by virtue of their high abundance and/or individual capacity to consume large numbers of CoTS [20], may be important in suppressing local CoTS densities. Cowan et al. [20] showed that a wide range of planktivorous damselfishes will feed on CoTS larvae, some of which (*D. aruanus* and *P. amboinensis*) have very high satiation limits. The capacity of these fishes to consume large numbers of CoTS larvae, combined with high densities of these fishes (as well as many other planktivorous fishes that may also readily feed on CoTS larvae) could be critical in limiting settlement rates [42,82] and thereby moderating the local severity of outbreaks. It is unlikely however, that these fishes could actually prevent an outbreak from ever occurring, unless either (i) they selectively target CoTS larvae (over other potentially more abundant prey) even at very low prey densities, and thereby prevent initiation of outbreaks; or (ii) their combined feeding capacity exceeds the very large number of CoTS larvae that can cause the rapid and pronounced onset of some outbreaks.

Early research on putative CoTS predators largely dismissed the importance of generalist reef predators, suggesting instead that important predators would have to be highly specialized (preferentially feeding on CoTS to the exclusion of almost all other potential prey) and feed on adult CoTS [80]. Importantly, if there are specialist predators that are highly effective in finding and killing CoTS even when they occur at very low densities, then it may be these predators that are key in preventing outbreaks from ever occurring [82]. However, effective CoTS predators must be sufficiently generalist to consume alternative prey [82] and thereby sustain themselves during non-outbreak periods when CoTS are scarce. If predation (or lack thereof) is a potential cause of CoTS outbreaks, it seems that we should also be focusing on predators that target pre-settlement, settlement, and post-settlement pre-reproductive stages [41,42]. Most notably, predation by benthic invertebrates on newly settled starfish appears, at present, to be the most significant bottleneck in their life-history [42], but this may be largely attributable to underestimates of predation rates on CoTS during other life-stages.

Quantitative data on predation rates is rare and in most cases comes from experimental studies that aim to determine maximum predation rates by specific organisms (e.g., [17,20,28,47,61]), or modelling efforts that predict the rate of predation needed to prevent outbreaks (e.g., [33,41,82,83]). Based on caging experiments, the triton shell *C. tritonis* is estimated to consume 0.7 adult starfish per week [28], however attacks are not always fatal [55] and this predator prefers to feed on other starfish when given a choice [28]. Furthermore, pre-fishing numbers of *C. tritonis* remain largely unknown and it is unlikely that this invertebrate was ever present in sufficiently high numbers to prevent outbreaks [84,85]. Starfish numbers are persistently low in areas where the corallimorph *P. hoplites* [61] or harlequin shrimp *H. picta* together with the lined fireworm *P. striata* [59] are abundant. Consumption rates of adult starfish (up to 340 mm diameter) by *P. hoplites* are estimated to be 29.5 g day^{-1} [61]. In the eastern Pacific, Glynn [59] reported that 5%–6% of the CoTS were being attacked by *H. picta* at any time, and 0.6% of the starfish population were preyed upon by both *H. picta* and *P. striata*. Approximately 50% of CoTS being attacked by *H. picta* ultimately died, compared to close to 100% for CoTS being attacked by both *H. picta* and *P. striata*, and these two predators are particularly effective in regulating numbers of juvenile starfish [59]. Keesing and Halford [17] measured significant mortality rates on post-metamorphic starfish (5.05% day^{-1} on 1-month old starfish and 0.85% day^{-1} on 4-month old starfish), due to predation by epibenthic fauna contained within dead coral rubble. This is much higher than the attack rate on juvenile starfish (1%–1.5%) that McCallum [40] suggested would be sufficient to limit the occurrence of CoTS outbreaks, based on demographic modelling. Cowan et al. [20] conducted a series of laboratory experiments, measuring maximal predation rates by a wide range of planktivorous damselfishes upon CoTS larvae, reporting consumption rates ranging from 14 larvae h^{-1} by *C. viridis*, up to 158 larvae h^{-1} by *D. aruanus*.

Ormond and Campbell [43] suggest that the predatory behaviour of large fishes (the pufferfish *A. hispidus*, and triggerfishes *P. flavimarginatus* and *B. viridescens*) may have the capacity to control densities of adult CoTS in the Red Sea, and may be capable of disbanding aggregations in their early stages. It is however, unlikely that these species would be effective in controlling CoTS populations across the entire Indo-Pacific region as they are not universally common [86].

However, triggerfishes (particularly, *Balistapus undulatus*) may nonetheless be important maintaining the structure of coral reef ecosystems, by predating on rock-boring urchins [87]. There may be other large predatory fishes capable of regulating densities of adult CoTS; Keesing and Halford [47] observed the pufferfish *A. stellatus* to be capable of consuming adult starfish (20 cm diameter) in less than 10 miin. Further, Ormond et al. [33] reported greater mean densities of lethrinids and large fish predators from the Red Sea, where no major outbreaks of CoTS were known to have occurred, compared to the GBR, where two cycles of large scale outbreaks had occurred [33]. Further, within the GBR, mean predatory fish densities were found to be reduced on reefs that were experiencing major outbreaks [33]. Fish species examined were commercially targeted or frequently caught as bycatch, and variations in population densities of predatory fish between sites was compatible with fisheries data on fishing intensity, therefore the pattern of difference in CoTS populations could be explained by differing fishing intensities between locations [33].

3.1. Sub-Lethal Injuries

Discussions to this point have focused on instances of whole animal or lethal predation, but sublethal (or partial) predation is often very apparent and well documented among echinoderms [88,89]. For CoTS, very high proportions of adults (up to 60%) have evidence of recent injuries (most apparent as missing arms), which is attributed to partial predation [15]. Even if the predation is not immediately fatal, sublethal attacks may still have an important influence on population dynamics. In the short term, open injuries and exposed internal organs may increase the likelihood of pathogenic infections and disease transmission among individuals [90–92] and can also increase susceptibility to further attacks [39]. Even if starfish effectively repair injuries caused by partial predation, effective declines in the size of individuals caused by sub-lethal predation will reduce food intake [88,93,94]. Crown-of-thorns starfish also regrow damaged or missing arms [28,95], which will require re-allocation of nutrients to regeneration, which could otherwise have been used for immune defence [11], reproduction or somatic growth [88,93]. The removal of arm tips, and consequent removal of the eyespot, may additionally result in reduced foraging efficiency due to the loss of vision, which is important for navigating between reef structures and locating prey [96].

There are strongly opposing views about the inferences of high incidence of partial predation in populations of crown-of-thorns starfish. In general, high incidence of sub-lethal predation has been considered to be generally reflective of high intensity of predation (e.g., [11,15]), such that high rates of partial mortality serve as a proxy for high levels of overall mortality. In the Philippines, Rivera-Posada et al. [11] showed that the incidence of injuries on CoTS was highest within a no-take marine reserve, and supporting information on the high diversity and abundance of reef fishes corroborate the idea that overall intensity of predation would likely be much higher inside versus outside of this reserve. Conversely, high incidence of partial mortality may reflect low intensity of predation pressure [16] because when predation is intense it would be expected that most predation events would result in complete mortality. As such, high incidences of starfish with partial injuries point to the strong regenerative capacity of starfish [14] and may suggest that predation is predominantly sub-lethal for crown-of-thorns starfish. Schoener [97] suggested that there is no reason to expect any relationship between rates of injuries versus rates of predation-driven mortality, because the efficiency of predation (the proportion of attacks that cause death) is independent of the attack rate or incidence of predation. For the asteroid *Asterias amurensis*, high density populations have been observed to be completely annihilated by an incursion of thousands of spider crabs (*Leptomithrax gaimardii*) that moved to shallow reefs in large numbers [89]. Additionally, where crab incursions involved fewer individuals, high rates of sub-lethal injury (~70% of starfish population injured) occurred independently of mass predator-driven mortality [89]. In this example, the high mobility of spider crabs can lead to overwhelming local impacts on starfish populations but this impact is highly variable across space and time [89]. Ultimately, a dedicated research project is needed to test the relationship between rates of partial versus complete predation mortality. However, in the absence of any empirical data on overall rates of predation mortality, the incidence of injuries serves as the best proxy to test for variation in vulnerability to predation among locations and with size of starfish [15].

A common trend among echinoderm species is for the incidence of sub-lethal injuries to decline with increasing body size [14,78,88,89,93,98] reflecting general declines in the vulnerability to predation. Accordingly, Messmer et al. [14] observed a clear linear decline in both rate and severity of predation with increasing starfish size. Rivera-Posada et al. [11] reported highest incidence of arm damage in the intermediate (11–20 cm diameter) size class however. This pattern may be explained by changes in behavioural and physical characteristics with increasing size, whereby intermediate sized individuals may have greatest exposure to predators, while smaller starfish (<10 cm) tend to remain hidden but are also more likely to be completely consumed [16]. Reduced incidence of sublethal predation in largest individuals ($\geq$21 cm) may be explained by increased length of spines [11] and satiation of predators following removal of a smaller proportion of the total body mass. Disparity between the trends observed in these studies may be a result of differing sample size or differing suites of predators between the locations on the GBR [14] versus in the Philippines [11].

Sub-lethal and/or 'trait-mediated' effects of predators [89] can include alteration of behaviour and spatial patterns in echinoderms, in addition to changes in adult abundances and local spatial distributions, which are important for the reproductive ecology of free-spawning marine invertebrates as they influence rates of fertilization success [99]. For CoTS, fertilization success may be close to 100% when spawning individuals are adjacent to each other [100], thus any predator that is capable of dissipating an aggregation, or sufficiently reducing adult density, is likely to have a significant impact on zygote production. Humans have been shown to indirectly alter spatial distributions of asteroids, leading to much higher rates of fertilization in human-driven hotspots of zygote production [97]. In this way, sub-lethal predatory effects, leading to reduced individual reproductive performance, in combination with alteration to spatial configurations at the time of spawning may impact on zygote production for asteroids [99], potentially influencing the occurrence of secondary outbreaks for CoTS [101].

3.2. Population Modelling

Population simulation models provide a means of exploring the possible role of predation (or other natural causes of mortality) in regulating populations of CoTS (e.g., [82]), especially given little or no empirical data on satiation levels and feeding efficiency of potential predators or predation rates on larval and juvenile CoTS. Simulation models indicate that changes to predation rates during the pre-reproductive, post-settlement stage may be particularly relevant in understanding the dynamics of CoTS populations [40,41,82]. Notably, given their exceptional reproductive capacity, small changes in proportional survivorship or settlement success of CoTS larvae will result in large changes in adult abundance. Accordingly, McCallum [41] argues that relatively subtle changes in the abundance of predators (e.g., caused by exploitation) and/or predation rates, will reduce the level of local recruitment required to overcome (or satiate) predators.

The potential ecological importance of predation as a regulatory factor upon CoTS populations depends largely upon the ability of predators to find and consume prey [102]. Quantification of the functional response, described by the intake rate of prey as a function of prey density [103], is a common method that provides insight into the dynamics of predator–prey systems [104–106]. Functional responses may be classified into three types. A type I response assumes a linear increase in the intake rate with increasing food density, generally up to a maximum value, beyond which the intake rate is constant [107]. Type II is characterized by a decelerating intake rate [107] and assumes that the predator is limited by its ability to process food [108,109]. Type III is associated with an accelerating intake rate, associated with prey switching behaviour (preferential consumption of the most common type of prey [110]), up to a saturation point [107]. Predators that exhibit prey-switching behaviour, feeding more frequently on CoTS in response to a marked increase in their local abundance, may be capable of dissipating an aggregation in its early stages [53]. In order to persist when population densities of CoTS are low, predators should be able to take a range of prey, exhibiting an increased feeding response in reaction to a rapid increase in the CoTS population [53]. Predators exhibiting these type II and type III functional responses [103] are typical of vertebrate predators, reinforcing the focus on fish, and are supported in McCallum's [15,82] population models. In laboratory feeding experiments on *Acanthaster* spp. larvae, planktivorous damselfishes exhibit primarily a type II functional response, indicating their capacity to consume sufficient larvae to suppress settlement rates when larvae are already scarce, thereby contributing to very low natural densities of CoTS recorded outside of outbreak periods [20]. However, very high densities of larvae, which are a necessary condition for the rapid and pronounced onset of outbreaks, are likely to swamp even the combined consumption capacity of all planktivorous reef fishes [40].

4. Conclusions

Crown-of-thorns starfish are vulnerable to predation from a wide range of coral reef organisms, and at all stages of their life cycle. Despite identification of potentially key predators, and groups of predators, natural predation rates in both outbreaking and non-outbreaking populations remain largely unknown. This considerably limits our understanding of the role of predation in structuring the population dynamics of CoTS and of approaches to managing their often-devastating impacts on coral reef ecosystems. Whilst predation is likely to be important in suppressing settlement rates and contributing to naturally low densities of CoTS, the initiation and spread of outbreaks cannot, at present, be definitively attributed to changes (presumably declines) in the abundance of predators and/or changes in predation rates (e.g., [111]). Babcock et al. [111] showed that there are likely to be multiple factors that contribute to outbreaks of CoTS, such that a diverse range of management strategies will be required to prevent, or reduce the occurrence, of outbreaks. Maximising the number and diversity of putative CoTS predators is nonetheless, a "no regrets" strategy to minimise the risk of CoTS outbreaks and increase the resilience of coral reef ecosystems, generally. In the meantime, further research into potential predators, as well as estimates of predation and mortality across all life-stages of CoTS, is still warranted. More specifically, rates of pre- and post-settlement predation should be explicitly compared along known gradients in abundance of putative CoTS predators (e.g., inside and outside of marine reserves with marked differences in the abundance and diversity of fishes that feed on CoTS). New technologies provide improved opportunities to explore spatial and temporal variation in the demographics of CoTS populations, for example, DNA screening of diets for large numbers of potential CoTS predators [44,112], plus increased potential to tag and track benthic species within reef environments [113]. Such novel approaches along with remote sensing techniques will provide new insights into changes in the population dynamics and/or environmental conditions during the onset of population outbreaks.

Acknowledgments: This research was supported by the National Environmental Science Programme Tropical Water Quality Hub, as well as Crown-of-Thorns Starfish Research Grants awarded to M.P., Z.L.C. and V.M. by Lizard Island Research Stations and the Australian Museum. Discussions with Peter Doherty and David Westcott were critical to the development of ideas put forward in this review.

Author Contributions: All authors contributed equally to the concept and layout of this review. Z.L.C. compiled and analysed the data; Z.L.C. and M.P. wrote the paper; S.L. and V.M. edited the paper.

Conflicts of Interest: The authors declare no conflict of interest.

References

1.   Howden, M.E.H.; Lucas, J.S.; McDuff, M.; Salathe, R. Chemical defences of *Acanthaster planci*. In *Crown-of-Thorns Starfish Seminar Proceedings*; Australian Government Publishing Service: Canberra, Australia, 1975; pp. 67–79.

2.   Barnett, D.; Dean, P.W.; Hart, R.J.; Lucas, J.S.; Salathe, R.; Howden, M.E.H. Determination of contents of steroidal saponins in starfish tissues and study of their biosynthesis. *Comp. Biochem. Physiol. Part B Comp. Biochem.* **1988**, *90*, 141–145. [CrossRef]

3. Shiomi, K.; Yamamoto, S.; Yamanaka, H.; Kikuchi, T. Purification and characterization of a lethal factor in venom from the crown-of-thorns starfish (*Acanthaster planci*). *Toxicon* **1988**, *26*, 1077–1083. [CrossRef]
4. Lucas, J.; Hart, R.; Howden, M.; Salathe, R. Saponins in eggs and larvae of *Acanthaster planci* (L.) (Asteroidea) as chemical defences against planktivorous fish. *J. Exp. Mar. Bio. Ecol.* **1979**, *40*, 155–165. [CrossRef]
5. Mackie, A.M.; Singh, H.T.; Fletcher, T.C. Studies on the cytolytic effects of seastar (*Marthasterias glacialis*) saponins and synthetic surfactants in the plaice *Pleuronectes platessa*. *Mar. Biol.* **1975**, *29*, 307–314. [CrossRef]
6. Shiomi, K.; Yamamoto, S.; Yamanaka, H.; Kikuchi, T.; Konno, K. Liver damage by the crown-of-thorns starfish (*Acanthaster planci*) lethal factor. *Toxicon* **1990**, *28*, 469–475. [CrossRef]
7. Shiomi, K.; Midorikawa, S.; Ishida, M.; Nagashima, Y.; Nagai, H. Plancitoxins, lethal factors from the crown-of-thorns starfish *Acanthaster planci*, are deoxyribonucleases II. *Toxicon* **2004**, *44*, 499–506. [CrossRef] [PubMed]
8. Yamaguchi, M. Early life histories of coral reef asteroids, with special reference to *Acanthaster planci* (L.). In *Biology and Geology of Coral Reefs*; Jones, O., Endean, R., Eds.; Academic Press: New York, NY, USA, 1973; pp. 369–387.
9. Bertness, M.D.; Garrity, S.D.; Levings, S.C. Predation pressure and gastropod foraging: A tropical-temperate comparison. *Evolution* **1981**, *35*, 995–1007. [CrossRef]
10. Pratchett, M.S.; Caballes, C.F.; Rivera-Posada, J.A.; Sweatman, H.P.A. Limits to understanding and managing outbreaks of crown-of-thorns starfish (*Acanthaster* spp.). *Oceanogr. Mar. Biol. Annu. Rev.* **2014**, *52*, 133–200.
11. Rivera-Posada, J.; Caballes, C.F.; Pratchett, M.S. Size-related variation in arm damage frequency in the crown-of-thorns sea star, *Acanthaster planci*. *J. Coast. Life Med.* **2014**, *2*, 187–195.
12. Chesson, P. Mechanisms of maintenance of species diversity. *Annu. Rev. Ecol. Syst.* **2000**, 343–366. [CrossRef]
13. Pekár, S.; Liíznarová, E.; Řezáč, M. Suitability of woodlice prey for generalist and specialist spider predators. *Ecol. Entomol.* **2015**, *41*, 123–130. [CrossRef]
14. Messmer, V.; Pratchett, M.; Chong-Seng, K. Variation in incidence and severity of injuries among crown-of-thorns starfish (*Acanthaster* cf. *solaris*) on Australia's Great Barrier Reef. *Diversity* **2016**, under review.
15. McCallum, H.I.; Endean, R.; Cameron, A.M. Sublethal damage to *Acanthaster planci* as an index of predation pressure. *Mar. Ecol. Prog. Ser.* **1989**, *56*, 29–36. [CrossRef]
16. Birkeland, C.; Lucas, J.S. *Acanthaster Planci: Major Management Problem of Coral Reefs*; CRC Press: Boca Raton, FL, USA, 1990.
17. Keesing, J.; Halford, A. Field measurement of survival rates of juvenile *Acanthaster planci*, techniques and preliminary results. *Mar. Ecol. Prog. Ser.* **1992**, *85*, 107–114. [CrossRef]
18. Zann, L.; Brodie, J.; Berryman, C.; Naqasima, M. Recruitment, ecology, growth and behavior of juvenile *Acanthaster planci* (L.) (Echinodermata: Asteroidea). *Bull. Mar. Sci.* **1987**, *41*, 561–575.
19. Cowan, Z.-L.; Dworjanyn, S.A.; Caballes, C.F.; Pratchett, M.S. Benthic predators influence microhabitat preferences and settlement success of crown-of-thorns starfish larvae (*Acanthaster* cf. *solaris*). *Diversity* **2016**, *8*. [CrossRef]
20. Cowan, Z.-L.; Dworjanyn, S.A.; Caballes, C.F.; Pratchett, M.S. Predation on crown-of-thorns starfish larvae by damselfishes. *Coral Reefs* **2016**, *35*, 1253–1262. [CrossRef]
21. Moran, P.J. The *Acanthaster* phenomenon. *Oceanogr. Mar. Biol.* **1986**, *24*, 379–480.
22. Endean, R. *Report on Investigations Made into Aspects of the Current Acanthaster planci (Crown of Thorns) Infestations of Certain Reefs of the Great Barrier Reef*; Queensland Department of Primary Industries (Fisheries Branch): Brisbane, Australia, 1969.
23. Estes, J.A.; Palmisano, J.F. Sea otters: Their role in structuring nearshore communities. *Science* **1974**, *185*, 1058–1060. [CrossRef] [PubMed]
24. Estes, J.A.; Smith, N.S.; Palmisano, J.F. Sea otter predation and community organization in the western Aleutian Islands, Alaska. *Ecology* **1978**, *59*, 822–833. [CrossRef]
25. Estes, J.A.; Duggins, D.O. Sea otters and kelp forests in Alaska: Generality and variation in a community ecological paradigm. *Ecol. Monogr.* **1995**, *65*, 75–100. [CrossRef]
26. Ling, S.D.; Johnson, C.R.; Frusher, S.D.; Ridgway, K.R. Overfishing reduces resilience of kelp beds to climate-driven catastrophic phase shift. *Proc. Natl. Acad. Sci. USA* **2009**, *106*, 22341–22345. [CrossRef] [PubMed]
27. Steneck, R.S.; Graham, M.H.; Bourque, B.J.; Corbett, D.; Erlandson, J.M.; Estes, J.A.; Tegner, M.J. Kelp forest ecosystems: Biodiversity, stability, resilience and future. *Environ. Conserv.* **2002**, *29*, 436–459. [CrossRef]

28. Pearson, R.G.; Endean, R. *A Preliminary Study of the Coral Predator Acanthaster planci (L.) (Asteroidea) on the Great Barrier Reef*; Queensland Department of Harbours and Marine: Brisbane, Queensland, Australia, 1969; Volume 3, pp. 27–55.

29. Fabricius, K.E.; Okaji, K.; De'ath, G. Three lines of evidence to link outbreaks of the crown-of-thorns seastar *Acanthaster planci* to the release of larval food limitation. *Coral Reefs* **2010**, *29*, 593–605. [CrossRef]

30. Brodie, J.; Fabricius, K.; De'ath, G.; Okaji, K. Are increased nutrient inputs responsible for more outbreaks of crown-of-thorns starfish? An appraisal of the evidence. *Mar. Pollut. Bull.* **2005**, *51*, 266–278. [CrossRef] [PubMed]

31. Potts, D.C. Crown-of-thorns starfish—Man-induced pest or natural phenomenon? In *The Ecology of Pests: Some Australian Case Histories*; Kitching, R., Jones, R., Eds.; Commonwealth Scientific and Industrial Research Organization: Melbourne, Australia, 1981; pp. 55–86.

32. Endean, R. Population explosions of *Acanthaster planci* and associated destruction of hermatypic corals in the Indo-West Pacific region. In *Biology and Geology of Coral Reefs*; Jones, O.A., Endean, R., Eds.; Academic Press: New York, NY, USA, 1973; pp. 389–438.

33. Ormond, R.; Bradbury, R.; Bainbridge, S.; Fabricius, K.; Keesing, J.; de Vantier, L.; Medlay, P.; Steven, A. Test of a model of regulation of crown-of-thorns starfish by fish predators. In *Acanthaster and the Coral Reef: A Theoretical Perspective*; Bradbury, R., Ed.; Springer: Berlin/Heidelberg, Germany, 1990; pp. 189–207.

34. Sweatman, H.P. Commercial fishes as predators of adult *Acanthaster planci*. In Proceedings of the 8th International Coral Reef Symposium, Panama City, Panama, 24–29 June 1996; Lessios, H.A., Macintyre, I.G., Eds.; Volume 1, pp. 617–620.

35. Mendonça, V.M.; Al Jabri, M.M.; Al Ajmi, I.; Al Muharrami, M.; Al Areimi, M.; Al Aghbari, H.A. Persistent and expanding population outbreaks of the corallivorous starfish *Acanthaster planci* in the Northwestern Indian Ocean: Are they through overfishing in coral reefs, or a response to a changing environment? *Zool. Stud.* **2010**, *49*, 108–123.

36. Dulvy, N.K.; Freckleton, R.P.; Polunin, N.V.C. Coral reef cascades and the indirect effects of predator removal by exploitation. *Ecol. Lett.* **2004**, *7*, 410–416. [CrossRef]

37. Sweatman, H. No-take reserves protect coral reefs from predatory starfish. *Curr. Biol.* **2008**, *18*, 598–599. [CrossRef] [PubMed]

38. Moran, P.J. Preliminary observations of the decomposition of crown-of-thorns starfish, *Acanthaster planci* (L.). *Coral Reefs* **1992**, *11*, 115–118. [CrossRef]

39. Glynn, P.W. An amphinoid worm predator of the crown-of-thorns sea star and general predation on asteroids in eastern and western pacific coral reefs. *Bull. Mar. Sci.* **1984**, *35*, 54–71.

40. McCallum, H.I. Effects of predation on organisms with pelagic larval stages: Models of metapopulations. In Proceedings of the 6th International Coral Reef Symposium, Townsville, Australia, 8–12 August 1988; Choat, J.H., Barnes, D., Borowitzka, M.A., Coll, J.C., Davies, P.J., Flood, P., Hatcher, B.G., Hopley, D., Hutchings, P.A., Kinsey, D., et al., Eds.; Volume 2, pp. 101–106.

41. McCallum, H.I. Effects of predation on *Acanthaster*: Age-structure metapopulation models. In *Acanthaster and the Coral Reef: A Theoretical Perspective*; Bradbury, R., Ed.; Springer: Berlin/Heidelberg, Germany, 1990; pp. 208–219.

42. Morello, E.B.; Plagányi, É.E.; Babcock, R.C.; Sweatman, H.; Hillary, R.; Punt, A.E. Model to manage and reduce crown-of-thorns starfish outbreaks. *Mar. Ecol. Prog. Ser.* **2014**, *512*, 167–183. [CrossRef]

43. Ormond, R.F.; Campbell, A.C. Formation and breakdown of *Acanthaster planci* aggregations in the Red Sea. In Proceedings of the 2nd International Coral Reef Symposium, Brisbane, Australia, 22 June–2 July 1973; Cameron, A.M., Cambell, B.M., Cribb, A.B., Endean, R., Jell, J.S., Jones, O.A., Mather, P., Talbot, F.H., Eds.; Volume 1, pp. 595–619.

44. Thomsen, P.F.; Willerslev, E. Environmental DNA—An emerging tool in conservation for monitoring past and present biodiversity. *Biol. Conserv.* **2015**, *183*, 4–18. [CrossRef]

45. Uthicke, S.; Doyle, J.; Duggan, S.; Yasuda, N.; Mckinnon, A.D. Outbreak of coral-eating crown-of-thorns creates continuous cloud of larvae over 320 km of the Great Barrier Reef. *Sci. Rep.* **2015**. [CrossRef] [PubMed]

46. Rivera-Posada, J.; Pratchett, M.S.; Aguilar, C.; Grand, A.; Caballes, C.F. Bile salts and the single-shot lethal injection method for killing crown-of-thorns sea stars (*Acanthaster planci*). *Ocean Coast. Manag.* **2014**, *102*, 383–390. [CrossRef]

47. Keesing, J.; Halford, A. Importance of postsettlement processes for the population dynamics of *Acanthaster planci* (L.). *Mar. Freshw. Res.* **1992**, *43*, 635–651. [CrossRef]

48. Boström-Einarsson, L.; Rivera-Posada, J. Controlling outbreaks of the coral-eating crown-of-thorns starfish using a single injection of common household vinegar. *Coral Reefs* **2016**, *35*, 223–228. [CrossRef]

49. Lucas, J. Environmental influences on the early development of *Acanthaster planci* (L.). In *Crown-of-Thorns Starfish Seminar Proceedings*; Australian Government Publishing Service: Canberra, Australia, 1975; pp. 109–121.

50. Sweatman, H.P.A. A field study of fish predation on juvenile crown-of-thorns starfish. *Coral Reefs* **1995**, *14*, 47–53. [CrossRef]

51. Birdsey, R. *Large Reef Fishes as Potential Predators of Acanthaster Planci: A Pilot Study by Alimentary Tract Analysis of Predatory Fishes from Reefs Subject to Acanthaster Feeding*; Report to the Great Barrier Reef Marine Park Authority: Townsville, Australia, 1988; unpublished.

52. Endean, R. Destruction and recovery of coral reef communities. In *Biology and Geology of Coral Reefs*; Jones, O., Endean, R., Eds.; Academic Press: New York, NY, USA, 1976; Volume 3, Biology 2; pp. 215–254.

53. Ormond, R.; Campbell, A.; Head, S.; Moore, R.; Rainbow, P.; Saunders, A. Formation and breakdown of aggregations of the crown-of-thorns starfish, *Acanthaster planci* (L.). *Nature* **1973**, *246*, 167–169. [CrossRef]

54. Owens, D. *Acanthaster planci* starfish in Fiji: Survey of incidence and biological studies. *Fiji Agric. J.* **1971**, *33*, 15–23.

55. Chesher, R.H. *Acanthaster planci: Impact on Pacific Coral Reefs*; Final report to U.S. Department of the Interior; Westinghouse Research Laboratories: Pittsburgh, PA, USA, 1969; p. 151.

56. Randall, J.E.; Head, S.M.; Sanders, A.P. L. Food habits of the giant humphead wrasse, *Cheilinus undulatus* (Labridae). *Environ. Biol. Fishes* **1978**, *3*, 235–238. [CrossRef]

57. Alcala, A.C. The sponge crab *Dromidiopsis dormia* as a predator of the crown of thorns starfish. *Silliman J.* **1974**, *21*, 174–177.

58. Wickler, W.; Seibt, U. Das Verhalten von *Hymenocera picta* Dana, einer Seesterne fressenden Garnele (Decapoda, Natantia, Gnathophyllidae). *Z. Tierpsychol.* **1970**, *27*, 352–368. [CrossRef]

59. Glynn, P.W. Acanthaster population regulation by a shrimp and a worm. In Proceedings of the 4th International Coral Reef Symposium, Manilla, Philippines, 18–22 May 1981; Gomez, E.D., Birkeland, C.E., Buddemeier, R.W., Johannes, R.E., Marsh, J.A., Jr., Tsuda, R.T., Eds.; 1982; Volume 2, pp. 607–612.

60. Brown, T.W. Starfish menaces coral reefs. *Hemisphere* **1970**, *14*, 31–36.

61. Bos, A.R.; Mueller, B.; Gumanao, G.S. Feeding biology and symbiotic relationships of the corallimorpharian *Paracorynactis hoplites* (Anthozoa: Hexacorallia). *Raffles Bull. Zool.* **2011**, *59*, 245–250.

62. Bos, A.R.; Gumanao, G.S.; Salac, F.N. A newly discovered predator of the crown-of-thorns starfish. *Coral Reefs* **2008**, *27*. [CrossRef]

63. Ciarapica, G.; Passeri, L. An overview of the maldivian coral reefs in Felidu and North Male Atoll (Indian Ocean): Platform drowning by ecological crises. *Facies* **1993**, *28*, 33–65. [CrossRef]

64. Babcock, R.C.; Milton, D.A.; Pratchett, M.S. Relationships between size and reproductive output in the crown-of-thorns starfish. *Mar. Biol.* **2016**, *163*, 1–7. [CrossRef]

65. Babcock, R.C. Spawning behaviour of *Acanthaster planci*. *Coral Reefs* **1990**, *9*, 124. [CrossRef]

66. Babcock, R.C.; Mundy, C.N. Reproductive biology, spawning and field fertilization rates of *Acanthaster planci*. *Aust. J. Mar. Freshw. Res.* **1992**, *43*, 525–534. [CrossRef]

67. Bailey, K.M.; Houde, E.D. Predation on eggs and larvae of marine fishes and the recruitment problem. *Adv. Mar. Biol.* **1989**, *25*, 1–83.

68. Fabricius, K.E.; Metzner, J. Scleractinian walls of mouths: Predation on coral larvae by corals. *Coral Reefs* **2004**, *23*, 245–248. [CrossRef]

69. Cowan, Z.-L.; Ling, S.D.; Dworjanyn, S.A.; Caballes, C.F.; Pratchett, M.S. Inter-specific variation in potential importance of planktivorous damselfishes as predators of *Acanthaster* sp. eggs. *Coral Reefs* **2016**, under review.

70. Keesing, J.K.; Wiedermeyer, W.L.; Okaji, K.; Halford, A.R.; Hall, K.C.; Cartwright, C.M. Mortality rates of juvenile starfish *Acanthaster planci* and *Nardoa* spp. measured on the Great Barrier Reef, Australia and in Okinawa, Japan. *Oceanol. Acta* **1996**, *19*, 441–448.

71. Roegner, G.C.; Mann, R. Early recruitment and growth of the American oyster *Crassostrea virginica* (Bivalvia: Ostreidae) with respect to tidal zonation and season. *Mar. Ecol. Prog. Ser. Oldend.* **1995**, *117*, 91–101. [CrossRef]

72. Almany, G.R.; Webster, M.S. The predation gauntlet: Early post-settlement mortality in reef fishes. *Coral Reefs* **2006**, *25*, 19–22. [CrossRef]

73. Chong-Seng, K.M.; Graham, N.A.J.; Pratchett, M.S. Bottlenecks to coral recovery in the Seychelles. *Coral Reefs* **2014**, *33*, 449–461. [CrossRef]

74. Glynn, P.W. Interactions between *Acanthaster* and *Hymenocera* in the field and laboratory. In *Proceedings: the 3rd International Coral Reef Symposium, Miami, Florida, USA*; Taylor, D.L., Ed.; University of Miami: Miami, FL, USA, 1977; Volume 1, pp. 209–216.

75. Wickler, W. Biology of *Hymenocera picta* Dana. *Micronesica* **1973**, *9*, 225–230.

76. Ling, S.D.; Johnson, C.R. Marine reserves reduce risk of climate-driven phase shift by reinstating size-and habitat-specific trophic interactions. *Ecol. Appl.* **2012**, *22*, 1232–1245. [CrossRef] [PubMed]

77. Glynn, P.W. Individual recognition and phenotypic variability in *Acanthaster planci* (Echinodermata: Asteroidea). *Coral Reefs* **1982**, *1*, 89–94. [CrossRef]

78. McClanahan, T.R.; Muthiga, N.A. Patterns of predation on a sea urchin, *Echinometra mathaei* (de Blainville), on Kenyan coral reefs. *J. Exp. Mar. Bio. Ecol.* **1989**, *126*, 77–94. [CrossRef]

79. Bonaviri, C.; Fernández, T.V.; Badalamenti, F.; Gianguzza, P.; Di Lorenzo, M.; Riggio, S. Fish versus starfish predation in controlling sea urchin populations in Mediterranean rocky shores. *Mar. Ecol. Prog. Ser.* **2009**, *382*, 129–138. [CrossRef]

80. Potts, D. Crown-of-thorns starfish—Man-induced pest or natural phenomenon? In *The Ecology of Pests: Some Australian Case Histories*; Kitching, R.E., Jones, R.E., Eds.; CSIRO: Melbourne, Australia, 1982; pp. 55–86.

81. Chesher, R.H. Destruction of Pacific corals by the sea star *Acanthaster planci*. *Science* **1969**, *165*, 280–283. [CrossRef] [PubMed]

82. McCallum, H.I. Predator regulation of *Acanthaster planci*. *J. Theor. Biol.* **1987**, *127*, 207–220. [CrossRef]

83. McCallum, H. Completing the circle: Stock-recruitment relationships and *Acanthaster*. *Mar. Freshw. Res.* **1992**, *43*, 653–662. [CrossRef]

84. Ormond, R.F.G.; Campbell, A.C. Observations on *Acanthaster planci* and other coral reef echinoderms in the Sudanese Red Sea. In *Symposia of the Zoological Society of London*; Academic Press: London, UK, 1971; Volume 28, pp. 433–454.

85. McClanahan, T.R. Kenyan coral reef-associated gastropod fauna: A comparison between protected and unprotected reefs. *Mar. Ecol. Prog. Ser.* **1989**, *53*, 11–20. [CrossRef]

86. Endean, R. *Acanthaster planci* infestations of reefs of the Great Barrier Reef. In *Proceedings: the 3rd International Coral Reef Symposium, Miami, Florida, USA*; Taylor, D.L., Ed.; University of Miami: Miami, FL, USA, 1977; Volume 1, pp. 185–191.

87. McClanahan, T.R.; Muthiga, N.A. Similar impacts of fishing and environmental stress on calcifying organisms in Indian Ocean coral reefs. *Mar. Ecol. Prog. Ser.* **2016**, *560*, 87–103. [CrossRef]

88. Lawrence, J.M.; Vasquez, J. The effect of sublethal predation on the biology of echinoderms. *Oceanol. Acta* **1996**, *19*, 431–440.

89. Ling, S.D.; Johnson, C.R. Native spider crab causes high incidence of sub-lethal injury to the introduced seastar *Asterias amurensis*. In *Echinoderms in a Changing World: Proceedings of the 13th International Echinoderm Conference, January 5–9 2009, University of Tasmania, Hobart Tasmania, Australia*; CRC Press: New York, NY, USA, 2012; pp. 195–201.

90. Rivera-Posada, J.A.; Pratchett, M.; Cano-Gómez, A.; Arango-Gómez, J.D.; Owens, L. Injection of *Acanthaster planci* with thiosulfate-citrate-bile-sucrose agar (TCBS). I. Disease induction. *Dis. Aquat. Organ.* **2011**, *97*, 85–94. [CrossRef] [PubMed]

91. Rivera-Posada, J.; Owens, L.; Caballes, C.F.; Pratchett, M.S. The role of protein extracts in the induction of disease in *Acanthaster planci*. *J. Exp. Mar. Bio. Ecol.* **2012**, *429*, 1–6. [CrossRef]

92. Caballes, C.F.; Schupp, P.J.; Pratchett, M.S.; Rivera-Posada, J.A. Interspecific transmission and recovery of TCBS-induced disease between *Acanthaster planci* and *Linckia guildingi*. *Dis. Aquat. Organ.* **2012**, *100*, 263–267. [CrossRef] [PubMed]

93. Lawrence, J.M. Arm loss and regeneration in Asteroidea (Echinodermata). *Echinoderm Res.* **1991**, *1992*, 39–52.

94. Lawrence, J.M. Energetic costs of loss and regeneration of arms in stellate echinoderms. *Integr. Comp. Biol.* **2010**, *50*, 506–514. [CrossRef] [PubMed]

95. Messmer, V.; Pratchett, M.S.; Clark, T.D. Capacity for regeneration in crown-of-thorns starfish, *Acanthaster planci*. *Coral Reefs* **2013**, *32*, 461. [CrossRef]

96. Sigl, R.; Steibl, S.; Laforsch, C. The role of vision for navigation in the crown-of-thorns seastar, *Acanthaster planci*. *Sci. Rep.* **2016**, *6*. [CrossRef] [PubMed]
97. Schoener, T.W. Inferring the properties of predation and other injury-producing agents from injury frequencies. *Ecology* **1979**, *60*, 1110–1115. [CrossRef]
98. Marrs, J.; Wilkie, I.C.; Sköld, M.; Maclaren, W.M.; McKenzie, J.D. Size-related aspects of arm damage, tissue mechanics, and autotomy in the starfish *Asterias rubens*. *Mar. Biol.* **2000**, *137*, 59–70. [CrossRef]
99. Ling, S.D.; Johnson, C.R.; Mundy, C.N.; Morris, A.; Ross, D.J. Hotspots of exotic free-spawning sex: Man-made environment facilitates success of an invasive seastar. *J. Appl. Ecol.* **2012**, *49*, 733–741. [CrossRef]
100. Benzie, J.A.H.; Black, K.P.; Moran, P.J.; Dixon, P. Small-scale dispersion of eggs and sperm of the crown-of-thorns starfish (*Acanthaster planci*) in a shallow coral reef habitat. *Biol. Bull.* **1994**, *186*, 153–167. [CrossRef]
101. Endean, R. *Acanthaster planci* on the Great Barrier Reef. In Proceedings of the 2nd International Coral Reef Symposium, Brisbane, Australia, 22 June–2 July 1973; Cameron, A.M., Cambell, B.M., Cribb, A.B., Endean, R., Jell, J.S., Jones, O.A., Mather, P., Talbot, F.H., Eds.; Volume 1, pp. 563–576.
102. Hassell, M.P. *The Dynamics of Arthropod Predator-Prey Systems*; Princeton University Press: Princeton, NJ, USA, 1978.
103. Holling, C. The functional response of predators to prey density and its role in mimicry and population regulation. *Mem. Entomol. Soc. Can.* **1965**, *97*, 5–60. [CrossRef]
104. Abrams, P.A. The effects of adaptive behavior on the Type-2 functional response. *Ecology* **1990**, *71*, 877–885. [CrossRef]
105. Buckel, J.A.; Stoner, A.W. Functional response and switching behaviour of young-of-the-year piscivorus bluefish. *J. Exp. Mar. Bio. Ecol.* **2000**, *245*, 25–41. [CrossRef]
106. Nilsson, P.A.; Ruxton, G.D. Temporally fluctuating prey and interfering predators: A positive feedback. *Anim. Behav.* **2004**, *68*, 159–165. [CrossRef]
107. Holling, C.S. The components of predation as revealed by a study of small-mammal predation of the European pine sawfly. *Can. Entomol.* **1959**, *91*, 293–320. [CrossRef]
108. Kaspari, M. Prey preparation and the determinants of handling time. *Anim. Behav.* **1990**, *40*, 118–126. [CrossRef]
109. Baker, D.J.; Stillman, R.A.; Smith, B.M.; Bullock, J.M.; Norris, K.J. Vigilance and the functional response of granivorous foragers. *Funct. Ecol.* **2010**, *24*, 1281–1290. [CrossRef]
110. Murdoch, W.W. Switching in general predators: Experiments on predator specificity and stability of prey populations. *Ecol. Monogr.* **1969**, *39*, 335–354. [CrossRef]
111. Babcock, R.C.; Dambacher, J.M.; Morello, E.B.; Plagányi, É.E.; Hayes, K.R.; Sweatman, H.P.A.; Pratchett, M.S. Assessing different causes of crown-of-thorns starfish outbreaks and appropriate responses for management on the Great Barrier Reef. *PLoS ONE* **2016**, *11*, e0169048. [CrossRef] [PubMed]
112. Redd, K.S.; Ling, S.D.; Frusher, S.D.; Jarman, S.; Johnson, C.R. Using molecular prey detection to quantify rock lobster predation on barrens-forming sea urchins. *Mol. Ecol.* **2014**, *23*, 3849–3869. [CrossRef] [PubMed]
113. MacArthur, L.D.; Babcock, R.C.; Hyndes, G.A. Movements of the western rock lobster (*Panulirus cygnus*) within shallow coastal waters using acoustic telemetry. *Mar. Freshw. Res.* **2008**, *59*, 603–613. [CrossRef]

![diversity logo] diversity

Communication

Crown-of-Thorns Starfish Larvae Can Feed on Organic Matter Released from Corals

Ryota Nakajima [1,*], Nobuyuki Nakatomi [2], Haruko Kurihara [3], Michael D. Fox [1],
Jennifer E. Smith [1] and Ken Okaji [4]

[1]   Scripps Institution of Oceanography, University of California San Diego, 9500 Gilman Drive,
     La Jolla, CA 92083-0202, USA; fox@ucsd.edu (M.D.F.); smithj@ucsd.edu (J.E.S.)
[2]   Faculty of Science and Engineering, Soka University, Hachioji, Tokyo 192-8577, Japan;
     nnakatomi@gmail.com
[3]   Faculty of Science, University of the Ryukyus, Nishihara, Okinawa 903-0213, Japan;
     harukoku@sci.u-ryukyu.ac.jp
[4]   Coralquest Inc., Atsugi, Kanagawa 243-0014, Japan; cab67820@pop06.odn.ne.jp
*   Correspondence: rnakajima@ucsd.edu; Tel.: +1-858-822-3271

Academic Editor: Morgan Pratchett
Received: 10 August 2016; Accepted: 28 September 2016; Published: 6 October 2016

Abstract: Previous studies have suggested that Crown-of-Thorns starfish (COTS) larvae may be able to survive in the absence of abundant phytoplankton resources suggesting that they may be able to utilize alternative food sources. Here, we tested the hypothesis that COTS larvae are able to feed on coral-derived organic matter using labeled stable isotope tracers (^{13}C and ^{15}N). Our results show that coral-derived organic matter (coral mucus and associated microorganisms) can be assimilated by COTS larvae and may be an important alternative or additional food resource for COTS larvae through periods of low phytoplankton biomass. This additional food resource could potentially facilitate COTS outbreaks by reducing resource limitation.

Keywords: *Acanthaster*; COTS; coral reefs; coral mucus; food limitation; isotope analysis; Japan

1. Introduction

Outbreaks of Crown-of-Thorns starfish (COTS), *Acanthaster* spp., can have devastating effects on coral reefs throughout the Pacific and Indian Oceans [1,2]. Determining the causes and spatial variability of COTS outbreaks has proven to be a major challenge for coral reef managers [2,3]. Efforts to determine the causes of COTS outbreaks on the Great Barrier Reef have identified anthropogenic eutrophication as an important correlate [3–6]. The increased phytoplankton concentrations that result from high inorganic nutrient concentrations in the water are thought to promote abnormally high survival rates of COTS larvae by acting as an important food resource [4–6]. However, most coral reefs are often considered oligotrophic systems with low phytoplankton biomass, which may help keep COTS populations stable in unperturbed conditions [4]. Maximal survival of COTS larvae has been directly linked to specific concentrations of food resources [6–9]. In the studies using natural phytoplankton assemblages as food sources, COTS larvae have been found to benefit from increasing food availability from 0.25 to 0.8 μg chl-*a* L^{-1}, with the highest growth rates observed at concentrations of >0.8 μg chl-*a* L^{-1} and resource limited mortality occurring below 0.25 μg chl-*a* L^{-1} [6,8]. Furthermore, Wolfe et al. [7] recently reported using a single microalgae species that phytoplankton levels of 1 μg chl-*a* L^{-1} were optimal for COTS larval development success.

On many coral reefs, typical concentrations of chl-*a* are approximately 0.2–0.6 μg chl-*a* L^{-1} (Table 1), which may be limiting to the survival of COTS larvae [6,8]. In Okinawa, chl-*a* concentrations often fall below 0.25 μg chl-*a* L^{-1}, the critical concentration for COTS larval survival [10–14].

However, local abundance of COTS suggests their larvae may be able to survive under low chl-*a* conditions, possibly by utilizing additional food sources [15,16]. Recent studies have reported that naturally-occurring COTS larvae were in the advanced developmental stages in the vicinity of coral communities [17–19], suggesting the utilization of organic matter derived from coral reefs by COTS larvae.

Table 1. Chlorophyll-*a* (Chl-*a*) concentrations in various coral reef waters. Some values were visually interpreted from figures. The values in parenthesis are the mean of several data in the study. GBR, Great Barrier Reef.

Site	Chl-*a* ($\mu g \cdot L^{-1}$)	Reference
Miyako Island (Okinawa, Japan)	0.10–0.15	[10]
Miyako Island (Okinawa, Japan)	0.1–0.4	[11]
Sesoko Island (Okinawa, Japan)	0.11–0.77 (0.45)	[12]
West coast of Okinawa Island (Japan)	<0.05–1.79 (0.17)	[13]
Ishigaki Island (Okinawa, Japan)	0.09–0.55	[14]
Princess Charlotte Bay (GBR, Australia)	0.06–0.28 (0.16)	[20]
Princess Charlotte Bay (GBR, Australia)	0.40	[21]
Cairns-Innisfail sector (GBR, Australia)	0.03–0.64 (0.25)	[20]
Wet Tropics (GBR, Australia)	0.70	[21]
Central GBR (Australia)	0.19–0.72 (0.38)	[22]
Whitsunday Islands (GBR, Australia)	0.31–1.21 (0.79)	[23]
Uvea Atoll (New Caledonia)	0.23	[24]
The Southwest lagoon (New Caledonia)	0.25–2.14 (0.60)	[25]
Maître Island (New Caledonia)	0.26–0.42 (0.30)	[26]
Tikehau Atoll (Tuamotu, French Polynesia)	0.17	[27]
Takapoto Atoll (Tuamotu, French Polynesia)	0.23	[27]
Takapoto Atoll (Tuamotu, French Polynesia)	0.21–0.23 (0.22)	[28]
Fakarava/Rangiroa Atolls (French Polynesia)	0.008–0.25	[29]
Ahe Atoll (French Polynesia)	0.08–0.85 (0.34)	[30]
Tioman Island (Malaysia)	0.20–0.24 (0.22)	[31]
Bidong Island (Malaysia)	0.28–0.30 (0.29)	[32]

Scleractinian corals release a large amount of organic matter in the form of mucus [33] as a result of basic metabolic activities and as a protective mechanism against various stresses [34]. Coral mucus particles contain carbohydrates, proteins, lipids, and numerous microorganisms [33,35]. When coral mucus particles become suspended in the water column it aggregates various organic particles (such as microbes and phytoplankton) and becomes more enriched in organic matter over time [33,36,37]. These coral mucus aggregates are considered to be one of the major contributors to the origin of particulate organic matter (POM) in reef waters [36,38]. These mucus aggregates have also been reported as an important food source for various reef animals, such as fish, zooplankton, and several benthic taxa, such as coral crabs and brittle stars [39]. Experimental evidence has shown that copepods and mysids, two common zooplankton taxa, directly utilize coral mucus aggregates as a food resource [40,41]. The release of mucus by corals increases with increasing ambient light and water temperature [42] as it originates from by-products of the photosynthesis of zooxanthellae [43]. Thus, mucus production is likely to be maximized in summer (July–August in Okinawa) which corresponds with the peak spawning of COTS (July in Okinawa). Coral mucus could, therefore, be an important food source for COTS larvae that develop on or near coral reefs during periods of low phytoplankton biomass.

Here, we examined whether COTS larvae are able to feed on coral-derived organic matter. Since understanding the key food sources of COTS larvae is critical for determining future recruitment of adults [7], it is important to investigate all possible food sources in the natural environment. If COTS larvae feed on organic matter released by corals, this would represent an additional food source not previously considered. This additional food resource may enhance the survival of COTS larvae and subsequent recruitment of adults in areas that are naturally low in phytoplankton abundance and/or during times of reduced phytoplankton biomass.

2. Materials and Methods

2.1. Collection of COTS Larvae

Adult COTS (ca. 25 cm diameter) were collected on shallow reefs around Onna Village (N26.508496; E127.854283), Okinawa, Japan, in June 2015. Individuals were immediately transferred to the flow-through indoor aquaria at the Sesoko Station, Tropical Biosphere Research Center (University of the Ryukyus). We refer to the individuals from the Pacific Ocean clade COTS used in this study as *Acanthaster* cf. *solaris* [44,45].

COTS larvae were obtained using the method of Birkeland and Lucas [1]. Several mL of 1-methyladenine solution (1×10^{-3} M) were added to a crystallization dish with mature ovary lobes from the collected adults to release the oocytes. The oocytes released were pipetted into a separate glass dish filled with filtered seawater. Two drops of dense spermatozoa suspension were then added and gently stirred to fertilize oocytes. Fertilized oocytes were repeatedly rinsed with filtered seawater and introduced into a 9 L cylindrical plastic container with filtered seawater. After approximately 24 h, actively swimming gastrula larvae were transferred to another container where they were reared in filtered seawater at 28 °C with enriched phytoplankton using *Dunaliella tertiolecta* at a cell density of 500 cells mL^{-1} (= 0.9–1.0 μg chl-*a* L^{-1}), until they reached the advanced bipinnaria stage. The algae *D. tertiolecta* (NRIA-0109) was provided by GeneBank of the Japan Fisheries Research and Education Agency (Yokohama, Japan). Daily seawater changes were carried out by gentle reverse filtration using a 60 μm mesh screen.

2.2. Collection of Coral Mucus

Colonies of the branching corals, *Acropora muricata* and *A. intermedia* were collected from the reefs of Sesoko Island, and raised in the outdoor aquaria facility at the Sesoko Station for more than a year. Two weeks before the experiment, the coral colonies were transferred to a 30 L flow-through outdoor aquarium tank. Fresh seawater was directly pumped from the reef to the aquaria at the flow rate of 2.5 L·min^{-1}.

The corals were labeled with enriched concentrations of stable carbon and nitrogen isotopes by incubating corals in isotopically-enriched seawater following the methods of Naumann et al. [46]. The water inflow was stopped at 1000 h, and the aquarium was treated with NaH^{13}CO$_3$ and Na^{15}NO$_3$ (Cambridge Isotope Laboratories, MA, USA, 98 atom %) to a final concentration of 20 mg·L^{-1} and 1 mg·L^{-1}, respectively. After a 5 h (at 1500 h) incubation, the flow-through seawater was resumed. The tank was aerated with a powerful air-pump for sufficient seawater agitation during each of the incubation periods. The labeling procedure was repeated for three days. On the final incubation day, labeled mucus was collected at 1600 h, one hour after the incubation period. To collect the mucus, corals were removed from the tank and washed with unlabeled filtered seawater [46]. The corals were then exposed to air and hung in direct sunlight to trigger mucus production [33]. The released mucus was collected in sterilized 50 mL Corning tubes. The collection of the labeled mucus was conducted twice, on July 18 and 25. The mucus was used immediately after each collection for the following feeding experiments. Triplicate subsamples of the labeled mucus from each collection period were placed in pre-weighed tin capsules and dried for 24 h at 60 °C, then stored in a desiccator until isotopic analysis was performed.

2.3. Feeding Experiments

Feeding experiments were conducted following each of the mucus collections using bipinnaria larvae. During the first feeding experiment, five-day-old (fertilized on 13 July) and 14-day-old (fertilized on 4 July) larvae were used (Experiment 1). For the second feeding experiment only five-day-old larvae (fertilized on 20 July) were used (Experiment 2). Prior to the experiments, actively-swimming larvae were gently siphoned out, concentrated, and rinsed thoroughly with GF/F (Whatman, NJ, USA) filtered seawater. These larvae were kept in GF/F filtered seawater for 24 h to empty their guts. Feeding incubations were conducted using 12 polycarbonate bottles filled with 1 L of GF/F filtered seawater. In Experiment 1, six bottles were used for five-day-old larvae and the other six for 14-day-old larvae. In each set of six bottles, three bottles treated with 2 mL of labeled mucus and the other three bottles were used as controls (without mucus). The feeding experiments contained approximately 300 starved larvae in each bottle. All bottles were incubated in the laboratory for 24 h keeping the water temperature at 28 °C. The bottles were rotated on a plankton wheel (Model II, Wheaton Instruments, NJ, USA) during the incubation to keep larvae and coral mucus particles in suspension. In Experiment 2, five-day-old larvae were introduced to all 12 bottles (300 individuals per bottle) and filled with 1 L of GF/F filtered seawater. Of these 12 bottles, six were used for the feeding treatment (with mucus) and the other six were used for the control treatment (without mucus). This second incubation was conducted for 48 h, and three bottles were collected from each treatment at 24 and 48 h.

After the incubations, the COTS larvae were collected on a 60 μm mesh screen and washed with unlabeled Milli-Q water (Whatman, NJ, USA) 10 times to remove remaining labeled external mucus material. The larvae were then transferred into 15 mL Corning tube full of Milli-Q water. These tubes were refrigerated until the larvae settled to the bottom. The settled larvae were gently pipetted out (this procedure allowed for the collection of all larvae within a sample in a small volume of water), then placed into a tin capsule, dried (40 °C, 24 h), and stored in a desiccator until isotopic analysis was performed. Due to low biomass, COTS larvae collected from triplicate bottles for each feeding treatment were pooled to ensure a good signal. Therefore, only four samples for each feeding treatment were analyzed. Samples were not acidified with HCl to remove possible inorganic carbon because the acid could have damaged the larvae and altered their isotopic signature.

2.4. Analysis

Carbon- and nitrogen-stable isotope ratios of the larvae were determined by elemental analysis/isotope ratio mass spectrometry (EA/IRMS) using a Flash EA1112-DELTA V PLUS ConFlo III System (Thermo Fisher Scientific, MA, USA) at SI Science Co., Ltd. (Saitama, Japan). The carbon and nitrogen isotopic ratios are expressed in δ notation (Vienna-PeeDee Belemnite limestone for carbon and atmospheric nitrogen for nitrogen) as the deviation from standards in parts per mill (‰) using the following equation: δX (‰) = $[(R_{sample}/R_{standard}) - 1] \times 1000$, where X is ^{13}C or ^{15}N and R is the ratio of $^{13}C/^{12}C$ or $^{15}N/^{14}N$. The analytical error was less than ± 0.13‰ for carbon and ± 0.61‰ for nitrogen. Significant differences (at the $p < 0.05$ level) of $\delta^{13}C$ and $\delta^{15}N$ values between control and mucus were determined using Student's *t*-test.

3. Results and Discussion

The average value ($\pm$ SE) of $\delta^{13}C$ and $\delta^{15}N$ of labeled coral mucus was 358.5‰ $\pm$ 92.7‰ and 1683.4 $\pm$ 112.1‰ for the first experiment and 707.4‰ $\pm$ 31.2‰ and 1983.8‰ $\pm$ 111.1‰ for the second experiment, respectively (Table 2). The $\delta^{13}C$ and $\delta^{15}N$ ratios of the labeled mucus showed higher values in the second experiment, likely because of label accumulation in coral tissue and/or the biological community in the mucous on the coral surface (e.g., bacteria and zooxanthellae) [46]. We measured the $\delta^{13}C$ and $\delta^{15}N$ values of raw coral mucus, which could contain bacteria and zooxanthellae. Therefore, we are not able to offer direct evidence of mucus labeling via coral tissue but rather the labeling of the aggregation of organic matter within the mucus particles. However, Naumann et al. [46] measured the $\delta^{15}N$ of coral mucus that was filtered through 0.2 μm filter to remove microorganisms, and showed coral mucus had been labeled successfully. Considering the $\delta^{15}N$ value in our labeled raw coral mucus (1.68‰–1.98‰) are similar to values reported in other studies ([46] ca. 2.00‰), we consider that the coral mucus in our experiment was successfully labeled.

Table 2. $\delta^{13}C$ and $\delta^{15}N$ signatures of labeled mucus released by *Acropora* corals. C/N ratios were determined as %C/%N of the samples. Bold values indicate mean ($\pm$ SE) of triplicate measurements.

Collection Day (mm/dd/yy)	$\delta^{13}C$ (‰)	$\delta^{15}N$ (‰)	C/N
07/18/2015	288.9	1562.5	6.7
	463.7	1783.9	7.7
	322.9	1703.7	7.2
	358.5 ± 53.5	**1683.4 ± 64.7**	
07/24/2015	690.6	2091.3	7.8
	688.1	1869.4	7.2
	743.3	1990.6	7.3
	707.4 ± 18.0	**1983.8 ± 64.1**	

After both the 24 and 48 h incubations, the COTS larvae in the mucus addition treatments exhibited highly-enriched $\delta^{13}C$ and $\delta^{15}N$ values compared to those in control treatments (Figure 1). The average value of $\delta^{13}C$ (40.7‰ ± 34.9‰) and $\delta^{15}N$ (133.4‰ ± 55.9‰) of larvae in the mucus treatments (n = 4) was significantly (P = 0.017 for $\delta^{13}C$ and 0.0040 for $\delta^{15}N$) higher than those in controls ($\delta^{13}C$, −16.8‰ ± 0.3‰; $\delta^{15}N$, 7.1‰ ± 0.2‰, n = 4) (Figure 1). Since $\delta^{13}C$ and $\delta^{15}N$ of the labeled mucus in the second experiment showed higher values, the following $\delta^{13}C$ and $\delta^{15}N$ values of COTS larvae also showed higher values compared to those in the first experiment. These results indicate that $\delta^{13}C$- and $\delta^{15}N$-labeled mucus are transferred into COTS larvae, providing further evidence for the potential use of organic matter derived from corals as a food source.

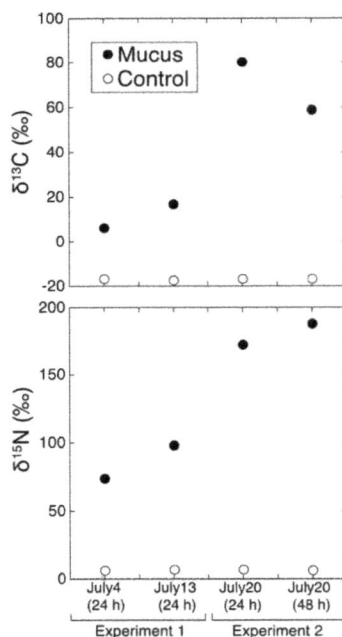

Figure 1. $\delta^{13}C$ and $\delta^{15}N$ signatures of COTS larvae after 24 h or 48 h of incubation with $\delta^{13}C$- and $\delta^{15}N$-labeled coral mucus and under control conditions. Closed and opened circles indicate fed larvae (with mucus) and unfed larvae (without mucus), respectively. Dates indicate the day of fertilization. Each plot comes from a single analysis.

This is the first study to show that COTS larvae can take up organic matter derived from corals. However, the raw coral mucus we provided to the COTS larvae may contain zooxanthellae and other microorganisms as mentioned above. Therefore, whether COTS larvae fed on whole coral mucus aggregates or selectively fed on the associated microorganisms remains unclear. Regardless, it is evident that the COTS larvae, in our feeding experiments, assimilated organic matter derived from corals, either directly (i.e., pure mucus), or indirectly via the associated microorganisms in/on the coral mucus aggregates (i.e., bacteria and zooxanthellae). It is also unclear whether COTS larvae fed on the particulate and/or dissolved fraction of the coral mucus, as we did not size-fractionate the coral mucus. Previous reports have shown that COTS larvae can utilize both POM and dissolved organic matter (DOM) [8,47,48]. Naumann et al. [46] also reported that coral associated epizoic acoelomorph *Wamioa* worms utilized both the particulate and dissolved fraction of mucus released by corals. Thus, it is possible that COTS larvae are capable of feeding on both POM (mucus particles) and DOM released by corals.

Previous studies that have observed high survival rates of COTS larvae even when exposed to low concentrations of phytoplankton, suggesting that they may utilize other food sources (e.g., [8,9]). For COTS larvae in near-reef waters, organic matter derived from corals may provide an important and previously underappreciated source of nutrition. Peak spawning for COTS tends to occur in summer (July in Japan). During this period, elevated water temperature and light intensity may also cause higher organic matter release by corals [42]. In Okinawa, numerous advanced COTS brachiolaria larvae were found near coral communities [19] and it appears that some populations of COTS larvae are retained in near-shore waters due to prevailing along-shore currents in the archipelago in summer [17]. If COTS larvae are entrained near-shore they are more likely to encounter coral-derived organic matter. When the reef water residence time is short and the net transport of water is offshore, some fraction of coral mucus may be physically transported offshore before this material settles or is consumed within the reef environment [39,49]. Thus, coral-derived organic matter may also provide an important food source to COTS larvae in pelagic habitats adjacent to coral reefs. Consequently, coral mucus and its aggregates may provide one of the energy pathways for the conversion of coral primary production to COTS larvae.

Coral-derived organic matter may be a particularly important resource for COTS larval under oligotrophic conditions typical of many coral reefs where nutrients and phytoplankton biomass are low. While phytoplankton can be ephemerally high in areas with internal tidal influence or with episodic upwelling events, coral-derived organic matter may be a more consistent source of nutrition to COTS larvae on most reefs. Further, these larvae may be able to extend their planktonic durations during times of low phytoplankton biomass using endogenous nutrients, or by increasing their capacity for food capture by extending ciliary bands [50]. Although it is not likely that coral mucus, alone, is a direct causal factor of COTS outbreaks, it does seem likely that coral-derived organic matter could be an important food source, especially during periods of low phytoplankton levels. In sum, while more data are needed to understand the quantitative importance of coral-derived organic matter to COTS population dynamics, our results suggest that the feeding ecology of this species is more complex than previously thought. Given the nutritional quality of coral mucus and associated material, it is not surprising that COTS larvae, as well as many other species, take advantage of this resource in an ecosystem where nutrients are often limiting.

Acknowledgments: We are grateful to three anonymous reviewers for their time and valuable suggestions to improve this article. We thank Sesoko Station staffs, M. Kitamura (Incorporated Foundation Okinawa Prefecture Environment Science Center) for his support on collection of COTS larvae, M. Okauchi and M. Tokuda (Japan Fisheries Research and Education Agency) for providing algal culture, and Y. Tadokoro for his help in sample collection. This study was made possible by a research project titled "Comprehensive management program of the Crown-of-Thorns Starfish outbreaks in Okinawa" supported by Nature Conservation Division, Okinawa Prefectural Government, by JSPS Fellowships for Research Abroad to the first author, and by Grant-in-Aid for JSPS Fellows (14J00884) to the second author.

Author Contributions: R.N. and N.N. conceived and designed the experiments; R.N., N.N. and K.O. performed the experiments; N.N. analyzed the data; H.K., N.N. and K.O. contributed reagents/materials/analysis tools; R.N., N.N., H.K., M.D.F., J.E.S. and K.O. wrote the paper.

Conflicts of Interest: The authors declare no conflict of interest.

References

1. Birkeland, C.; Lucas, J. *Acanthaster Planci: Major Management Problem of Coral Reefs*; CRC Press: Florida, FL, USA, 1990.
2. Pratchett, M.; Caballes, C.; Rivera Posada, J.; Sweatman, H. Limits to understanding and managing outbreaks of crown-of-thorns starfish (*Acanthaster* spp.). *Oceanogr. Mar. Biol. Annu. Rev.* **2014**, *52*, 133–200.
3. De'ath, G.; Fabricius, K.E.; Sweatman, H.; Puotinen, M. The 27-year decline of coral cover on the Great Barrier Reef and its causes. *Proc. Natl. Acad. Sci. USA* **2012**, *109*, 17995–17999. [CrossRef] [PubMed]
4. Lucas, J. Quantitative studies of feeding and nutrition during larval stages of the coral reef asteroid *Acanthaster planci* (L.). *J. Exp. Mar. Biol. Ecol.* **1982**, *65*, 173–193. [CrossRef]
5. Brodie, J.; Fabricius, K.; De'ath, G.; Okaji, K. Are increased nutrient inputs responsible for more outbreaks of crown-of-thorns starfish? An appraisal of the evidence. *Mar. Pollut. Bull.* **2005**, *51*, 266–278. [CrossRef] [PubMed]
6. Fabricius, K.E.; Okaji, K.; De'ath, G. Three lines of evidence to link outbreaks of the crown-of-thorns seastar *Acanthaster planci* to the release of larval food limitation. *Coral Reefs* **2010**, *29*, 593–605. [CrossRef]
7. Wolfe, K.; Graba-Landry, A.; Dworjanyn, S.A.; Byrne, M. Larval starvation to satiation: Influence of nutrient regime on the success of *Acanthaster planci*. *PLoS ONE* **2015**, *10*, 1–17. [CrossRef] [PubMed]
8. Okaji, K.; Ayukai, T.; Lucas, J.S. Selective feeding by larvae of the crown-of-thorns starfish, *Acanthaster planci* (L.). *Coral Reefs* **1997**, *16*, 47–50. [CrossRef]
9. Okaji, K. Feeding Ecology in the Early Life Stages of the Crown-of-Thorns Starfish, *Acanthaster planci* (L.). Ph.D. Thesis, James Cook University, Brisbane, Australia, June 1996.
10. Ferrier-Pagès, C.; Gattuso, J.P. Biomass, production and grazing rates of pico- and nanoplankton in coral reef waters (Miyako Island, Japan). *Microb. Ecol.* **1998**, *35*, 46–57. [CrossRef] [PubMed]
11. Casareto, B.E.; Charpy, L.; Blanchot, J.; Suzuki, Y.; Kurosawa, K.; Ihikawa, Y. Photorophic prokaryotes in Bora Bay, Miyako Island, Okinawa, Japan. *Proc. 10th Int. Coral Reef Symp.* **2006**, *31*, 844–853.
12. Tada, K.; Sakai, K.; Nakano, Y.; Takemura, A.; Montani, S. Size-fractionated phytoplankton biomass in coral reef waters off Sesoko Island, Okinawa, Japan. *J. Plankton Res.* **2001**, *25*, 991–997. [CrossRef]
13. Kinjyo, K.; Yamakawa, E. Survey of chlorophyll distribution. *A Report on the Comprehensive Management Program of the Crown-of-Thorns Starfish Outbreaks in Okinawa, 2014*; Okinawa Prefectural Government, Department of Environmental Affairs, Nature Conservation and Afforestation Promotion Division: Okinawa, Japan, 2014; pp. 99–112.
14. Fukuoka, K.; Shimoda, T.; Abe, K. Community structure and abundance of copepods in summer on a fringing coral reef off Ishigaki Island, Ryukyu Islands, Japan. *Plankt. Benthos Res.* **2015**, *10*, 225–232. [CrossRef]
15. Olson, R. In situ culturing of larvae of the crown-of-thorns starfish *Acanthaster planci*. *Mar. Ecol. Prog. Ser.* **1985**, *25*, 207–210. [CrossRef]
16. Olson, R. In situ culturing as a test of the larval starvation hypothesis for the crown-of-throns starfish, Acanthaster planci. *Limnol. Oceanogr.* **1987**, *32*, 895–904. [CrossRef]
17. Nakamura, M.; Kumagai, N.H.; Sakai, K.; Okaji, K.; Ogasawara, K.; Mitarai, S. Spatial variability in recruitment of acroporid corals and predatory starfish along the Onna coast, Okinawa, Japan. *Mar. Ecol. Prog. Ser.* **2015**, *540*, 1–12. [CrossRef]
18. Uthicke, S.; Doyle, J.; Duggan, S.; Yasuda, N.; McKinnon, A.D. Outbreak of coral-eating Crown-of-Thorns creates continuous cloud of larvae over 320 km of the Great Barrier Reef. *Sci. Rep.* **2015**, *5*, 16885. [CrossRef] [PubMed]
19. Suzuki, G.; Yasuda, N.; Ikehara, K.; Fukuoka, K.; Kameda, T.; Kai, S.; Nagai, S.; Watanabe, A.; Nakamura, T.; Kitazawa, S.; et al. Detection of a high-density brachiolaria-stage larval population of crown-of-thorns sea star (*Acanthaster planci*) in Sekisei Lagoon (Okinawa, Japan). *Diversity* **2016**, *8*, 9. [CrossRef]
20. McKinnon, A.; Duggan, S.; De'ath, G. Mesozooplankton dynamics in nearshore waters of the Great Barrier Reef. *Estuar. Coast. Shelf Sci.* **2005**, *63*, 497–511. [CrossRef]

21. Fabricius, K.; De'ath, G.; McCook, L.; Turak, E.; Williams, D.M. Changes in algal, coral and fish assemblages along water quality gradients on the inshore Great Barrier Reef. *Mar. Pollut. Bull.* **2005**, *51*, 384–398. [CrossRef] [PubMed]

22. Furnas, M.J.; Mitchell, A.W.; Gilmartin, M.; Revelante, N. Phytoplankton biomass and primary production in semi-enclosed reef lagoons of the central Great Barrier Reel Australia. *Coral Reefs* **1990**, *9*, 1–10. [CrossRef]

23. Van Woesik, R.; Tomascik, T.; Blake, S. Coral assemblages and physico-chemical characteristics of the Whitsunday Islands: Evidence of recent community changes. *Mar. Freshw. Res.* **1999**, *50*, 427–440. [CrossRef]

24. Le Borgne, R.; Rodier, M.; Le Bouteiller, A.; Kulbicki, M. Plankton biomass and production in an open atoll lagoon: Uvea, New Caledonia. *J. Exp. Mar. Biol. Ecol.* **1997**, *212*, 187–210. [CrossRef]

25. Rochelle-Newall, E.J.; Torréton, J.P.; Mari, X.; Pringault, O. Phytoplankton-bacterioplankton coupling in a subtropical South Pacific coral reef lagoon. *Aquat. Microb. Ecol.* **2008**, *50*, 221–229. [CrossRef]

26. Torréton, J.P.; Rochelle-Newall, E.; Pringault, O.; Jacquet, S.; Faure, V.; Briand, E. Variability of primary and bacterial production in a coral reef lagoon (New Caledonia). *Mar. Pollut. Bull.* **2010**, *61*, 335–348. [CrossRef] [PubMed]

27. Charpy, L.; Dufour, P.; Garcia, N. Particulate organic matter in sixteen Tuamotu atoll lagoons (French Polynesia). *Mar. Ecol. Ser.* **1997**, *151*, 55–65. [CrossRef]

28. Sakka, A.; Legendre, L. Carbon budget of the planktonic food web in an atoll lagoon (Takapoto, French Polynesia). *J. Plankton Res.* **2002**, *24*, 301–320. [CrossRef]

29. Ferrier-Pagès, C.; Furla, P. Pico- and nanoplankton biomass and production in the two largest atoll lagoons of French Polynesia. *Mar. Ecol. Prog. Ser.* **2001**, *211*, 63–76. [CrossRef]

30. Charpy, L.; Rodier, M.; Fournier, J.; Langlade, M.J.; Gaertner-Mazouni, N. Physical and chemical control of the phytoplankton of Ahe lagoon, French Polynesia. *Mar. Pollut. Bull.* **2012**, *65*, 471–477. [CrossRef] [PubMed]

31. Nakajima, R.; Yoshida, T.; Othman, B.; Toda, T. Biomass and estimated production rates of metazoan zooplankton community in a tropical coral reef of Malaysia. *Mar. Ecol.* **2014**, *35*, 112–131. [CrossRef]

32. Nakajima, R.; Tsuchiya, K.; Nakatomi, N.; Yoshida, T.; Tada, Y.; Konno, F.; Toda, T.; Kuwahara, V.S.; Hamasaki, K.; Othman, B.H.R.; et al. Enrichment of microbial abundance in the sea-surface microlayer over a coral reef: Implications for biogeochemical cycles in reef ecosystems. *Mar. Ecol. Prog. Ser.* **2013**, *490*, 11–22. [CrossRef]

33. Wild, C.; Huettel, M.; Klueter, A.; Kremb, S.G. Coral mucus functions as an energy carrier and particle trap in the reef ecosystem. *Nature* **2004**, *428*, 66–70. [CrossRef] [PubMed]

34. Brown, B.E.; Bythell, J.C. Perspectives on mucus secretion in reef corals. *Mar. Ecol. Prog. Ser.* **2005**, *296*, 291–309. [CrossRef]

35. Wild, C.; Naumann, M.; Niggl, W.; Haas, A. Carbohydrate composition of mucus released by scleractinian warm- and cold-water reef corals. *Aquat. Biol.* **2010**, *10*, 41–45. [CrossRef]

36. Huettel, M.; Wild, C.; Gonelli, S. Mucus trap in coral reefs: Formation and temporal evolution of particle aggregates caused by coral mucus. *Mar. Ecol. Prog. Ser.* **2006**, *307*, 69–84. [CrossRef]

37. Naumann, M.S.; Richter, C.; El-Zibdah, M.; Wild, C. Coral mucus as an efficient trap for picoplanktonic cyanobacteria: Implications for pelagic-benthic coupling in the reef ecosystem. *Mar. Ecol. Prog. Ser.* **2009**, *385*, 65–76. [CrossRef]

38. Hata, H.; Kudo, S.; Yamano, H.; Kurano, N.; Kayanne, H. Organic carbon flux in Shiraho coral reef (Ishigaki Island, Japan). *Mar. Ecol. Prog. Ser.* **2002**, *232*, 129–140. [CrossRef]

39. Nakajima, R.; Tanaka, Y. The role of coral mucus in the material cycle in reef ecosystems: Biogeochemical and ecological perspectives. *J. Jpn. Coral Reef Soc.* **2014**, *16*, 3–27. [CrossRef]

40. Richman, S.; Loya, Y.; Slobodkin, L. The rate of mucus production by corals and its assimilation by the coral reef copepod Acartia negligens. *Limnol. Oceanogr.* **1975**, *20*, 918–923. [CrossRef]

41. Gottfried, M.; Roman, M.R. Ingestion and incorporation of coral-mucus detritus by reef zooplankton. *Mar. Biol.* **1983**, *72*, 211–218. [CrossRef]

42. Naumann, M.S.; Haas, A.; Struck, U.; Mayr, C.; el-Zibdah, M.; Wild, C. Organic matter release by dominant hermatypic corals of the Northern Red Sea. *Coral Reefs* **2010**, *29*, 649–659. [CrossRef]

43. Crossland, C. In situ release of mucus and DOC-lipid from the corals Acropora variabilis and Stylophora pistillata in different light regimes. *Coral Reefs* **1987**, *6*, 35–42. [CrossRef]

44. Vogler, C.; Benzie, J.; Lessios, H.; Barber, P.; Wörheide, G. A threat to coral reefs multiplied? Four species of crown-of-thorns starfish. *Biol. Lett.* **2008**, *4*, 696–699. [CrossRef] [PubMed]

45. Haszprunar, G.; Spies, M. An integrative approach to the taxonomy of the crown-of-thorns starfish species group (Asteroidea: *Acanthaster*): A review of names and comparison to recent molecular data. *Zootaxa* **2014**, *3841*, 271–284. [CrossRef] [PubMed]

46. Naumann, M.S.; Mayr, C.; Struck, U.; Wild, C. Coral mucus stable isotope composition and labeling: Experimental evidence for mucus uptake by epizoic acoelomorph worms. *Mar. Biol.* **2010**, *157*, 2521–2531. [CrossRef]

47. Hoegh-Guldberg, O.; Ayukai, T. *Assessment of the Role of Dissolved Organic Matter and Bacteria in the Nutrition of Crown-of-Thorns Starfish Larva*; Australian Institute of Marine Science: Townsville, Australia, 1992.

48. Hoegh-Gulberg, O. Is *Acanthaster planci* able to utilise dissolved organic matter (DOM) to satisfy the energetic requirements of larval development? In *The Possible Causes and Consequences of Outbreaks of the Crown-of-Thorns Starfish*; Great Barrier Reef Marine Park Authority Workshop: Townsville, Australia, 1992; pp. 37–54.

49. Wyatt, A.S.J.; Lowe, R.J.; Humphries, S.; Waite, A.M. Particulate nutrient fluxes over a fringing coral reef: Source—Sink dynamics inferred from carbon to nitrogen ratios and stable isotopes. *Limnol. Oceanogr.* **2013**, *58*, 409–427. [CrossRef]

50. Wolfe, K.; Graba-Landry, A.; Dworjanyn, S.A.; Byrne, M. Larval phenotypic plasticity in the boom-and-bust crown-of-thorns seastar, *Acanthaster planci*. *Mar. Ecol. Prog. Ser.* **2015**, *539*, 179–189. [CrossRef]

Article

The Influence of Water Currents on Movement Patterns on Sand in the Crown-of-Thorns Seastar (*Acanthaster* cf. *solaris*)

Robert Sigl * and Christian Laforsch *

Department Animal Ecology I, University of Bayreuth and BayCEER, Universitaetsstr. 30,
95447 Bayreuth, Germany
* Correspondence: Robert.Sigl@uni-bayreuth.de (R.S.); Christian.Laforsch@uni-bayreuth.de (C.L.);
  Tel.: +49-921-55-2648 (R.S.); +49-921-55-2650 (C.L.)

Academic Editor: Morgan Pratchett
Received: 9 October 2016; Accepted: 17 November 2016; Published: 25 November 2016

Abstract: Outbreaks of the coral-eating crown-of-thorns seastar (*Acanthaster* cf. *solaris*) threaten coral reefs of the Indo-Pacific. Movement patterns may play an important role in the spread of outbreak populations, but studies investigating adult movement behavior are scarce. It remains unknown if *Acanthaster* cf. *solaris* orientates in inter-reef areas using chemical, visual, or mechanical cues (e.g., water currents) or which trigger is used for the onset of movement. We investigated the movement patterns of adult starved, fed, and blinded *A.* cf. *solaris* on sand at two sites with different unidirectional water current strengths. We found that the movement direction of the seastars in strong currents was downstream, whereas movement in weaker currents was random and independent from the current direction. However, the directionality of movement was consistently high, independent of the nutritional state, its visual abilities, or current strength. Starved *A.* cf. *solaris* started to move significantly faster compared to fed individuals. Therefore, starvation might trigger the onset of movement. Our findings indicate that navigation of *A.* cf. *solaris* in inter-reef areas is inefficient. Movements between reefs may be random or current-dependent and finding a new reef from a distance subject to chance, unless it is only few meters away.

Keywords: movement; migration; outbreak; *Acanthaster* cf. *solaris*; water current; orientation; navigation; rheotaxis; crown-of-thorns seastar

---

1. Introduction

Coral eating crown-of-thorns seastars (*Acanthaster* cf. *solaris*) are abundant inhabitants of many coral reefs. Their specialized diet and frequent occurrence in high-density populations have caused extensive damage to coral reefs in the Indo-Pacific region for several decades [1]. During population outbreaks thousands of adult individuals may appear on a reef and move across it in feeding-bands [2]. Over time, local coral resources get depleted and the moving seastars reach the end of a reef or face a channel. In these situations *A.* cf. *solaris* may leave its preferred reef substrate and move on the bare sandy bottom in the search of reef areas with sufficient coral cover [3–6]. Although the spatial scales and individual numbers of migrations are not yet known, migrating adult *A.* cf. *solaris* may subsequently contribute to the spread of outbreak populations or cause other reefs to be infected [7,8]. Therefore, it is of considerable interest what triggers the onset of their movement and what cues might be used by *A.* cf. *solaris* to direct their movement and orientate themselves between reefs, on sandy substrate bare of suitable food.

Several variables have been proposed to influence the movements of *A.* cf. *solaris*: density of corals, exposure to wave action, temperature, time of day, light, and type of substratum, but also age or size, condition, or nutritional state [9–11]. Most of these variables, however, only influence the speed of seastar movement or cause the seastar to actively avoid unpleasant areas. The direction of movement, however, may be guided by the presence of visual cues [12–14], chemical cues (from food [15], predators [16], or conspecifics [17]), or water currents [18]. The visual sense of *A.* cf. *solaris* was shown to be involved in navigation on sand towards reef structures from a distance of several meters [13,14]. To successfully navigate between reefs, however, the reef structures have to be in sight, limiting the use of visual cues for long distance orientation on large, bare, sandy areas [14]. It is commonly assumed that *A.* cf. *solaris* navigate using their well-developed chemical sense [19]. However, chemoreception in seastars has only proven its functionality over short distances [20]. If chemoreception is used from a distance, then chemical cues need to be carried towards the seastar by water currents. The seastars may then follow a gradient of prey odors towards their source [21]. For *A.* cf. *solaris* navigating towards its coral food on the reef using chemoreception may be difficult, as the cues from corals are omnipresent and the flow patterns may be intricate and turbulent [14,22]. However, localization of prey using chemical cues is most effective in subtidal environments and where currents flow in one direction [20], and this may be the case in inter-reef areas. Seastars are also known to combine chemotaxis with their ability to perceive the mechanical stimulation by currents (rheotaxis). In the presence of chemical cues from food they may show positive rheotaxis (moving against the current) [22–24] or cross-current movement [18]. Stronger currents may thereby enhance the capability of seastars to locate their prey [18]. In this context, the responsiveness to chemical cues from food is frequently increased by starvation [25–28] and also results in a more consistent movement behavior [18]. This may help *A.* cf. *solaris* to overcome even larger distances between reefs where no food is available.

We aimed to investigate the potential of adult *A.* cf. *solaris* to navigate between reefs. In particular, we tried to identify cues, which may be used for orientation and factors, which trigger the onset of movement. We therefore analyzed the movement patterns of starved, fed and blinded *A.* cf. *solaris* in the field on sandy substrates under different strengths of water currents. Depending on the senses (vision, chemoreception, or mechanoreception) *A.* cf. *solaris* might use for orientation, we expected to see differences in movement patterns between treatments, such as an orientation of the seastars towards the reef or along water currents. In particular, we expected starved seastars to show the most directional movement and an orientation towards the reef (food source), while fed seastars may be less directed in their movement, but still being oriented towards the reef (seeking food and shelter). Blinded seastars should either not differ in their movement patterns from starved or fed, if the chemical sense is primarily used for orientation between reefs, or should show different movement patterns, if the visual sense is used for orientation.

2. Materials and Methods

2.1. Study Site

Two sites on the island Mo'orea, Society Islands, French Polynesia (Figure 1a,c), named Temae (in the northwest; 17°29′52.43″ S; 149°45′28.61″ W) and Maharepa (in the north; 17°28′52.79″ S; 149°48′56.99″ W) were chosen for the experiments. Both areas consisted of sandy ground with few interspersed living coral heads. At both sites the water carried towards the seastars had to pass the reef crest before entering the sandy experimental area and may, therefore, have carried chemical cues from corals.

Temae was situated in a large lagoon that was delimited by a sandy beach to the west and otherwise surrounded by fringing or patch reefs on each side. The starting point was located approximately 50 m from the beach and 370 m from the surrounding atoll reef crest at a water depth of about 2.5 m. A consistent, relatively strong current (>0.15 m·s^{-1}), whose direction was independent from the tides was running from north to south, more or less parallel to the beach (Figure 1a,b).

Maharepa was located at a sandflat behind a backreef in about 3–5 m depth. To the south of the sandflat the ground dropped into a channel, which was approximately 25 m deep. The incoming tides produced a current, approximately 40% weaker than in Temae, that ran over the fringing reef towards to shore. We timed the experiments with the incoming tides so that they produced a consistent current over the reef perpendicular to the shore (Figure 1c,d).

Although one location was chosen in the lagoon and the other at the backreef, both areas provided similar conditions as both were downstream of an outer reef (for Maharepa this accounts only for the incoming tide), both had comparable water depths (2.5 m vs. 3–5 m), and at both the ground was covered to almost 100% by sand.

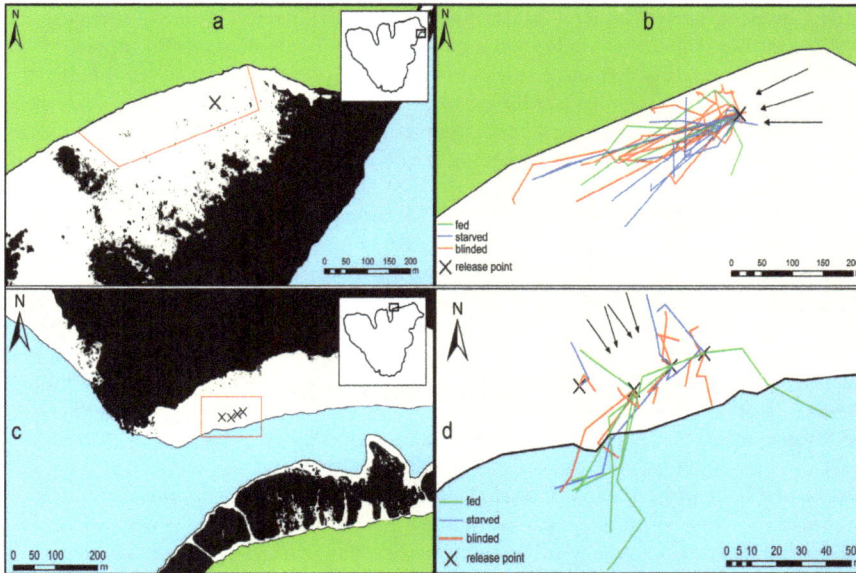

Figure 1. Experimental sites and movement tracks of *Acanthaster* cf. *solaris* on sand. The green-colored area is land and the beige-colored areas consist of sandy substrate at water depths of 0–5 m. The blue-colored area represents water depths of >5 m. The black line represents the shoreline or the division of water depth greater or smaller than 5 m. Black areas represent reef areas with >30% hard substrate. (**a**) Overview of the experimental site Temae. The red line indicates the area that was systematically searched for *A.* cf. *solaris* six times a day. The box in the upper right shows the position of the experimental site around the Island of Mo'orea; (**b**) movement tracks of *A.* cf. *solaris* in Temae (strong current) recorded during a 2.5-week period. Arrows indicate the direction of the current; (**c**) overview of the experimental site Maharepa. The red box indicates the area that was systematically searched for *A.* cf. *solaris* hourly for 3 h. The box in the upper right shows the position of the experimental site around the Island of Mo'orea; and (**d**) movement tracks of *A.* cf. *solaris* in Maharepa (weaker current) recorded during 3 h over a period of three days. Arrows indicate the direction of the current.

2.2. General Experimental Procedure

The experiments took place in November and December 2013. *A.* cf. *solaris* were collected at several locations around Mo'orea and transported to the R.B. Gump South Pacific Research Station. Eleven individuals each (diameter ~25–45 cm) were randomly assigned to the treatments 'starved', 'fed', and 'blinded' for each of the two experimental sites. Individuals of the "starved" treatment were kept without food in large plastic bins supplied with seawater from the ocean for 3–6 weeks. Individuals for the "fed" and "blinded" treatment were collected 1–4 days prior to the experiment and transported to the station. Individuals in the 'blinded' treatment were anaesthetized using 3.5%

magnesium-chloride hexahydrate (Mg$_2$Cl × 6H$_2$O) mixed with seawater (procedure adapted from Messenger et al. [29]) prior to dissection. Subsequently the terminal ossicle harboring the terminal tube foot containing the compound eye was carefully removed using a pair of scissors one day prior to the experiments.

Depending on the experimental site the seastars were transported either by car (Temae) or by boat (Maharepa) in large coolers filled with seawater (duration of transport approximately 25 min and 15 min, respectively). *A.* cf. *solaris* were marked individually using numbered plastic tags (~12 × 15 mm pieces of flagging tape) pulled over an aboral spine. The seastars were released in a shelter (size approximately 50 × 50 × 30 cm) built out of dead coral rock that was collected within a radius of 150 m. The shelter was used to allow the seastars to start moving freely without influencing their initial movement direction and to provide them with a safe hiding place so that the onset of movement was not influenced by handling and exposure. At the same time this design had the disadvantage that not all individuals actually had moved out of the shelters after 1–3 h, therefore, different sample sizes emerged for the observations on movement directions. Additionally, not all seastars could be recaptured after 3 h.

In Temae one shelter was constructed and three seastars (one individual from each treatment) were placed in the shelter on each day a trial was conducted. In total, 11 trials were conducted during 2.5 weeks (Supplementary Materials Figure S1) resulting in 33 individuals released during this period of time. The tagged individuals were not collected after release to enable the recording of their movement over the period of 2.5 weeks. The experiment always started at 10 a.m. and the shelters were visited hourly until 1 p.m. Moreover, they were also checked once in the afternoon (3–4 p.m.) and in the evening (9–10 p.m.), in order to record how long *A.* cf. *solaris* individuals were staying in the shelters, if they had not left after 3 h. Additionally, at these times a large area around the release point was searched systematically by snorkeling transects parallel to the shore using a compass and underwater landmarks and the positions of the seastars were recorded (Figure 1a).

The experiment in Maharepa consisted of 10 trials, which were performed during three consecutive days. Four shelters were built in a line parallel to the reef of which only three were used in the first two days and the fourth only at the third day of the experiment. After each trial the released individuals were collected and transported back to the research station. The shelters were slightly smaller than the one built in Temae, measuring approximately 30 × 30 × 30 cm. To prevent potential effects of higher densities of conspecifics in the area as compared to Temae, shelters were set up approximately 20 m apart. This ensured that individuals were not able to detect each other by visual means [14]. In addition, the direction of the prevailing current should have prevented the seastars released in each shelter to sense each other chemically because it ran perpendicular to the row of shelters. The release procedure was identical to that described in Temae, placing one individual of each treatment in one shelter, with the only exception that all three to four shelters were equipped with seastars quickly one after another. *A.* cf. *solaris* movements were tracked using GPS (GPSMap 60 CSx; Garmin International, Inc., Olathe, KS, USA) hourly for 3 h after they were released.

The displacement and the direction of movement were analyzed using ArcMap (ArcGIS 10.2.1, ESRI, Redlands, CA, USA) and the ArcMET tool [30]. The water current direction was determined to the nearest 5° using a compass and an underwater plastic flag and checked during the experiments to ensure it was not changing. The strength of the current was determined each day prior to the start of the experiment using a neutrally buoyant plastic piece allowed to drift underwater for 2 m and a stopwatch. The mean of three such measurements was considered as the current velocity.

2.3. Data Analysis

To test for a preferred movement direction in *A.* cf. *solaris* the compass headings obtained after one and three hours, and several days of movement, were analyzed using the Rao's spacing test. This test was applied because the data were partly diametrically bimodal and violated the von Mises distribution assumption [31]. The directionality of movement (a measure of how close the movement resembles a straight path) was analyzed by calculating the $D:W_{all}$ value [32], where D is the shortest distance from the starting point to the end position of the seastar (after three hours of movement) and W_{all} is the total distance travelled during one observation, thus, the sum of the distances travelled after one, two, and three hours. Moving in a straight line means displaying perfect directional movement and would, therefore, result in a $D:W_{all}$ value of 1. The smaller the value gets, the less directional the movement is. A $D:W_{all}$ value of >0.7 is considered to be 'highly directional', a value of >0.5 as 'partly directional' and a value of <0.5 is 'undirected' movement [32,33]. The data on the duration of stay in shelters violated the normality and homogeneity of variances assumptions. Hence, to test for a general difference between treatments in the duration of stay in shelters a Kruskal-Wallis test was applied. Single treatments were then compared using Mann-Whitney tests.

3. Results

The current produced by the tides in Maharepa was generally weaker than the tide-independent current present in Temae (Tables 1 and 2). The movement tracks of the seastars in Temae during the 2.5-week period are shown in Figure 1b and the ones in Maharepa, recorded during 3 h, in Figure 1d. Most of the individuals from the fed treatment in Temae stayed for more than three hours in the shelter provided; therefore, no data on movement patterns could be obtained on them during this time. However, their movement patterns could be obtained from the observation for several days (Section 3.3). The mean distances covered by *A.* cf. *solaris* during 1 and 3 h of movement, from which the directionality of movement was calculated, are shown in Table 3. The observation of the seastars during all experiments revealed that they always actively moved using their tube feet, even in strong currents, and were not passively transported by them.

3.1. Movement Patterns after One Hour

In Temae (strong current, see Table 1) starved *A.* cf. *solaris* showed significant preference for a common direction of movement and blinded *A.* cf. *solaris* tended towards a common directional preference that approximately resembled the direction of the currents present (255°–270°). For fed individuals a statistical analysis was not possible, as only one individual had left the shelter after 1 h. In Maharepa (weak current, see Table 1) no common preference for a certain direction could be detected in any of the treatments (Table 1). The water currents in Maharepa came from between 320° and 340° and no general rheotactic orientation could be detected in the movement tracks of the seastars.

Table 1. Mean movement direction of *Acanthaster* cf. *solaris* after one hour and test for a preferred direction (Rao's spacing test) at two sites with respective water current strengths. Note that the circular mean direction is not very meaningful when there is no preferred direction in the seastars' movement (when Rao's spacing test is not significant). * Refers to significant test $p < 0.05$.

Site	Treatment	N	Circular Mean Direction	Mean Current Velocity [m·s⁻¹]	Rao's Spacing Test p-Value
Temae	blinded	8	232.9°	0.163 ± 0.076	0.056
Temae	starved	6	251.3°	0.163 ± 0.076	0.026 *
Temae [1]	fed	1	265.8°	0.163 ± 0.076	-
Maharepa	blinded	10	242.0°	0.096 ± 0.029	0.400
Maharepa	starved	5	275.9°	0.096 ± 0.029	0.506
Maharepa	fed	6	230.7°	0.096 ± 0.029	0.333

[1] Only one individual could be observed in this treatment, therefore no mean ± SD or statistical results are given.

3.2. Movement Patterns after Three Hours

The directionality of movement of individual *A.* cf. *solaris* was high ($D:W_{all}$ value ≥ 0.7), regardless of site or treatment. Starved and blinded *A.* cf. *solaris* in Temae showed a significant preferred direction of movement that resembled the direction of the currents present (255°–270°).

In contrast, no preferred direction could be observed in blinded, starved or fed *A.* cf. *solaris* at the location with weaker currents i.e., Maharepa (Table 2).

Table 2. Mean movement direction of *Acanthaster* cf. *solaris* after 3 h, test for a preferred direction (Rao's spacing test) and D:W$_{all}$ value at two sites with respective water current strengths. Note that the circular mean direction is not very meaningful when there is no preferred direction in the seastars movement (when Rao's spacing test is not significant). * Refers to significant test $p < 0.05$.

Site	Treatment	N	D:W$_{all}$ ± SD	Circular Mean Direction	Mean Current Velocity [m·s^{-1}]	Rao's Spacing Test p-Value
Temae	blinded	6	0.83 ± 0.27	257.0°	0.183 ± 0.062	0.004 *
Temae	starved	4	0.89 ± 0.17	258.8°	0.183 ± 0.062	0.011 *
Temae [1]	fed	1	0.98	257.4°	0.183 ± 0.062	-
Maharepa	blinded	10	0.70 ± 0.18	230.9°	0.096 ± 0.029	0.500
Maharepa	starved	4	0.85 ± 0.15	319.7°	0.096 ± 0.029	0.818
Maharepa	fed	5	0.79 ± 0.07	196.9°	0.096 ± 0.029	0.074

[1] Only one individual could be observed in this treatment, therefore no mean ± SD or statistical results are given.

Table 3. Mean distance covered after one and three hours of *A.* cf. *solaris*.

Site	Treatment	N 1st h	Mean Displacement ± SD [m] after 1 h	N 3rd h	Mean Displacement (D) ± SD [m] after 3 h
Temae	blinded	8	16.7 ± 10.5	6	49.8 ± 27.7
Temae	starved	6	13.4 ± 5.6	4	26.3 ± 13.4
Temae [1]	fed	1	11.8	1	27.9
Maharepa	blinded	10	7.2 ± 5.7	10	15.7 ± 10.2
Maharepa	starved	5	12.2 ± 11.3	4	29.7 ± 17.3
Maharepa	fed	6	27.4 ± 7.9	5	49.8 ± 8.9

[1] Only one individual could be observed in this treatment, therefore no mean ± SD or statistical results are given.

3.3. Movement Patterns after Several Days in Temae

The analysis of the positions of *A.* cf. *solaris* from all treatments after more than 3 h of movement revealed a significant common directional preference (Rao's spacing test: fed ($N = 11$): $U = 244$, $p < 0.001$, circular mean direction: 245°; starved ($N = 8$): $U = 248.2$, $p = 0.002$, circular mean direction: 235°; blinded (11): $U = 208$, $p = 0.001$, circular mean direction: 248°) to move downstream (Figure 1b, Table S1). Only one individual from the blinded treatment and one from the starved treatment had its last recorded position upstream of the release point. The amount of time the movement of single individuals could be tracked was between 3 h and 11 days, however, most individuals could not be followed for longer than six days. Additionally, the tracking time was shortened, as some individuals did not leave the shelter for several hours or days (see Section 3.4). For an overview of release and tracking times, and numbers and distances moved, please see Figure S1.

3.4. Duration of Stay in a Shelter

A. cf. *solaris* from different treatments significantly differed in the time they stayed in the shelter (Kruskal-Wallis test; $p = 0.013$). Starved *A.* cf. *solaris* stayed for a significantly shorter time in the shelter than fed ones (Mann-Whitney-Test; $U = 27.5$, $p = 0.028$). Fed individuals remained in the shelter for 28 ± 11 h. All except one fed individual stayed longer than one hour and one fed individual stayed for five days in the shelter. In contrast, the starved individuals left the shelter after 5 ± 2 h and half of the starved *A.* cf. *solaris* left the shelter less than one hour after their release. Blinded individuals did not differ in their duration of stay from starved ones (Mann-Whitney-Test; $U = 51.0$, $p > 0.05$), but differed significantly from fed ones, too (Mann-Whitney-Test; $U = 21.5$, $p = 0.008$, Figure 2).

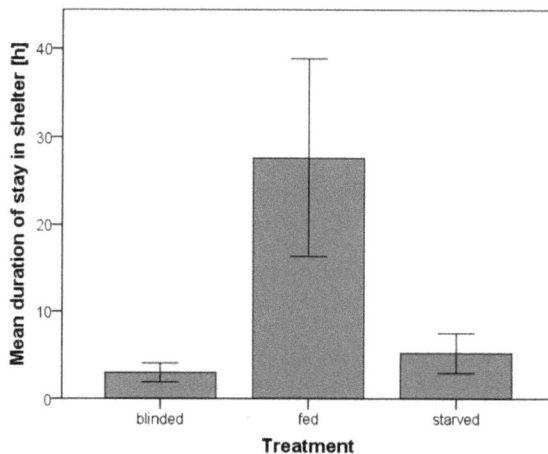

Figure 2. Mean duration of stay in a shelter of *Acanthaster* cf. *solaris*. Error bars represent the standard error.

4. Discussion

A trigger for *A.* cf. *solaris* to start its movements between reefs may be its state of nourishment. Our experiment showed that starved individuals left a safe shelter almost immediately after they were placed there, indicating that a poor nutritional state induces migration towards a new food patch. In contrast, fed individuals stayed there, sometimes for days. Blinded individuals were also in a good nutritional state, but started as early as starved individuals, a behavior that contradicts the hypothesis that nourishment induces movement. However, blinded individuals were obviously not behaving like starved or fed individuals. Some of them ended up very close to the shore (in Temae) and when they were recaptured they were often moving on sand, while others had found a small reef block to hide or feed. This supports the finding that vision is playing an important role in the orientation of *A.* cf. *solaris* [13,14]. At the same time, the lack of visual information may have caused their early departure, as they may not have recognized being in a safe location, in the shade of a shelter.

The analysis of the movement patterns after the seastars had left the shelter indicate that there is a relationship between the strength of the water current and the direction of *A.* cf. *solaris* movements on sand. Surprisingly, in strong current *A.* cf. *solaris* followed its direction and showed negative rheotaxis independent of their nutritional state or their ability to see, even after several days of observation. This behavior is uncommon as often seastars show positive rheotaxis [22,34–36], even in the absence of chemical cues in the water [24]. Castilla and Crisp [34] found a reversal of the normally observed positive rheotaxis in *Asterias rubens* in the laboratory to be caused by the following factors: a sudden increase of the sea water temperature, a reduction of the sea water salinity below 25‰ S, a drop in the oxygen tension below 4.18 mL O_2/L, a pH of less than 6.9, and long periods of captivity under starvation. Although water parameters were not recorded in the study area, such dramatic changes are highly unlikely to have occurred during the experiments, especially given that no storms, heavy rains or swells took place. In addition, strong surge, which is suggested to alter *A.* cf. *solaris* movements [37], can be excluded as an influencing factor, as experiments were only performed during very calm seas. Seastars may show negative rheotaxis in the presence of a cue that indicates potential predators or harmful conditions [34]. However, this is also highly unlikely to have caused the observed movement patterns as such reactions have only been shown as a response to an immediate predator cue present in the water [17,38,39] and, therefore, the immediate presence of a predator, which was not observed during the time of the experiment.

A factor that indeed might have caused the negative rheotaxis observed in the present study was starvation. However, blinded individuals also showed negative rheotaxis in strong currents, which is not directly attributable to their nutritional state. Additionally, instead of showing negative rheotaxis, other seastars even enhance their positive responsiveness to cues when starved [18,27,40]. It might be assumed that the sand did not provide enough gripping surface to the tube feet and the seastars might, therefore, have been passively transported by the current rather than actively moving with it [3]. However, the observations made during the experiments show that the seastars moved actively using their tube feet and could still change their direction of movement to the side and that several individuals could also move against the current. One remaining explanation for the observed downstream movement is that the seastars actively move with the current if it is too strong to save energy. The fact that *A.* cf. *solaris* showed movement in random directions and no general downstream movement in weaker currents support this hypothesis. In even weaker currents (mean velocity: 0.054 m·s^{-1}) and on shorter timescales, also, no influence of currents on movement direction in *A.* cf. *solaris* was observed [32]. The only other field study that investigated movement patterns of tagged *A.* cf. *solaris* was conducted on a solid reef structure, not like the present one on sandy substrate. Movement patterns here were random, as in the present study during weak currents, but no clear measurements of current strengths were made stating only that currents moved strongly in 'both directions parallel to the shore' [5]. Such inconsistent movement directions as a response to a constant cue were also often found in chemoreception-studies of other seastars (reviewed by Sloan and Campbell [23]). Still, in the present study, random movement directions were consistently shown over all treatments during weak currents. This suggests that either no cue for orientation could be detected by *A.* cf. *solaris* or the cues were detected but the seastars did not react to them.

The probability that chemical cues from coral food were present in the water was very high, as in both experimental sites the water carried towards the seastars first had to pass the reef crest. Although, the water was not tested for chemical traces of corals, the question is why *A.* cf. *solaris* was not moving towards the reef in the vicinity at all, considering that especially starved individuals should try to reach feeding grounds. This challenges the ability of *A.* cf. *solaris* to successfully navigate between reefs and maybe even the functionality of chemoreception for long distance navigation. At the same time, the fact that blinded seastars followed the same inconsistent movement in weak currents as fed and starved ones implies that visual cues, which have been proven to be important in short-distance navigation [14], were also not responsible for the observed movement patterns at long distances. The observation that some of the individuals in Maharepa had started moving into the deep water of the channel south to the release point, and that others in Temae had moved into shallower water in the direction of the beach, does rather imply that movement on larger scales between reefs may be random. Additionally, directed navigation or orientation seems to be limited to the very presence of a distinct chemical or visual cue only a few meters away. Still, in both experiments, movement of single individuals was generally highly directional, which may prevent re-encountering already traversed areas [33]. Such a high directionality of movement has already been shown for *A.* cf. *solaris*, although in shorter time scales and on artificial substrate [32]. For movements between reefs this may be beneficial, as the distance covered is maximized, however, it still needs to move in the right direction to find a reef. A directed movement towards the reef was only shown in one individual from each treatment in Maharepa, which underpins that *A.* cf. *solaris* is not efficient in finding reefs from a distance.

5. Conclusions

A. cf. *solaris* in strong currents showed downstream movement, in contrast to weaker currents where movement was random; however, movement was not directed towards the reef in both experiments. This indicates that the senses of *A.* cf. *solaris* are generally not well suited to locate distant reefs. It remains unknown if *A.* cf. *solaris* uses specific cues to direct its movement between reefs, although the trigger for the onset of movement may well be a poor nutritional state. Importantly, the negative rheotaxis (movement with the current) shown might protect certain reefs from being invaded by *A.* cf. *solaris*, when currents between reefs are strong and consistently flow into one direction causing the seastar to move parallel to, or away from, the reef.

Supplementary Materials: The following are available online at www.mdpi.com/1424-2818/8/4/25/s1, Table S1: Compass directions after several hours or days of *A.* cf. *solaris* in Temae, Figure S1. Overview of release and tracking time and date and distances moved in-between by *A.* cf. *solaris* in Temae.

Acknowledgments: Financial support for Robert Sigl by the Cusanuswerk is gratefully acknowledged. We want to thank Maximilian Schweinsberg and the staff of the Richard B. Gump South Pacific Research Station for their help. This publication was funded by the German Research Foundation (DFG) and the University of Bayreuth in the funding program Open Access Publishing.

Author Contributions: R.S. and C.L. conceived, designed and performed the experiments; R.S. analyzed the data; R.S. and C.L. wrote the paper.

Conflicts of Interest: The authors declare no conflict of interest. The founding sponsors had no role in the design of the study; in the collection, analyses, or interpretation of data; in the writing of the manuscript, and in the decision to publish the results.

References

1. Pratchett, M.S.; Caballes, C.F.; Rivera-Posada, J.; Sweatman, H.P.A. Limits to understanding and managing outbreaks of crown-of-thorns starfish (*Acanthaster planci* spp.). In *Oceanography and Marine Biology: An Annual Review*; Hughes, R.N., Hughes, D.J., Smith, I.P., Eds.; CRC Press: Boca Raton, FL, USA, 2014; pp. 133–200.
2. Birkeland, C.; Lucas, J. *Acanthaster Planci: Major Management Problem of Coral Reefs*; CRC Press: Boca Raton, FL, USA, 1990.
3. Chesher, R.H. Destruction of Pacific corals by the sea star *Acanthaster planci*. *Science* **1969**, *165*, 280–283. [CrossRef] [PubMed]
4. Endean, R. Report on Investigations Made into Aspects of the Current *Acanthaster planci* (crown-of-thorns) Infestations of Certain Reefs of the Great Barrier Reef. Available online: http://trove.nla.gov.au/work/24663525?q&versionId=29777393 (accessed on 29 October 2016).
5. Branham, J.M.; Reed, S.; Bailey, J.H.; Caperon, J. Coral-eating sea stars *Acanthaster planci* in Hawaii. *Science* **1971**, *172*, 1155–1157. [CrossRef] [PubMed]
6. Moran, P.J. Preliminary observations of the decomposition of crown-of-thorns starfish, *Acanthaster planci* (L.). *Coral Reefs* **1992**, *11*, 115–118. [CrossRef]
7. Branham, J.M. Crown of thorns on coral reefs. *Bioscience* **1973**, *23*, 219–226. [CrossRef]
8. Kayal, M.; Vercelloni, J.; Lison de Loma, T.; Bosserelle, P.; Chancerelle, Y.; Geoffroy, S.; Stievenart, C.; Michonneau, F.; Penin, L.; Planes, S.; et al. Predator crown-of-thorns starfish (*Acanthaster planci*) outbreak, mass mortality of corals, and cascading effects on reef fish and benthic communities. *PLoS ONE* **2012**, *7*, e47363. [CrossRef] [PubMed]
9. Moran, P. The *Acanthaster* phenomenon. *Oceanogr. Mar. Biol. Annu. Rev.* **1986**, *24*, 379–480.
10. Ormond, R.F.; Campbell, A.C. Formation and breakdown of *Acanthaster planci* aggregations in the Red Sea. In Proceedings of the Second International Coral Reef Symposium, Brisbane, Australia, 22 June–2 July 1973; Cameron, A.M., Cambell, B.M., Cribb, A.B., Endean, R., Jell, J.S., Jones, O.A., Mather, P., Talbot, F.H., Eds.; The Great Barrier Reef Committee: Brisbane, Australia; pp. 595–619.
11. Barham, E.G.; Gowdy, R.W.; Wolfson, F.H. *Acanthaster* (Echinodermata, Asteroidea) in the Gulf of California. *Fish. Bull.* **1973**, *71*, 927–942.
12. Garm, A.; Nilsson, D. Visual navigation in starfish: First evidence for the use of vision and eyes in starfish. *Proc. R. Soc. B Biol. Sci.* **2014**, *281*. [CrossRef] [PubMed]
13. Petie, R.; Hall, M.R.; Hyldahl, M.; Garm, A. Visual orientation by the crown-of-thorns starfish (*Acanthaster planci*). *Coral Reefs* **2016**, *35*, 1139–1150. [CrossRef]
14. Sigl, R.; Steibl, S.; Laforsch, C. The role of vision for navigation in the crown-of-thorns seastar, *Acanthaster planci*. *Sci. Rep.* **2016**, *6*. [CrossRef] [PubMed]
15. Barnes, D.; Brauer, R.; Jordan, M. Locomotory response of *Acanthaster planci* to various species of coral. *Nature* **1970**, *228*, 342–344. [CrossRef] [PubMed]
16. Castilla, J.C.; Crisp, D.J. Responses of *Asterias rubens* to olfactory stimuli. *J. Mar. Biol. Assoc. UK* **1970**, *50*, 829–847. [CrossRef]

17. Campbell, A.; Coppard, S.; D'Abreo, C.; Tudor-Thomas, R. Escape and aggregation responses of three echinoderms to conspecific stimuli. *Biol. Bull.* **2001**, *201*, 175–185. [CrossRef] [PubMed]

18. Rochette, R.; Hamel, J.; Himmelman, J. Foraging strategy of the asteroid *Leptasterias polaris*: Role of prey odors, current and feeding status. *Mar. Ecol. Prog. Ser.* **1994**, *106*, 93–100. [CrossRef]

19. Ormond, R.; Campbell, A.C.; Head, S.H.; Moore, R.J.; Rainbow, P.R.; Saunders, A.P. Formation and breakdown of aggregations of the crown-of-thorns starfish, *Acanthaster planci* (L.). *Nature* **1973**, *246*, 167–169. [CrossRef]

20. Sloan, N. Aspects of the feeding biology of Asteroids. *Oceanogr. Mar. Biol. Annu. Rev.* **1980**, *18*, 57–124.

21. Atema, J. Chemoreception in the sea: Adaptations of chemoreceptors and behaviour to aquatic stimulus conditions. *Symp. Soc. Exp. Biol.* **1985**, *39*, 387–423. [PubMed]

22. Sloan, N.; Northway, S. Chemoreception by the Asteroid *Crossaster papposus* (L.). *J. Exp. Mar. Biol. Ecol.* **1982**, *61*, 85–98. [CrossRef]

23. Sloan, N.; Campbell, A. Perception of food. In *Echinoderm Nutrition*; Jangoux, M., Lawrence, J., Eds.; Balkema: Rotterdam, The Netherlands, 1982; pp. 3–23.

24. Drolet, D.; Himmelman, J.H. Role of current and prey odour in the displacement behaviour of the sea star *Asterias vulgaris*. *Can. J. Zool.* **2004**, *82*, 1547–1553. [CrossRef]

25. Brauer, R.; Jordan, M. Triggering of the stomach eversion reflex of *Acanthaster planci* by coral extracts. *Nature* **1970**, *228*, 344–346. [CrossRef] [PubMed]

26. Valentinčič, T. Food finding and stimuli to feeding in the sea star *Marthasterias glacialis*. *Neth. J. Sea Res.* **1973**, *7*, 191–199. [CrossRef]

27. Ribi, G.; Jost, P. Feeding rate and duration of daily activity of *Astropecten aranciacus* (Echinodermata: Asteroidea) in relation to prey density. *Mar. Biol.* **1978**, *45*, 249–254. [CrossRef]

28. McClintock, J.B.; Lawrence, J.M. Ingestive conditioning in *Luidia clathrata* (Say) (Echinodermata: Asteroidea): Effect of nutritional condition on selectivity, teloreception, and rates of ingestion. *Mar. Behav. Physiol.* **1984**, *10*, 167–181. [CrossRef]

29. McCurley, R.S.; Kier, W.M. The functional morphology of starfish tube feet: The role of a crossed-fiber helical array in movement. *Biol. Bull.* **1995**, *188*, 197–209. [CrossRef]

30. Wall, J. *Movement Ecology Tools for ArcGIS (ArcMET)*. Available online: www.movementecology.net (accessed on 12 August 2014).

31. Jammalamadaka, S.R.; SenGupta, A. *Topics in Circular Statistics*; World Scientific Publishing: Singapore, 2001.

32. Mueller, B.; Bos, A.; Graf, G.; Gumanao, G. Size-specific locomotion rate and movement pattern of four common Indo-Pacific sea stars (Echinodermata; Asteroidea). *Aquat. Biol.* **2011**, *12*, 157–164. [CrossRef]

33. Scheibling, R. Optimal foraging movements of *Oreaster reticulatus* (L.) (Echinodermata: Asteroidea). *J. Exp. Mar. Biol. Ecol.* **1981**, *51*, 173–185. [CrossRef]

34. Castilla, J.C.; Crisp, D.J. Responses of *Asterias rubens* to water currents and their modification by certain environmental factors. *Neth. J. Sea Res.* **1973**, *7*, 171–190. [CrossRef]

35. Fenchel, T. Feeding biology of the sea-star *Luidia sarsi* Düben & Koren. *Ophelia* **1965**, *2*, 223–236.

36. Nickell, T.D.; Moore, P.G. The behavioural ecology of epibenthic scavenging invertebrates in the Clyde Sea area: Laboratory experiments on attractions to bait in moving water, underwater TV observation in situ and general conclusions. *J. Exp. Mar. Biol. Ecol.* **1992**, *159*, 15–35. [CrossRef]

37. Goreau, T.F.; Lang, J.C.; Graham, E.A. Structure and ecology of the Saipan Reefs in relation to predation by *Acanthaster planci* (Linnaeus). *Bull. Mar. Sci.* **1972**, *22*, 113–152.

38. Montgomery, E.M.; Palmer, A.R. Effects of body size and shape on locomotion in the bat star (*Patiria miniata*). *Biol. Bull.* **2012**, *222*, 222–232. [CrossRef] [PubMed]

39. Van Veldhuizen, H.; Oakes, V. Behavioral responses of seven species of Asteroids to the Asteroid predator, *Solaster dawsoni*. *Oecologia* **1981**, *48*, 214–220. [CrossRef]

40. McClintock, J.B.; Lawrence, J. Characteristics of foraging in the soft-bottom benthic starfish *Luidia clathrata* (Echinodermata: Asteroidea): Prey selectivity, switching behavior, functional responses and movement patterns. *Oecologia* **1985**, *66*, 291–298. [CrossRef]

diversity

MDPI

Article

Using Long-Term Removal Data to Manage a Crown-of-Thorns Starfish Population

Masako Nakamura [1,*], Yoshimi Higa [2], Naoki H. Kumagai [3] and Ken Okaji [4]

[1] School of Marine Science and Technology, Tokai University, Shimizu, Shizuoka 424-8610, Japan
[2] Onna Village Fisheries Cooperative, Onna, Okinawa 904-0414, Japan; mozuku@pony.ocn.ne.jp
[3] National Institute for Environmental Studies, Tsukuba, Ibaraki 305-8506, Japan; nh.kuma@gmail.com
[4] Coralquest Inc., Atsugi, Kanagawa 243-0014, Japan; cab67820@pop06.odn.ne.jp
*   Correspondence: mnakamura@tsc.u-tokai.ac.jp; Tel.: +81-54-334-0411

Academic Editors: Morgan Pratchett and Sven Uthicke
Received: 21 September 2016; Accepted: 14 November 2016; Published: 18 November 2016

Abstract: Background: Removal programs are effective strategies for short-term management of Crown-of-Thorns Starfish (*Acanthaster* spp.) populations, especially on a small scale. However, management programs are costly, and, in order to be effective, they must be based on local *Acanthaster* spp. population dynamics. We have developed simple models to predict the annual number of removable *A*. cf. *solaris* along the Onna coast of western central Okinawa Island, where chronic outbreaks have continued for several decades. Methods: The Onna coastal area was divided into five sectors, and annual abundance of small *A*. cf. *solaris* individuals was used to predict the total number of removable individuals of a cohort in each sector. Three models were developed, based on size class data collected by the Onna Village Fisheries Cooperative (OVFC) for 2003–2015, according to possible patterns of recruitment and adult occurrence. Using the best-fit models selected for each of the five sectors, the number of individuals that potentially escaped removal was calculated. Results: Best-fit models were likely to differ among the five sectors instead of small differences in the coefficients of determination. The models predict differences in the number of residual starfish among sectors; the northernmost sector was predicted to have a high number of residuals and the potential density of *A*. cf. *solaris* in the sector exceeded the outbreak criterion. Conclusions: These results suggest how to allocate resources to reduce the population of *A*. cf. *solaris* along the Onna coast in 2016. The OVFC implemented a control program for *A*. cf. *solaris* based on three model predictions.

Keywords: *Acanthaster planci*; coral; conservation; statistical model

---

1. Introduction

Outbreaks of the Crown-of-Thorns Starfish, *Acanthaster* spp., have been one of the major threats to coral communities in the Indo-Pacific region for more than a half-century [1]. A recent report showed that the Great Barrier Reef (GBR) has suffered heavy predation by *A*. cf. *solaris* for at least 27 years, resulting in a 42% loss of coral cover [2]. In a simulation, a 0.89% yr^{-1} increase of coral cover was predicted in the absence of *A*. cf. *solaris*, despite the impacts of other disturbances to corals in the GBR [2]. This study highlighted the catastrophic influence of *Acanthaster* predation on coral communities. Similar to the GBR, other areas in the Indo-Pacific region have been damaged by unusually high densities of *Acanthaster* spp.: Countries within the Coral Triangle [3–5]: the Ryukyu Islands [6,7], Philippines [8], French Polynesia [9,10], and islands in the Indian Ocean [11].

Understanding causal factors of outbreaks of *Acanthaster* spp. is the principal weapon in the fight to prevent future coral loss. Decades of studies have determined that the major factor could be larval supply, which fluctuates drastically with phytoplankton availability [12]. However, neither of the factors controlling juvenile survival, nor other factors influencing outbreaks of *Acanthaster* spp., are fully understood [1]. Alternatively, direct control activities for removing adult *Acanthaster* spp. are commonly implemented in many localities in order to protect coral communities. Hand-collection and/or injection of lethal chemicals are the major methods of locally controlling *Acanthaster* spp. locally [7,13,14]. These control methods are usually ineffective if conducted as one-shot management attempts, because outbreaks of *Acanthaster* spp. normally occur for at least several years [6]. Therefore, continuing control activities of *Acanthaster* spp. could be proactive measures against future crises. However, removal programs are costly; thus, it is important to predict population dynamics of *Acanthaster* spp. to wisely allocate budget in order to develop management plans. If the programs are conducted without such plan, a budget might run out before resulting in a significant impact on the number of *Acanthaster* spp. That science-based budgeting is necessary to effectively control populations of *Acanthaster* spp. In Japan, local government budgets are based on a single-fiscal-year system, and are allocated according to the previous year's request. Because of this system, management programs were halted when the budget ran out, despite a population explosion of *Acanthaster* spp. [6]. Therefore, control methods must be strategically planned, with adequate budgets, based on accurate models of demographics of *Acanthaster* spp.

Quantification of densities and distributions of juvenile *Acanthaster* spp. are considered to be effective at predicting outbreaks [15,16]. Monitoring juvenile population dynamics could help to predict outbreaks ~2 years ahead because *Acanthaster* spp. reaches adulthood (~20 cm) in two years ([17,18], Okaji unpublished data). Such field observations might contribute to effective management programs and might enable adequate budgeting by local governments [15,16].

A long-term *A.* cf. *solaris* removal program has been conducted for more than 30 years along the west coast of Okinawa Island, Japan, by the Onna Village Fisheries Cooperative (OVFC, [7]). OVFC has been removing *A.* cf. *solaris* individuals along the Onna coast by hand-collection [7]. Removal data revealed that *A.* cf. *solaris* is widely occurring in the coral reef ecosystem of the Onna coast, suggesting that *A.* cf. *solaris* is likely to maintain chronically high densities in the area [7]. Moreover, in 2003, *A.* cf. *solaris* were sorted into 5-cm size classes. Size class data suggested that multiple, successive recruitment events could potentially drive high *A.* cf. *solaris* population densities along the Onna coast [7]. As such, OVFC's removal data helps in understanding local *A.* cf. *solaris* population dynamics.

In this study, applying the idea of juvenile monitoring, we developed models to predict the number of removable *A.* cf. *solaris* individuals, which can be removed during the removal programs of OVFC, along the Onna coast using the OVFC data.

2. Materials and Methods

Size-class data of removed *A.* cf. *solaris* were collected beginning in 2003 at Maeda, Maeganaku, Minami Onna, Seragaki, and Afuso on the coast of Onna Village (Figure 1) by the OVFC. *A.* cf. *solaris* were manually removed by snorkeling and using tongs and other devices. The areas of removal were at 1–5 m depth during the daytime with 1–2 h surveys of both mornings and afternoons on selected days. The size of removed *A.* cf. *solaris* was determined with a diameter from the tip of one arm to the tip of the opposite arm. *A.* cf. *solaris* were divided into nine 5-cm size classes from 0–<5 cm to ≥40 cm, and individuals ≥40 cm were excluded from the analysis because any individuals of this size were found for 2003–2015.

We developed a model to predict the number of removable *A.* cf. *solaris* in the five sectors along the Onna coast using the size-class data of removed *A.* cf. *solaris* of the OVFC. In our model, the number of small individuals (<15 cm), estimated to be about one-year-old, was considered as the explanatory variable and used to predict the total number of removable individuals of that cohort. The growth rate of *Acanthaster* spp. in laboratory and field observations indicate that *Acanthaster* spp. reaches 3–15 cm in the first year, 18–25 cm in the second year, and 25–40 cm in the third year ([17,18], Okaji unpublished data). To choose an appropriate explanatory variable for the model, we considered three size classes of explanatory variables: <5 cm, <10 cm, and <15 cm. However, only individuals of

<10 cm and <15 cm were chosen as a potential explanatory variable because the number of removed individuals <5 cm in diameter was low until 2010 (~10 individuals per year collected along the Onna coast). In the analyses, we used complete cohort data, which comprised individuals of the 1st year (<15 cm), 2nd year (15–< 25 cm), and of the 3rd year (25–< 40 cm). For example, when building up a complete cohort, size-class data were rearranged because size-class data for 2003, for example, consists of the 1st year individuals born in 2002, the 2nd year individuals born in 2001, and 3rd year individuals born in 2000, but a complete cohort of 2003 consists of removed individuals of <15 cm in 2004, of 15–< 25 cm in 2005, and of 25–< 40 cm in 2006. Therefore, we acquired complete cohort data from 2002 to 2012, which were used for the analyses.

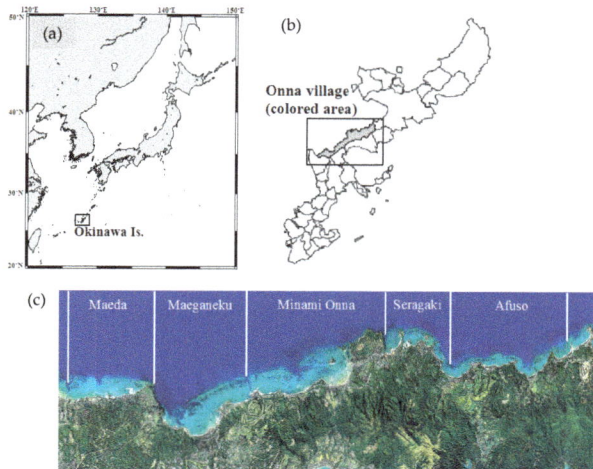

Figure 1. Study site maps: (**a**) Okinawa Island, Japan; (**b**) Onna Village on Okinawa Island; (**c**) five sectors, which are distinguished with channels, along the Onna coast.

We first constructed linear models to predict cohorts of *A*. cf. *solaris* as a function of the number of individuals of the 1st year (i.e., the size class <10 cm or <15 cm, selected as a potential explanatory variable, as above) at each sector along the Onna coast, as follows: Cohort = First year individuals + First year individuals x Sector. This model analyzed the relationship between the numbers of first-year *A*. cf. *solaris* removed (as First year individuals) and total numbers of removable individuals of the same cohort (as Cohort), and how the relationships differed among the five sectors (as First year individuals x Sector). The Akaike information criterion (AIC) was used to estimate an optimum explanatory variable for the model; the size class <10 cm or <15 cm. With the optimum variable, we examined the best-fit model for each of the five sectors along the Onna coast.

In selecting the best-fit model, we assumed three different cases for COTS population dynamics along the Onna coast with simple mathematical models; (1) linear, (2) logarithmic, and (3) exponential ones. These accord to *Acanthaster* juvenile and adult behavior. *Acanthaster* juveniles are negatively phototaxic, hiding in the shaded areas, e.g., crevices and holes of reefs, and are thus nocturnally active [19]. The negative phototaxis and nocturnality lessen with growth and the increase in density [19]. In addition, *Acanthaster* spp. move to shallow areas for the spawning [20]. According to these behavioral characteristics of *Acanthaster* spp., Case 1: The linear model indicates that larvae consistently settled, developed, and were eventually removed in areas where OVFC conducted *A*. cf. *solaris* removal. Case 2: The logarithmic model indicates that the areas of settlement have a number of places for hiding for *A*. cf. *solaris*. In this context, "places for hiding" contains two meanings; large area with many hiding places for juveniles, and large area where fishermen could not patrol the whole area due to limited time and manpower, thus, *A*. cf. *solaris* could escape from removal. This model could show a maximum number of removable individuals with limited time and manpower. Case 3: The exponential model indicates that areas of settlement have relatively less places to hide from fishermen, that is, relatively narrow areas, where fishermen could cover all areas when patrolling. Therefore, *A*. cf. *solaris* appear easily in shallower areas while growing, and then, eventually, be removed. Therefore, we selected the best-fit model for each of the five sectors. We first estimated and tested coefficients for the three models. The estimated total removable individuals of a cohort were then calculated using the estimated coefficients for each of the three models. Residuals were tested for normality with the Shapiro-Wilk test. Then we selected the models showing the largest coefficients of determination among the fitted models (best-fit) for each of the five sectors, provided that the residuals indicated data normality. For the best-fit models, 95% confidence intervals were also calculated. With the best-fit models, we predicted the total numbers of removable and non-removable individuals (difference between the predicted total removable individuals and the already-removed individuals of the 1st and/or the 2nd year), for the cohorts 2013 and 2014.

3. Results and Discussion

Size-frequency distributions of removed *A*. cf. *solaris* demonstrated abundance maxima for all five sectors along the Onna coast over 13 years. The abundant size classes were consistently from <15 cm to <25 cm for all sectors (Figure 2), indicating successive recruitments in the five sectors along the Onna coast over many years [7,21]. Size class abundance data also demonstrate the importance of larval recruitment in maintaining the *A*. cf. *solaris* population along the Onna coast.

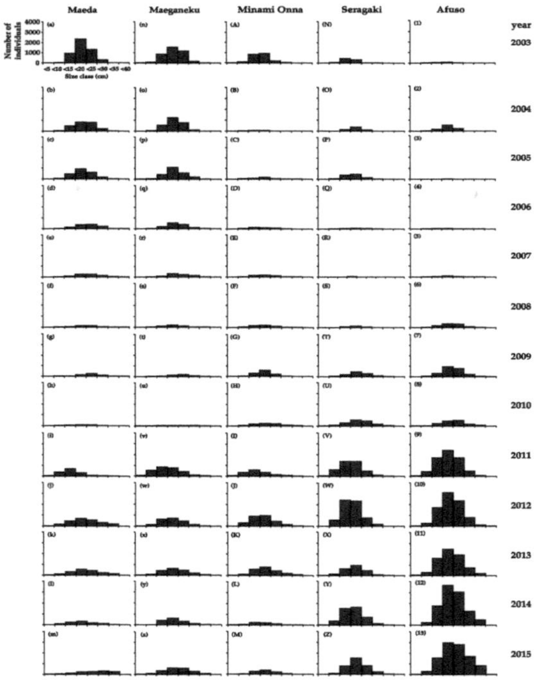

Figure 2. Size-frequency distributions of *Acanthaster* cf. *solaris* for five sectors along the Onna Coast, Okinawa, Japan, over 13 years. (**a–m**), Maeda; (**n–z**), Maeganeku; (**A–M**), Minami Onna; (**N–Z**), Seragaki; (**1–13**), Afuso.

Annual fluctuation in removed *A.* cf. *solaris* demonstrated a south-to-north change of sectors of more abundant *A.* cf. *solaris* along the coast (Figure 2). Maeda (Figure 2a–m) and Maeganeku (Figure 2n–z), the two most southern sectors, showed relatively high removal numbers in 2003 and a gradual decrease until 2010, following this, there was another increase in 2011. Minami Onna showed an abrupt decline from 2003 to 2004 and then a gradual increase to 2012 (Figure 2A–M). The numbers of removed starfish at Seragaki (Figure 2N–Z) and Afuso (Figure 21–13) were relatively low until 2008, followed by an increase from 2009. This pattern could be related to northeastward onshore currents that prevail along the Onna coast during the *A.* cf. *solaris* larval dispersal period [22].

Removal data suggested the importance of juveniles for *A.* cf. *solaris* population maintenance along the Onna coast. Based on the linear model, numbers of both <10 cm and <15 cm individuals accurately predicted the total number of removable individuals belonging to a given cohort (Tables 1 and 2). However, the model, using only starfish <15 cm as an explanatory variable, was a better model according to AIC (Tables 1 and 2). Using <15 cm individuals as an explanatory variable, best-fit models were likely to differ among sectors, instead of having small differences in the coefficients of determination (Tables 3 and 4, Figure 3 and Figure S1). The relationship between the number of first year starfish and the total removable individuals belonging to a cohort in Afuso was optimally fitted to the logarithmic model. At Seragaki, Maeganeku, and Maeda, the linear model provided the best fit. Minami Onna was likely to be described by the exponential model. Small differences in the coefficients of determination suggested a necessity for continuation of removal programs in order to accumulate more data. Moreover, the present study demonstrated that we could predict the number of removable individuals of *A.* cf. *solaris* in a certain area using a simple and low-cost calculation using long-term removal size-class data.

Table 1. Parameter estimates of the predictive model for the total number of removable individuals in the same cohort (first to third year individuals) as a function of the number of <10 cm individuals designated as first-year *Acanthaster* cf. *solaris* removed at each sector along the Onna Coast, Japan. AF = Afuso, MO = Minami Onna, MG = Maeganeku, MD = Maeda. *p*-value: significance level based on *t*-value.

	Estimate	Std. Error	*t*-value	*p*-value
Intercept	1035	208.5	4.961	8.83×10^{-6}
1st year class (<10 cm)	6.355	1.486	4.277	8.74×10^{-5}
1st year class (<10 cm)_AF	14.81	2.321	6.384	5.98×10^{-8}
1st year class (<10 cm)_MO	−3.278	2.810	−1.167	0.249
1st year class (<10 cm)_MG	−2.946	2.454	−1.201	0.236
1st year class (<10 cm)_MD	−2.421	2.974	−0.814	0.420
AIC	948.1			

Table 2. Parameter estimates of the predictive model for the total number of removable individuals in the same cohort (first to third year individuals) as a function of the number of <15 cm individuals designated as first-year *Acanthaster* cf. *solaris* removed at each sector along the Onna Coast, Japan. AF = Afuso, MO = Minami Onna, MG = Maeganeku, MD = Maeda. *p*-value: significance level based on *t*-value.

	Estimate	Std. Error	*t*-value	*p*-value
Intercept	637.4	161.6	3.943	25.5×10^{-5}
1st year class (<15 cm)	1.680	0.243	6.927	8.61×10^{-9}
1st year class (<15 cm)_AF	2.552	0.325	7.841	3.35×10^{-10}
1st year class (<15 cm)_MO	−0.475	0.493	−0.964	0.340
1st year class (<15 cm)_MG	0.267	0.475	0.562	0.577
1st year class (<15 cm)_MD	0.367	0.533	0.688	0.495
AIC	905.8			

Table 3. Coefficient of determination for three models for the five sectors along the Onna coast, Japan. The largest values are in bold. n.a. indicates the model that did not ensure normality.

Sector	Models		
	Linear	Logarithmic	Exponential
Afuso	n.a.	**0.808**	0.791
Seragaki	**0.753**	0.716	0.549
Minami Onna	0.720	0.687	**0.723**
Maeganeku	**0.779**	0.739	0.567
Maeda	**0.911**	0.897	0.783

Table 4. Best model for all sectors along the Onna coast for predicting total removable cohort (Y) membership based upon the number of individuals <15 cm (X).

Sector	Best Model
Afuso	$Y = 2218 \log (X) - 8484$
Seragaki	$Y = 1.615 X + 758.4$
Minami Onna	$Y = 422.2 \times 1.001^X$
Maeganeku	$Y = 1.955 X + 631.2$
Maeda	$Y = 2.277 X + 470.0$

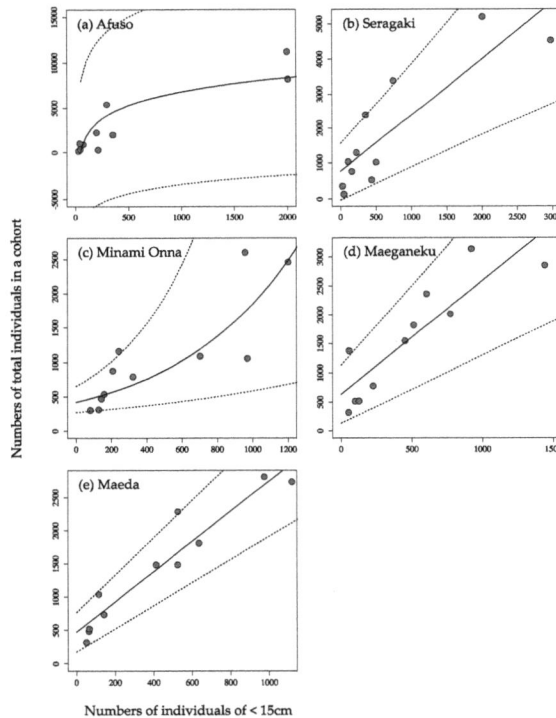

Figure 3. Best-fit model with 95% confidence intervals for the relationship between the numbers of removed individuals <15 cm and the total number of removed individuals in the same cohort of *Acanthaster* cf. *solaris* for five sectors along the Onna coast, Okinawa, Japan. (**a**) Afuso; (**b**) Seragaki; (**c**) Minami Onna; (**d**) Maeganeku; (**e**) Maeda.

Using these models, the total predicted removable individuals (Figure S2) and the predicted removable individuals that escaped removal were calculated for the cohorts 2013 and 2014 (Table 5). There were potentially uncaptured individuals in the 2013 cohort in Afuso and Maeda after two years of removal and in the 2014 cohort in all sectors after one year of removal. This implies that two years of concerted eradication efforts might not be sufficient to eradicate *A*. cf. *solaris* in Afuso and Maeda. According to the predictive models, control programs should allocate more effort in Maeda and especially in Afuso, which might have relatively larger hiding places for *A*. cf. *solaris* from the OVFC members than the other sectors. Indeed, the model predicted that as many as 6350 individuals from the 2014 cohort could remain in Afuso (Table 5). Additionally, the area of removal in Afuso is about 238.8 ha, which suggests there could be 26.5 *A*. cf. *solaris* ha^{-1} in Afuso, exceeding the outbreak criterion of 15 individuals ha^{-1} [23].

Table 5. The number of individuals that escaped removal, predicted from the best-fit models for the cohorts 2013 and 2014 for five sectors along the Onna coast, Okinawa, Japan.

Sector	Number of Individuals that Escaped Removal	
	Cohort 2013	Cohort 2014
Afuso	577	6350
Seragaki	0	1315
Minami Onna	0	357
Maeganeku	0	1057
Maeda	508	644

The OVFC's efforts enable the prediction of the number of uncaptured *A*. cf. *solaris* at each of the five sectors along the Onna coast using predictive models. Therefore, the OVFC can plan how to allocate removal efforts along the coast in order to further reduce the number of starfish in January 2016. The OVFC implemented a control program based on these models since January 2016 with the main objective being to remove remaining adults before the July spawning season [24]. Such effort could reduce successful fertilization in subsequent seasons and potentially control the number of *A*. cf. *solaris* [14].

4. Conclusions

Predictive models based on long-term removal data could contribute to planning a strategic management program for *A*. cf. *solaris* along the Onna coast. Our data suggest that simple models can be applied to local control efforts. Five-centimeter size class data could be useful for selecting the best predictive model. These models are useful as a rough standard for *A*. cf. *solaris* management programs, and should facilitate budgetary planning for *A*. cf. *solaris* control.

Supplementary Materials: The following are available online at www.mdpi.com/1424-2818/8/4/24/s1, Figure S1: Relationship between the numbers of individuals <15 cm and of total individuals in the same cohort of *Acanthaster* cf. *solaris* for five sectors along the Onna coast, Okinawa, Japan, with three different models. Black line, linear model; blue line, logarithmic model; red line, exponential model, Figure S2: Total removable individuals of cohort 2013 (Red) and 2014 (Blue) predicted from the best-fit models for the relationship between the numbers of individuals of <15 cm and total individuals in the same cohort of *Acanthaster* cf. *solaris* for five sectors along the Onna coast, Okinawa, Japan with observed values (Grey).

Acknowledgments: We are very grateful to the members of the Onna Village Fisheries Cooperative. We thank Steven D. Aird for editing our manuscript.

Author Contributions: Y.H. and K.O. conceived and designed the experiments; M.N. and N.H.K. analyzed the data; M.N. wrote the paper.

Conflicts of Interest: The authors declare no conflicts of interest.

References

1. Pratchett, M.S.; Caballes, C.F.; Rivera-Posada, J.A.; Sweatman, H.P.A. Limits to understanding and managing outbreaks of Crown-of-Thorns starfish (*Acanthaster* spp.). *Oceanogr. Mar. Biol.* **2014**, *52*, 133–200.

2. De'ath, G.; Fabricius, K.E.; Sweatman, H.; Puotinen, M. The 27-year decline of coral cover on the Great Barrier Reef and its causes. *Proc. Natl. Acad. Sci. USA* **2012**, *109*, 17995–17999. [CrossRef] [PubMed]

3. Pratchett, M.S.; Schenk, T.J.; Baine, M.; Syms, C.; Baird, A.H. Selective coral mortality associated with outbreaks of *Acanthaster planci* L. in Bootless Bay, Papua New Guinea. *Mar. Environ. Res.* **2009**, *67*, 230–236. [CrossRef] [PubMed]

4. Baird, A.H.; Pratchett, M.S.; Hoey, A.S.; Herdiana, Y.; Campbell, S.J. *Acanthaster planci* is a major cause of coral mortality in Indonesia. *Coral Reefs* **2013**, *32*, 803–812. [CrossRef]

5. Plass-Johnson, J.G.; Schwieder, H.; Heiden, J.; Weiand, L.; Wild, C.; Jompa, J.; Ferse, S.C.A.; Teichberg, M. A recent outbreak of crown-of-thorns starfish (*Acanthaster planci*) in the Spermonde Archipelago, Indonesia. *Reg. Environ. Chang.* **2015**, *15*, 1157–1162. [CrossRef]

6. Yamaguchi, M. *Acanthaster planci* infestations of reefs and coral assemblages in Japan: A retrospective analysis of control efforts. *Coral Reefs* **1986**, *5*, 23–30. [CrossRef]

7. Nakamura, M.; Okaji, K.; Higa, Y.; Yamagawa, E.; Mitarai, S. Spatial and temporal population dynamics of the crown-of-thorns starfish, *Acanthaster planci*, over a 24-year period along the central west coast of Okinawa Island, Japan. *Mar. Biol.* **2014**, *161*, 2521–2530. [CrossRef]

8. Bos, A.R. Crown-of-thorns outbreak at the Tubbataha Reef UNESCO World Heritage Site. *Zool. Stud.* **2010**, *49*, 124.

9. Adjeroud, M.; Michonneau, F.; Edmounds, P.J.; Chancerelle, Y.; Lison de Loma, T.; Penin, L.; Thibaut, L.; Vidal-Dupiol, J.; Salvat, B.; Galzin, R. Recurrent disturbances recovery trajectories, and resilience of coral assemblages on a South Central Pacific reef. *Coral Reefs* **2009**, *28*, 775–780. [CrossRef]

10. Kayal, M.; Vercelloni, J.; de Loma, T.L.; Bosserelle, P.; Chancerelle, Y.; Geoffroy, S.; Stievenart, C.; Michonneau, F.; Penin, L.; Planes, S.; et al. Predator crown-of-thorns starfish (*Acanthaster planci*) outbreak, mass mortality of corals, and cascading effects on reef fish and benthic communities. *PLoS ONE* **2012**, *7*, e47363. [CrossRef] [PubMed]

11. Roche, R.C.; Pratchett, M.S.; Carr, P.; Turner, J.R.; Wagner, D.; Head, C.; Sheppard, C.R.C. Localized outbreaks of *Acanthaster planci* at an isolated and unpopulated reef atoll in the Chagos Archipelago. *Mar. Biol.* **2015**, *162*, 1695–1704. [CrossRef]

12. Fabricius, K.E.; Okaji, K.; De'ath, G. Three lines of evidence to link outbreaks of the corwn-of-thorns seastar *Acanthaster planci* to the release of larval food limitation. *Coral Reefs* **2010**, *29*, 593–605. [CrossRef]

13. Bostrom-Einarsson, L.; Rivera-Posada, J. Controlling outbreaks of the coral-eating crown-of-thorns starfish using a single injection of common household vinegar. *Coral Reefs* **2016**, *35*, 223–228. [CrossRef]

14. Bos, A.R.; Gumanao, G.S.; Mueller, B.; Saceda-Cardoza, M.M.E. Management of crown-of-thorns sea star (*Acanthaster planci* L.) outbreaks: Removal success depends on reef topography and timing within the reproduction cycle. *Ocean Coast Manag.* **2013**, *71*, 116–122. [CrossRef]

15. Yokochi, H. The early detection of *Acanthaster* outbreaks by monitoring the algae-eating juvenile starfish. *Midoriishi* **1998**, *9*, 12–14. (In Japanese)

16. Research Institute for Subtropical Regions. *Manual for Monitoring of Juvenile Crown-of-Thorns Starfish*; Research Institute for Subtropical Regions: Naha, Japan, 2006; p. 29. (In Japanese)

17. Yamaguchi, M. Growth of juvenile *Acanthaster planci* (L) in the laboratory. *Pac. Sci.* **1974**, *28*, 123–138.

18. Lucas, J.S. Growth, maturation and effects of diet in *Acanthaster planci* (L.) (Asteroidae) and hybrids reared in the laboratory. *J. Exp. Mar. Biol. Ecol.* **1984**, *79*, 129–147. [CrossRef]

19. Birkeland, C.; Lucas, J.S. *Acanthaster planci: Major Management Problems of Coral Reefs*; CRC Press, Inc.: Florida, FL, USA, 1990.

20. Babcook, R.C.; Mundy, C.N. Reproductive biology, spawning and field fertilization rates of *Acanthaster planci*. *Aust. J. Mar. Freshw. Res.* **1992**, *43*, 525–534. [CrossRef]

21. Pratchett, M.S. Dynamics of an outbreak population of *Acanthaster planci* at Lizard Island, northern Great Barrier Reef (1995–1999). *Coral Reefs* **2005**, *24*, 453–462. [CrossRef]

22. Nakamura, M.; Kumagai, N.H.; Sakai, K.; Okaji, K.; Ogasawara, K.; Mitarai, S. Spatial variability in recruitment of acroporid corals and predatory starfish along the Onna coast, Okinawa, Japan. *Mar. Ecol. Prog. Ser.* **2015**, *540*, 1–12. [CrossRef]

23. Moran, P.J.; De'ath, G. Estimates of the abundance of the crown-of-thorns starfish *Acanthaster planci* in outbreaking and non-outbreaking populations on reefs within the Great Barrier Reef. *Mar. Biol.* **1992**, *113*, 509–515. [CrossRef]

24. Yasuda, N.; Ogasawara, K.; Kajiwara, K.; Ueno, M.; Oki, K.; Taniguchi, H.; Kakuma, S.; Okaji, K.; Nadaoka, K. Latitudinal differentiation in the reproduction patterns of the crown-of-thorns starfish *Acanthaster planci* through the Ryukyu Island Archipelago. *Plankton Benthos Res.* **2010**, *5*, 156–164. [CrossRef]

diversity

Article

Citric Acid Injections: An Accessible and Efficient Method for Controlling Outbreaks of the Crown-of-Thorns Starfish *Acanthaster* cf. *solaris*

Alexander C. E. Buck [1,*], Naomi M. Gardiner [1] and Lisa Boström-Einarsson [1,2]

1   Marine Biology and Aquaculture, College of Science and Engineering, James Cook University,
    Townsville, QLD 4811, Australia; naomi.gardiner@jcu.edu.au (N.M.G.);
    lisa.bostromeinarsson@my.jcu.edu.au (L.B.-E.)
2   Australian Research Council Centre of Excellence for Coral Reef Studies, James Cook University,
    Townsville, QLD 4811, Australia
*   Correspondence: alexander.buck@my.jcu.edu.au; Tel.: +61-04-2313-7575

Academic Editors: Sven Uthicke and Michael Wink
Received: 10 October 2016; Accepted: 3 December 2016; Published: 10 December 2016

Abstract: Outbreaks of the crown-of-thorns starfish (*Acanthaster* cf. *solaris*, COTS) are one of the primary causes of coral decline in the Indo-Pacific region. Effective methods to control COTS outbreaks may therefore be one of the most direct and immediate ways to reduce coral loss. However, the cost and logistical challenges associated with current control methods have undermined the effectiveness of many control efforts. In this study, we tested the feasibility of using powdered citric acid, which is widely available and low-cost, as an injection chemical for COTS control. We tested what combination of concentration, number of injections, volume, and water type were most efficient at killing COTS. All COTS injected in two or four sites died, irrespectively of the concentration of citric acid used, while single injections failed at reaching 100% mortality. The fastest combination was the injection of 150 g·L^{-1} citric acid solution in four injection sites (5 mL per site), which killed the starfish in 26.4 ± 4 h. These results suggest that injections of powdered citric acid are an effective, economical, and widely available alternative to current COTS control methods.

Keywords: COTS; outbreak; control methods; pest control; coral reefs; injections

1. Introduction

Periodic outbreaks of *Acanthaster* cf. *solaris* (crown-of-thorns starfish, COTS) represent one of the single biggest threats to tropical coral reefs in the Indo-Pacific [1,2]. The corallivorous starfish can cause extensive damage when present at outbreak densities on coral reefs. For example, a large outbreak in Guam reduced live coral cover by more than 90% over a period of 2–3 years [3]. On Australia's Great Barrier Reef (GBR), mean live coral cover has halved between 1985 and 2012, and ~40% of that coral loss is attributed to COTS [4,5]. Preventing and managing COTS outbreaks is therefore a major priority for environmental science and resource management, and may be the most immediate and effective mechanism to prevent ongoing coral loss across the Indo-Pacific [6]. Management of most anthropogenic disturbances that threaten coral reefs (e.g., climate-induced disturbances and declining water quality) require large-scale interventions and international policy changes [7]). In contrast, management of localised outbreaks of COTS are potentially feasible and may significantly increase the resilience of coral reef communities [5]. However, finding efficient, cheap, and safe methods that can be applied at large scales and across developed and developing nations has proven challenging [8].

Historically, a wide range of methods have been employed to cull COTS [8], however, most have been inefficient or damaging to the marine environment. Currently, large-scale control programs for

COTS involve a single injection of a bile salt solution, which effectively kills COTS in less than 24 h [9, 10]. However, bile salts can be expensive, are only accessible through specialised suppliers, and carry quarantine restrictions that impede international operations [11]. Thus, for many communities in remote areas of the Indo-Pacific that experience recurrent COTS outbreaks, it can be difficult to obtain bile salts. To overcome these challenges, recent research has been investigating the potential use of alternate chemicals that are cheap, readily available, easy to deploy, safe to both the marine environment and to humans, and equally efficient at killing COTS as bile salts. To date, injections with cooking salt solution, vinegar, and lime juice have been demonstrated to be both lethal to COTS and safe for the environment [12–17]. However, not all of these chemicals are viable for large-scale control programs in remote coral reef regions. For instance, the high quantities of cooking salt and limes required during extensive control programmes may be challenging to transport in remote locations [16]. Limes are not readily available in all seasons nor in all locations, which may increase their price. In addition, juice extraction is labour-intensive, adding labour costs to control programmes, and the juice may deteriorate if not rapidly used [15]. Instead, this study examines the viability of using powdered citric acid, which can be purchased locally from grocery stores, is inexpensive, readily available, and has a long shelf life.

For a large-scale COTS control programme to be efficient, it is critical that any method employed achieves 100% COTS mortality. Achieving complete mortality depends not only on the efficiency of the injected chemical, but also on the technique used when administering the injection. Recent studies have revealed the importance of the number of injections [12,15], the size of the needle [10,12], location of injection site [10], and the volume or concentration injected [12,13,15,18]. For example, a single injection of 20 mL of lime juice did not kill 100% of treated COTS, while splitting the volume between two injection sites did [15]. Furthermore, a single 20 mL injection of vinegar with a 16-gauge (1.2 mm inner diameter) needle resulted in 100% mortality, but using a 4 mm diameter needle reduced mortality to 87% [12]. The ideal injection method hinges on finding the optimal combination of techniques, using chemicals readily available in the local area.

The aim of this study was, therefore, to investigate the potential of using powdered citric acid as an efficient and cost-effective method to cull COTS. To achieve this, we aimed to determine: (1) the lowest effective concentration and number of injection sites needed to obtain 100% mortality, (2) the most effective injection volume, and (3) the efficiency of untreated seawater compared to distilled water as solvent for the citric acid solution. Finally, we compared the efficiency of citric acid injections to other chemical products in order to provide management recommendations.

2. Materials and Methods

2.1. Collection Site and Maintenance Conditions

The study was carried out at Lizard Island (14°40′S, 145°28′E), Northern GBR, Queensland, Australia in February 2015. Adult *A.* cf. *solaris* were collected from reefs around Lizard Island and transported to Lizard Island Research Station in plastic aquaria (64 × 41 × 40 cm, max. 20 COTS per aquaria) with aerators. Then, COTS were allowed to acclimatise for at least 24 h in two large aerated holding tanks (160 cm diameter × 50 cm deep) with flow-through ambient seawater. Injured and weak specimens were discarded. Specimens were then measured from the tip of one randomly selected arm to the tip of the diametrically opposite arm (mean size 28.2 ± 0.4 cm SE) and placed individually in aquaria (40 × 30 × 25 cm) with flow-through ambient seawater.

This study was conducted in accordance with James Cook University ethical guidelines and the Queensland Animal Care and Protection Act 2001.

2.2. Experiment 1: Concentration and Number of Injection Sites

To test whether different concentrations and number of injection sites influenced the effectiveness in killing COTS, we conducted a factorial experiment. We tested three different concentrations of citric acid in seawater solution (90, 120, and 150 g·L^{-1}) crossed with three different numbers of injection sites (one, two, and four injection sites, hereafter I.S.). Six replicate COTS were injected in each of the nine treatment combinations (three citric acid concentrations × three injection site levels) using a total of 54 individual COTS. In addition, five COTS per I.S. treatment level were injected with 20 mL of seawater to control for the injection itself. All treatments were run simultaneously, except the three 90 g·L^{-1} treatments and the 4 I.S., 120 g·L^{-1} treatments, where COTS were injected in bunches of three (each still in individual tanks). The second bunch was injected ~24 h after the first one to keep the start of the experiment at a similar time of day, therefore minimizing any variation in water temperature.

A 16-gauge stainless steel needle mounted on a 25 mL disposable syringe was chosen to perform the injection in order to minimise puncture size and leakage postinjection [12]. Injections were performed at the base of the arms [10], targeting the hydrovascular system of the starfish [19]. Single injections were administered at a randomly selected arm, double injections at opposing arms, and four injections in different quartiles. For each treatment, a total of 20 mL of solution was evenly split between injection sites (one I.S. treatment = 20 mL injected in one arm, two I.S. treatments = 10 mL per arm, and four I.S. treatments = 5 mL per arm).

The temporal progression of the treatments was recorded every 4 h (three observations per day) for up to one week or until death. The clinical signs recorded were: (1) no response vs hyperactivity, (2) matting of spines, (3) swelling and loss of turgor, (4) appearance and increased production of mucus, (5) appearance of bacterial films, (6) loss of any arms or splitting, (7) immobility, and (8) death [cf. 9]. The primary response variables measured were mortality, time to immobility, and time to death, also recorded every 4 h (three observations per day). Immobility was defined as the inability to cling to the walls of the tank or move [12], while death was determined when all tube feet completely stopped moving [10,20,21]. Time to immobility was recorded because it represents the ecological death of the starfish, due to their inability to feed when immobile. In some cases, time to death was delayed by up to 38 h past the time of immobility as individual tube feet remained motile.

2.3. Experiment 2: Volume

A second experiment was performed to evaluate whether increasing the injected volume could improve the efficacy of the single injection method in achieving 100% mortality. Here, six COTS were injected with 30 mL of 120 g·L^{-1} solution via two simultaneous injections at the base of one arm. Injections of 15 mL of solution from two syringes were performed simultaneously, with the needles held approximately 5 mm apart (considered one I.S.). Two needles were needed to accommodate the increased volume, as larger syringes were not readily available. A maximum of 30 mL was used, as it is the capacity of the most commonly used gun used in the field to inject sodium bisulphate and bile salts [10]. The effect of 30 mL injections on mortality, time to immobility, and time to death was compared to the effect of 20 mL injections of 120 g·L^{-1} at one I.S. (performed in Experiment 1).

2.4. Experiment 3: Seawater vs Distilled Water

To compare the effect of the solvent on citric acid efficacy, we compared COTS mortality and response times when using seawater as the solvent versus distilled water. Distilled water was used to test whether a hypoosmotic reaction induced by low salinity and lower pH of the solvent would have increased the efficacy of the citric acid solution, increased percentage of mortality, and/or reduced time to immobility or to death. Distilled water was expected to induce an osmotic shock in COTS tissues, thereby increasing the efficiency of the citric acid injections, in two ways. First, by accentuating acidosis, because distilled water (pH 7.0) is less alkaline than seawater (pH 7.5–8.4) and should lower COTS coelomic pH after citric acid injection. Second, by causing a hypoosmotic shock, because COTS are unable to tolerate drastic changes in internal salinity or osmotic pressure [13,16]. Distilled water was used as the solvent in one treatment (one I.S., 20 mL of 120 g·L^{-1} citric acid solution) repeated on six individuals. The effect of distilled water on mortality, time to immobility, and time to death was compared to the effect of 20 mL injections of 120 g·L^{-1} at one I.S. (performed in Experiment 1).

2.5. Statistical Analysis

For Experiment 1, differences between times to immobility and to death were analysed using two-way fixed factor analysis of covariance (ANCOVAs). Because only treatments that reached 100% mortality were included in these models, none of the single injection site treatments were included in the statistical tests for Experiment 1. Additionally, for the 150 g·L^{-1}, 4 I.S., 20 mL seawater treatment, specific hours of immobility and death were not attained because all COTS reached immobility and died overnight. Therefore, death was assumed to have occurred at the same time as immobility, and both times were scored at the following morning observation. For all analyses, assumptions of normality among residuals were analysed using the Shapiro–Wilk test (time to immobility: W = 0.787, $p < 0.001$; time to death: W = 0. 772, $p < 0.001$). Homogeneities of variances between concentrations and number of injection sites were analysed with Levene's test. Time to immobility and to death measurements were subsequently log transformed to meet assumptions of normality: log(time to immobility): W = 0.958, $p = 0.317$; log(time to death): W = 0.950, $p = 0.118$. The dependent variables were time to immobility and to death (hours), analysed separately, and the independent fixed factors were the concentrations of the solution (three levels: 90, 120, 150 g·L^{-1}) and the number of injection sites (two levels: two and four I.S.). To control for the effect of body size, we used the diameter of the starfish as covariate. Because the interaction terms among size, concentration, and number of injection sites were not significant for time to immobility and to death, the full models were rerun without the interaction terms to increase the power of the tests. Tukey's post hoc tests were used to analyse statistical differences between concentration groups and I.S. treatments.

For Experiments 2 and 3, differences between times to immobility and to death (dependent variables) were analysed separately using one-way ANCOVAs with volume and water type as independent fixed parameters, while using the diameter of the starfish as covariate. Assumptions of normality were analysed with the Shapiro–Wilk test of normality (Experiment 2, time to immobility: W = 0.978, $p = 0.954$, time to death: W = 0.901, $p = 0.226$; Experiment 3, time to immobility: W = 0.876, $p = 0.144$, time to death: W = 0 .878, $p = 0.152$) and homogeneity of variances were analysed with Levene's tests. No transformations were required. Because the interaction terms between size and volume and size and water type were not significant for time to immobility and to death, the full models were rerun without the interaction terms to increase the power of the tests. Statistical analyses were conducted using *TIBCO Spotfire S+*® 8.2 Programmer's Guide, *TIBCO* Software Inc. (Palo Alto, CA, United States) Technical Support.

3. Results

3.1. Experiment 1: Concentration and Number of Injection Sites

All treatments using either two or four injection sites had 100% mortality, regardless of the citric acid concentration (Figure 1) or of COTS size (Table 1). Concentration significantly affected time to immobility (Table 1), with the 120 g·L^{-1} treatments being almost twice as fast (32.2 ± 8.2 h SEM) than the 90 g·L^{-1} treatments (69.9 ± 18.7 h). In contrast, the concentration of citric acid did not have a significant impact on time to death (Table 1), although the 150 g·L^{-1} treatments were, on average, 1.8 times faster than the 90 g·L^{-1} ones (respectively, 46.8 ± 7.4 h and 84.4 ± 18.7 h). Time to death was instead significantly affected by the number of injection sites (Table 1), as COTS with more injections died faster. On average, four injections halved the time to death compared to two injections (44.1 ± 6.7 h and 79.2 ± 12.1 h, respectively). In contrast, time to immobility was not affected by number of injection sites (Table 1), although four injections immobilised COTS twice as fast as two injections (respectively, 32.8 ± 6.2 h and 59.6 ± 12.8 h). The 150 g·L^{-1}, four I.S. treatment was the quickest overall at killing COTS (26.4 ± 4 h), and was 4.3 times faster than the 90 g·L^{-1}, double injection. The latter was the slowest treatment that still achieved 100% mortality and took 112.8 ± 31.8 h (Figure 1).

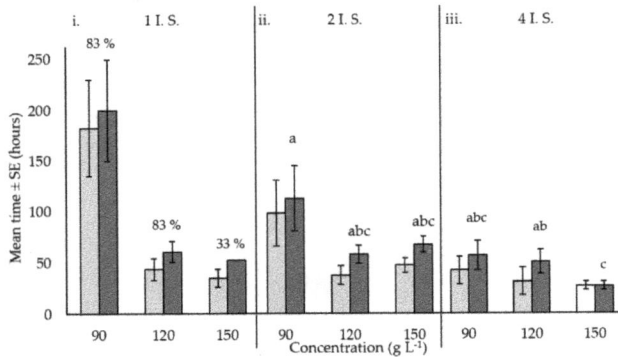

Figure 1. The effects of citric acid concentration and the number of injections. Mean time to immobility (□) and to death (■) ± standard error for *Acanthaster* cf. *solaris* injected with 20 mL of citric acid and seawater solution at concentrations of 90, 120, and 150 g·L^{-1} in one (**i**), two (**ii**) and four (**iii**) injection sites (I.S.). Numbers above bars represent total percent mortality; where no numbers are shown, mortality was equal to 100%. For the 150 g·L^{-1}, 4 I.S. treatment, time to immobility was assumed to equal time to death (□). Six replicates per treatment combination were used. Letter notations above bars (a, b, c) indicate Tukey's post-hoc groupings between injection site treatments and concentrations for time to death. Different letters indicate significant differences among treatments.

When a single injection was administered, mortality depended on the concentration used. At 90 and 120 g·L^{-1}, mortality was 83% (five out of six starfish died), while at 150 g·L^{-1} mortality decreased to 33% (only two out of six starfish died, Figure 1). However, increasing the concentration reduced time to immobility by 5-fold and time to death 4-fold. At 90 g·L^{-1}, COTS were immobile in 182.8 ± 47.1 h and died in 200 ± 50 h, while at 150 g·L^{-1} they were immobile in 35 ± 8.5 h and died in ~50 h.

Behaviour and Macroscopic Progression

Immediately after injection with citric acid, COTS showed symptoms of stress through hyperactivity, swelling, and increased mucus production. Approximately one day after injection, mucus production, matting of the spines, and localised necrosis on at least one arm were common symptoms. Additionally, many specimens, particularly those treated with higher concentrations, were already partly immobile or dead and others had split in half or thirds. In some cases, tissue decomposition progressed over the central disc, exposing internal organs. After two days, the first dense colonies of bacteria started forming, creating orange-red films around decomposing parts of COTS. Matting of the spines was usually body-wide, and some spines had been dropped or were bleached. Starfish motility was usually very low, with most animals completely immobile or dead. Three to four days after injection, tissue necrosis and bacterial decomposition progressed. By this time, most COTS were considered unrecoverable, completely immobile, or dead. However, in some cases where the starfish had split, one half was completely dead and decomposing while the other half, or a few roaming arms, remained alive for up to 10 d postinjection, before eventually dying. These individuals (eight) were scored as survivors in our analyses, because they were still alive during the observation period.

3.2. Experiment 2: Volume

Increasing the volume of citric acid solution from 20 to 30 mL did not increase the percentage of mortality in the starfish injected once with the 120 g·L^{-1} solution, which remained at 83% (five out of six COTS died in both treatments) (Figure 2). However, increasing the volume accelerated the mean time to immobility by around ~20%, from 43.8 ± 10.3 h (20 mL) to 37.2 ± 10.1 h (30 mL), although this difference was not statistically significant (Table 1). Time to death was also not significantly affected by changes in the injected volume (Table 1), although higher volumes resulted in a 27% decrease in time to death, passing from 60.7 ± 10.3 h (20 mL) to 47.8 ± 9.3 h (30 mL) (Figure 2).

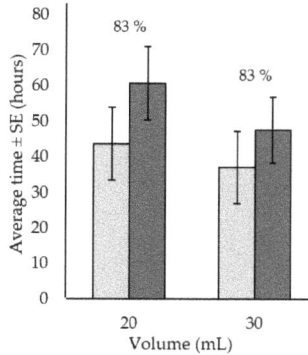

Figure 2. The effect of volume on response to citric acid injections. Mean time to immobility (□) and to death (■) ± standard error for *Acanthaster* cf. *solaris* injected in one site with either 20 mL or 30 mL of 120 g·L^{-1} citric acid and seawater solution. Numbers above bars represent total percent mortality for each treatment. Six replicates per treatment were used.

3.3. Experiment 3: Water Type

Seawater appeared to be a better solvent than distilled water, as it resulted in (1) higher mortality and (2) ~30% reduction in time to immobility and to death, although these differences were not significant (Figure 3, Table 1). Seawater killed five out of six COTS (or 83%), while distilled water only killed four out of six COTS (or 67%). Additionally, the seawater treatment caused immobility in 43.8 ± 10.3 h, while the distilled water solvent took 59.4 ± 22.5 h (Figure 3). Similarly, COTS injected with seawater solution died in 60.7 ± 10.3 h, while those injected with the distilled water solution died in 77.8 ± 22.6 h.

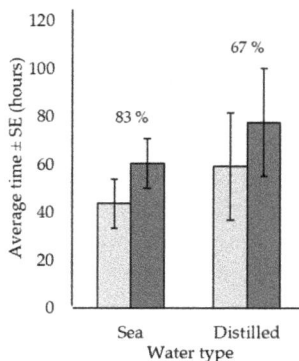

Figure 3. The effect of solvent on responses to citric acid injections. Mean time to immobility (□) and to death (■) ± standard error for *Acanthaster* cf. *solaris* injected in one site with 20 mL of 120 g·L^{-1} citric acid and either seawater or distilled water solution. Numbers above bars represent total percent mortality for each treatment. Six replicates per treatment were used.

Table 1. Results of the analysis of covariance (ANCOVA) on time to immobility and to death for *Acanthaster* cf. *solaris* injected with citric acid. Experiment 1 tests the response to acid concentration (90, 120, or 150 g·L^{-1}) and number of injection sites (two or four). Experiment 2 tests the effect of injection volume (20 mL or 30 mL). Experiment 3 tests the effect of distilled water or seawater as solvent for the citric acid. Body diameter was used as covariate in all models. Data for Experiment 1 were log transformed. * indicates significance at P < 0.05.

Source	DF	SS	MS	F	P
Experiment 1: two-way ANCOVA					
Log(Time to Immobility)					
Size	1	0.20	0.20	0.18	0.677
Concentration	2	8.94	4.47	3.91	0.031 *
Injection sites	1	1.74	1.74	1.53	0.226
Error	30	34.23	1.14		
Log(Time to Death)					
Size	1	0.01	0.01	0.02	0.882
Concentration	2	1.35	0.68	2.49	0.100
Injection sites	1	2.82	2.82	10.37	0.003 *
Error	30	8.16	0.27		
Experiment 2: one-way ANCOVA					
Time to Immobility					
Size	1	196.60	196.60	0.34	0.579
Volume	1	23.44	23.44	0.04	0.847
Error	7	4070.46	581.49		
Time to Death					
Size	1	709.08	709.08	1.43	0.270
Volume	1	96.74	96.74	0.20	0.671
Error	7	3454.80	493.54		
Experiment 3: one-way ANCOVA					
Time to Immobility					
Size	1	61.44	61.44	0.05	0.836
Water type	1	800.55	800.55	0.61	0.465
Error	6	7878.63	1313.11		
Time to Death					
Size	1	367.11	367.11	0.30	0.603
Water type	1	1241.10	1241.10	1.01	0.352
Error	6	7333.14	1222.19		

4. Discussion

This study demonstrated that injections of citric acid powder and seawater solution represent an efficient, economical, and easy-to-use method to cull crown-of-thorns starfish (COTS). We found that injecting 20 mL of seawater and citric acid powder at a concentration of 90–150 g·L^{-1} in two or four opposing arms effectively kills 100% of COTS, with times to death comparable to that of bile salts, vinegar, and lime juice (Table 2). Although time to death was higher for the two injections method (~58 h, [120 g·L^{-1} citric acid) compared to four (~26 h, [150 g·L^{-1} citric acid]), we argue that the reduced handling time makes the two injections method the most cost and time effective for operations in the field.

Mortality rates were significantly affected by both the number of injections and the concentration used. Indeed, for the single injection treatments, increasing the concentration from 90–120 g·L^{-1} to 150 g·L^{-1} reduced mortality from 83% to 33%. This concentration-related decrease in mortality may have occurred because the high localised dose of acid induced rapid, local necrosis of the injected area, with consequent loss of one or more arms, while the rest of the body survived. In contrast, all the multiple injection treatments reached 100% mortality, regardless of the concentration used, perhaps because the acidic solution was more evenly spread throughout the starfishes' bodies. While the precision of mortality rates within each of our treatments is quite low (with six replicates, the error is 1/6 or 16.6%) multiple injections were conclusively effective at culling all COTS thus treated. All 36 specimens treated with multiple injections died (18 double and 18 quadruple injections, with a combined sampling error for multiple injections of 5.5%).

For the treatments that reached 100% mortality (double and quadruple injections), time to immobility significantly decreased with increasing concentration, dropping from 70 ± 19 h for the 90 g·L^{-1} to 37 ± 5 h for the 150 g·L^{-1} solution. Contrarily, time to death was not affected by concentration, but by the number of injections: four injections halved the time to death compared to two (respectively, 44 ± 7 h and 79 ± 12 h). However, due to the low number of replications for each treatment in this study, specific times to immobility and to death should only be used as guidelines for which is the most effective (fastest) method, and not as accurate indicators of true times to immobility and to death.

Another determinant factor affecting times to immobility and to death in this study was the incredible resistance of some roaming arms, which in some cases survived for up to 10 days postinjection. However, although COTS can regenerate from extensive tissue loss [22,23], it is unlikely that these roaming parts would have caused further coral loss, because of the absence of central disc and pyloric caeca, and the increased predation caused by chemoattraction of predators to injured COTS [24,25]. Therefore, we consider the lag time in reaching death, found herein, to be inconsequential.

Because none of the single injection treatments of Experiment 1 reached 100% mortality, we evaluated whether using distilled water as solvent or increasing the injected volume could improve the efficiency of citric acid single injections. Distilled water was expected to induce an osmotic shock in COTS tissues, thus increasing the efficiency of the citric acid injections. On the contrary, it reduced mortality and slowed time to immobility and to death. This may have happened because the hypoosmotic conditions caused by distilled water activated the opening of the water vascular system in attempt to restore the physiological osmolarity [26], leading to the exchange of water with the environment and consequently flushing out the citric acid. In contrast, COTS injected with seawater may not have had the possibility of doing so, lacking the hypoosmotic triggering factor. Therefore, we argue that distilled water should not be used as a solvent for the citric acid injections. Similarly, increasing the volume of a single injection of citric acid from 20 to 30 mL did not increase the percent mortality, although it moderately reduced time to immobility and to death. Likewise, a single injection of 10, 15, and 20 mL of lime juice failed to achieve 100% mortality and is therefore not an effective method to cull COTS [15]. We conclude that single injections of citric acid should not be performed on COTS in the field, given that mortality with single injection methods only reached 83%.

COTS death by citric acid injections is most likely induced by chronic pH stress caused by the low pH of the solution injected, which ranged between ~1.6 and 1.7. Similar mechanisms of death were proposed for vinegar (pH 2.2) and lime juice (pH 1.8) [12,15]. Indeed, echinoderms are poor acid–base regulators [27], and citric acid, like acetic acid, is both water-soluble and lipid-soluble, so it can easily perfuse into COTS tissues, where the low tissue pH causes protein degeneration and tissue necrosis [12,17].

This study showed that injection with citric acid is an efficient way of culling COTS, and thus of potentially controlling localised outbreaks. But why should this method be used over other current alternatives like sodium bisulphate, bile and cooking salts, vinegar, or lime juice? Compared to the single injection of bile salts [10] and vinegar [12], this method may be slightly more time consuming, because, from the data available, at least two injections per starfish are required to achieve 100% mortality. Nevertheless, citric acid has several characteristics that make it a valid alternative to those control methods (Table 2). First, compared to both sodium bisulphate and bile salts, it is generally available for purchase from a variety of stores. Indeed, rapid intervention, which is crucial to the successful control initiatives against COTS outbreaks [1,28,29], is possible only with a reliable and easy access to the chemical product. However, due to quarantine restrictions and high importation cost, bile salts may be inaccessible for some remote island communities that experience recurrent COTS outbreaks throughout the Indo-Pacific. Secondly, citric acid has a favourable ecological profile (compared, for example, to sodium bisulphate) and is unlikely to accumulate in soil or sediment, as it is rapidly degraded by naturally occurring bacteria [30–32]. Additionally, predators feeding upon decomposing COTS would not be expected to suffer from ingestion of acidic tissues. Investigations on the environmental side effects of similar natural acidic products (lime juice and vinegar) have found no evidence of an impact on other marine organisms [12,15,17]. Third, in contrast to vinegar and lime juice, citric acid is easily transportable as a lightweight powder which can be mixed on site with seawater, reducing transportation costs and volumes on land. It also has a long shelf life, allowing storage in remote areas where fresh citrus juice may not be available or rapidly deteriorate. Additionally, compared to extracting juice from fresh limes, using powdered citric acid is far less labour intensive and more readily available (limes are seasonal and their price can vary greatly).

Until effective measures to prevent COTS outbreaks are found, having a collective workforce and/or volunteers "adopt a reef" may be the most efficient way of deploying COTS treatments and thereby preventing further mass coral predation. However, some important considerations need to be made. Firstly, permits need to be obtained from relevant local authorities. Secondly, safety guidelines for operators are needed to avoid spiking hazards from COTS spines and syringe needles, and for correct handling of citric acid powder to avoid skin, eye, and respiratory inflammations. Thirdly, further studies should aim at developing a single injection protocol for citric acid and investigate the efficiencies of injecting citric acid in different parts of COTS bodies (i.e., distal and medial portion of the arm and central disc). Finally, considering that the objective of COTS control programmes is prevention of extensive coral mortality and not the eradication of the species, injections should be carried out only if more than four to five COTS are counted during a 15 min swim, considered active outbreaking density [11]. If an outbreak is identified, citric acid solution [~90–150 g·L^{-1}] can be easily prepared by mixing the solid powder with seawater; 20 mL of solution can then be administered with double or quadruple injections using any syringe attached to a veterinary 16 Ga × 1/2" needle, both easily available from any chemist store. It is important to note, however, that we caution against the use of citric acid until large-scale field trials have been undertaken. Citric acid is a promising alternative to existing COTS control techniques due to the ease of access, transport, storage, handling, and delivery compared to current methods. As such, it provides a new option to combat COTS outbreaks that is especially useful in remote locations and developing countries.

Table 2. Comparison of the chemicals currently available for injections of *Acanthaster* cf. *solaris* (crown-of-thorns starfish (COTS)).

	Absolute Lethal Dose (LD$_{100}$)	Time to Death	Advantages	Disadvantages
Sodium bisulphate	Multiple injections of up to 180 mL [33] of 140 g·L^{-1} solution [8,9]	Unreported	-Highly effective	-Multiple injections required -Potent oxygen scavenger [8,9]
Bile salts	1 × 10 mL injection of 8 g·L^{-1} solution [10]	~28 h [10]	-Single injection -No known environmental side effects	-Not readily available in remote areas -Quarantine restrictions on access ~0.05 to 0.29 USD per injection [10,15,18,33]
Cooking salt	2 × 10 mL injections of 400 g·L^{-1} solution [13,16]	~48 h [16]	-Readily available -No known environmental side effects ~0.05 USD per COTS [13]	-High quantities required (8 kg/1000 COTS) -Solution preparation requires heating -Precipitation and crystallization [16]
Vinegar	2 × 10 mL injections [12,18] or 1 × 25 mL injection [15]	~30 h [15], ~40 h [12]	-Single injection -Readily available -No known environmental side effects ~0.05 USD per COTS [15]	-High quantities required (20–25 L/1000 COTS)
Lime juice	2 × 10 mL injections [15,18]	~20 h [15]	-No known environmental side effects [15]	-High quantities required (20 L/1000 COTS) -Laborious process for juice extraction -Seasonal and not ubiquitously cheap -Perishable
Powdered Citric acid [a]	2 × 10 mL or 4 × 5 mL injections of 90–150 g·L^{-1} solution	~26 h [b]	-Readily available, long shelf life -No known environmental side effects [a] ~0.05 USD per COTS [a] -Easily transportable (180–300 g/1000 COTS)	Multiple injections required

[a] This study. [b] Four injections of 150 g·L^{-1} solution of citric acid.

Acknowledgments: This study was funded by a grant to Lisa Boström-Einarsson from the Ian Potter Foundation 50th Anniversary COTS Control Grant to the Australian Museum's Lizard Island Research Station (LIRS), and by James Cook University. We wish to thank Robert Streit for helping in the field and laboratory work, Anne Hoggett and Lyle Vail for their assistance at the Lizard Island Research Station and Rie Hagiara for the statistical help.

Author Contributions: A.B., L.B.E. and N.G. conceived and designed the experiments; A.B and L.B.E. performed the experiments; A.B analysed the data; A.B. wrote the majority of the paper; L.B.E. and N.G. edited the manuscript.

Conflicts of Interest: The authors declare no conflict of interest. The founding sponsors had no role in the design of the study; in the collection, analyses, or interpretation of data; in the writing of the manuscript, and in the decision to publish the results

References

1. Birkeland, C.; Lucas, J. *Acanthaster planci: Major Management Problem of Coral Reefs*; CRC press: Boca Raton, FL, USA, 1990.
2. Pratchett, M.S.; Caballes, C.F.; Rivera-Posada, J.A.; Sweatman, H.P.A. Limits to understanding and managing outbreaks of crown-of-thorns starfish (*Acanthaster* spp.). *Oceanogr. Mar. Biol. Ann. Rev.* **2014**, *52*, 133–200.
3. Chesher, R.H. Destruction of Pacific corals by the sea star *Acanthaster planci*. *Science* **1969**, *165*, 280–283. [CrossRef] [PubMed]
4. Osborne, K.; Dolman, A.M.; Burgess, S.C.; Johns, K.A. Disturbance and the dynamics of coral cover on the Great Barrier Reef (1995–2009). *PLoS ONE* **2011**, *6*, e17516. [CrossRef] [PubMed]
5. De'ath, G.; Fabricius, K.E.; Sweatman, H.; Puotinen, M. The 27–year decline of coral cover on the Great Barrier Reef and its causes. *Proc. Nat. Acad. Sci.* **2012**, *109*, 17995–17999. [CrossRef] [PubMed]
6. Rivera-Posada, J.; Owens, L.; Caballes, C.F.; Pratchett, M.S. The role of protein extracts in the induction of disease in *Acanthaster planci. J. Exp. Mar. Biol. Ecol.* **2012**, *429*, 1–6. [CrossRef]
7. Hughes, T.P.; Graham, N.A.; Jackson, J.B.; Mumby, P.J.; Steneck, R.S. Rising to the challenge of sustaining coral reef resilience. *Trends Ecol. Evol.* **2010**, *25*, 633–642. [CrossRef] [PubMed]
8. Rivera-Posada, J.; Pratchett, M.S. A review of existing control efforts for *Acanthaster planci*; limitations to successes. Report to the Department of Sustainability, Environment, Water, Population & Communities, NERP, Tropical Environmental Hub: Townsville, Australia, 2012. Available online: http://citeseerx.ist.psu.edu/viewdoc/download?doi=10.1.1.721.3890&rep=rep1&type=pdf (accessed on 20 April 2015).
9. Rivera-Posada, J.; Caballes, C.F.; Pratchett, M.S. Lethal doses of oxbile, peptones and thiosulfate-citrate-bile-sucrose agar (TCBS) for *Acanthaster planci*; exploring alternative population control options. *Mar. Poll. Bull.* **2013**, *75*, 133–139. [CrossRef] [PubMed]
10. Rivera-Posada, J.; Pratchett, M.S.; Aguilar, C.; Grand, A.; Caballes, C.F. Bile salts and the single-shot lethal injection method for killing crown-of-thorns sea stars (*Acanthaster planci*). *Ocean Coast. Manag.* **2014**, *102*, 383–390. [CrossRef]
11. Crown-of-thorns starfish control guidelines. Great Barrier Reef Marine Park Authority: Townsville, Australia, 2014. Available online: http://hdl.handle.net/11017/2874 (accessed on 20 April 2015).
12. Boström-Einarsson, L.; Rivera-Posada, J. Controlling outbreaks of the coral-eating crown-of-thorns starfish using a single injection of common household vinegar. *Coral Reefs* **2016**, *35*, 223–228. [CrossRef]
13. De Dios, H.Y.; Sotto, B.F.; Dy, D.T.; Ilano, A.S. Response of *Acanthaster planci* (Echinodermata: Asteroidea) to hypersaline solution: Its potential application to population control. *Galaxea* **2015**, *17*, 23–30. [CrossRef]
14. Introduction to the Control of COTS by Acetic Acid Injection. Kuroshio Biological Research Foundation: Kochi, Japan, 2012. Available online: https://chushikoku.env.go.jp/to_2012/data/0530aa_2.pdf (accessed on 23 June 2015).
15. Moutardier, G.; Gereva, S.; Mills, S.C.; Adjeroud, M.; Beldade, R.; Ham, J.; Kaku, R.; Dumas, P. Lime Juice and Vinegar Injections as a Cheap and Natural Alternative to Control COTS Outbreaks. *PLoS ONE* **2015**, *10*, e0137605. [CrossRef] [PubMed]
16. Rivera-Posada, J.; Owens, L. Osmotic shock as alternative method to control *Acanthaster planci. J. Coastal Life Med.* **2014**, *2*, 99–106. [CrossRef]
17. Yamamoto, T.; Otsuka, T. Experimental validation of dilute acetic acid solution injection (*Acanthaster planci*). *Naturalistae* **2013**, *17*, 63–65.

18. Dumas, P.; Gereva, S.; Moutardier, M.; Ham, J.; Kaku, R. Collective action and lime juice fight crown-of-thorns starfish outbreaks in Vanuatu. *SPC Fish. Newsl.* **2015**, *146*, 47–52. Available online: http://www.spc.int/ DigitalLibrary/Doc/FAME/InfoBull/FishNews/146/FishNews146_47_Dumas.pdf (accessed on 23 June 2015).

19. Nichols, D. The water-vascular system in living and fossil echinoderms. *Palaeontology* **1972**, *15*, 519–538.

20. Rivera-Posada, J.; Pratchett, M.S.; Cano-Gómez, A.; Arango-Gómez, J.; Owens, L. Injection of *Acanthaster planci* with thiosulfate-citrate-bile-sucrose agar (TCBS). I. Disease induction. *Dis. Aquat. Organ.* **2011**, *97*, 85–94. [CrossRef] [PubMed]

21. Pratchett, M.S.; Rivera-Posada, J.; Aguilar, C.; Caballes, F.C.; Grand, A. *Efficacy of Oxbile for Controlling Outbreak Populations of Acanthaster Planci on the Great Barrier Reef*; Great Barrier Reef Marine Park Authority: Townsville, QLD, Australia, 2007; pp. 1–8.

22. Sweatman, H.; Butler, I. An experimental investigation of the ability of adult crown-of-thorns starfish to survive physical damage. In *The Possible Causes and Consequences of Outbreaks of the Crown-of-Thorns Starfish, Proceedings of a Workshop series N° 18, Townsville, QLD, Australia, 10 June 1992*; Engelhardt, U., Lassing, B., Eds.; Great Barrier Reef Marine Park Authority: Townsville, Australia, 1993; pp. 71–82.

23. Messmer, V.; Pratchett, M.S.; Clark, T. Capacity for regeneration in crown of thorns starfish, *Acanthaster planci*. *Coral Reefs* **2013**, *32*, 461. [CrossRef]

24. Glynn, P.W. *Acanthaster* population regulation by a shrimp and a worm. In Proceedings of the Fourth International Coral Reef Symposium, Manila, Phillippines, 1981; Gomez, E.D., Birkeland, C.E., Buddemeier, R.W., Johannes, R.E., Marsh, J.A., Tsuda, R.T., Eds.; Marine Sciences Center: Quezon City, Phillippines, 1982; Volume 2, pp. 607–612.

25. Glynn, P.W. An amphinomid worm predator of the crown-of-thorns sea star and general predation on asteroids in eastern and western Pacific coral reefs. *Bull. Mar. Sci* **1984**, *35*, 54–71.

26. Santos-Gouvea, I.A.; Freire, C.A. Effects of hypo-and hypersaline seawater on the microanatomy and ultrastructure of epithelial tissues of *Echinometra lucunter* (Echinodermata: Echinoidea) of intertidal and subtidal populations. *Zool. Stud.* **2007**, *46*, 203–215.

27. Wittmann, A.C.; Pörtner, H.O. Sensitivities of extant animal taxa to ocean acidification. *Nat. Clim. Chang.* **2013**, *3*, 995–1001. [CrossRef]

28. Yamaguchi, M. *Acanthaster planci* infestations of reefs and coral assemblages in Japan: A retrospective analysis of control efforts. *Coral Reefs* **1986**, *5*, 23–30. [CrossRef]

29. Bos, A.R.; Gumanao, G.S.; Mueller, B.; Saceda-Cardoza, M.M. Management of crown-of-thorns sea star (*Acanthaster planci* L.) outbreaks: Removal success depends on reef topography and timing within the reproduction cycle. *Ocean Coast. Manage.* **2013**, *71*, 116–122. [CrossRef]

30. Human and Environmental Risk Assessment on ingredients of Household Cleaning Products. Substance: Citric Acid and Salts (CAS# 77–92–9; 5949–29–1; 6132–04–3). 2005. Available online: http://www.heraproject. com/files/37-f-05-hera_citricacid_version1_april05.pdf (accessed on 23 June 2015).

31. Hoyt, H.L.; Gewanter, H.L. Citrate. In *Detergent*; De Oude, N.T., Ed.; Springer: Berlin/Heidelberg, Germany, 1992; Volume 3, pp. 229–242. [CrossRef]

32. OECD SIDS. Citric Acid CAS N°77–92–9. 2001. Available online: http://www.inchem.org/documents/ sids/sids/77929.pdf (accessed on 23 June 2015).

33. Great Barrier Reef Marine Park Authority. Crown-of-thorns starfish control guidelines. GBRMPA: Townsville, Australia, 2014. Available online: http://www.gbrmpa.gov.au/__data/assets/pdf_file/0006/185298/COTS-control-guidelines.pdf (accessed on 23 June 2015).

MDPI AG

St. Alban-Anlage 66

4052 Basel, Switzerland

Tel. +41 61 683 77 34

Fax +41 61 302 89 18

http://www.mdpi.com

Diversity Editorial Office

E-mail: diversity@mdpi.com

http://www.mdpi.com/journal/diversity